PROPERTIES AND INTERACTIONS OF INTERPLANETARY DUST

ASTROPHYSICS AND SPACE SCIENCE LIBRARY

A SERIES OF BOOKS ON THE RECENT DEVELOPMENTS
OF SPACE SCIENCE AND OF GENERAL GEOPHYSICS AND ASTROPHYSICS
PUBLISHED IN CONNECTION WITH THE JOURNAL
SPACE SCIENCE REVIEWS

VOLUME 119
PROCEEDINGS

PROPERTIES AND INTERACTIONS OF INTERPLANETARY DUST

PROCEEDINGS OF THE 85TH COLLOQUIUM OF THE INTERNATIONAL ASTRONOMICAL UNION, MARSEILLE, FRANCE, JULY 9–12, 1984

Edited by

R. H. GIESE

Ruhr-University Bochum,
Department of Physics and Astronomy: Extraterrestrial Physics,
Bochum, F.R.G.

and

P. LAMY

Laboratory for Space Astronomy, Marseille, France

SPRINGER-SCIENCE+BUSINESS MEDIA, B.V.

Library of Congress Cataloging in Publication Data

CIP

International Astronomical Union. Colloquium (85th: 1984: Marseille, France)
Properties and interactions of interplanetary dust.

(Astrophysics and space science library; v. 119)
Includes indexes.
1. Cosmic dust–Congresses. I. Giese, R. H. (Richard H.)
II. Lamy, P. L. (Philippe L.) III. Title. IV. Series.
QB791.I56 1984 523.2 85-18281

ISBN 978-94-010-8912-8 ISBN 978-94-009-5464-9 (eBook)
DOI 10.1007/ 978-94-009-5464-9

Originally published by D. Reidel Publishing Company, Dordrecht, Holland
Softcover reprint of the hardcover 1st edition 1985

TABLE OF CONTENTS

PREFACE	xi
PICTURE	xiv
LIST OF PARTICIPANTS	xvii
ORGANIZING COMMITTEES	xix
NECROLOGY	xxi

I. ZODIACAL LIGHT AND F-CORONA: OBSERVATIONS

WEINBERG J L: Zodiacal Light and Interplanetary Dust (IR) 1

NIKOLSKY G, KOUTCHMY S, LAMY Ph L, NESMJANOVICH I A: Photographic Observations of the Inner Zodiacal Light Aboard Saliout 7 7

MISCONI N Y, WEINBERG J L: Ground-Based Observations of Near Ecliptic Zodiacal Light Brightness 11

PFLEIDERER J, LEUPRECHT G: An Isophote Map of the Zodiacal Light in V 17

TOLLER G N, WEINBERG J L: The Change in the Near-Ecliptic Zodiacal Light Brightness with Heliocentric Distance 21

MAUCHERAT A, LLEBARIA A, GONIN J C: Zodiacal Light, Gegenschein and Sky Background 27

FELDMAN P D, CEBULA R P, TENNYSON P D, HENRY R C: Recent Rocket Observations of the Ultraviolet Spectrum of the Zodiacal Light 31

HONG S S, MISCONI N Y, VAN DIJK M H H, WEINBERG J L, TOLLER G N: A Search for Small Scale Structures in the Zodiacal Light 33

GIOVANE F, WEINBERG J L, MANN H M, OLIVER J P: An Observational Search for the Schuerman Dust Arcs 39

HAUSER M G, GAUTIER T N, GOOD J, LOW F J: IRAS Observations of the Interplanetary Dust Emission (IC) 43

ISOBE S, TANABE H, HIRAYAMA T, KOMA Y, SOEGIJO J, BABA N: Balloon Observation of the F-Corona at the 1983 Total Solar Eclipse (IC) 49

MAIHARA T, MIZUTANI K, HIROMOTO N, TAKAMI H, HASEGAWA H: A Balloon Observation of the Thermal Radiation from the Circumsolar Dust Cloud in the 1983 Total Eclipse 55

MUKAI T: On the Solar Dust Ring(s) 59

KOUTCHMY S, LAMY P L: The F-Corona and the Circum-Solar Dust Evidences and Properties (IR) 63

SHESTAKOVA L I: Interpretation of F-Corona Radial Velocity Field 75

SHCHEGLOV P V, SHESTAKOVA L I, AJMANOV A K: Interferometric Observation of the F-Corona Radial Velocities Field Between 3 and 7 $R_\odot$ 77

EAST I R, REAY N K: The Motion of Interplanetary Dust Particles 81

ROBLEY R, BÜCHER A, KOUTCHMY S, LAMY Ph : Doppler Shifts Measurements of the Zodiacal Light at the Pic Midi Observatory 85

II. INTERPLANETARY DUST: SPACE AND GROUND STUDIES

ZOOK H A, LANGE G, GRÜN E, FECHTIG H: The Interplanetary Micrometeoroid Flux and Lunar Primary and Secondary Microcraters (IR) 89
KESSLER D J: Explorer 46 Meteoroid Bumper Experiment: Earth Orbital Debris Interpretation 97
BARSUKOV V L, NAZAROVA T N: On the Earth's Dust Envelope 99
GRÜN E, FECHTIG H, KISSEL J: Orbits of Interplanetary Dust Particles Inside 1 AU as Observed by Helios 105
GÖLLER J R, GRÜN E: Calibration of the Galileo/ISPM Dust Detectors with Iron Particles 113
SINGER S F, STANLEY J E, KASSEL P: The LDEF Interplanetary Dust Experiment 117
FECHTIG H, HÖRZ F, IGENBERGS E, JESSBERGER E, KUCZERA H, LANGE G, PAILER N, SUTTON S, SWAN P, WALKER R, ZINNER E: Measurements of the Elemental and Isotopic Composition of Interplanetary Dust Collected on LDEF 121
WALKER R M, ZINNER E: Prospects for Cosmic Dust Experiments on the Planned Reflight of LDEF 127
KUCZERA H, IGLSEDER H, WEISHAUPT U, IGENBERGS E: Acoustic Penetration and Impact Detector for Micrometeoroid and Space Debris Application 129
CAREY W C, McDONNELL J A M, DIXON D G: An Empirical Penetration Equation for Thin Metallic Films Used in Capture Cell Techniques 131
LAMY P L: Some Remarks on the Density of Interplanetary Dust Grains 137
KOUTCHMY S, LAMY P L: A Coronascanner for a Sun Oriented Space Mission 141

III. INTERPLANETARY DUST: LABORATORY STUDIES

BROWNLEE D E: Collection of Cosmic Dust: Past and Future (IR) 143
FAHEY A, McKEEGAN K D, SANDFORD S A, WALKER R M, WOPENKA B, ZINNER E: Complementary Laboratory Measurements of Individual Interplanetary Dust Particles 149
WALLENWEIN R, BLANK H, JESSBERGER E K, TRAXEL K: Proton Induced X-Ray Emission (PIXE) Analysis of Meteoritic Microsamples 157
BORGHESI A, BUSSOLETTI E, COLANGELI L: Physical Properties of Submicronic Carbonaceous Particles Candidate as Cosmic Dust 159
PENG H C, ZHUANG S J, LIU Z K, YU Z: A Discover of Iron Spherules with Golden Ni-Fe Core and Silicate Spherules with Golden Ni-Fe Inclusion in Deep-Sea Sediments 163
RAISBECK G M, YIOU F, KLEIN J, MIDDLETON R, BROWNLEE D: $^{26}Al/^{10}Be$ in Deep Sea Spherules as Evidence of Cometary Origin 169
YAMAKOSHI K, HONMA K: Chemical and Isotopic Compositions of Refractory Elements in Deep Sea Spherules 175

YAMAKOSHI K: Cosmic Ray Exposure Age Determinations of Cosmic Spherules from Marine Sediments 179
PARKIN D W: Cosmic Spherules, Asteroid Collisions and the Solar Constant 183
OZIMA M, TAKAYANAGI M, ZASHU S, AMARI S: Observation of high $^{3}He/^{4}He$ Ratio in Ocean Sediments - Evidence for Extraterrestrial Material Contamination 185
KVIZ Z: Cometary Dust and the Periodicity of Rainfall Singularities 191

IV. OPTICAL STUDIES OF DUST

ZERULL R H: Laboratory Investigations and Optical Properties of Grains (IR) 197
DUMONT R, LEVASSEUR-REGOURD A C: Remote Sensing of the Zodiacal Cloud along Secants to Earth's Orbit (IC) 207
HONG S S: A Method for Deriving the Mean Volume Scattering Phase Function for Zodiacal Dust 215
WEISS-WRANA K, GIESE R H, ZERULL R H: Microwave and Laser Facilities to Determine Scattering and Colour Signatures Related to the Physical Properties of Dust Particles 219
WEISS-WRANA K, GIESE R H, ZERULL R H: Experimental Results on Scattering by Irregular and Meteoritic Dust Particles Related to Photopolarimetry of Zodiacal Light and Comets 223
GUSTAFSON B A S: Laboratory Results on Polarization Properties of Elongated Particles and Comparisons to Dust in the Tail of Comet IKEYA-SEKI (1965 VIII) 227
BLIEK P, LAMY P L, COURTES G: Preliminary Results of a Dust Scattering Experiment 231
MUKAI S, MUKAI T: Influence of Surface Roughness on Scattering Patterns 235
LAMY PH L, PERRIN J M: Zodiacal Light and the Spatial Density of Interplanetary Grains 239
PERRIN J M, LAMY P L: Optical Properties of Rough Grains: A Theoretical Study 245
SCHIFFER R, THIELHEIM K O: Wavelength Dependence of the Zodiacal Light 249
GIESE R H , KINATEDER G , KNEISSEL B, RITTICH U: Optical Models of the Three Dimensional Distribution of Interplanetary Dust 255
CRIFO J F: Equilibrium Temperatures of H_2O, CO_2 and NH_3 Ice Grains in the Solar System 261
PEGOURIE B, PAPOULAR R: The Optical Properties of Circumstellar and Interstellar Dust in the Mid-IR 267
SANDFORD S A, WALKER R M: Laboratory Infrared Transmission Measurements of Interplanetary Dust and Implications for Remote Observations of Cosmic Particles 273
STRONG I B, BROWNLEE R R, FARNUM E H, HUEBNER W F, KUNKLE T D, STEPHENS J R, BODE M F: Plans for Release of Simulated Interplanetary Materials Into Low Earth Orbit 275

V. RELATIONSHIPS TO COMETS AND METEOROIDS

LIU C P, KIMURA H: On the Bimodal Nature of the Particle-Size-Distribution Function of Cometary Dust 279
KIMURA H, LIU C P, JOCKERS K: On the Dynamic Behaviour of Dust Striae Observed in Comets West 1976VI, and Mrkos 1957V 283
ZOOK H A, FERNANDEZ J A, GRÜN E: Selection Effects Against Small Comets 287
ZOOK H A, POTTER A E: Optical Detection of Large Meteoroids in Space 293
STEEL D I, BAGGALEY W J: Meteoroid Orbits Determined by Southern Hemisphere Radar 299

VI. DUST - PLASMA INTERACTIONS

FAHR H J, RIPKEN H W: Dust - Plasma - Gas Interactions in the Heliosphere (IR) 305
IP W H, KIMURA H, LIU C P: Interaction of the Cometary Dust with the Solar Wind and Cometary Plasma 325
QUEMERAIS A, SEIGNAC A, PRIOL M, LEFEVRE J: Optical and Photoelectric Properties of Materials of Astrophysical Interest 329
LAMY P L, LEFEVRE J, MILLET J M, LAFON J P: Electrostatic Charge of Interplanetary Dust Grains: New Results 335
LAFON J P, MILLET J M: On the Electrical Effects of the Presence of Fluff on the Surface of Cosmic Dust Grains 341
MILLET J M, LAFON J P: Secondary Emission from Interplanetary Dust Grains 347
RÖSSLER K, EICH G: Computer Calculation of Collision Cascades by Energetic Particles Penetrating Dust Grains 351
RÖSSLER K: Laboratory Simulation of Chemical Interactions of Accelerated Ions with Dust and Ice Grains 357
IBADOV S: On Two Mechanisms of X-Ray Generation in Comets 365

VII. DYNAMICS OF INTERPLANETARY DUST

LEINERT C: Dynamics and Spatial Distribution of Interplanetary Dust 369
MISCONI N Y, RUSK E T: The Gravitational Zones of Influence of the Planets Acting on Small Celestial Bodies 377
GUSTAFSON B A S: A New Approach to Evaluate Planetary Perturbations on a Cloud of Dust in Low Eccentricity Heliocentric Orbits 381
GUSTAFSON B A S: Planetary Perturbations: Effects of the Shape of a Cloud of Dust in Circular Heliocentric Orbits 385
BURKHARDT G: Dynamics of Dust Particles in the Solar System 389
DERMOTT S F, NICHOLSON P D, BURNS J A, HOUCK J R: An Analysis of IRAS' Solar System Dust Bands 395

GRÜN E, ZOOK H A, FECHTIG H, GIESE R H: Mass Input into and Output from the Meteoritic Complex 411
BARGE P, PELLAT R, MILLET J: Lorentz Scattering of Interplanetary Dust Grains 417
BELKOVICH O I, POTAPOV I N: Possible Distributions of some Orbital Elements of Interstellar Particles in the Solar System 421

IR = Invited Review
IC = Invited Contribution

FINAL DISCUSSION 425
SUMMARY AND CONCLUSION OF THE COLLOQUIUM 429
INDEX OF AUTHORS 435
INDEX OF SUBJECTS 437

PREFACE

Investigation of the interplanetary dust cloud is characterized by contributions from quite different methods and fields, such as research on zodiacal light, meteors, micrometeoroids, asteroids, and comets. Since the earth's environment and interplanetary space became accessible to space vehicles these interrelations are clearly evident and extremely useful. Space measurements by micrometeoroid detectors, for example, provide individual and eventually detailed information on impact events, which however are limited in number and therefore restricted in statistical significance. On the other hand, zodiacal light measurements involve scattered light from many particles and therefore provide global information about the average values of physical properties and spatial distribution of interplanetary grains. Additional knowledge stems from lunar samples and from dust collections in the atmosphere and in deep sea sediments. All these sources of complementary information must be put together into a synoptical synthesis. This also has to take into account dynamical aspects and the results of laboratory investigations concerning physical properties of small grains. Such considerable effort is not merely an academic exercise for a few specialists interested in the solar dust cloud. Since this same cloud exclusively allows direct in-situ acess to investigate extraterrestrial dust particles over a wide range of sizes and materials, it provides valuable information for realistic treatment of dust phenomena in other remote cosmic regions such as in dense molecular clouds, circumstellar dust shells, and even protostellar or protoplanetary systems.

After the famous first meeting at Honolulu (1967) it has become a tradition of the "interplanetary dust community" to periodically review on an interdisciplinary level the progress and state of the art: l.c. the meetings at Heidelberg (1975) and Ottawa (1979). This time the Colloquium on "Properties and Interactions of Interplanetary dust" at Marseille (1984) was organized by IAU Commission No. 21 (Light of the Night Sky). It was cosponsored by IAU commissions No.15, 22, 49, and by the relevant Section ISC.B-1 of COSPAR. The timing took into account that earthbound observations and data from Helios, Pioneer and other spacecraft were analyzed to a large extent and that the community is stepping forward towards a synopsis, combining optical and in-situ data with laboratory measurements, collected sample analysis and advanced modelling. With respect to the years 1985/86 (apparition of P/Halley) it seemed very appropriate and timely to provide a well founded, up-to-date synopsis of interplanetary dust. It was first envisaged to have the colloquium at Graz in combination with the COSPAR General Assembly 1984, which partially involves the same scientists. Since, due to funding problems, it was not possible to combine the IAU Colloquium directly with the meeting at Graz we gratefully accepted the invitation of the LAS (Laboratoire d' Astronomie Spatiale, CNRS, Director Prof. Courtes) to have the colloquium at Marseille immediately after COSPAR. It was a good choice. Due to cooperation from COSPAR, which shifted its similar Topical Meeting to the Marseille program and by the engagement of all participants, the meeting became an exciting international workshop, involving 78 scientists of 14 nations, including China, which recently joined the IAU. Here is the place especially to thank the Local Organizing Commitee for their excellent organization, and to acknowledge the support of IAU, CNRS, CNES, LAS, and of the Universite de Provence, which provided their facilities for the meeting.

It was decided to publish the proceedings in a comprehensive form and to leave for later detailed presentations to appropriate journals. The rather limited space of about 4 or 10 pages for contributed papers or invited reviews, respectively, puts severe requirements of conciseness on the authors, who succeeded in condensing the highlights of their achievements into the very limited space. We acknowledge with thanks this high degree of cooperation in most cases. From the proceedings of former dust colloquia it is known that such a book, which presents the complete state of the art by short articles of the original investigators is an inestimable tool for daily work: Students at Bochum named the Heidelberg Proceedings "The Bible"! As a compromise between a high standard of language, uniform presentation and fast publication the editors refrained from sending papers to english speaking colleagues for linquistic perfection or to retype all contributions. However, when they were asked for or when they realized major problems, they changed the manuscripts and sent them for approval to the authors. To save time, minor changes or formal adaptions were made without contacting the authors. We apologize and accept responsibility for any mistake which might be brought in by this procedure. On the other hand - with the exception of the panel discussion - all discussion remarks were to be handed out directly to the authors, who were responsible and free to incorporate in their camera ready manuscripts aspects raised in the discussion or to include formal answers. Certaintly each paper does not reflects the view of the editors or of the majority of investigators attending the meeting. It is, however, an attribute of colloquia to stimulate progress in research by sometimes new or even exotic ideas which are presented to the scientific community for discussion and helpful criticism.

Due to the manifold interrelationships between methods and problems there is no "king's way" to organize a book on interplanetary dust. Nor is there an importance to the order of presentation. In this proceedings we followed mainly the order of presentation during the meeting. We start with observations of zodiacal light and F-corona (I) followed by space and ground studies of interplanetary dust (II). Investigations of interplanetary dust particles by laboratory studies (III) has become an important tool of research on dust. In a similar way optical observations provide further progress only if they are followed by laboratory investigations, theoretical work and modelling. This type of work was taken care for in the section "Optical studies of dust (IV)". Comets and meteoroids (V) were treated as far as they are related to interplanetary dust. An important aspect of dust particles in the heliospheric medium are dust-plasma interactions (VI). Finally, if all topics referred to above are involved, one arrives at a realistic treatment of the dynamics of interplanetary dust (VII), which will lead to an understanding of the complicated interactions between sources, sinks, development of interplanetary dust particles and to an insight as to why the interplanetary dust cloud is as we find it today. An outlook to future problems and recommended investigations in the panel discussion (VIII) and a summary of the present status as it results from the meeting is presented in the conclusions (IX).

Bochum and Marseille, February 1985

Richard H. Giese
Philippe L. Lamy

IAU colloquium n°85

1	Courtés	14	Mukai	27	Mateshivili	40	van Dijk
2	Lefèvre	15	Kneißel	28	Nazarova	41	Mrs Kuczera
3	Göller	16	Walker	29	Giese	42	Peng
4	Killinger	17	Schiffer	30	Mrs Harmitt (Secretary)	43	Liu
5	Kimura	18	Raisbeck	31	Mrs Miethe (Secretary)	44	Mrs Giese
6	Fechtig	19	Tanabe	32	Grün	45	Zook
7	Kuczera	20	Lamy	33	Dumont	46	East
8	Pegourie	21	Perrin	34	Hong	47	Brownlee
9	Burns	22	Weinberg	35	Maurel (Projectionist)	48	Piercey
10	Yamakoshi	23	Strong	36	Kessler	49	Mrs Petit (Secretary)
11	Zerull	24	Weiss-Wrana	37	Mrs Bethoux (Secretary)	50	Zhuang
12	Zinner	25	Parkin	38	Hanner	51	Hughes
13	Singer	26	Millet	39	Mrs Dumont	52	Mulholland
						53	Miss (Dumont's Dog)

List of Participants

BARGE P, Observatoire de Marseille, 13248 Marseille Cedex 4, France
BLIEK P, Laboratoire d'Astronomie Spatiale, 13012 Marseille, France
BORGHESI A, Gruppo di fisica cosmica, Lecce, Italie
BROWNLEE D E, University of Washington, Seattle, WA 98195, U.S.A.
BUCHER A, Observatoire du Pic du Midi, 65200 Bangeres de Bigorre, France
BURKHARDT G, Astron. Rechen-Institut, 6900 Heidelberg, F.R.G.
BURNS J A, Cornell University, Ithaca, N.Y. 14853, U.S.A.
BUSSOLETTI E, Universita di Lecce, 73100 Lecce, Italie
CHIAPPETTA P, Centre de Physique Theorique, 13288 Marseille Cedex 2, France
COURTES G, Laboratoire d'Astronmie Spatiale, 13012 Marseille, France
CRIFO J F, L.P.S.P., 91371 Verrieres le Buisson, France
DOLLFUS A, Observatoire de Paris, 92190 Meudon, France
DUMONT R, Observatoire de Bordeaux, 33270 Floirac, France
EAST I, Dept. of Physics and Astronomy, London WCIE6BT, U.K.
FAHR H J, Universität Bonn, 5300 Bonn, F.R.G.
FECHTIG H, Max-Planck-Institut f. Kernphysik, 6900 Heidelberg, F.R.G.
FELDMAN P F, Johns Hopkins University, Baltimore, MD 21218, U.S.A.
GIESE R H, Ruhr-Universität Bochum, 4630 Bochum 1, F.R.G.
GÖLLER J R, Max-Planck-Institut f. Kernphysik, 6900 Heidelberg, F.R.G.
GRUN E, Max-Planck-Institut f. Kernphysik, 6900 Heidelberg, F.R.G.
GUSTAFSON B A S, University of Florida, Gainesville, FL 32601, U.S.A.
HANNER M A, Jet Propulsion Laboratory, Pasadena, CA 91109, U.S.A.
HONG S S, Seoul National University, Seoul, Corea
HUGHES S W, Dept. of Physics, Sheffield S3 7RH, U.K.
JESSBERGER E K, Max-Planck-Institut f. Kernphysik, 6900 Heidelberg, F.R.G.
JONES J, University of Western Ontario, Ontario, NGA 3k7, Canada
KESSLER D, NASA-Johnson Space Center, Houston, TX 75058, U.S.A.
KILLINGER R, Ruhr-Universität Bochum, 4630 Bochum 1, F.R.G.
KIMURA H, Max-Planck-Institut f. Aeronomie, 3411 Katlenburg-Lindau, F.R.G.
KNEISSEL B, Ruhr-Universität Bochum, 4630 Bochum 1, F.R.G.
KOEBERL C, Inst. for Analytical Chemistry, A-1094 Vienne, Austria
KOUTCHMY S, Institut d'Astrophysique, 75014 Paris, France
KUCZERA H, Technische Universität München, 8000 Munich 2, F.R.G.
KVIZ Z, University of New South Wales, Broken Hill, NSW 2880, Australia
LAFON J P, Observatoire de Meudon, Meudon 92190 Principal Cedex, France
LAMY P L, Laboratoire d'Astronomie Spatiale, 13012 Marseille, France
LEFEVRE J, Faculte des Sciences, 44072 Nantes, France
LEINERT C, Max-Planck-Institut für Astronomie, 6900 Heidelberg, F.R.G.
LEUPRECHT G, Institut für Astronomie, A-6020 Innsbruck, Austria
LEVASSEUR-REGOURD A C, C.N.R.S., Verrieres le Buisson, France
LILLIE C F, TRW, Redondo Beach, CA 90178, U.S.A.

LIU C P, Max-Planck-Institut für Aeronomie, 3411 Katlenburg-Lindau, F.R.G.
LOKANADHAM B, Osmania University, Hyderabad, India
LUMME K, University of Helsinki, SF-00130 Helsinki 13, Finland
MANDEVILLE J C, ONERA / CERT-DERTS, Toulouse 31055 Cedex, France
MATESHIVILI G G, Abastumani Astrophys. Obs., 383762 Abastumani Georgie, U.S.S.R.
MAUCHERAT J, Laboratoire d'Astronomie Spatiale, 13012 Marseille, France
MC DONNELL J A M, Space Science Laboratory, Canterbury, Kent CT2 7NT, U.K.
MERCER R D, ISSR, Inc., Silver Spring, MD 20906, U.S.A.
MILLET J, Laboratoire d'Astronomie Spatiale, 13012 Marseille, France
MUKAI T, Kanazawa Inst. of Technology, Ishikawa, Japan
MULHOLLAND J D, University of Texas, Austin, TX 78712, U.S.A.
NAZAROVA T N, USSR Acad. of Sciences, 117975 GSP Moscow V-334, U.S.S.R.
OZIMA M, Max-Planck-Institut für Kernphysik, 6900 Heidelberg, F.R.G.
PARKIN D W, University of Bath, Bath, U.K.
PEGOURIE B, CEN-Saclay - Service d'Astrophysique, 91191 Gif-sur-Yvette, France
PENG H C, First Institut of Oceanography, Quindao, P.R.China
PERRIN J M, Laboratoire d'Astronomie Spatiale, 13012 Marseille, France
PIERCEY R B, University of Florida, Gainesville, FL 32601, U.S.A.
RAISBECK G, Laboratoire Rene Bernas, 91406 Orsay, France
RÖSSLER K, Institut für Chemie, 5170 Jülich, F.R.G.
SCHIFFER R, Universität Kiel, 2300 Kiel, F.R.G.
SCHWEHM G, ESTEC, Noordwijk, The Netherlands
SINGER F, 1600 S. EADS ST., Arlington, VA 22202, U.S.A.
STRONG I B, Los Alamos Laboratory, Los Alamos M 87545 New Mexico, U.S.A.
TANABE H, Tokyo Astronomical Observatory, Mitaka, Tokyo 181, Japan
VAN DIJKM H H, Free University, Amsterdam, The Netherlands
VERLHAC A, C.N.R.S., 91370 Verrieres le Buisson, France
WALKER R M, McDonnell Center for the Space Sciences, St. Louis, MO 63130, U.S.A.
WALLENWEIN R, Max-Planck-Institut für Kernphysik, 6900 Heidelberg, F.R.G.
WEINBERG J L, University of Florida, Gainesville, FL 32601, U.S.A.
WEISS-WRANA K, Fasanenweg 39, 7312 Kirchheim/Teck, F.R.G.
YAMAKOSHI K, University of Tokyo, Tokyo 188, Japan
ZARNECKI J C, University of Kent, Canterbury, Kent CT2 7NT, U.K.
ZERULL R H, Ruhr-Universität Bochum, 4630 Bochum 1, F.R.G.
ZHUANG S J, Ministry of Metallurgical Industry, Guilin, Guangxi Prov., P.R.China
ZINNER E, McDonnell Center for the Space Sciences, St. Louis, MO 63130, U.S.A.
ZOOK H, NASA Johnson Space Center, SN3 Houston, TX 77058, U.S.A.

IAU COLLOQUIUM NO 85:
PROPERTIES AND INTERACTIONS OF INTERPRLANETARY DUST

SPONSORS

Commission 21, Commissions 15, 22, 49 of the IAU
COSPAR
Supporting Institutions: CNES, CNRS, INAG, Universite de Provence

SCIENTIFIC ORGANIZING COMMITTEE

R.H. Giese, Ruhr-Universität Bochum, Bochum, F.R.G. (Chairman)
O.I. Belkovich, Engelhardt Astronomical Observatory, Kazan, U.S.S.R.
D.E. Brownlee, University of Washington, Seattle, U.S.A.
G. Courtes, Laboratoire d'Astronomie Spatiale, Marseille, France
H.J. Fahr, Universität Bonn, Bonn, F.R.G.
E. Grün, Max-Planck-Institut f.Kernphysik, Heidelberg, F.R.G.
S. Koutchmy, Institut d'Astrophysique, Paris, France
P.L. Lamy, Laboratoire d'Astronomie Spatiale, Marseille, France
J.A.M. McDonnell, Space Science Laboratory, Canterbury, U.K.
T. Mukai, Kanazawa Inst. of Technology, Ishikawa, Japan
V. Vanysek, Charles University Prague, Prague, Chechoslovakia
R.M. Walker, McDonnell Center Space Sciences, St.Louis, U.S.A.
J.L. Weinberg, University of Florida, Gainesville, U.S.A.

LOCAL ORGANIZING COMMITTEE

Laboratoire d'Astronomie Spatiale, Marseille, France:
P.L. Lamy (Chairman)
P. Bliek
A. Maucherat
J. Millet
J.M. Perrin

C. L. Hemenway
(1920 - 1982)

On November 18, 1982, Curtis Hemenway unexpectedly passed away.

He went to school at Rutgers University, where he received his MS and Ph.D. in physics.

From 1949-56 he was at Union College at Schenectady, N.Y. as Assistant Professor and Associate Professor.

In 1956 he was asked to take over the Dudley Observatory, Albany, N.Y., which, at that time, was located in its old castle-like building at Albany, N.Y. There he was the director until 1979.

From 1964-1979 he was also Professor of Astronomy and Space Science at the State University of New York, Albany (SUNYA) and from 1964-71 he served as the Chairman of the Department of Astronomy and Space Science at SUNYA.

His scientific peak activities were in the 60ties and 70ties. He pioneered a variety of new techniques for collecting extraterrestrial dust particles. The revolutionary nature of these experiments and Hemenway's encouragement of outsiders to participate in these experiments catalyzed a rapid growth of the cosmic dust research: Already in 1959 he flew the famous Venus Flytrap rocket experiment to collect particles in the atmosphere. He developed balloon borne collectors, including a superpressured balloon, which travelled around the globe. The first satellite-borne collectors for the Gemini program were also performed by Hemenway. In 1973 he was Principal Investigator of several dust collections on Skylab. Already at that time one could see space debris, but nobody was aware of this fact at that time.

Hemenway's scientific results were partly unexpected and partly controversial. Only since we have the sound basis of the lunar sample results, we can estimate the reality in a better way. He has already collected fluffy particles, but at that time it was not known that these fluffy particles were the ones we were looking for.

He was a member of the IAU, active in Commission 22. His activities within COSPAR were excellent. In 1968 he founded the Cosmic Dust Panel IIC and was its first chairman until 1973. Owing to his activities, the small but strong dust community has vigorously succeeded during that time.

Hemenway's final years were tragic. The department of Astronomy and Space Science at SUNYA was dissolved and the activities at the Dudley Observatory were not funded any longer. This terminated his activities. Disappointed, but still defending his scientific convictions, he went back to the state where he had, originally, come from: to Maine. There he lived in his summer house near the ocean. Since he was a sailor, he was always proud of his sail boat named "star dust".

His last visit with the international dust community was when he came to Ottawa/Canada in May 1982 to the COSPAR-meeting.

H. Fechtig

G.M. NIKOLSKY (1929-1982)

Professor Guennady M. Nikolsky, who passed away in his 54th year was a prominent russian astrophysicist, member of the IAU. His sudden disappearance occuring at the summit of his creative life is a great loss for the astronomical community. G.M. Nikolsky, was born in Rostov at Don River on Sept. 28, 1929. He obtained a degree in astrophysics from the Astronomical Department of Kiev State University in 1953, where he took up his first job as a scientific assistant to Professor S.K. Vsekhsvyatsky. In 1956-58 he joined the Astrophysical Institute at Alma-Ata where he studied the zodiacal light, planets, and the airglow. Since 1958 up to his last day he worked at IZMIRAN, near Moscow, where he created the Laboratory of Solar Activity which became famous to all experts in Solar Physics. G.M. Nikolsky was a brilliant experimentator and observer. He designed the world largest coronograph, of the Lyot type, with a 535mm objective diameter and a new type of magnetograph. Prof. Nikolsky was generally recognized as an authority in solar total eclipse observations. His contribution in promoting space research have been remarkable. He had been the P.I. of the Artificial Solar Eclipse Experiment realized at the joint Soviet-American mission Apollo-Soyuz in 1975. In 1982, with the Soviet-French crew aboard Salyut-7, Prof. Nikolsky had been on the Soviet part of the PCN experiment. Unique color photographs of the zodiacal light and the ionospheric glow were obtained then. Lately, G.M. Nikolsky had developed ideas on new and original experiments and studies which he had no time to fulfill.

Gennady Nikolsky was a brilliant and open person. He had many friends, not only in his homeland but also among solar astrophysicists abroad although he had never the opportunity to leave his country. The untimely death of Prof. G.M. Nikolsky is an unmeasurable lost for his disciples, friends, colleagues, to all those who had the privilege to share his friendship.

S. Koutchmy

D. W. SCHUERMAN (1943 - 1982)

Dr. Donald W. Schuerman was just 38 when an automobile accident prematurely ended his promising career as an astrophysicist. He was a member of the IAU, COSPAR, and the American and Royal Astronomical Societies, and he had already made major contributions to the subjects of this IAU Colloquium in his relatively short seven years in this field. Dr. Schuerman was born in Cincinnati, Ohio on July 13, 1943. He obtained his advanced degree in physics and astronomy from the University of Rochester in 1968, after which he was manager of the NASA/Johnson Space Center UV rocket program. In 1972 he joined Dudley Observatory in Albany, New York where he worked on a Skylab coronagraph designed to measure atmospheric aerosols and contaminant particulates around the spacecraft. This experiment introduced him to some of the many problems involving light scattered by small particles. In 1975 he joined the Space Astronomy Laboratory at which he worked in Albany and, later, in Gainesville, Florida until his death in 1982. While there he worked on theoretical topics and on space (Skylab, Pioneer 10/11, ISPM, Space Shuttle, ESA/ Giotto mission to Halley) and laboratory experiments. Included among his activities:

- light scattering by irregularly shaped particles (he directed the program in microwave analog measurements of small particles);
- studies of spacecraft induced atmospheres (particulates associated with manned spacecraft - Skylab, Space Shuttle);
- he defined and developed the science concept for the ISPM zodiacal light/background starlight experiment and for the optical probe experiment (HOPE) which will be used for measurements of Comet Halley from the ESA/Giotto probe;
- he derived the general inversion of the zodiacal light brightness integral and used it with Pioneer 10/11 data to provide the first direct information on the properties and distribution of zodiacal dust beyond 1 AU;
- with Dr. Frank Giovane he developed the data reduction methodology

used to derive an all-sky map of background starlight as seen from beyond the asteroid belt;
- he predicted that large scale interplanetary dust arcs may be associated with each planet, those of Jupiter and Saturn being the most predominant.

In his last years he became increasingly convinced that optics, dynamics, and (IR) emission must continue to "come together" if we are to fully understand the interplanetary dust complex. His convictions are strongly confirmed by the recent zodiacal emission results from IRAS.

Although Dr. Schuerman was a brilliant and quick theoretician, he also understood and worked comfortably with instrumentation. He was also expert at metal and woodworking. Muzzle-loading guns and the associated woodsmanship was a major hobby for him as it is for his father and for his son. Don Schuerman went all out in his science, in his love of family and friends, and in his love of life itself. His scientific work ethic, contagious smile, everready wit, and zest for life will remain forever with his family, colleagues, and friends.

J. L. Weinberg

ZODIACAL LIGHT AND INTERPLANETARY DUST

J. L. Weinberg
Space Astronomy Laboratory, University of Florida
1810 NW 6th Street
Gainesville, Florida 32609
USA

ABSTRACT. Recent results on zodiacal light are used to show that optics, dynamics, and infrared must be considered *together* to properly and fully characterize the interplanetary dust complex.

Zodiacal light observations have been widely used to infer the large scale properties of the interplanetary dust - with mixed results. Zodiacal light is certainly more stable (and better understood) than the published literature would suggest over the 25 years that it has been studied by this writer. The interested reader can trace this history in the Proceedings of this Colloquium's predecessor meetings (Honolulu 1967, Heidelberg 1975, Ottawa 1979), in triennial Reports on Astronomy of IAU Commission 21, and in various reviews (e.g.: Leinert 1975; Weinberg and Sparrow 1978; Fechtig, Leinert, and Grün 1981).

There have been intensive ground and space observations over the past two decades together with laboratory and theoretical studies, and, as we entered the 1980's, there was general agreement that:

- zodiacal light has solar color from the near UV to the near IR, except for a slight reddening in sky regions observed near the sun
- zodiacal light brightness is relatively smooth and remarkably stable over times as long as a solar cycle (Burnett 1976; Dumont and Levasseur-Regourd 1978; Leinert, et al. 1982a)
- zodiacal light brightness decreases monotonically with heliocentric distance R and is negligible beyond the asteroid belt (Hanner, et al.; 1974, 1976)
- zodiacal light is partially plane polarized with its electric vector perpendicular to the scattering plane, except for regions at large elongations where there is polarization reversal (electric vector parallel to scattering plane).

Recent results present a somewhat different, more complex picture of the interplanetary dust. Representative of this changing "view" are the recent, exciting observations by IRAS. Figure 1 shows IRAS 4-color data on total thermal emission for representative ecliptic pole-to-pole scans. The zodiacal emission peaks near the ecliptic, is the dominant source at 12, 25, and 60μm, and appears to be the brightest diffuse

R. H. Giese and P. Lamy (eds.), Properties and Interactions of Interplanetary Dust, 1–6.

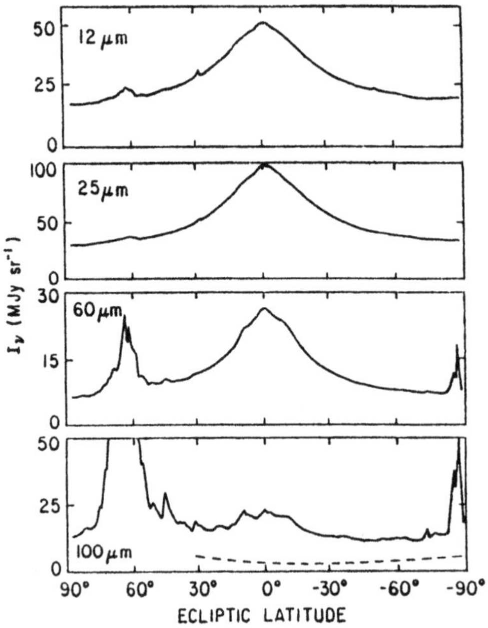

Figure 1. Ecliptic pole-to-pole profiles of total sky brightness at each of the IRAS wavelengths - 12, 25, 60, 100μm - at elongation 91.1° (Hauser, et al. 1984). The enhancement near ecliptic latitude 60° corresponds to the large Magellanic Cloud. Structure near the ecliptic and at 10° on either side corresponds to the zodiacal dust emission bands.

source away from the Milky Way even at 100μm. The sensitive, high resolution IRAS telescope (Neugebauer, et al. 1984a) found the zodiacal emission to be up to a factor of 4 brighter than previously published measurements at 11μm and 20μm (Price, et al.; 1980, 1982). In addition, IRAS found widespread structure in the IR sky background in all four wavelength bands, including an IR "cirrus" in the 100μm data (Low, et al. 1984). Particularly striking was the discovery of relatively narrow dust emission bands in and near the ecliptic for all ecliptic longitudes (Neugebauer, et al. 1984b; Low, et al. 1984). Gautier, et al. (1984) and Hauser, et al. (this volume) find this dust to be at approximately 2.5AU and suggest an asteroidal origin (see, also, Dermott, et al. 1984), compared to the prevailing wisdom that comets are the principal source of the main body of zodiacal dust. At almost the same time, Hong, et al. (this volume) found small scale structure in the elongation dependence of visible wavelength Gegenschein observations from Pioneer 10 and from the ground. In an analysis of additional Pioneer 10 observations, Toller and Weinberg (this volume) found similar "structures" at the same elongations above and below the ecliptic and which reproduced at different heliocentric distances.

Pioneer 10/11 and Helios 1/2 were the first space probes to carry photometric experiments designed to measure zodiacal light from different locations in the solar system: Pioneer for R beyond 1AU and Helios from 1AU to 0.31AU from the sun. Outward-looking measurements from Pioneer found zodiacal light brightnesses for R>2.8AU to be negligible compared to the background starlight (Schuerman, et al. 1977). The R-dependence of dust number density was found from Helios data to vary as $R^{-1.3}$ (Link, et al. 1976; Leinert, et al. 1981), whereas the Pioneer data suggest a steeper decrease for R>1AU (Weinberg and Sparrow 1978; Schuerman 1980a; Toller and Weinberg, this volume). Hong (A&A, in press) suggests that the spatial distribution may be closer to R^{-1} for regions closer than 0.3AU to the sun and (personal communication) that dynamical considerations and the R-dependence of zodiacal light brightness may argue against a single power-law exponent.

Zodiacal light polarization is sensibly zero at the antisun. At other sky positions the orientation of the plane of zodiacal light polarization is either perpendicular or parallel to the scattering plane (i.e., to the direction to the sun/antisun). [By definition this is referred to as positive or negative polarization, respectively.] This is illustrated in Figure 2 for two of ten visible wavelengths observed from Skylab (Weinberg, et al. 1976). The polarization directions are symmetric about the antisun except for distortion caused by Milky Way polarization which, in general, has a different polarization direction. The change from positive to negative polarization or polarization reversal is evident in both colors at large elongations. No evidence was found for time changes in direction or amount of polarization (see, also, Leinert and Planck 1982). Although not readily apparent in the data shown here, the Skylab data are consistent with the result of Weinberg and Mann (1968) that the position of zero polarization or neutral point moves closer to the sun with increasing wavelength. At IR wavelengths there may be more than one such polarization reversal at large elongations and even a reversal in near-sun regions if there are interstellar grain size particles there (Hong, personal communication). As shown by Beard (1984), IR polarization measurements near the sun also "very effectively determine the size of interplanetary dust or place significant limits on the size." [This writer is not aware of the existence of suitable IR polarization measurements.] Sparrow, et al. (1976) and Weinberg and Hahn (1980) found the polarized brightness of zodiacal light to have solar color at five moderate elongation positions, suggesting that the degree of polarization in this region is independent of wavelength between 4000Å and 8200Å. In contrast, Helios results show significant color differences in both brightness and polarization (Leinert, et al; 1981, 1982b). Zodiacal light brightness was found to be redder than the sun, with the reddening decreasing with increasing elongation, and independent of R. The former supports earlier results that the particles primarily responsible for zodiacal light are tens to hundreds of microns in diameter. Polarization degree was found to increase systematically with decreasing wavelength (i.e., from V to B to U) and to be significantly higher at Helios aphelion (1AU) than at perihelion (0.3AU). No explanation was found for the latter.

Zodiacal light brightness Z is a function of both the density n and nature (scattering cross section σ) of the dust. The traditional method of analysis (separation) involves the following two assumptions:

1. dust density is a power of heliocentric distance: $n \propto R^{-\nu}$
2. the nature of the dust is independent of location; i.e., the same type of particle is arbitrarily distributed throughout the solar system.

With these assumptions, observations at different R make it possible to solve for n and σ (Hanner and Leinert 1972). With the second assumption, the brightness Z can be written as $Z(\varepsilon,R) = R^{\alpha}f(\varepsilon,R)$, where ε is elongation and α is an index independent of ε. To test this assumption, Schuerman (1980a) fit a power law of the form $Z \propto R^{\alpha}$, $\alpha = -1-\nu$, to Pioneer data on brightness Z versus ε for 12 different heliocentric distances R. If the aforementioned assumptions are correct, a *single* value of α should apply at all ε. Figure 3 shows the combined result of the 12 (ε) log Z versus log R data sets and of other independent Pioneer data: a systematic

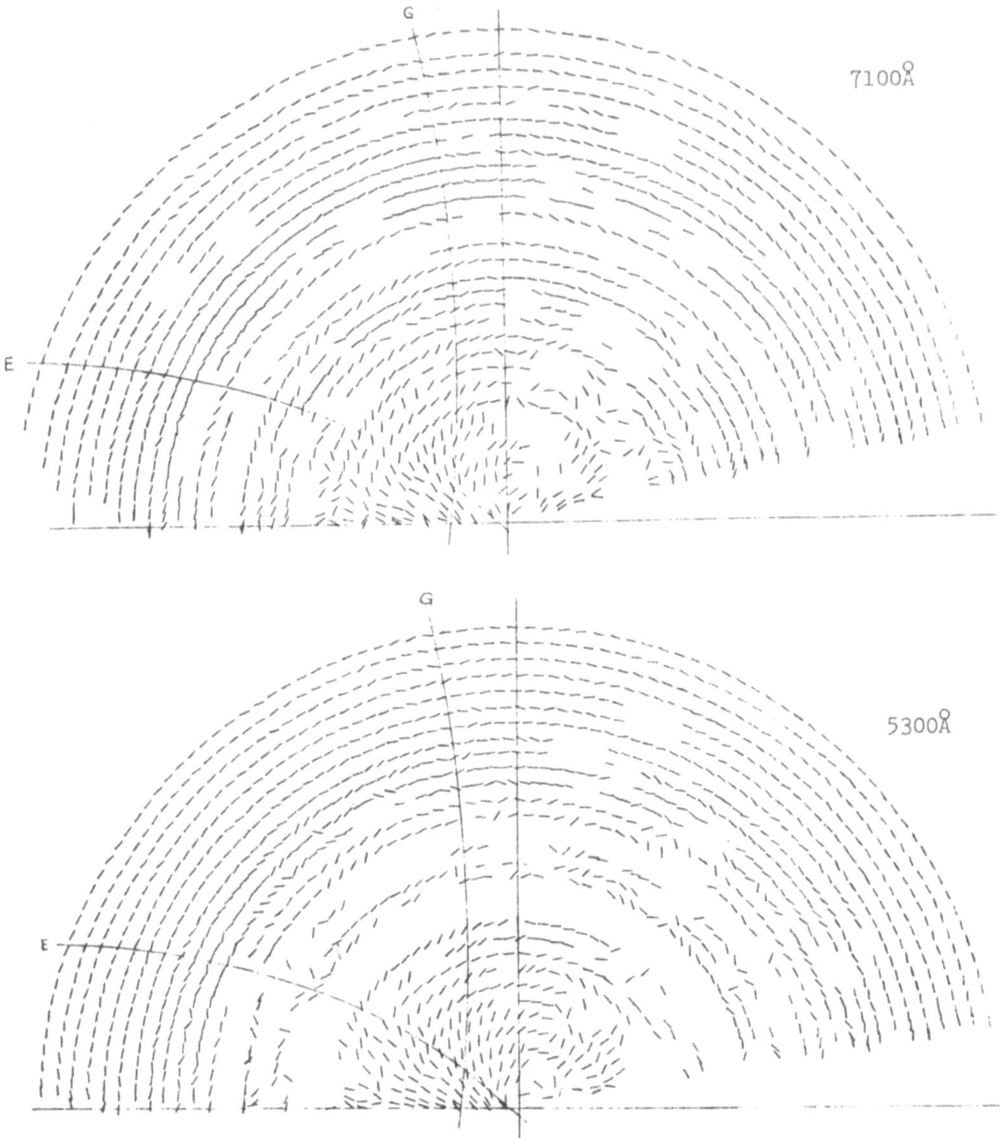

Figure 2. Measured orientations of the plane of polarization (electric vector) taken from Skylab in June 1973. The antisun is at the intersection of the straight lines, and E and G correspond to the positions of the ecliptic and galactic equators, respectively. The outermost data are approximately 70° from the sun. Gaps are primarily due to telemetry data dropouts or noisiness due to bright stars.

decrease in α towards $\varepsilon=100^{\circ}$. Schuerman's tentative conclusion was that one or both assumptions are incorrect; i.e., *the particles cannot be the same throughout the solar system.* It then follows that the plot of α

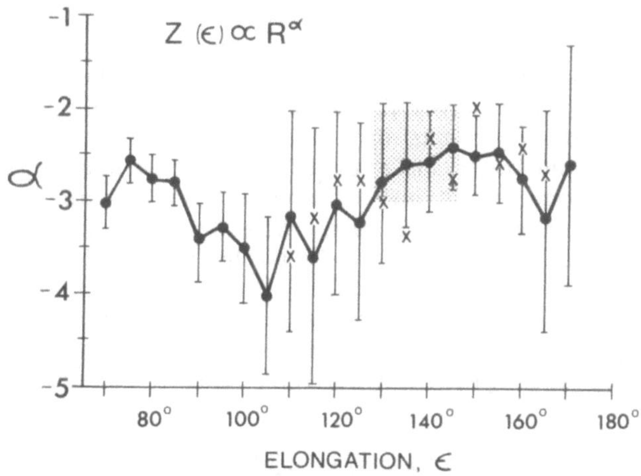

Figure 3. Angular dependence of the index α in brightness $Z \propto R^{\alpha}$, $\alpha = -1-\nu$ (Schuerman 1980a). Data points marked by x's refer to in-ecliptic Pioneer data presented earlier (Schuerman 1979a). The shaded rectangle approximates the earlier and independent Pioneer results of Hanner, et al. (1976). If the scattering properties of the dust are independent of R, α should be constant.

versus ε must contain information concerning the distribution of different types of particles, with the maximum information only able to be obtained from inversion techniques (Dumont 1972, 1973; Schuerman 1979b; others). [D. W. Schuerman's untimely death in May 1982 prevented his completing analysis of the same data with inversion methods.] Changes in polarization with R observed from Helios (see earlier) and from Pioneer 10 (Clarke and Weinberg, unpublished) may be a result of an inhomogeneous dust distribution.

From analysis of the positions of maximum zodiacal light brightness, Misconi and Weinberg (1978) and Misconi (1980) suggest that there may not be a single plane of dust concentration but a "multiplicity" of planes associated with the orbital planes of the planets. Can long-term gravitational perturbations by the planets "shepherd" particles in this way? ... In a search for mechanisms which might produce azimuthal asymmetries in the zodiacal cloud, Schuerman (1980b) found that there may be large scale dust arcs associated with the sun, planets, and the classical L_4 and L_5 Lagrangian points. Special observing geometries, techniques, and timing are needed to search for such arcs from the ground (Giovane, et al., this volume). The demonstrated existence of the arcs and their relatively small particles would add a new dimension to interplanetary dust dynamics. Could such particles contribute to the aforementioned dependence of α on ε? ... IRAS data at different wavelengths refer to particles at different temperatures and therefore at different R. Does the Pioneer data contain information on the same particles? ... As noted earlier, zodiacal scattering is remarkably constant. Are the recently discovered zodiacal emission bands similarly constant? ... Misconi (1977) showed that the F-corona and inner zodiacal light arise primarily from particles physically close to the sun. Brightness, polarization, color, thermal emission, and spectroscopy (doppler shifts and line/band material "signatures") of the F-corona/ inner zodiacal light from several solar radii out to approximately 20° contains *unique* information on the optical and physical nature of the particles, the physics/dynamics/chemistry of particles near the sun, the transition from solar corona (dust and electrons) to zodiacal light, and gas-dust interactions during solar storms. The sensible limit for ground observations of zodiacal light is 30° from the sun, with the result that *ground observations never see the effects of dust closer than 0.5AU to the sun*; i.e., this region can only be observed from space. ... The results and the questions in these few pages are just part of the unfolding picture connecting optics, dynamics, and infrared.

REFERENCES

Beard, D.B.: 1984, Astron. Astrophys. 132, 317-320.
Burnett, G.G.: 1976, Lecture Notes in Physics 48, 53-57.
Dermott, S.F., et al.: 1984, Nature 312, 505-509.
Dumont, R.: 1972, Compt. Rend. Acad. Sci. Paris 275, 765-768.
Dumont, R.: 1973, Planetary Space Sci. 21, 2149-2155.
Dumont, R., Levasseur-Regourd, A.C.: 1978, Astron. Astrophys. 64, 9-16.
Fechtig, H., et al.: 1981, Landolt-Börnstein 2, sub-vol. 3, 228-243.
Gautier, T.N., et al.: 1984, Bull. Amer. Astron. Soc. 16, 442.
Hanner, M., Leinert, C.: 1972, Space Research XII, 445-455.
Hanner, M., et al.: 1974, J. Geophys. Res. 79, 3671-3675.
Hanner, M., et al.: 1976, Lecture Notes in Physics 48, 29-35.
Hauser, M.G., et al.: 1984, Astrophys. J. 278, L15-L18.
Leinert, C.: 1975, Space Sci. Rev. 18, 281-339.
Leinert, C., et al.: 1981, Astron. Astrophys. 103, 177-188.
Leinert, C., Planck, B.: 1982, Astron. Astrophys. 105, 364-368.
Leinert, C., et al.: 1982a, Astron. Astrophys. 110, 111-114.
Leinert, C., et al.: 1982b, Astron. Astrophys. 110, 355-357.
Link, H., et al.: 1976, Lecture Notes in Physics 48, 24-28.
Low, F.J., et al.: 1984, Astrophys. J. 278, L19-L22.
Misconi, N.Y.: 1977, Astron. Astrophys. 61, 497-504.
Misconi, N.Y.: 1980, in Proc. IAU Symp. 90, eds. Halliday/McIntosh, 49-53.
Misconi, N.Y., Weinberg, J.L.: 1978, Science 200, 1484-1485.
Neugebauer, G., et al.: 1984a, Astrophys. J. 278, L1-L6.
Neugebauer, G., et al.: 1984b, Science 224, 14-21.
Price, S.D., et al.: 1980, Astron. J. 85, 765-771.
Price, S.D., et al.: 1982, Astron. J. 87, 131-140.
Schuerman, D. W., et al.: 1977, Bull. Amer. Astron. Soc. 9, 313.
Schuerman, D. W.: 1979a, Space Research XIX, 447-450.
Schuerman, D. W.: 1979b, Planetary Space Sci. 27, 551-556.
Schuerman, D. W.: 1980a, in Proc. IAU Symp. 90, 71-74.
Schuerman, D. W.: 1980b, Astrophys. J. 238, 337-342.
Sparrow, J.G., et al.: 1976, Lecture Notes in Physics 48, 45-51.
Weinberg, J.L., Mann, H.M.: 1968, Astrophys. J. 152, 665-666.
Weinberg, J.L., et al.: 1976, Bull. Amer. Astron. Soc. 8, 502.
Weinberg, J.L., Sparrow, J.G.: 1978, in Cosmic Dust, ed. McDonnell, 75-122.
Weinberg, J.L., Hahn, R.C.: 1980, in Proc. IAU Symp. 90, 19-22.

PHOTOGRAPHIC OBSERVATIONS OF THE INNER ZODIACAL LIGHT ABOARD SALIOUT 7.

G. Nikolsky[(1)], S. Koutchmy [(2)], P.L. Lamy[(3)]
I.A. Nesmjanovich[(4)]
(1) Izmiran, Moscow
(2) Institut d'Astrophysique, Paris
(3) Laboratoire d'Astronomie Spatiale, Marseille
(4) Kiev University, Kiev

ABSTRACT. Color photographs of the inner Zodiacal Light were obtained aboard Saliout 7 with Kodak Ekta 400. Absolute calibrations were performed in-Flight with attenuated Sun light. The photometric analysis was carried out in three colors and corrections were introduced for the OI emission of the F layers. The brightness of the Zodiacal Light along the ecliptic is found in good agreement with published data while its color is found slightly redder than the Sun. Its out-of-ecliptic variation coincides neither with the fan nor the ellipsoidal model; the photometric axis is found near the projection of the orbital plane of Venus.

1 - INTRODUCTION AND OBSERVATION.

As part of a continuying program to study the inner Zodiacal Light which remains poorly known, we present here the results of an observation obtained during the joint french-soviet flight in June 1982, aboard the Saliout 7 orbital station. The so-called PCN (Photographie du Ciel Nocturne) experiment consist in a Nikon 35mm camera with a 58mm f/1.2 Nikkor lense (Belmahdi and Koutchmy, 1982). Three improvements were introduced in comparison with our past observations, (Gretchko et al.,1982, Lamy et al., 1980):

i) a two-axis mount allows to perform manual corrections by visually tracking a star in the field of an auxilliary finder and thus, to offset any linear displacement; however, the rotation of the field cannot be corrected with the device;

ii) a new sensitometer whose key component is a square matrix of 6x6 densities ranging from 0 to 3.5, which can be mounted directly in front of the film and is directly illuminated by attenuated (10^{-8}) solar light (Koutchmy,1981); problems of color correction are therefore enterely avoided and each film is routinely calibrated;

iii) the kodak Ektachrome 400 ASA color film has been selected as the a-posteriori photometric analysis in three colors is now well mastered (Koutchmy, 1978; Lamy and Koutchmy, 1982).

The present photograph, see Koutchmy and Nikolsky, 1983, was taken by Cosmonaut V.A. Djanibekov on June 30, 1982 at 16h U.T. (before sunrise)

R. H. Giese and P. Lamy (eds.), Properties and Interactions of Interplanetary Dust, 7–10.

the solar depression being - 19°. It is a two minutes exposure which unfortunately shows the rich star field - down to a magnitude of 8.5 - of the Hyades and Pleiades regions. The ecliptic plane was well oriented, making a 70° angle with the local horizon and the Zodiacal Light is conspicuously seen in the elongation interval 23 to 35°. The planet Mercury is burried in the bright atmospheric arc while Venus saturates a large disk at the center of the field. Most part of this region could not be observed by the Helios probe (Leinert et al., 1981) because of its pointing 16° away from the ecliptic plane, even when it reached 0.3 AU. The resolution of 2 to 10 min of arc of the present photographic observation represents also an improvement in comparison with the 1° resolution of Helios.

2 - PROCESSING.

The results presented here are based on seven photometric scans having a resolution of 0.1° and judiciously choosen so as to reveal the most interesting features of the Zodiacal Light (Nesmjanowich, 1984). Densities were successively recorded through the three filters (blue, green and red) of the Kodak color separation system. Relative photometry was carried out in the three bandpass which were further corrected for the differences with the Johnson system with a final accuracy of ± 1.5%. Absolute photometry was performed in the green bandpass using 38 stars of the Hyades cluster. It offers a well-established photometric sequence which resulted in a ± 20% accuracy for the absolute photometry of the Zodiacal Light. The background of unresolved stars (V>8.5) was substracteed using the tables of Roach and Gordon (1973) and represents a contribution of less than 10%. No correction was introduced for the galactic component but this will be done in the final processing. However the more intense OI emission from the F-layers at 200-250 km height was carefully studied via its green line (5577 Å) and mainly its red lines (6300 and 6364 Å) which altogether are four times more intense than the green line. Their measured contributions of 40 S10 for the former and 160 S10 for the latter are non-negligible and may introduce an error in the determination of the color of the Zodiacal Light if they are not accounted for.

3 - RESULTS

The intensities of the Zodiacal Light in the plane of maximum brightness for the B, V and R bandpass are given in Fig. 1.

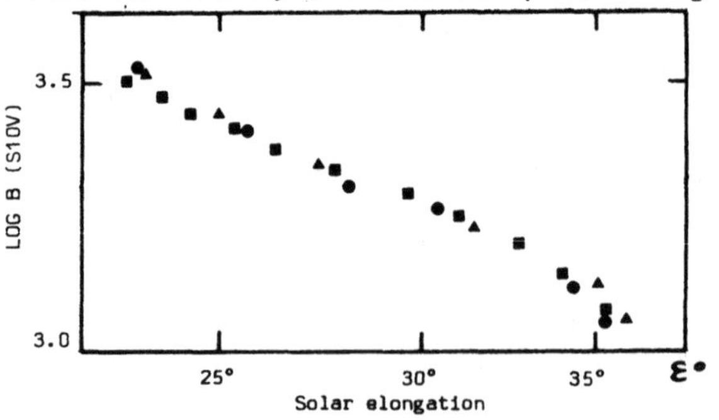

Fig. 1: Brightness of the Zodiacal Light for the B(■), the V(▲) and the R(●) bandpass.

and are in satisfactory agreement with published results (e.g., Leinert, 1975). The slope in the log-log scale is - 2.04 ± 0.1 somewhat smaller than the value of - 2.30 of the model of Koutchmy and Lamy (this volume) in this range of elongation. The color is slightly redder than the Sun, the color index being:

$$\frac{B(4500)}{B(5500)} = \frac{B(5500)}{B(6300)} = 0.94 \pm 0.1$$

in agreement with a now classical result (see Weinberg, this volume). Fig. 2 shows calibrated isophotes which illustrate the two-dimensional aspect of the Zodiacal cloud. The out-of-ecliptic variation is specifically illustrated in Fig.3 and is shown to agree neither with the fan nor the ellipsoidal models.

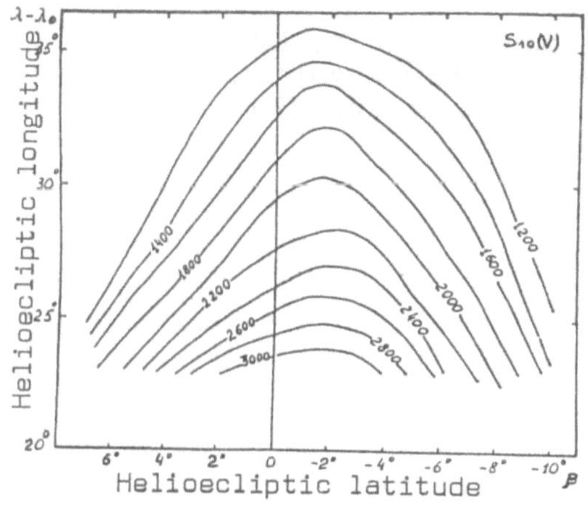

Fig. 2: The isophote in helioecliptic coordinates in units of S_{10} (V)

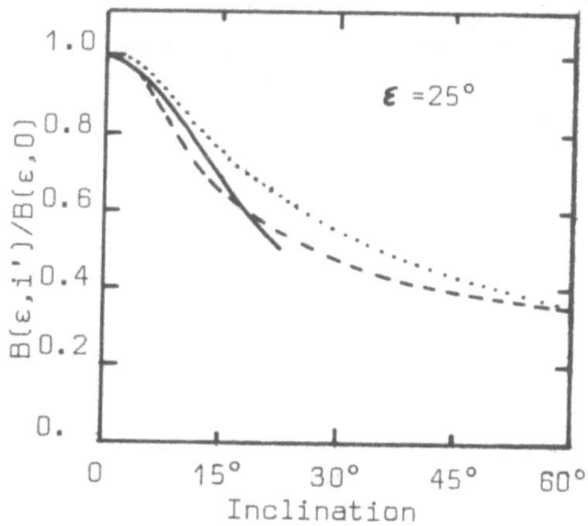

Fig. 3: The out-of-ecliptic variation of the zodiacal light at ε = 25° (solid line) to be compared with the fan (dotted line) and the ellipsoidal (broken line) models.

Finally, the axis of symmetry of the isophotes was found not to coincide exactly with the projection of the orbital plane of Venus, a result which may refine the observation of Misconi and Weinberg (1978).

AKNOWLEDGEMENTS: The PCN experiment was carried out in the frame of French-Soviet cooperation for space research by CNES, France and Inter-Cosmos, USSR involving the participation of many researchers and technicians whose collaboration is acknowledged.

REFERENCES

Belmahdi,M. and Koutchmy,S.: 1982, L'Astronomie, 96, p. 454.
Gretchko,G.M., Divari,N.B , Nikolsky,G.M. and Romanenko,Y.V.: 1982, Pisma Sov. Astron. Zhurnal, 8, 52.
Koutchmy,S.: 1978, Proc. ESO Workshop "Modern Techniques in Astronomical Photography" (Eds R.M. West and J.L. Heudier), p. 225.
Koutchmy,S.: 1981, Astronomical Photography 1981 (Eds, J.L. Heudier and M.E. Sim), p.11.
Koutchmy,S. and Nikolsky,G.M.: 1983, Sky and Tel. 65, 1, 23.
Lamy,P.L. and Koutchmy,S.: 1982, Proc. ESO Workshop "On the Need for Coordinated ground -based observations of Halley'sComet", (Eds, P. Véron and M. Festou), p. 243.
Lamy,P.L., Llebaria,A. and Koutchmy,S.: 1980, Solid Particles in the Solar System (Eds, I. Halliday and B.A. Mc Intosh), p. 37.
Leinert,C.: 1975, Space Sci. Rev. 18, p. 281.
Leinert,C., Richter,I., Pitz,E., Planck,B.: 1981, Astron. Astrophys. 103, p. 177.
Misconi,N.Y. and Weinberg,J.L.: 1978, Science 200, p. 1484.
Nesmjanovich,I.A.: 1984, Dissertatsia, Kiev Univ.-Izmiran.
Roach,F.E. and Gordon,J.L.: 1983, The Light of the Night Sky (ed. D. Reidel).

GROUND-BASED OBSERVATIONS OF NEAR ECLIPTIC ZODIACAL LIGHT BRIGHTNESS

Nebil Y. Misconi and J.L. Weinberg
Space Astronomy Laboratory, University of Florida
1810 NW 6th Street
Gainesville, Florida 32609
USA

ABSTRACT. Ground-based observations of the evening zodiacal light taken by Weinberg and Mann from Mt. Haleakala, Hawaii, during March 1966 are used to derive a table of zodiacal light brightnesses at high spatial resolution (as little as 0.5° in differential ecliptic longitude $\lambda-\lambda_{\odot}$ and 1.0° in ecliptic latitude β) over the region $29.5° < \lambda-\lambda_{\odot} < 56°$, $-30° < \beta < 30°$. Significant differences are found in the brightness distributions above and below the ecliptic plane.

Brightness, polarization, color and angular dependence of the light of the night sky were systematically observed by Weinberg and Mann from Mt. Haleakala, Hawaii between 1965 and 1969 (Weinberg and Mann, 1967). One of the several observing techniques that was used involved scanning the main cone of the zodiacal light over a range of 160° in azimuth centered on the ecliptic, beginning (evening) or ending (morning) with the onset of astronomical twilight. A multicolor photopolarimeter scanned back and forth in azimuth at 2.5 deg/sec, incrementing elevation in 1° steps between 5° and 24°. A sample of these data, taken in March 1966 at 5080Å, is reduced to isolate the zodiacal light. Additional data and full details of the observations, calibration, and data reduction will be presented elsewhere.

Data below elevation 10° are omitted here due to difficulties in the atmospheric corrections. The measured brightnesses were converted to absolute units ($S_{10}(V)_{G2V}$) by reference to a calibrated (by NBS Fritz Peak Observatory), 17.8-cm diameter ^{14}C-activated phosphor source. The source was placed over the objective before and after each night's observations, filling both the aperture and the 3° diameter field of view (FOV). Bright stars were used to obtain an independent absolute calibration, the two methods agreeing to better than 5 percent.

Extinction corrections were made with coefficients derived from observations of bright stars using the same instrument. Atmospheric scattering corrections followed the method outlined by Weinberg (1964). The brightness contributed by "resolved" stars in each FOV were subtracted using a special merged star catalog developed for each color used with this instrument. Background starlight was subtracted using data obtained from Pioneer 10 observations beyond the asteroid belt, where the zodi-

R. H. Giese and P. Lamy (eds.), Properties and Interactions of Interplanetary Dust, 11–15.

acal light was found to be vanishingly small compared to the background starlight (Hanner, et al., 1974). The airglow continuum contribution was obtained by using a typical zenith value of 43 $S_{10}(V)$ (Roach and Smith, 1964) multiplied by the van Rhijn function at each elevation for a plane parallel atmosphere (Weinberg, 1964). The band width and off-band rejection of the 5080Å filter were such that the airglow green line (5577Å) emission can be ignored. Tables 1 and 2 give the derived near-ecliptic zodiacal light brightnesses, at northern and southern ecliptic latitudes, respectively, for differential ecliptic longitudes from 29.5° to 56.0°. Missing data in the Tables correspond to regions in the Milky Way and/or with bright stars, where the separation of components is uncertain, or is a result of the timing and spatial coverage of the observing program. Other observations, not yet reduced in this manner, will be used to fill in most of the data gaps. The full body of data will be used to depict the topology of zodiacal light as seen from 1 AU and to model the large scale spatial distribution of the dust - in and out of the ecliptic.

Some of the Table 1 and 2 data are plotted in Figure 1 together with Dumont's ground observations (Levasseur-Regourd and Dumont, 1980) and with observations from the Helios 16° photometer (Leinert, et al., 1982). The Haleakala brightness data have higher spatial resolution than any other published data, they separately show data north and south of the ecliptic and thereby contain information on the positions of maximum brightness ("photometric axis") in March 1966, and, at Hong's suggestion, are not smoothed in order to avoid masking possible brightness structure. The Haleakala data increases more rapidly toward the sun and falls off more slowly away from the ecliptic than the Dumont data. Possible reasons for these differences are still being evaluated. Two other things are evident in the Haleakala data: (1) the zodiacal light is brighter above than below the ecliptic (i.e., the photometric axis or symmetry plane is displaced 1-2° north of the ecliptic), and (2) brightness structures first found by Hong at large elongations (this volume) were subsequently also found here. Similar structures are also seen in the zodiacal light data from Pioneer 10 (Toller and Weinberg, this volume).

ACKNOWLEDGEMENTS

This research received partial support from NASA grant NAGW-0126 and from the U.S. Air Force Office of Scientific Research grant 83-0107.

REFERENCES

Hanner, M.S., Weinberg, J.L., DeShields, L.M., Green, B.A., and Toller G.N., J. Geophys. Res. 79, 3671, 1974.

Leinert, C., Richter, I., Pitz, E. and Hanner, M., Astron. Astrophys. 110, 355, 1982.

Levasseur-Regourd, A.C. and Dumont, R., Astron. Astrophys. 84, 277, 1980.

Roach, F.E. and Smith, L.L., NBS Technical Note 214, 1964.

Weinberg, J.L., Ann. Astrophys. 27, 718, 1964.

Weinberg, J.L. and Mann, H.M., in Proc., Symp. on the Zodiacal Light and the Interplanetary Medium, NASA SP-150, ed. J.L. Weinberg, p.3, 1967.

$\lambda-\lambda_{\odot}$ / β°	0	1	2	3	4	5	6	7	8	9	10	11	12	13	14	15	16	17	18	19	20	21	22	23	24	25	26	27	28	29	30	β° / $\lambda-\lambda$
56																		193	201													56
55.5															291	246	248															55.5
55												316	289	278						200												55
54.5							393	371	344	336	326							209	208									123				54.5
54	434	418	395	427	434	406								292	278	263	243										125					54
53.5										376			344												154	152						53.5
53			456	414	413	406	378	372	379														156	171								53
52.5	449	455																			214	188										52.5
52																		249	250	223				168	154							52
51.5															289	269	248					197	197									51.5
51											350	367	340	307						240	217											51
50.5					480	442	442	420	421	383							276	261	250					178	171							50.5
50	508	500	507	502									348	343	315	299						203	194				176					50
49.5									443	425	386	351							341	332	240				227	212						49.5
49		539	522	532	541	506	472	446								317	302	289					214	219								49
48.5	566												386	364	328							268										48.5
48							506	481	473	446	424						330	295	264	258					193	198						48
47.5	598	609	607	567	543	536								399	359	346								187								47.5
47											448	441	412							210			260			212	188					47
46.5			613	629	609	577	554	532	503	492							316		275		271	255			211					168	156	46.5
46	632	640											422	385	359	330												166	148			46
45.5								562	518	483	479	438						288	286	290						253	204				196	45.5
45	662	656	662	660	617	597	571									344	325							249	255				262	219		45
44.5												488	448	438	402						268	262	253					231				44.5
44				697	657	629	613	600	573	512								344	333	299						247	234					44
43.5	688	712	720											454	407	389	356						300	272	262					156	160	43.5
43									629		590									368	341	306					218	198	188			43
42.5	763	783	773	730	714	679	685	643																	244	228						42.5
42													555	504	474	423	391	384	344			316	275	249					213			42
41.5						789	759	703	676	613	572	526							353	328	333						252	236				41.5
41	888	863	841	857	824										448	439	379	413						296	285	269					236	41
40.5										654		572	534	483							343	346	302						252	246		40.5
40	894	893	899	852	824	829	829	763	677								464	390	391							306	305	305				40
39.5													585	582										342	320						269	39.5
39						878	849	880	797	765	678	622									417	368						300	291	272		39
38.5	996	1016	987	923	933											522	515	482	463	446					324		318					38.5
38												675	621	589	571							391	381	352							290	38
37.5		1124	1090	1051	997		911	853											478	468	450							304	303	308		37.5
37	1111														614	564	528	508							374	357	328					37
36.5								885	855	812	798	756	693									479	440	384		385				340	336	36.5
36	1194	1180	1184	1132	1090	1053	1006											541	526	497	475								362			36
35.5												768	732	696	650	612	585							438				386				35.5
35			1294	1260	1187	1134	1109	1019	936	896	852									563	530	496	464		405	402	420			390	401	35
34.5	1338	1369													726	709	687	679	604										379			34.5
34									1065	1007	931	880	830	802									515	482			467	439				34
33.5	1507	1491	1488	1418	1354	1285	1192	1127											639	605	546								447	444	400	33.5
33														816	779	728	712	656							534	499	476	474				33
32.5							1347	1190	1179	1070	1028	960	894								657	599	580	560								32.5
32	1647	1648	1602	1560	1515	1383											777	726	664	664								481	466	435		32
31.5											1165	1106	1033	937	922										567	547	530					31.5
31	1882	1874	1814	1712	1649	1601	1510	1421	1321	1210											709	689	632	593								31
30.5															993	936	861	840	781	737												30.5
30							1707	1608	1508	1349	1288	1229	1147	1097																		30
29.5	2145	2135	2070	1977	1902	1794																										29.5

Table 1. Zodiacal light brightness at 5080Å in $S_{10}(V)_{G2V}$ units as a function of differential ecliptic longitude $\lambda-\lambda_{\odot}$, and at northern ecliptic latitudes, β.

λ-λ☉ \ β°	0	-1	-2	-3	-4	-5	-6	-7	-8	-9	-10	-11	-12	-13	-14	-15	-16	-17	-18	-19	-20	-21	-22	-23	-24	-25	-26	-27	-28	-29	-30	β° / λ-λ☉
56																							140	135	128							56
55.5																					166	154					123					55.5
55																	205	178														55
54.5												250	245	217	224	219						153	158	161								54.5
54	434	425	404	393	376	362	351	316	304	296	264								176	170	165						136					54
53.5															228	225	215	186								141						53.5
53									334	320	278	273	257	240								161										53
52.5	449	452	433	418	407	388	366	347											166	168	152					150						52.5
52																	216	200													124	52
51.5																					174	171	175							96		51.5
51																	233	221	213	189						135	110					51
50.5											341	324	314	308	264	262							180	147	148					134	125	50.5
50	508	495	495	483	458	420	426	401	384	372										186	183	171						142	129			50
49.5																249	234	229	217						150	174	157					49.5
49								398	378	369	358											191	180	180						134	133	49
48.5	566	544	530	494	467	466	437												212	216	193							141	143			48.5
48															278	272	264	244							174	152	146					48
47.5	598	578	582	551	540	550	514	473	424	392	371	359										213	208	197						132	133	47.5
47																		284	259	237	218						139	151	147			47
46.5										408	393	388	366	347	293	293	281							219	202	184						46.5
46	632	627	602	560	554	528		452	439											228	213	216	211							148	139	46
45.5														363	338	301	292	266	247								166	144	133			45.5
45	622	643	623	610	573	560	506	502	484	440	410	441	426											183	195	178						45
44.5																				241	221	219	200						161	150	149	44.5
44													391	309	341	322	302	269	256							210	197	179				44
43.5	688	683	653	627	621	587	537	522	501	456	429	368											227	225	212							43.5
43																		301	304	270	260	240							201	193	182	43
42.5	763	769	740	740	712		602	560	538	494	458	450		421	420	390	363									188	221	221				42.5
42																						302	274	262	199							42
41.5															436	404	386	372	338	298	293								182	173	173	41.5
41	888	859	847	787	741	675	666	640	619	576	544	497	472	440											233	209	201	187				41
40.5													474	447	434	417	379	337	308	292	282	267	255	237								40.5
40	894	890	884							568																			198	165	157	40
39.5				847	833	739	705		627																228	220	201	185				39.5
39																				319	282	257		248								39
38.5	996	978	905	871	831	796	741	731	651	614	580	530	523	480	437	410	379	368	338									220	215			38.5
38																						302	267	272	262	257	227					38
37.5														522	508	460	438	397	382	369	336											37.5
37	1111	1085	1041	1003	941	895	844	795	750	703	664	632	562																			37
36.5																						326	304	310	279							36.5
36	1194	1167	1157	1104	1040	983	962	830	766	769	688	642	594	558	526	480	439	398		408	348										243	36
35.5																										340	320	277	282	242		35.5
35																		490	453	431	415	394	378	347	356							35
34.5	1338	1314	1280	1208	1195	1115	1035	938	876	849	780	751		634	604	561	481															34.5
34																																34
33.5	1507	1481	1454													629	616	546	507	503	449	427	391	396								33.5
33				1407	1312	1222	1124	1067			876		773		696															291	277	33
32.5																						434	390	379	362	359	349	327	288			32.5
32	1647	1613	1514	1476	1375	1297	1246	1136	1065	977	902	861	798	747	694	625	584	562	512	511	469											32
31.5																														327	314	31.5
31	1882																				569	516		457	435	416	395	390	352			31
30.5		1844	1776	1716	1626	1524	1426	1306	1187	1114	1055	967	917	879	829	732	699	669	625	604												30.5
30																												449	413	373	380	30
29.5	2145	2090	2017	1900	1791	1669	1571	1458	1342	1270	1183	1080	995	967	912	835	789	740	725	685	646	589	567	535	496	491	471					29.5

Table 2. Zodiacal light brightness at 5080Å in $S_{10}(V)_{G2V}$ units as a function of differential ecliptic longitude, $\lambda-\lambda_{\odot}$, and at southern ecliptic latitudes, β.

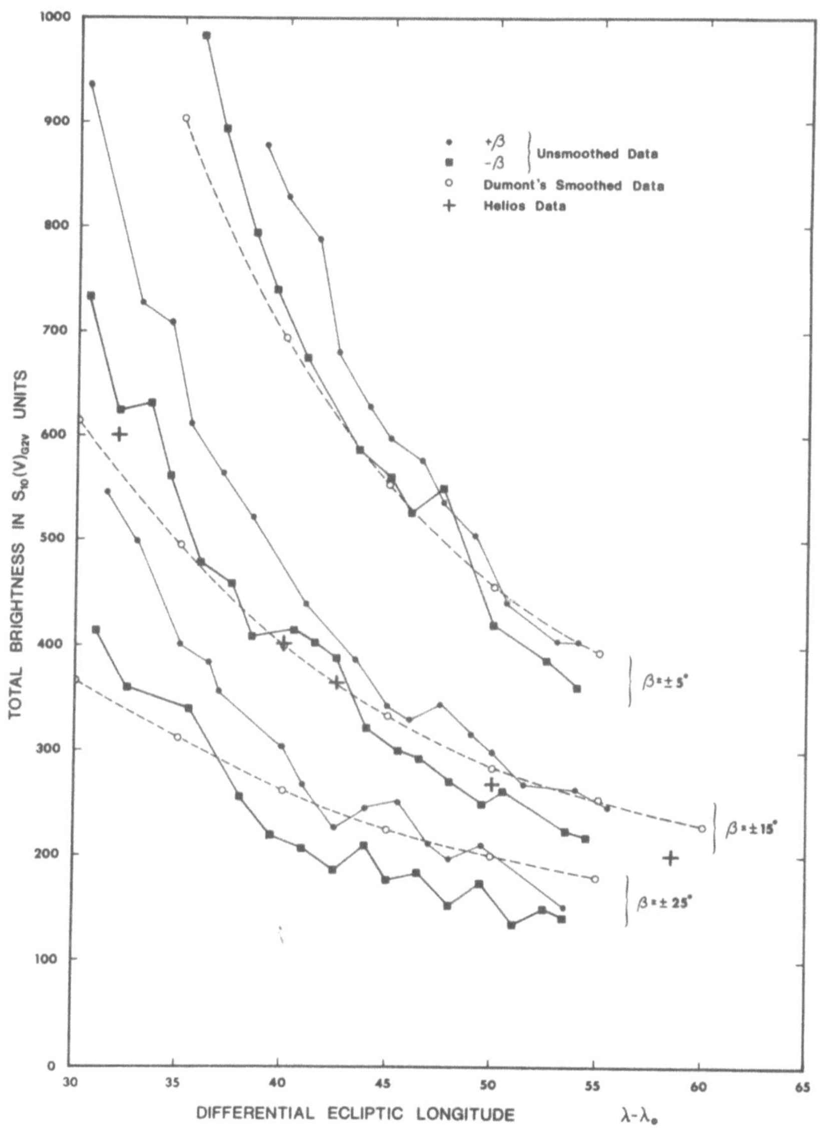

Figure 1. Zodiacal light brightness as a function of differential ecliptic longitude, $\lambda-\lambda_{\odot}$, for ecliptic latitudes, β, of ± 5°, ±15°, and ±25°. Dumont (o) and Helios (+) data are shown for comparison.

AN ISOPHOTE MAP OF THE ZODIACAL LIGHT IN V

J. Pfleiderer and G. Leuprecht
Institute of Astronomy
University of Innsbruck
Austria

ABSTRACT: The reduction method and the results of a photoelectric surface photometry of the Zodiacal Light at a wavelength of 555 nm are presented. A comparison with published results is also given.

The data were obtained on Rockdale Mountain, South Africa, during 1961 and 1962 (for details see Pfleiderer and Mayer 1971). The instrument was a refractor (D = 125 mm, f = 750 mm) with a RCA 1P21 photomultiplier and a GG11 Schott filter (2 mm). The effective wavelength was about 555 nm, the field of view 1.1 square degrees. The data were recorded on strip charts and later on digitized with the semi-automatic digitizer of the Innsbruck computer centre. During the digitization, bright stars down to about 7^m were removed individually from the data.

Contributions to the observed brightnesses come from the Zodiacal Light (ZL), the Milky Way (MW), the airglow (AG), and from light scattered in the lower atmosphere (SL). We have assumed the AG and the SL both to depend on the zenith distance and time but to be independent of azimuth. They can be rather effectively removed by subtracting a background constant from each observed almucanteral scan. During the iteration process described below, the constants were adjusted to give a smooth function of zenith distance and time and also to give consistent results for repeated observations of the same sky areas.

This procedure leaves open, however, one constant for each of the four contributions to the observed brightness. For example, the minimum brightness, which we have observed and set arbitrarily to zero during the reduction, can be assumed to consist mainly of the minimum MW (far outside of the galactic plane), plus the minimum of the ZL, plus the minimum AG, plus the minimum SL (= Rayleigh scattering). These individual contributions cannot be found from our measurements alone but only from outside information.

After subtraction of the background, the remaining brightness -corrected for extinction- consists of MW plus ZL. These contributions can be separated by the relative motion of MW and ZL during the 7 months of observations.

R. H. Giese and P. Lamy (eds.), Properties and Interactions of Interplanetary Dust, 17–20.

We used an iterational process: The ZL (or the MW) was found by subtracting the MW (or ZL) from the observed brightness (minus background) whereever the MW (or ZL) was estimated to be fainter than S_o. The result was used for the estimate of ZL (or MW) in the next iteration. Convergence of the iteration proved to be quite good. We choose S_o = 100 S10V. We found however that the choice is rather uncritical.

The last step of the reduction is the choice of the constants mentioned above. Our best estimate is as follows: The minimum observed brightness was about 95 S10V (zenith distance 30°). According to Landold-Börnstein, at least 30% of this should be AG (28 S10V). The average extraterrestrial brightness (MW + ZL) is about 100 S10V, the average extinction is about twice the zenith extinction , or 25%. About half the scattered light reaches the ground, or roughly 13 S10V. The zenith region being darker than the average, we estimate a SL contribution of about 9 S10V. Finally, Roach and Megill quoted about 28 S10V for the faintest parts of the MW, of which about 24 S10V reach the ground. The minimum ZL is therefore about 95 - 28 - 9 - 24 = 34 S10V observed or 39 S10V extraterrestrial.

The resulting map, averaged over 5×5 degrees, is given in table 1 and compared to other maps in table 2. The minimum at (180,50) is at lower latitudes than that found by LD, and also than the relative minimum at $|\epsilon'|$= 180° found by Classen (1976) in B. A relative minimum occurs at (100-130, 45-65). It was not found by FHLT, and its presence is only marginally supported by the map of LD. Its position coincides, however, approximately with that of the absolute minimum found by Classen in B. It seems that the brightness distribution of the ZL in the darkest regions is still an unsettled question.

While our pole brightness and the brightness change along the ecliptic are in reasonable agreement with FHLT and LD, our intensities in the ecliptic plane are definitely lower. This cannot be solely due to a wrong choice of our zero point. However, part of the discrepancy can be removed by the assumption that our estimate of minimum AG and minimum MW were too high. This would mean that our pole brightness becomes higher than found by anybody else except Wolstencroft and Rose (1967).

This work was partly supported by the Austrian Fonds zur Förderung der Wissenschaftlichen Forschung, project no. 3843. The cooperation of E. Mravlag during the first reduction stages is gratefully acknowledged.

Table 1: Surface brightness of the Zodiacal Light in S10V

|ε'| \ |ß|	0	5	10	15	20	25	30	35	40	45	50	55	60	65	70	75	80
0															58	54	64
5													125	83	53	54	63
10													108	75	59	58	63
15												130	101	68	62	57	63
20										147	128	97	84	64	60	58	62
25										139	111	88	77	66	62	62	62
30									159	115	88	82	77	73	61	64	62
35								155	134	117	81	81	81	67	63	62	62
40							168	147	128	96	86	83	76	67	61	59	63
45						170	163	130	102	93	85	75	71	69	59	59	64
50	425	400	293	257	203	181	146	115	101	91	82	77	69	65	62	66	62
55	392	355	287	246	199	162	138	99	99	91	79	78	69	64	60	64	62
60	325	300	256	213	187	152	111	98	93	79	76	76	69	64	58	66	65
65	272	239	211	184	160	150	143	99	86	78	77	78	67	60	61	62	63
70	231	211	174	146	127	124	115	102	84	71	68	66	60	57	59	61	64
75	199	193	160	140	109	100	96	94	84	73	65	58	59	60	55	57	65
80	171	167	143	113	95	89	84	81	80	76	60	58	61	62	57	61	62
85	158	150	132	109	89	79	76	74	68	65	63	67	63	62	59	63	63
90	148	138	119	102	87	74	68	66	63	64	63	58	56	57	58	66	62
95	137	129	115	97	86	76	67	61	60	58	55	53	54	55	56	61	63
100	129	121	106	94	82	74	64	62	60	53	53	50	53	55	53	56	64
105	119	110	101	91	79	72	66	63	61	54	50	45	51	52	53	61	64
110	110	101	93	85	75	67	64	63	63	57	54	57	53	50	57	60	62
115	99	96	86	79	71	69	65	65	63	54	54	61	57	50	53	59	65
120	89	88	79	74	67	67	63	60	59	51	52	57	56	61	55	60	64
125	89	85	78	71	66	63	60	60	58	53	52	55	55	57	58	59	67
130	89	85	77	73	66	61	60	60	57	56	58	52	54	55	55	59	63
135	87	81	75	73	64	59	58	60	61	55	63	55	52	54	55	60	63
140	84	78	74	71	63	61	59	58	56	58	62	59	57	52	50	57	62
145	83	79	76	71	66	63	59	59	56	57	61	63	57	53	48	56	63
150	81	80	76	70	62	62	58	62	58	56	62	64	56	49	50	56	61
155	79	80	77	69	64	59	57	62	61	57	59	57	54	53	49	56	63
160	82	81	77	72	64	61	59	61	62	59	56	54	52	51	50	57	64
165	84	81	80	75	66	64	63	64	62	60	51	50	47	54	52	57	64
170	91	85	79	76	70	67	63	62	57	54	46	50	48	53	53	61	62
175	102	93	82	78	71	66	59	58	55	50	43	49	48	48	49	61	62
180	115	99	84	78	71	65	57	52	49	44	39	45	44	48	49	57	59

The surface brightness at an ecliptic latitude of 85 to 90 degree remains 63 S10V during all helioecliptic longitudes.
ε' ecliptic longitude with respect to the sun ($\lambda-\lambda_{\odot}$).
ß ecliptic latitude.

Table 2: Comparison with other maps.

Authors	FHLT	LD	P	R	PL
Measured from	balloon	ground	ground		ground
Wavelength [nm] or spectral region	500	502	542.5	V	555
Surface Brightness $B(\|\varepsilon'\|,\|\beta\|)$ in S10V:					
Ecliptic pole	60	63	-	78	63
B(60,0)	-	394	392	500	325
B(100,0)	186	175	211	220	129
B(140,0)	130	139	-	164	84
B(180,0)	166	180	-	205	115
B(180,0)-B(140,0)	36	41	-	41	31
B(100,0)-B(140,0)	56	36	-	56	45
B(100,0)-B(180,0)	20	-5	-	15	14
Minimum ($\|\beta\|=0^o$)	130	139	-	164	79
$\|\varepsilon'\|$ of this minimum	150°	142°	-	140°	155°

FHLT Frey, Hofmann, Lemke, Thum, Astron. Astrophys. 36, 447 (1974)
LD Levasseur-Regourd, Dumont, Astron. Astrophys. 84, 277 (1980)
P Peterson, Astrophys. J. 133, 668 (1961)
R Roach, Astron. J. 77, 887 (1972)
PL Pfleiderer, Leuprecht (present paper)

REFERENCES

Classen, C.: PhD-Thesis Bonn (1976)
Landolt-Börnstein: Astronomy and Astrophysics, Voigt H.H., Ed., Springer Verlag (1965)
Pfleiderer, J., Mayer, U.: Astron. J. 76, 691 (1971)
Roach, F.E., Megill, L.R.: Astrophys. J. 133, 228 (1961)
Wolstencroft, R.D., Rose, L.J.: Astrophys. J. 147, 271 (1967)

THE CHANGE IN NEAR-ECLIPTIC ZODIACAL LIGHT BRIGHTNESS WITH HELIOCENTRIC DISTANCE

G. N. Toller and J. L. Weinberg
Space Astronomy Laboratory, University of Florida
1810 NW 6th Street
Gainesville, Florida 32609
U.S.A.

ABSTRACT. Background starlight observed by the Pioneer 10 Imaging Photopolarimeter from beyond the asteroid belt is used to isolate zodiacal light in Pioneer observations at heliocentric distances R between 1 and 3 AU. Near-ecliptic zodiacal light brightness data in the range 65° to 180° elongation ε are used to depict changes in the shape of the zodiacal light with ε and R and are compared to the corresponding views seen from the Earth and from the Helios 1 and 2 spacecraft.

The heliocentric dependence of zodiacal light can only be observed directly by a space probe, the first such probe measurement being made by Pioneer 10. During the cruise phase of this outer solar system mission, sky brightness and polarization were mapped using a photopolarimeter (Weinberg, et al.; 1973, 1974). At elongations greater than 90°, the contribution of zodiacal light decreased to negligible levels beyond 3 AU (Hanner, et al.; 1974, 1976). As part of an analysis of additional Pioneer 10 data, Schuerman, et al. (1977) pointed out that there is no evidence for the zodiacal light being absent beyond 3 AU - only that it becomes vanishingly small compared to the background starlight (integrated starlight, diffuse galactic light, extragalactic background light). These studies are extended to smaller elongations and, for the first time, zodiacal light brightnesses are tabulated as a function of ε, β, and R (Table 1).

Since zodiacal light is negligible at large R, observations in these regions have been used to derive maps of background starlight over the sky (Toller, 1981; Weinberg, 1981). These data and brightnesses due to discrete ($m_V < 6.5$) stars are subtracted from Pioneer 10 observations between 1 and 3 AU to derive the 4400Å ε,β,R zodiacal light topology given in Table 1. The three entries in each ε,R box correspond to data at ecliptic latitudes β = +10°, 0°, -10° from top to bottom, respectively. Empty boxes indicate regions not yet analyzed. Blank areas within a box denote regions where the data was either not analyzed or is uncertain due to the presence of a bright star in the field of view. Additional Pioneer 10 blue data and the corresponding Pioneer 10 red (6400Å) data will be similarly analyzed and published elsewhere.

R. H. Giese and P. Lamy (eds.), Properties and Interactions of Interplanetary Dust, 21–25.

Figure 1a illustrates the change in zodiacal light brightness, I, with R for $\beta = 0°$. It confirms the falloff of zodiacal light to undetected levels at R > 2.8 AU for outward-looking directions, as noted by Schuerman, et al. (1977). $R^{-2.6}$ is the best fit to the run of relative brightness with R (Figure 1b), a steeper decline than the $R^{-2.3}$ found by Link, et al. (1976) and Leinert, et al. (1981) from Helios 1/2 observations between 0.31 and 1.0 AU. Weinberg and Sparrow (1978) suggested that this difference is probably the result of different viewing directions and spacecraft positions for Pioneer 10 and Helios 1/2. Figure 2 depicts zodiacal light brightness versus elongation for several values of R and β. Pioneer 10 observations from R = 1.011 AU, at .01 AU below the ecliptic, and in a plane parallel to the ecliptic are 20% below the ground-based, in-ecliptic observations taken by Dumont (Levasseur-Regourd and Dumont, 1980). 3% of this can be attributed to the .011 AU difference in R between the Pioneer and ground data, with approximately 10% more due to Pioneer's position below the plane of symmetry of the dust (Hanner, et al., 1976). Helios data at 0.984 AU normalized to 1 AU (Leinert, et al., 1981) also falls below the ground data in the range $\varepsilon > 90°$. There is brightness "structure" in the Pioneer data (Figure 2) that appears to reproduce at the same elongations above and below the ecliptic *and at different heliocentric distances* (i.e., seen through different parts of the zodiacal cloud). Similar ε,β structures were discovered by Hong in Pioneer 10 Gegenschein data and in ground observations from Mt. Haleakala, Hawaii (Hong, et al., this volume).

This research was supported by U.S. Air Force Office of Scientific Research grant 83-0107.

REFERENCES

Hanner, M.S., Sparrow, J.G., Weinberg, J.L., Beeson, D.E.: 1976, in Lecture Notes in Physics, No. 48, 29 (Springer-Verlag).
Hanner, M.S., Weinberg, J.L., DeShields, L.M., Green, B.A., Toller, G.N.: 1974, J. Geophys. Res., **79**, 3671.
Leinert, C., Richter, I., Pitz, E., Planck, B.: 1981, Astron. Astrophys., **103**, 177.
Levasseur-Regourd, A.C., Dumont, R.: 1980, Astron. Astrophys., **84**, 277.
Link, H., Leinert, C., Pitz, E., Salm, N.: 1976, in Lecture Notes in Physics, No. 48, 24 (Springer-Verlag).
Schuerman, D.W., Weinberg, J.L., Beeson, D.E.: 1977, B.A.A.S., **9**, 313.
Toller, G.N.: 1981, Dissertation, State Univ. of N.Y. at Stony Brook.
Weinberg, J.L., Hanner, M.S., Mann, H.M., Hutchison, P.B., Fimmel, R.: 1973, Space Res. **XIII**, 1187.
Weinberg, J.L., Hanner, M.S., Beeson, D.E., DeShields, L.M., Green, B.A.: 1974, J. Geophys. Res., **79**, 3665.
Weinberg, J.L., Sparrow, J.G.: 1978, in Cosmic Dust, 75 (Wiley-Interscience).
Weinberg, J.L.: 1981, Sky and Tel., **61**, 114.

ELONGATION, ε (DEGREES)	HELIOCENTRIC DISTANCE, R (AU): 1.011	1.150	1.171	1.349	1.453	1.652	1.861	2.294	2.413	2.641	2.939
65											4
70								55 63 71		25	2 5
75								40 44 50	33	19 32	4 1 5
80							36 35	33 29 35	25 23	15 22 23	4 3 2
85							25 25 33	26 25 22	20 19 15	7 15 12	1 2 3
90						30 22 24	23 16 19	18 19 18	18 17 13	8 10	0 1 1
95						27 23 27	21 17 21	15 19 19	16 14 13	4 8 9	0 0 0
100					 27	25 24 26	24 25 21	13 16 8	11 13 14	7 4 10	1 0 0
105					40 37 36	31 27 27	25 20 20	10 10 12	9 12 13	5 6	0 0 0
110				 42	41 34 34	36 33 32	22 22 17	10 9	11 4 9	5 7 7	0 0 0
115	122			45 42 49	49 37 36	32 33 27	22 26	8 10	10 11 10	5 6 4	0 0 0
120	109 126			47 41 45	46 45 35	30 32 29	23 25 20	7 6 11	6 11 9	1 6 3	2 0

Table 1 (part 1). Zodiacal light brightness at 4400Å in $S_{10}(V)_{G2V}$ units as a function of elongation (angular distance from the sun) and heliocentric distance R. The three entries in each ε,R box correspond to ecliptic latitudes β = +10°, 0°, -10° from top to bottom, respectively.

HELIOCENTRIC DISTANCE, R (AU)

ELONGATION, ε (DEGREES)	1.011	1.150	1.171	1.349	1.453	1.652	1.861	2.294	2.413	2.641	2.939
	108		69	48	47	32	24	11	13	3	0
125				50	43	37	17	11	13	5	0
		59	57	44	40	31	19	12			
	117	73	73	55	46	29	23	7	13	0	1
130		63	64	52	46	34	10	8	10	0	0
		62	64	45	38	30	20	13			
	113	68	68	53	45	29	20	15	5	1	0
135		68	64	52	43		23	8	0	0	
		68	60	43	37	26	20	8			
	116	73	76	60	42	29	25	7	8		
140	117	74	69	53	48		22	8	6		
		68	65	48	41	28	18	15			
	118	73	78	56	47	34	23	9	5		
145	120	73	79	56	45	35	26				
	101	68	67	46	41	29	17				
	118	83	83	54	52	33	31	6			
150	114	81	80	52	39	35					
	104	73	67	48	45	33					
	117	82	80	59	52	41	23				
155	118	83	80		51	33					
	103	72	72	50	43	28					
	115	89	90	60	48	35	22				
160	119	81	84	59	49						
		79		53	44						
	119	91	86	60	55	43					
165	130	98	90	61	56						
	111	79	78	51	43						
	122										
170	134	107	90	65	57						
	120	83	83	56							
175	138	109	100	70							
180	146	111	108	70							

Table 1 (part 2). Zodiacal light brightness at 4400Å in $S_{10}(V)_{G2V}$ units as a function of elongation (angular distance from the sun) and heliocentric distance R. The three entries in each ε,R box correspond to ecliptic latitudes β = +10°, 0°, −10° from top to bottom, respectively.

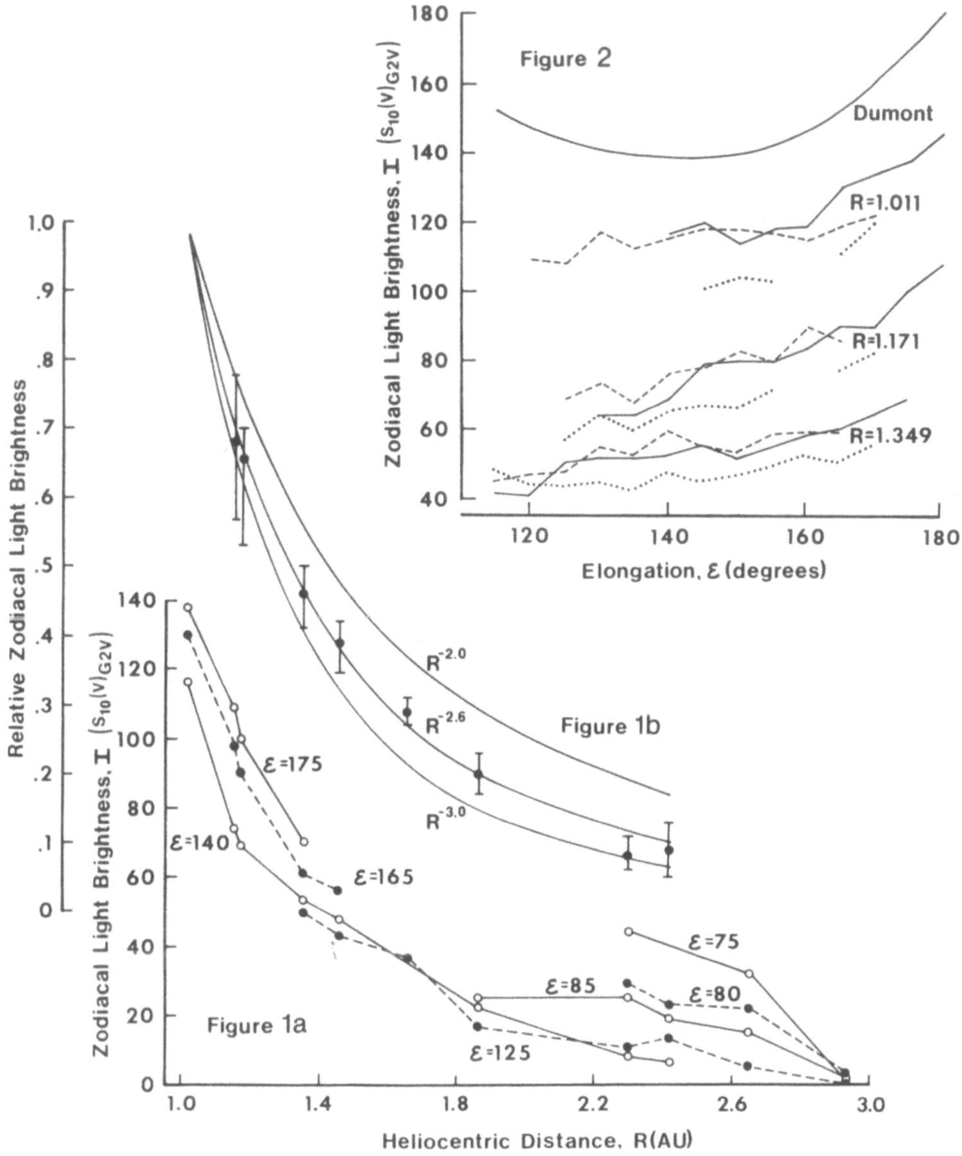

Figure 1a. Pioneer 10 zodiacal light brightness versus heliocentric distance for $\beta = 0^\circ$ and for several elongations.

Figure 1b. Relative zodiacal light brightness versus heliocentric distance. Circles are an average over several elongations, and error bars denote the range of relative brightness for all elongation angles. Power law curves are included for comparison.

Figure 2. Near-ecliptic zodiacal light brightness at large elongations for several heliocentric distances. Observations by Dumont at 1 AU are included for comparison. Solid lines indicate $\beta = 0^\circ$, dashed lines $\beta = +10^\circ$, and dotted lines $\beta = -10^\circ$.

ZODIACAL LIGHT, GEGENSCHEIN AND SKY BACKGROUND

A. MAUCHERAT
A. LLEBARIA
J.C. GONIN
Laboratoire d'Astronomie Spatiale
Marseille

ABSTRACT. Results from photometers on board the D2B satellite (1975) for the whole sky in ultraviolet and blue light, are combined to generate a very well defined empirical representation of the zodiacal light and Gegenschein. The sky background is then obtained by subtraction of this model from the data.

1. INSTRUMENTATION

The observations were obtained with two photometers on board the D2B satellite launched in september 1975. Its orbit had an altitude of 500 - 700 km, an inclination of 37° with respect to the equator and a period of 96 min. The satellite was pointed at the sun with an accuracy of ± 30' during the day and spin stabilized during the night. The spin rate was $1°.5\ s^{-1}$ in the average (Cruvellier, 1970).

2. OBSERVATION OF THE ZODIACAL LIGHT

The data come from the ELZ photometer whose optical axis is pointed at a constant solar elongation of 90° and whose main characteristics are λ =3100 Å, $\Delta\lambda$ = 500 Å and field-of-view = 1° x 2°.8.

The photometric scans, in a plane perpendicular to the earth-sun direction, record the same sky field every six months.

In order to separate the sky background from the zodiacal light, we make use of a very sensitive zero method based on the residue:

$$R(\lambda,\beta) = S(\lambda,\beta,t + 6\ \text{months}) - S(\lambda,\beta,t)$$

as a function of λ and β.

This residue is non zero because of
1) the inclination of the zodiacal cloud with respect to the ecliptic plane and
2) the variation of the earth-sun distance; it is further unrelated to the skybackground (see figures 1 and 3).

The residue can be written in terms of the two components I_{sky} and I_{ZL}:

R. H. Giese and P. Lamy (eds.), Properties and Interactions of Interplanetary Dust, 27–30.

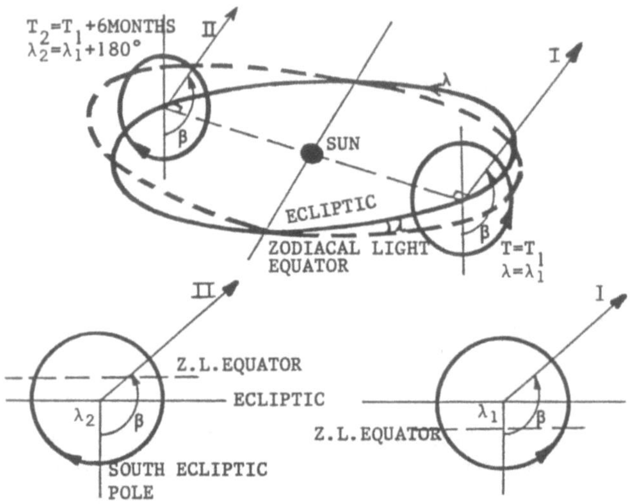

Figure 1. optical axis scans the same sky field every six months, whose planes are perpendicular to the anti-solar directions: λ_I and $\lambda_{II}=\lambda_I+180°$

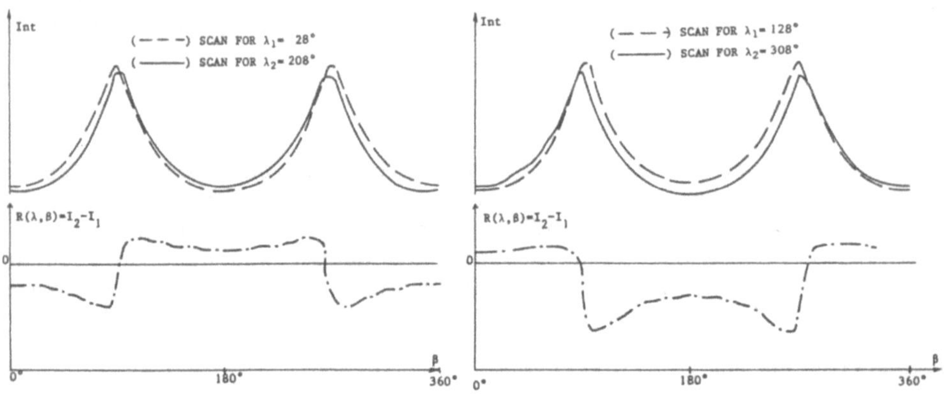

Figure 2. two examples of residues (— · — ·) for two longitudes, computed with models. The south ecliptic pole direction gives the zero of β axis; 180° is the north ecliptic pole.

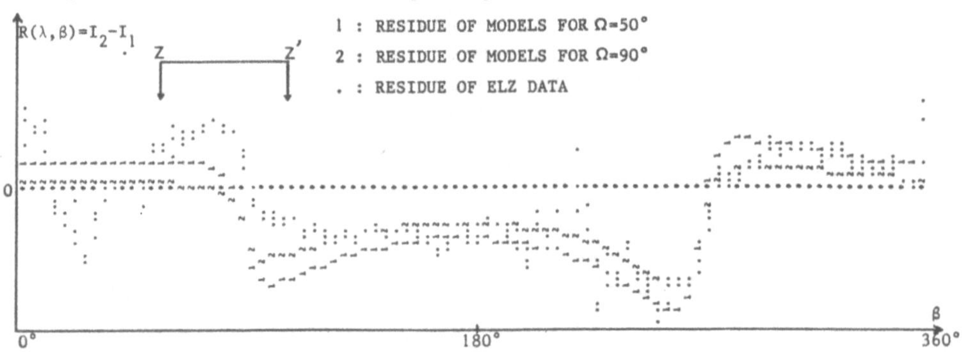

Figure 3. examples of residue for two models computed with two values of ascending nodes, compared with data, for $\lambda_I=128°$ and $\lambda_{II}=128°+180°=308°$
ZZ'= zone where intensity of the zodiacal light is higher for $\lambda_{II}=308°$ than for $\lambda_I=128°$

$$R(\lambda,\beta) = \underbrace{I_{sky}(\lambda,\beta) + I_{2ZL}(\lambda,\beta)}_{II} - \underbrace{\left[I_{sky}(\lambda,\beta) + I_{1ZL}(\lambda,\beta)\right]}_{I}.$$

Residues obtained from the observational data were compared with calculated residues using various models proposed by Leinert et al. (1976).

The fan model of the form:

$$I = K\int_0^{\infty} \frac{r_1^{-\nu}}{r^2}\ \sigma_1(\Theta)\ \exp(-w\ |\sin^{u}\beta_{\odot}|)dx$$

was found to generate intensities and residues $R(\lambda,\beta)$ whose variations match exceedingly well those of the observational data and residues over the period 1975–1976.

We adopted a phase function $\sigma_1(\Theta)$ derived from the work of Mujica et al. (1979):

$\sigma_1\ (\Theta) = 0.00021\ (\Theta - 90°.0)^2 + 1.$

The nominal values of the parameters of the above model as well as the inclination i and the ascending node Ω of the plane of symmetry, and the corresponding error bars were determined by bracketing the observational residues within two extreme calculated residues.

Figure 3 shows an example of variation of the residue for two values of Ω.

We used skyfields where I is critically sensitive to ν, w, u, i and Ω separatly, and we obtained:

$\nu = 1.0 \pm 0.1$	$i = 2°.0 \pm 0°.5$
$w = 4.0 \pm 0.2$	$\Omega = 70°.0 \pm 10°.0$
$u = 1.3 \pm 0.1$	

slightly different from values found by Leinert et al. (1976). We present in figure 4 the profile of the zodiacal cloud along an ecliptic meridian for $\lambda = 90°$

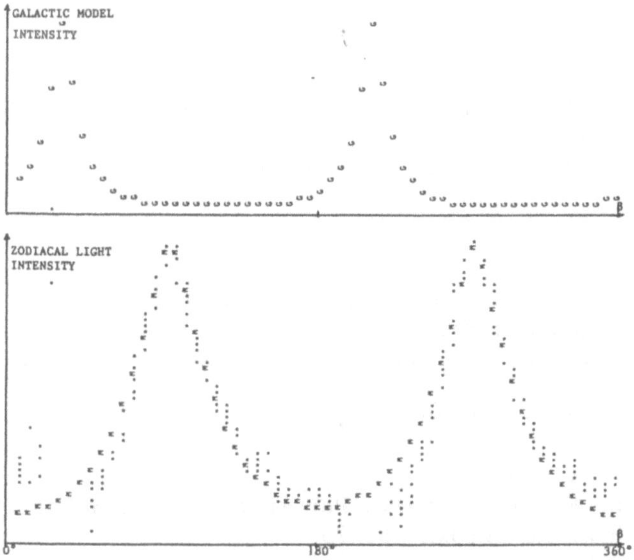

Figure 4. data for λ=90°, minus a galactic model, compared to the zodiacal light model (M). G is the galactic model for this scan.

3. GEGENSCHEIN

The data came from the ERC photometer which scanned the antisolar direction at $\lambda = 4500$ Å, $\Delta\lambda = 1000$ Å in a circle of 9° radius with a resolution of 1° x 1°.

Using the Pioneer 10 results for the sky background (Weinberg, 1981), we obtained nine gegenschein maps for nine ecliptic longitudes over the period 1975-1976.

By applying the zodiacal light model obtained perpendicular to the antisolar direction, with a new phase function: $\sigma_2\ (\Theta) = \sigma_1\ (\Theta) + \sigma'\ (\Theta)$

$$\sigma'\ (\Theta) = 0 \text{ for } \Theta < 177°$$
$$\sigma'\ (\Theta) = 0.02\ (\Theta - 90°.0)^2 - 3.49\ (\Theta - 90°.0) + 152.25 \text{ for } 177° < \Theta < 180°$$

we found that it is necessary to enhance the model in a 3° circle centered at the antisolar point as already suggested by Misconi (1981). The map for $\lambda = 185°$ is shown in figure 5.

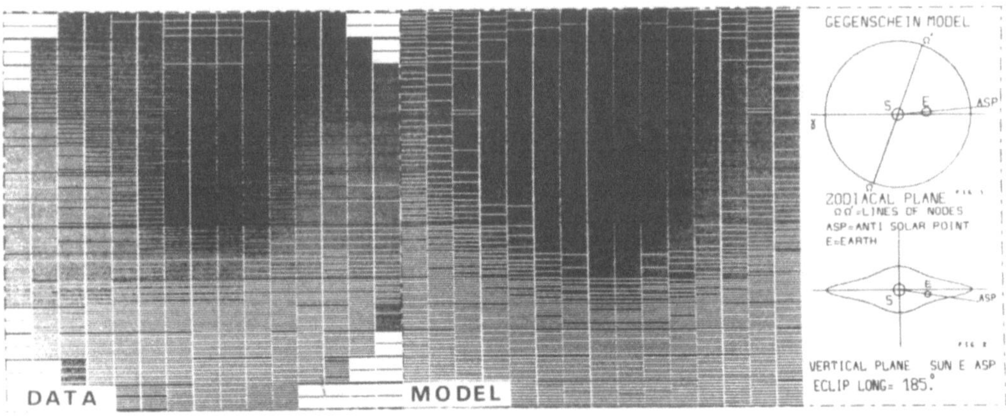

Figure 5. comparison of gegenschein map obtained for λ=185°, computed with ERC data, with the gegenschein model in the same conditions. The enhancement near antisolar direction is clearly visible.

REFERENCES

Cruvellier, P.: 1970, Proceedings of Summer School, Space Optics ed. A. Maréchal and G. Courtes, Gordon and Breach

Leinert, C., Link, H., Pitz, E., Giese, R.H.: 1976, Astron. Astrophys. 47 p. 221

Maucherat-Joubert, M., Cruvellier, P., Deharveng, J.M.: 1979, Astro. Astrophsy. 74 p. 218

Misconi, N.: 1981, Icarus, 47 p. 265

Mujica, A., Lopez, G., Sanchez, F.: 1980, Solid Particles in the Solar System (eds, J. Halliday and B.A. Mc Intosh) p. 75

Weinberg, J.L.: 1981, Sky and Telescope 61, p. 114

RECENT ROCKET OBSERVATIONS OF THE ULTRAVIOLET SPECTRUM OF THE ZODIACAL LIGHT

P.D. Feldman, R.P. Cebula,
P.D. Tennyson, and R.C. Henry
Department of Physics and Astronomy
The Johns Hopkins University
Baltimore, MD 21218
U.S.A.

ABSTRACT

Observations of the spectrum of the zodiacal light in the wavelength range 1800-3000 Å have been obtained in two recent sounding rocket experiments. The September 1979 observation was made close to the ecliptic plane at elongation angles of 20° and 28° and with a spectral resolution of 25 Å, while in April 1981 the elongation was 125° (inclination 60°), and the spectral resolution 50 Å. Both observations confirm the slight reddening in the ultraviolet reported previously and show the same dependence on elongation as in the visible. There appears to be some structure in the ultraviolet albedo of the dust grains near 2800 Å but the exact nature of this structure cannot be determined at the low spectral resolution of the present observations. Detailed descriptions of the two observations has been given by Cebula and Feldman (1982) and Tennyson (1983), respectively.

REFERENCES

Cebula, R.P., Feldman, P.D.: 1982, Astrophys. J., **263**, 987-992.
Tennyson, P.D.: 1983, Ph.D. Dissertation, The Johns Hopkins University, chapter 7.

R. H. Giese and P. Lamy (eds.), Properties and Interactions of Interplanetary Dust, 31.

A SEARCH FOR SMALL SCALE STRUCTURES IN THE ZODIACAL LIGHT

S. S. Hong*, N. Y. Misconi, M. H. H. van Dijk[1],
J. L. Weinberg, and G. N. Toller
Space Astronomy Laboratory, University of Florida
1810 NW 6th Street, Gainesville, Florida 32609, USA
[1]Free University, Amsterdam, the Netherlands

ABSTRACT. Gegenschein observations from Pioneer 10 were found to have brightness structures with an amplitude of about 10% and a period of several to ten degrees in elongation. A search is made for such structures in high angular resolution ground-based observations from Mt. Haleakala, Hawaii. A new empirical method is used to correct for atmosphere-originated radiation. Background starlight is subtracted using Pioneer 10 observations from beyond the asteroid belt. Preliminary analysis of the ground data also indicates the presence of small amplitude structures in the brightness distribution.

This is a first report on our search for small scale structures in the zodiacal light brightness. By structures we mean small amplitude spatial fluctuations in the brightness distribution. Such structures might be expected from inhomogeneities in the spatial distribution of the zodiacal dust and/or structure in the scattering phase function. There have been a number of observational hints for the existence of such structures. Ground-based observations of zodiacal light brightness by Weinberg (1963) show an enhancement at elongation 135°, which also has a counterpart in the polarized brightness (Weinberg, 1964; Frey, et al., 1974). Other space and ground-based observations of zodiacal light have also shown irregularities in the brightness distribution (Sparrow and Ney, 1972; Dumont and Sánchez, 1975). Despite these observational indications, structures in the zodiacal light have generally been ignored for two reasons. First, there was a lack of detailed information on the distribution of background starlight over the sky. Second, for ground observations it is difficult to determine the airglow continuum and to correct for airglow and astronomical diffuse background that is scattered into the telescope's field of view by the Earth's atmosphere.

Figure 1 shows Pioneer 10 near-Earth observations of zodiacal light brightness as a function of differential ecliptic longitude $\lambda-\lambda_\odot$ for various ecliptic latitudes. Connection of the points in Figure 1

*On leave of absence from Seoul National University, Korea.

R. H. Giese and P. Lamy (eds.), Properties and Interactions of Interplanetary Dust, 33–37.

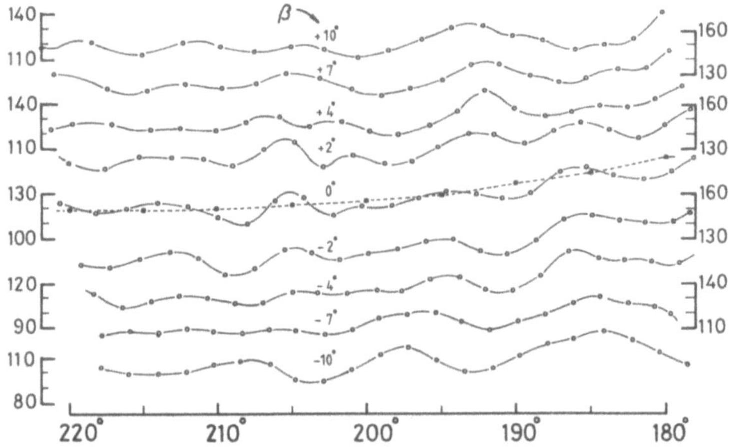

Figure 1. Zodiacal light brightness distribution near anti-solar point as seen from Pioneer 10 on March 14, 1972 at 1.011 AU from the sun. Abscissa represents differential ecliptic longitude λ-λ⊙ and ecliptic latitude β is given to each curve.

reveals a persistent pattern of fluctuations having 10% amplitude 7 to 10 degree period. When the fluctuations are smoothed out, the underlying average distribution has the same differential ecliptic longitude dependence as Dumont's ground observations (Levasseur-Regourd and Dumont, 1980); the filled circles connected by dashes are their ecliptic values multiplied by a constant factor 0.86. Because the amplitude of the structure is not much greater than the noise level of the observations and the Pioneer's instantaneous field of view is approximately 2.3° by 2.3° , we decided to examine selected higher angular resolution ground-based observations obtained by Weinberg and Mann.

One night's observations, 21/22 August 1968 at Mt. Haleakala, Hawaii, are particularly well-suited for this investigation. The sky was scanned repeatedly over the entire range (360°) of azimuth at elevations 10°(05)45° using a 1.7°-radius field of view. Eleven such groups of scans were made at each of wavelengths 5080 Å and 5300 Å. Observations at 5080 Å are analyzed here.

Because the same sky positions were observed at different times through different air masses, the following correction scheme for atmosphere-originated radiation was devised specifically for this body of data. The observed brightness (B) of a sky position includes the zodiacal light (ZL), background starlight (BS), resolved starlight (ST), airglow emission (AG), and the airglow and astronomical radiation diffusely scattered into the field of view (SCT). Putting the airglow emitting layer outside the scattering atmosphere and denoting the optical path length of the atmosphere at elevation h by t(h), we have

$$B = (ZL+BS+ST)\exp[-t(h)] + AG(h)\exp[-t(h)] + SCT(h). \qquad (1)$$

Similarly, when the same sky position is viewed at h´, we have

$$B' = (ZL+BS+ST)\exp[-t(h')] + AG(h')\exp[-t(h')] + SCT(h'). \quad (2)$$

$X(h) \equiv AG(h) + SCT(h)\exp[t(h)]$ is the amount of airglow emission and total scattered light which must be removed to obtain ZL+BS+ST. From equations (1) and (2) we obtain

$$X(h) - X(h') = B\exp[t(h)] - B'\exp[t(h')]. \quad (3)$$

Figure 2 illustrates how X(h) increases with increasing atmospheric extinction exp [t(h)]. Values on the left ordinate represent the excess brightness of X(h) over X(45°) in $S_{10}(V)_{\odot}$ units. Because the Haleakala Observatory is located at latitude 20.7 degrees north, the elevation 20 degree scans include the north celestial pole, enabling us to determine X(45°) and giving an absolute scale for X(h) as shown on the right ordinate. Spreads of the data points shown by vertical bars in Figure 2 are clear indications of spatial and temporal variations of the atmosphere-originated radiation. To minimize the azimuthal dependence of X(h) due to spatial variations of astronomical background, we only include regions more than 30° away from the galactic plane.

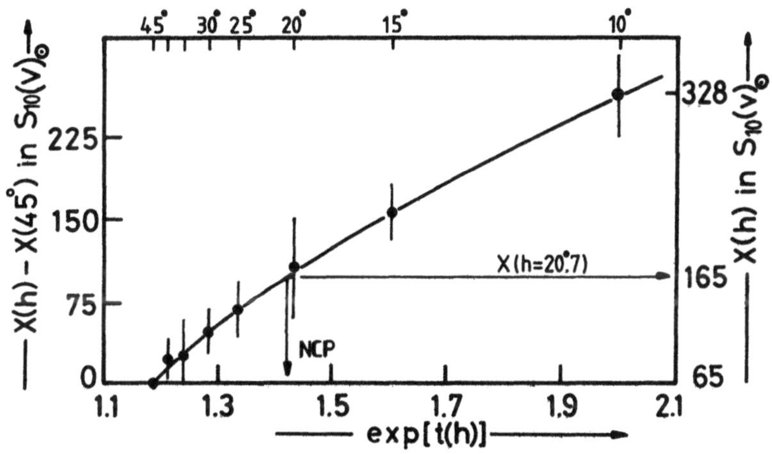

Figure 2. Elevation dependence of the atmospheric correction term X(h). To set up an absolute scale for X(h), 64 and 73 $S_{10}(V)_{\odot}$ were used for the brightnesses of background starlight at the north celestial pole and zodiacal light there at the time of observation, respectively.

The zodiacal light is determined by using X(h) (the solid line in Figure 2) to correct for airglow continuum emission and atmospheric scattering, by subtracting Pioneer 10 determinations of background starlight BS (Toller, 1981), and by subtracting the contribution ST by stars brighter than visual magnitude 6.5 from Bexp[t(h)]. The resulting brightness distribution is shown, in Figure 3, as a function of $\lambda - \lambda_{\odot}$

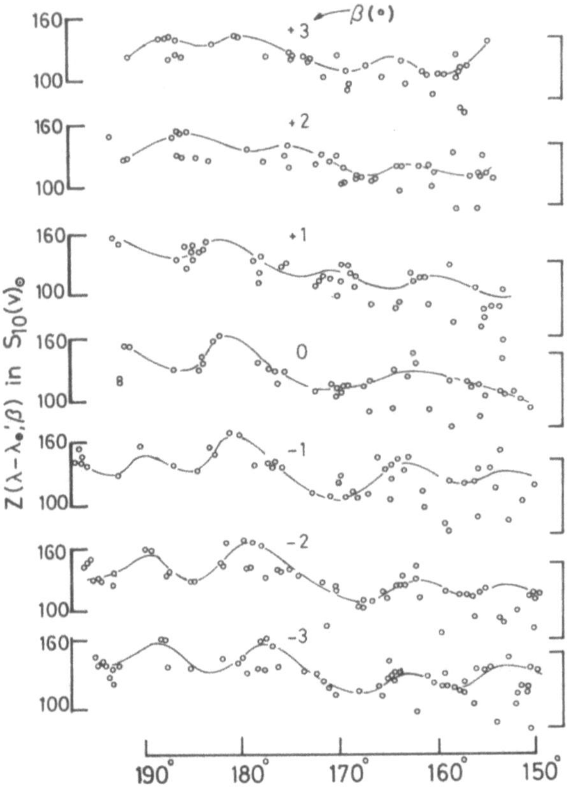

Figure 3. Zodiacal light brightness distribution near anti-solar point as observed from Mt. Haleakala, Hawaii on 21/22 August, 1968. Abscissa represents the differential ecliptic longitude and ecliptic latitude β is given to each curve.

for various ecliptic latitudes. A 10% amplitude and 10 degree periodicity of the structures are evident in the ground-based observations, and the structures persist at different latitudes. In order to confirm the structures seen in this preliminary study, we intend to refine the absolute calibrations, examine polarization observations, and extend coverage in elongation, wavelength and season. Nevertheless, the consistency in overall characteristics of the structures that has emerged from both the space and ground-based observations strongly suggests that the zodiacal light has fine structure which could provide new information on the spatial distribution and/or optical properties of the interplanetary dust particles.

This work was supported by US Air Force Office of Scientific Research grant 83-0107.

M.H.H.v.D. acknowledges support by the Netherlands Organization for the Advancement of Pure Science (Z.W.O.), and also acknowledge travel support from the Leids Kerkhoven-Bosscha Fonds.

REFERENCES

Dumont, R., Sánchez, F.: 1975, Astron. Astrophys. **38**, 405.
Frey, A., Hoffmann, D., Lemke, D., Thum, C.: 1974, Astron. Astrophys. **36**, 447.
Levasseur-Regourd, A. C., Dumont, R.: 1980, Astron. Astrophys. **84**, 227.
Sparrow, J. G., Ney, E. P.: 1972, Ap. J. **174**, 705.
Toller, G. N.: 1981, PhD Dissertation, State University of New York at Stony Brook.
Weinberg, J. L.: 1963, PhD Dissertation, University of Colorado.
______________: 1964, Ann. d'Astrophys. **27**, 718.

DISCUSSIONS

Dumont: The patchy and sometimes rapidly evolutive distribution of the airglow continuum is known to be a main difficulty in ground-based zodiacal light observations. Part of the structure that you report in the Gegenschein could be ascribed to these airglow fluctuations. To what extent can your reduction method take into accound such telluric effects?

Hong: As scatter in Figures 2 and 3 indicates, airglow is certainly a problem. But, for the following reasons, we do not think the structure seen in this study is due to airglow variation. The pattern of structures persists at widely different ecliptic latitudes. The structure seen in the Pioneer data is consistent, at least in overall characteristics, with the one seen in the ground data. All night monitoring of brightness at the north celestial pole at Mt. Haleakala typically does not show rapid changes at 5080 Å and 5300 Å, which is in agreement with the 5000 Å result from the balloon-borne observations by Frey, et al. (1974). From extensive multicolor observations of celestial pole brightness we hope to derive a time-dependent correction for the atmospheric term, thereby reducing the scatter in Figure 3.

AN OBSERVATIONAL SEARCH FOR THE SCHUERMAN DUST ARCS

F. Giovane, J. L. Weinberg, H. M. Mann, and J. P. Oliver
Space Astronomy Laboratory, University of Florida
1810 NW 6th Street
Gainesville, Florida 32609
U.S.A.

ABSTRACT. Large-scale interplanetary dust arcs were predicted by Schuerman (1980) to be associated with the planets. An arc, if one exists, would produce asymmetries on opposite sides of the Gegenschein during the Earth's two passages through the arc each year. A brief description is given of Gegenschein/arc observing constraints, of future viewing opportunities, and of the results of a test program conducted from Mt. Haleakala, Hawaii, in Jan/Feb 1984 when the Earth encountered the arc associated with Saturn.

In the course of performing dust dynamics studies associated with the International Solar Polar Mission (ISPM) Zodiacal Light Experiment (Schwehm, et al., 1981) and with analysis of Pioneer 10/11 data, Schuerman (1980) found that a previously unsuspected solar-system phenomenon might exist: interplanetary dust arcs which could span the entire solar system. An arc, hereafter called Schuerman Dust Arc or SDA, was predicted to be associated with each planet - those of Jupiter and Saturn being predominant. His prediction was based on the results of recalculating the restricted three-body problem to include radiation pressure. The combined effects of sunlight and gravity extend the classical equilibrium (Lagrangian) points L_4 and L_5 into a circular arc extending from L_4 through the sun to L_5 - an arc made up of "trapped" .01 μm to 1.0 μm size particles. For particles in this size range, the radiation pressure force is not negligible compared to the gravitational force. Such grains are referred to as β particles. Perturbing and/or destructive forces such as the Lorentz force resulting from the interplanetary magnetic field, solar wind interaction, and gravitational effects of other planets were not included in Schuerman's treatment. Thus, the stability of the arcs has not been definitively established. Schuerman estimated that for a typical value of $\beta = 0.57$, particles in the Jupiter-sun and Saturn-sun arc systems have characteristic lifetimes T of 14,700 and 50,000 years, respectively. Since β particles now exist in the interplanetary medium and the gradual spiraling into the sun due to the Poynting-Robertson effect occurs on a time scale on the order of T, then the replenishment mechanism, regardless of details as to its source, must be of sufficiently short time scale to replenish the interplanetary

R. H. Giese and P. Lamy (eds.), Properties and Interactions of Interplanetary Dust, 39–42.

dust and therefore the "leaking" SDA material as well.

We had planned to look for azimuthal asymmetries in the interplanetary dust complex using observations of the ISPM Zodiacal Light Experiment (ZLE). During its two periods of solar polar passage, at 1.7 AU above and below the ecliptic, isophotes of zodiacal light would have shown if the dust complex is truly azimuthally symmetric. And gross enhancements in the spatial density of dust along arcs associated with Jupiter and Saturn should have been detectable. When NASA's ISPM spacecraft, including the ZLE, was cancelled, we turned to a program of ground-based observations in search of the SDAs. Our plan was to view an arc when it is encountered by the Earth (twice per year) and look along the arc to take advantage of the longer optical path (Figure 1). The requirement to observe tangentially to an arc and to include the brightest portion means that we will be looking at and near the Gegenschein. An SDA, if one exists, would produce a subtle, low-light level asymmetry to one side of the Gegenschein during one observing period with reversed asymmetry approximately six months later.

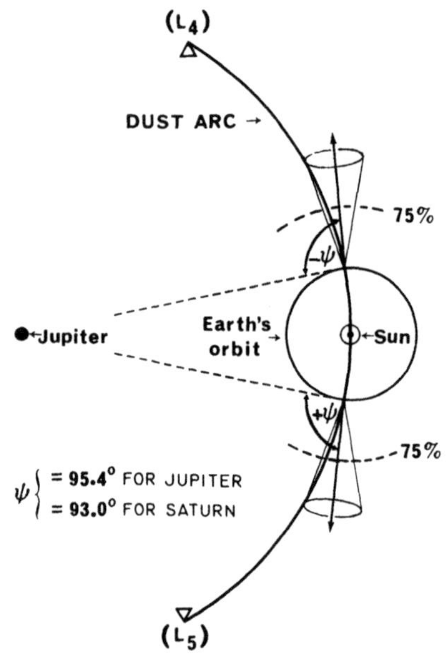

Figure 1. Viewing the Gegenschein when the Earth is in Jupiter's dust arc. The dashed curves labelled 75% indicate that 3/4 of the Gegenschein originates from particles within 1 AU of the Earth (Misconi, 1977).

The Gegenschein itself - and, especially, an SDA-associated asymmetry in the Gegenschein - has formidable observing constraints from the ground: no moon or twilight, wide separation from the horizon and from the Milky Way, a low latitude site to minimize airglow line and continuum emission, and a high altitude stable atmosphere with relatively few aerosols. Since the Gegenschein is an extended diffuse source, it is necessary to map a relatively large region of the sky, typically 40° by 40° centered on the antisun, in order to properly characterize and isolate the angular extent of the Gegenschein/SDA complex. A multicolor photopolarimeter was used during a Saturn SDA "window" in January/February 1984 from Mt. Haleakala, Hawaii, to test the observing methodology: number of wavelengths, timing, sky coverage, data rates. The method most suitable for mapping the Gegenschein/SDA region involves scanning back and forth in azimuth at a fixed elevation (preferably 40° or more) over a period of time sufficient to measure 20° east and west of the antisun position. Holding the elevation constant as the astronomical sky drifts through the azimuthal scan minimizes variations in airglow continuum and atmospheric scattering and extinction associated with changing air mass. It should be noted that knowledge of the absolute value of these atmospheric effects is not crucial in the search for SDAs; it is only necessary that these effects be minimized and that horizontal inhomogeneities be averaged out.

The condition required for the Earth to be in position to observe an SDA can be specified in terms of the angular distance ψ, as viewed from the Earth, between the antisolar point and the planet (Figure 1). For Jupiter and Saturn these angular distances are 95.4° and 93.0°, respectively. Figure 2 illustrates five ecliptic longitude tracks during 1985-86: the antisolar point, $\psi = \pm 95.4°$ (Jupiter) and $\psi = \pm 93.0°$ (Saturn). Times when the antisolar point position crosses the Jupiter- and Saturn-related positions establish when the Earth is in the arc and the geometry is suitable for observing. These (8) crossings are denoted in Figure 2 by open circles, and the conditions of these observing times are given in Table 1. Only four of the eight possible observing periods are considered good or excellent for observing SDAs during 1985-1986; three of these are associated with the Saturn SDA.

Table 1. SDA Observing Times and Conditions, 1985-86.

crossing	date	arc	age of moon	qualitative estimate
1	14 Feb 1985	Saturn	6 days	good
2	30 Apr 1985	Jupiter	10	fair
3	18 Aug 1985	Saturn	2	excellent
4	7 Nov 1985	Jupiter	24	good
5	25 Feb 1986	Saturn	15	poor
6	6 Jun 1986	Jupiter	28	poor
7	30 Aug 1986	Saturn	24	good
8	13 Dec 1986	Jupiter	11	poor

An absolute minimum of two observing sessions, with the Earth on opposite sides of the sun, is required to establish the existence of one SDA system. Gegenschein mappings should also be made at times when the Earth is not in the arc, for use as a "control". If no SDA is detected, multicolor maps would be obtained of brightness and polarization near the antisun - a region of high information content on the properties of interplanetary dust.

This research was supported by U.S. Air Force Office of Scientific Research grant 83-0107.

REFERENCES

Misconi, N. Y.: 1977, Astron. Astrophys., **61**, 497.
Schuerman, D. W.: 1980, Ap. J., **238**, 337.
Schwehm, G. H., Giese, R. H., Giovane, F., Schuerman, D. W., and Weinberg, J. L.: 1981, in Adv. Space Res., **1**, 121.

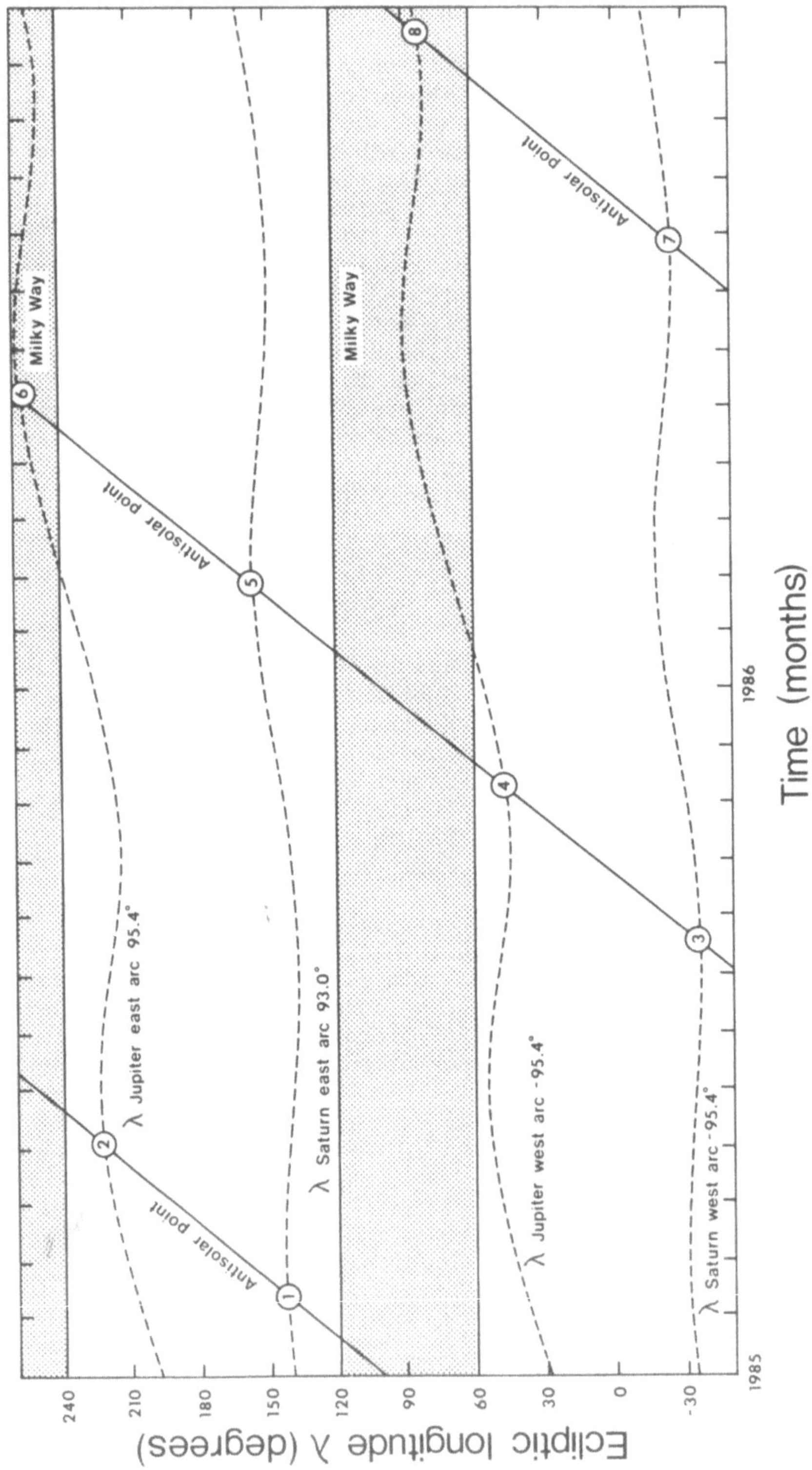

Figure 2. Schuerman dust arc observing opportunities (crossings of antisolar point positions and Jupiter/Saturn arc positions), 1985-86. See, also, Table 1.

IRAS* OBSERVATIONS OF THE INTERPLANETARY DUST EMISSION

M. G. Hauser
Goddard Space Flight Center
Greenbelt, Maryland 20771

T. N. Gautier and J. Good
Jet Propulsion Laboratory
Pasadena, California 91011

F. J. Low
University of Arizona
Tucson, Arizona 85721

ABSTRACT. The Infrared Astronomical Satellite (IRAS) has completed a sensitive, highly redundant survey of the full sky in four broad photometric bands at 12, 25, 60, and 100 micrometers wavelength. The survey measured interplanetary dust emission over elongation angles ranging from 60 to 120 degrees. Bright emission from the main cloud is consistent with optically thin blackbody emission. The grains are evidently quite black, with an "apparent albedo" of about 0.07. The data show clear evidence for deviation of the dust symmetry surface from the ecliptic plane. Surprising bands of emission were discovered near the ecliptic plane and about ten degrees on either side of it. The heliocentric distance of this material, suggested to be asteroidal in origin, is inferred to be about 2.5 AU from both color temperature and parallax measurements.

1. MISSION DESCRIPTION

The IRAS satellite was launched on Jan. 26, 1983, and operated successfully until depletion of its stored superfluid helium on Nov. 22, 1983 (Neugebauer et al. 1984a). The primary purpose of the mission was to conduct an unbiased survey of the full celestial sphere in four broad photometric bands at 12, 25, 60, and 100 micrometers wavelength. Approximately 60 percent of the mission life was devoted to the survey, giving redundant coverage of 98% of the celestial sphere. The satellite

*The Infrared Astronomical Satellite (IRAS) was developed and operated by the Netherlands Agency for Aerospace Programs (NIVR), the United States National Aeronautics and Space Administration (NASA), and the United Kingdom Science and Engineering Research Council (SERC).

R. H. Giese and P. Lamy (eds.), Properties and Interactions of Interplanetary Dust, 43–48.

was placed in a 900 km altitude, circular orbit at 99 deg inclination with a twilight nodal crossing. This is a Sun-synchronous orbit: precession of the orbital plane keeps the Sun-Earth line roughly normal to the orbital plane. The survey strategy was to scan the sky at approximately the orbital pitch rate at a fixed solar elongation, completing all or part of a small or great semi-circle scan (depending on elongation) on each side of the Earth.

The survey instrument contained a linear-staggered array of photoconductive detectors mounted at the focal plane of a 57-cm diameter Cassegrain telescope (Neugebauer et al. 1984a). The detectors in each wavelength band were arranged in two independent modules separated in the in-scan direction so that two independent measurements were obtained on each sky region in an interval of several seconds as the telescope scanned the sky. The full focal plane assembly viewed a 1/2-deg wide region in the cross-scan direction. The scan track was offset cross-scan by 1/4 degree on successive orbits, assuring that each sky region was seen on consecutive orbits to confirm the reality of the measurements. During the first half year of the survey, each sky position was scanned on a second pair of orbits about a week after the initial pair to permit discrimination of solar system objects.

For studies of the interplanetary dust emission (hereafter, zodiacal emission or ZE), data from all detectors in a given wavelength band for a given sky position were averaged together and then averaged over 1/2 deg in the in-scan direction to provide low resolution samples of high signal-to-noise ratio. The detector signals were dc-coupled, providing intensity measurements on all spatial scales larger than the instantaneous field of view. The electronic baselines were found to be very stable on time scales of hours, as evidenced by the reproducibility of the data on successive orbits. Calibration of the zero point and scale factor was accomplished using procedures briefly discussed by Neugebauer et al. (1984a) and Hauser et al. (1984).

In terms of Sun-referenced coordinates, each survey scan provided data at a fixed elongation with inclination varying uniformly with time. During the first half year of the survey (Feb. 9 to Aug. 26), all ecliptic longitudes were redundantly observed at elongation angles varying between about 80 and 100 degrees. A 70% complete second survey was achieved between Aug. 26 and Nov. 22 including elongation angles from 60 to 120 degrees. The net result is an extensive self-confirming data set containing roughly 8000 partial-circle scans, most crossing the ecliptic plane, and many extending from one ecliptic pole to the other.

2. RESULTS AND INTERPRETATION

The data from single scans clearly show smoothly varying emission peaking near the ecliptic plane in all bands; this emission is the dominant large-scale feature at 12 and 25 μm (Hauser et al. 1984). This large-scale emission shows a number of interesting features, including high brightness at all wavelengths, irregularity of the profiles near the ecliptic plane, differences in brightness between the ecliptic poles, and displacement of the symmetry axis of the profile from the ecliptic plane. Some implications of these features are discussed below. It should first be emphasized that there is no clear evidence of significant large-scale galactic emission more than 20 degrees or so from the galactic plane except at 100 μm. A procedure for estimating the ZE component of the 100 μm emission is given by Hauser et al. (1984): in what follows, we deal only with the ZE component at all wavelengths.

2.1 Brightness and Spectrum of ZE: Grain Albedo and Size

The ZE brightness at 12 and 25 μm exceeds that of previously published rocket data at comparable wavelengths by amounts varying from 50% to a factor of 3-4 (Briotta 1976; Price, Murdock, and Marcotte 1980). An immediate implication is low dust albedo. Comparing the spectrally integrated emission in the ecliptic plane at elongation 90 deg with the integrated scattered intensity in the same direction, based upon a visual brightness of 200 $S_{10}(V)$ (Levasseur-Regourd and Dumont 1980), shows the emission to be about a factor of 13 brighter. The "apparent albedo" is thus $\sim$0.07. Adopting an anisotropic scattering function implies a larger albedo, but the particles are evidently quite black.

The spectrum of the ZE is, within the IRAS calibration uncertainties, consistent with optically thin blackbody emission, though a grain radiative efficiency proportional to frequency cannot be ruled out. The implication is a significant population of grains with radii $a > 16$ μm. The blackbody grain temperature toward the ecliptic pole, 275±57 K, is at or slightly below the greybody temperature at 1 AU of 298 K.

2.2 Angular Distribution of the ZE: Dust Spatial Distribution

No attempt has yet been made to invert the brightness distribution to obtain the spatial distribution of the dust. However, the brightness profiles in the ecliptic plane from elongation 68 to 103 deg and in the plane at elongation 90 deg have been compared with the emission model tabulated by Frazier (1977) based on a dust density distribution determined from visual observations of zodiacal light:

$$n(r,z) \sim (1/r) \exp[-2.6 (z/r)^{1.3}],$$

where z is distance normal to the symmetry plane and r is the radial distance from the Sun in that plane. A tentative conclusion is that the infrared observations are generally consistent with dust radial and z distributions determined from previous studies.

An important characteristic of the dust distribution is its symmetry plane, which can provide clues to the dynamics and perhaps origin of the dust. Two characteristics of the IRAS data relate fairly directly to the symmetry plane: the difference in brightness at the ecliptic poles at any given time, and the ecliptic latitude about which the scan profiles are symmetric. Both of these quantities show an annual variation as a result of the Earth's motion with respect to the symmetry plane.

The difference in brightness between the ecliptic poles can be determined from single half-orbit IRAS scans extending from one ecliptic pole to the other at 90 deg elongation. For small inclination of the symmetry plane, the annual variation of this difference is expected to be sinusoidal. The data show the expected sinusoidal variation clearly in all bands except 100 μm, where galactic sources make the measurement of the difference more difficult. The ecliptic longitude of the ascending node of the symmetry plane is about 78 deg (Hauser and Gautier 1984), slightly smaller than the 87 deg value reported by Leinert et al. (1980) from the Helios measurements. The fractional difference in polar brightness varies annually by 40% peak-to-peak. This variation implies an inclination of the symmetry plane with respect to the ecliptic plane of about 2.5 degrees.

The time variation of the apparent ecliptic latitude of the symmetry axis of each scan across the ecliptic plane has been examined in a preliminary fashion. Comparison of the data obtained at elongation angles between 80 and 100 deg with expectations for a model in which the dust is concentrated in a Sun-centered circle in the symmetry plane, and for models with more realistic dust spatial distributions, yields symmetry plane parameters of about 1.5 deg inclination, 55 deg ascending node (Hauser and Gautier 1984). Comparing these results with those of ground and space-based visible zodiacal light studies lends support to the suggestion that the symmetry "plane" is really a more complicated surface (Misconi 1980), with parameters resembling those of the Venus orbit in the region interior to 1 AU observed by Helios, and parameters more nearly those of the Martian orbit exterior to 1 AU.

2.3 The Interplanetary Dust Bands

The most surprising feature in the ZE data noticed thus far is structure in the latitude profiles near the ecliptic. Though this structure is a small perturbation on the main shape of the profiles, it has been found to be very reproducible from scan to scan. The structure generally takes the form of shoulders about 10 deg on either side of the ecliptic plane and a flattened top at the ecliptic plane. In order to enhance the visibility of these features, the scan data were filtered so as to suppress the very large scale gradients. When presented as a contour plot, the data then clearly show ridges at the ecliptic and about 10 deg on either side (Neugebauer et al. 1984b). These features represent about 5% of the total emission intensity near the ecliptic. Similar filtering of the data from the entire survey shows the ridges to persist all around the sky, leading to their description as "bands" of zodiacal dust about the ecliptic (Low et al. 1984). From the relative intensities of these bands, Low et al. inferred color temperatures in the range 165 - 200 K, implying heliocentric distances (for grey particles) in the range 2.2 - 3.2 AU. Examination of the apparent separation of the outer bands in the data from the second half year clearly shows systematic variation as elongation changes from 60 to 120 deg, as expected from parallax considerations for material at a few AU. Initial analysis of these data yield an inclination for the orbits of the outer band material with respect to their symmetry plane of about 8.1 deg, with a heliocentric distance of 2.5 AU, quite consistent with the color temperature determination (Gautier, Hauser, and Low 1984).

These bands of emission are interpreted as arising from dust in inclined orbits in the vicinity of the main asteroid belts. If, for example, orbital inclinations of about 10 deg are preferentially populated, then the fact that particles spend more time near the points of extreme displacement from the symmetry plane than close to it will cause enhancements in the observed emission near 10 degrees. The bands are particularly intriguing because they may represent a clear example of an asteroidal, as opposed to cometary, origin for at least part of the interplanetary dust. Dermott et al. (1984) have proposed a detailed model for the bands based upon the orbital properties of Hirayama's asteroid families, which do include groupings near 10 deg and 0 deg inclination. This model includes specific predictions for additional consequences, e.g., variations in brightness and band separation with longitude, which will have to be compared closely with the data.

3. CONCLUSION

The IRAS survey has yielded an extensive new set of observations of the interplanetary dust. The results discussed here are based upon preliminary examination of a small fraction of the available data. Final data reduction with a consistent calibration scheme is still in progress. Detailed analysis of the IRAS data along with previous observations should contribute substantially to our understanding of the nature of the interplanetary dust, its distribution, and its origins.

ACKNOWLEDGMENTS

We thank our many colleagues whose labors contributed to the IRAS mission, and particularly the dedicated staff of the IRAS Scientific Data Analysis System at the Jet Propulsion Laboratory for rendering the data into forms needed for this investigation. We are grateful to J. Weinberg for presenting this paper at the Colloquium. T. N. Gautier was supported by an NRC Resident Research Associateship.

REFERENCES

Briotta, D. A. 1976, Ph.D. thesis, Cornell University.
Dermott, S. F., Nicholson, P. D., Burns, J. A., and Houck, J. R. 1984, preprint (submitted to Nature).
Frazier, E. N. 1977, Proc. SPIE, 124, 139.
Gautier, T. N., Hauser, M. G., and Low, F. J. 1984, Bull. Am. Astr. Soc., 16, 442.
Hauser, M. G. et al. 1984, Ap. J. (Letters), 278, L15.
Hauser, M. G. and Gautier, T. N. 1984, Bull. Am. Astr. Soc., 16, 495.
Leinert, C., Hanner, M., Richter, I., and Pitz, E. 1980, Astr. Ap., 82, 328.

Levasseur-Regourd, A. C. and Dumont, R. 1980, Astr. Ap., 84, 277.
Low, F. J. et al. 1984, Ap. J. (Letters), 278, L19.
Misconi, N. Y. 1980, Solid Particles in the Solar System, I. Halliday and B. A. McIntosh (eds.), 49-53.
Neugebauer, G. et al. 1984a, Ap. J. (Letters), 278, L1.
Neugebauer, G. et al. 1984b, Science, 224, 14.
Price, S. D., Murdock, T. L., and Marcotte, L. P. 1980, A.J., 85, 765.

RESPONSE TO QUESTIONS SUBMITTED FOLLOWING PAPER DELIVERED FOR THE AUTHORS BY J. WEINBERG

LAMY: The method of residues used by J. Maucherat for analyzing the D2B ZL data (i.e., comparing the same field of view at 6 months interval) turned out very powerful. Could it be used to analyze the IRAS data or part of the data?

AUTHORS: The IRAS extended emission data are being reduced separately for each sky coverage so that such comparisons can be carried out.

LEVASSEUR-REGOURD: Would the 20% variation in intensity between the first and the second semester observations mainly be due to the position of the symmetry plane--better, the surface of maximum brightness--with respect to the ecliptic plane?

AUTHORS: The inclination of the symmetry surface with respect to the ecliptic plane is evidently the main cause of the effect. Another potential contributor, which would not show up in the polar differences, is the eccentricity of the Earth's orbit. Search for evidence of this contribution awaits complete reduction of the data with a consistent baseline calibration.

FECHTIG: A possible interpretation for low albedo is the following: as the beer can experiment on Pioneer 10/11 has shown, the dust particles at $r > 5$ AU are moving eccentrically ($e \sim 0.99$) and/or have random inclinations. Therefore, the particles are on typical cometary orbits, and are hence young (time since release from comet nucleus). Young cometary particles have organic mantles (Greenberg) which turn black when irradiated in space (Lanzerotti et al.). Therefore, particles in the outer solar system have extremely low albedos.

GRÜN: With what radial and z dependence of the dust density distribution are the IRAS data compatible? What particle size distribution is implied by the IRAS data?

AUTHORS: The brightness distributions for a small sample of the IRAS data are roughly compatible with the spatial distributions inferred from visible light studies, as indicated above. Further modelling is in progress to determine whether the spatial distributions can be more tightly constrained. Implications for the particle size distribution have not yet been investigated.

BALLOON OBSERVATION OF THE F-CORONA AT THE 1983 TOTAL SOLAR ECLIPSE

S. Isobe, H. Tanabe, T. Hirayama
Tokyo Astronomical Observatory, Mitaka, Tokyo 181, Japan

Y. Koma
The Institute of Space and Aeronautical Science,
Meguro-ku, Tokyo 153, Japan

J. Soegijo
Indonesian National Institute of Aeronautics and Space,
Jalan Laks, L.R.E. Martadinata 166, Bandung, Indonesia

and

N. Baba
Department of Applied Physics, Hokkaido University,
Sapporo 060, Japan

ABSTRACT. A balloon observation of the F-corona in visual and infrared regions was carried out by Japanese and Indonesian teams at the total solar eclipse on June 11, 1983, in Java, Indonesia. For the visual observation, a SIT television camera, with 4 interference filters (5300Å, 6000Å, 7200Å and 8000Å) and a 45°-step rotating polarizer, was used. The camera measured brightness distributions in a sky area of 5°×5° centered at the eclipsed sun at each polarizer position for each filter. In this paper, a part of results, which are the brightness and polarization distributions in a half area of the 6000Å picture, is shown.

1. INTRODUCTION

At the total solar eclipse on June 11, 1983, in Indonesia, we have made a photo-polarimetry of the F-corona in both visual and infrared regions by using a balloon under a cooperation between Japanese and Indonesian teams. Our main purpose was to find precise position of the boundary of the dust free zone, where the interplanetary dust approaching to the sun sublimates due to temperature increase.

The use of a balloon is not only needed for infrared measurement in order to avoid strong absorption by the lower atmosphere, but also much advantageous for visual observation in reducing the background sky brightness as shown in Fig. 1 (from Saito (1959)).

This paper is a report on the circumstances of our balloon observation and result of the visual region measurement reduced from a part of data until the present.

R. H. Giese and P. Lamy (eds.), Properties and Interactions of Interplanetary Dust, 49–54.

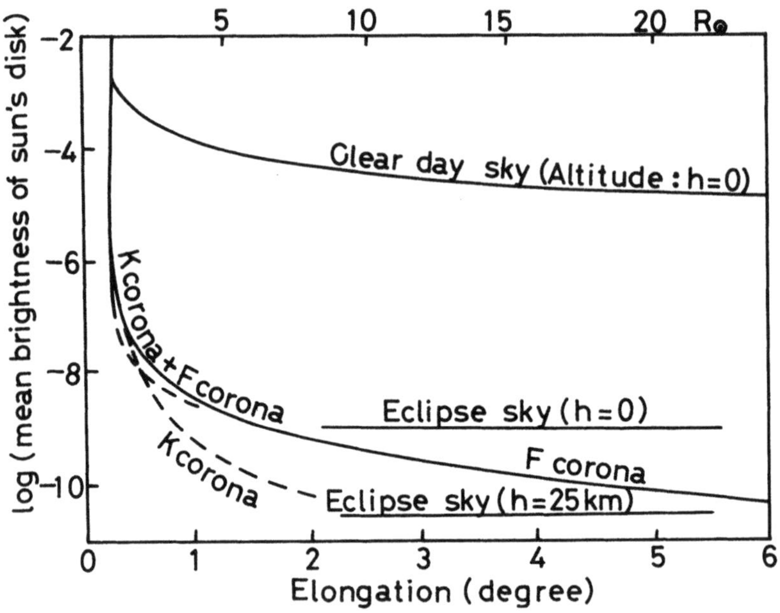

Figure 1. Brightness of the corona and sky (from Saito (1959)).

2. PREPARATORY PHASE

Detailed prediction and explanation of the total solar eclipse on June 11, 1983 had been given by Fiala and Lukac (1982).

In 1980 the Japanese team began to study the feasibility of a balloon observation at the eclipse in Indonesia, and the Japan-Indonesia cooperative plan started in 1981.

The balloon and payloads; photometers, telemeter, attitude control and remote control systems, had been prepared in Japan. Performance test of these instruments was made by a balloon flight in Japan in September 1982.

In Indonesia, a new balloon launching base was constructed at Watukosek, East Java by the Indonesian National Institute of Aeronautics and Space. Ground equipments; launcher, gas injection system, receiving and transmitting system, have been provided.

In the tropics, usually, the stratospheric wind direction alternates between easterly and westerly with a period of 26 months (see Labitzke, 1982), and a prediction (Yamanaka, (1983)) showed that June 1983 would be just in its reversal period. However, from our small rubber balloon flights made during three weeks prior to the eclipse day, we found that the wind direction is favorable (easterly) for our eclipse observation.

3. INSTRUMENTS

A. SIT (Silicon Intensifier Target) television camera was used for the visual region observation. In front of the objective lens (50mm aperture, F/2), a shutter, a filter-wheel and a 45°-step rotating polarizer are installed. The camera takes pictures, one consists of 256×256 pixels, of the brightness distribution in a sky area of 5°×5° centered at the eclipsed sun at each polarizer position (1.8 sec/position) with exchanging four interference filters of 5300Å, 6000Å, 7200Å and 8000Å (15 sec/filter). To eliminate the effect from the bright inner corona, the camera has an occulting disk at its direct focus, which is a sputtered metal on a thin optical glass inclined at 45° to the optical axis. The radius of the occulting disk corresponds to $3.5R_{\odot}$ ($R_{\odot}$: sun's radius).

Both infrared and visual photometers were set on a same platform in a gondola. The elevation angle of the platform can be changed by remote control while watching a monitor-display at Watukosek Base. By an attitude control system developed by Nishimura et al. (1984), the direction of the gondola can be kept to the sun's azimuth, which changes gradually, with referring to the direction of geomagnetic line of force.

All the data were recorded in a video tape-recorder on board, and a part of them was transmitted to Watukosek Base through the telemeter.

4. OBSERVATION

The balloon of $15,000m^3$ with payload of 150kg was launched at 0713 LT on June 11. Fig. 2 is its construction. The balloon reached the level altitude of 30.8km at 0837 LT, then drifted to west-south-west direction with a speed of 80km/h. The trajectory is shown in Fig. 3. The SIT camera was used for monitoring the sun during the partial phase of the eclipse through a small iris and a density filter.

The balloon encountered shadow of the moon at 1128 LT at 30.5km height over a position of about 40km east-south-east from Jogjakarta. Duration of totality at the balloon was 3^m50^s. Immediately after the 2nd contact, observation was started by remote control. The instruments worked properly throughout the totality under attitude control within ±8 arc seconds.

After the observation, the payload was recovered safely by a parachute.

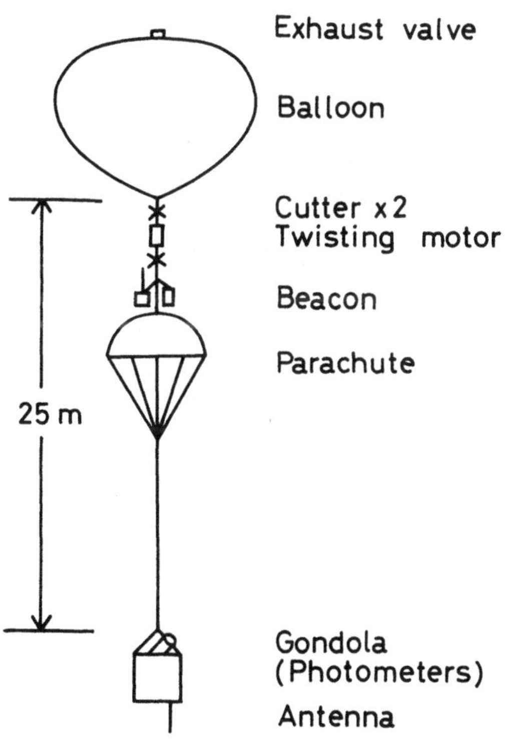

Figure 2. Construction of the Balloon.

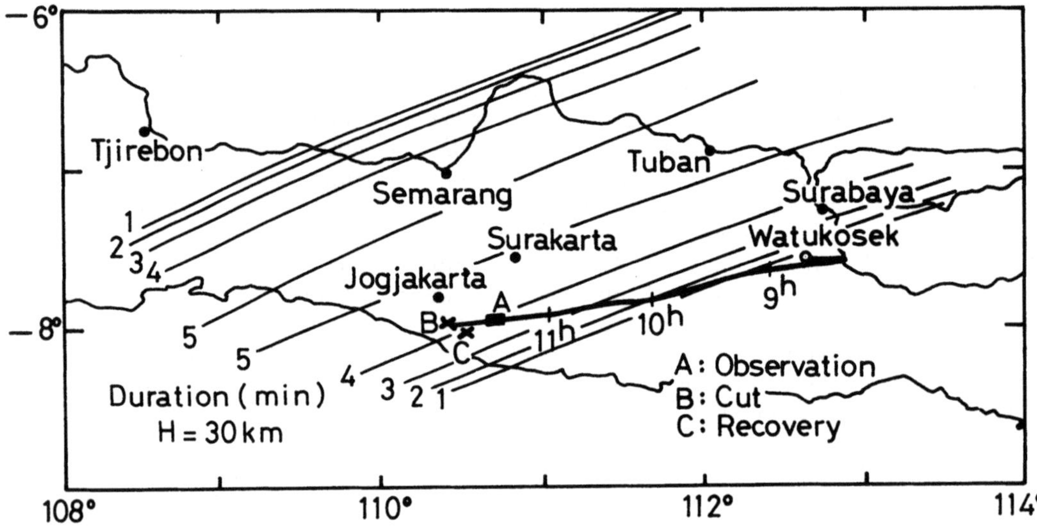

Figure 3. Trajectory of the balloon, June 11, 1983.
Durations of the totality in the eclipse zone are indicated.

5. METHOD OF REDUCTION

Before full reduction of all the data using a computer, we took several tens sample points in a picture, and analyzed their data with careful check and necessary corrections at the following each process. (1) Inter-calibration among sensitivities of the pixels was made by using data of picture of a uniformly illuminated surface. (2) From Fig. 1 and from the discussion by Blackwell et al. (1967), we assumed that the background sky brightness at 30.5km is negligible compared to the F-corona brightness inside of our 5°×5° picture. (3) Polarization effect by the 45° glass plate (occulting disk) set at the direct focus was measured in the laboratory. The ratic k, of the transmitted intensities of linearly polarized light, of which electric vectors are parallel; $I_{\parallel}$, and perpendicular; $I_{\perp}$, to the plane of incidence, is $k = I_{\perp} / I_{\parallel} = 0.827$. (4) The Stokes parameters; I,Q,U,V, at each sample point were obtained with the least-squares method using equations;

$$\begin{aligned} 2kI + 2kQ &= I(0°) \\ (1+k)I + (1+k)\cdot\tan 2\chi\cdot Q &= I(45°) \\ 2I - 2Q &= I(90°) \\ (1+k)I - (1+k)\cdot\tan 2\chi\cdot Q &= I(135°), \end{aligned} \qquad (1)$$

where I(θ) is the observed intensity at polarizer position θ. We assumed here that V=0, and position angle χ, of the F-corona polarization is perpendicular to the radial vector from the sun, which gives relation of $U=Q\cdot\tan 2\chi$. (5) Relative brightness; B, of each sample point was obtained from $B=\Sigma I(\theta)$, θ=0°, 45°, 90°, 135°. (6) We did not separate the K-corona component, which is usually small (<1/10) compared to the F-corona component outside of $5R_{\odot}$.

6. RESULTS

We show here our results only for the left half area of the 6000Å picture, which have been obtained until the present time.

Fig. 4 is an isophote map of the relative brightness. Contour lines were drawn with log B, and B was expressed in an arbitrary scale. Since the data within $5R_{\odot}$ from the sun showed a gradual saturation, we plotted only the brightness outside of $5R_{\odot}$. The contours in Fig. 4 are concentric circles slightly elongated toward the ecliptic. However, we can not determine the precise direction of the axis of elongation only from this map.

The slope of the brightness decrease with increase of the heliocentric distance in the equatorial region

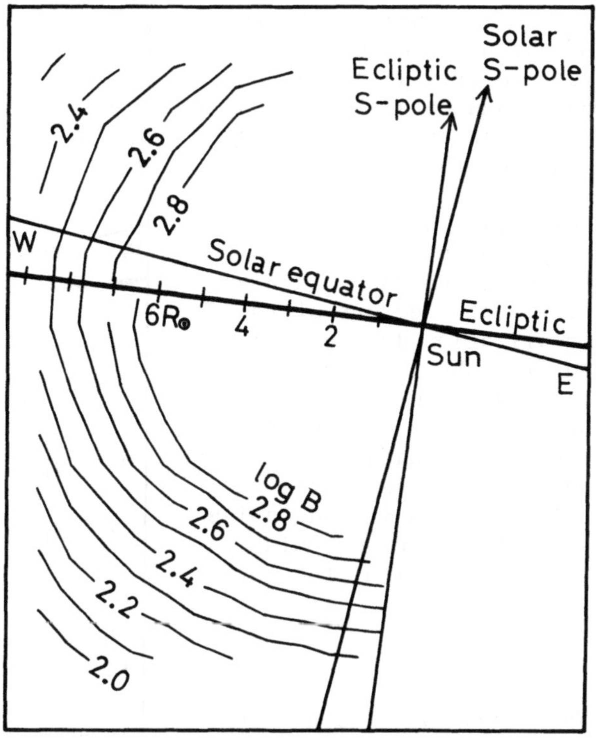

Figure 4. 6000Å relative brightness distribution of the F+K corona.

TABLE 1. Positions, Heliocentric distances ($R/R_{\odot}$) and Polarization degrees (p, in per cent) of the Sample Points.

$\lambda-\lambda_{\odot}$	β	$R/R_{\odot}$	p	$\lambda-\lambda_{\odot}$	β	$R/R_{\odot}$	p
0°65	2°08	8.27	9	1°89	-1°30	8.73	16
0.70	1.69	6.95	4	1.62	2.21	10.40	6
0.75	1.30	5.72	6	1.67	1.82	9.38	9
1.11	-1.41	6.83	18	1.72	1.43	8.51	8
1.31	-1.38	7.23	9	1.77	1.05	7.82	6
1.04	2.13	9.00	9	2.08	-1.28	9.29	17
1.09	1.74	7.81	6	2.22	-0.86	9.07	19
1.14	1.35	6.73	5	2.28	-1.25	9.88	16
1.19	0.97	5.83	10	2.00	2.26	11.48	12
1.50	-1.36	7.69	11	2.03	2.07	11.01	14
1.38	0.99	6.48	11	2.06	1.87	10.57	10
1.42	2.18	9.90	15	2.11	1.48	9.80	14
1.47	1.79	8.83	4	2.16	1.10	9.21	11
1.53	1.41	7.89	5	2.47	-1.23	10.48	20
1.58	1.02	7.14	11	2.50	-1.42	10.92	16
1.63	0.63	6.65	9				

agrees in general with those of the previous observers, particularly with that of Saito and Hata (1964) observed in an approximately same phase of the solar cycle.

The positions and polarization degrees of the sample points are listed in Table 1, and shown in Fig. 5. Probable errors of the polarization degree estimated from the least squares calculation of equations (1) are within ±3 per cents. Because the data of the points in the ecliptic and polar regions include noise slightly large for polarization calculation, they are not shown here.

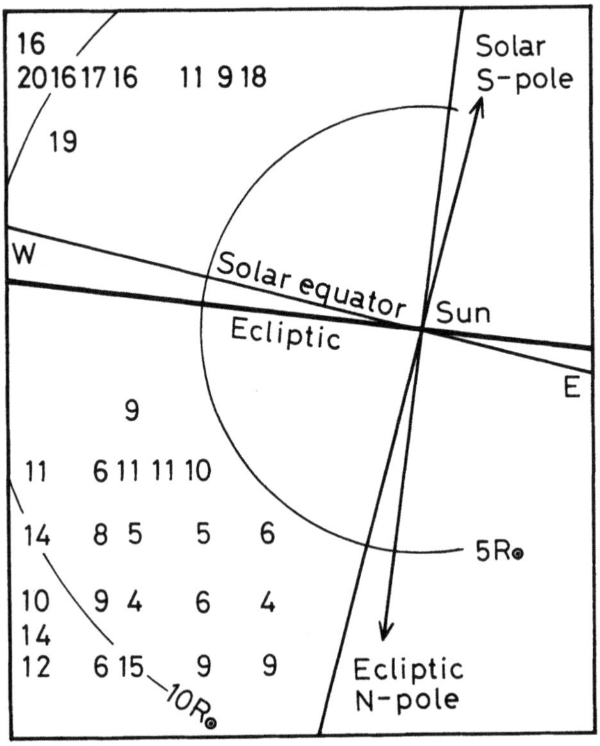

Figure 5. 6000Å polarization of the F+K corona (in per cent).

We acknowledge the governments of Japan and Indonesia for their great support to our cooperative project.

REFERENCES

Blackwell, D.E., Dewhirst, D.W. and Ingham, M.F., 1967, Advances in Astronomy and Astrophysics, 5, (ed. Z. Kopal), Academic Press, New York and London, p.2.

Fiala, A.D. and Lukac, M.R., 1982, U.S. Nav. Obs. Circ., No. 165.

Labitzke, K., 1982, J. Meteor. Soc. Japan, 60, 124.

Nishimura, J., Koma, Y. and Ohta, S., 1984, Proc. of the 14th ISTS, Tokyo, to be published.

Saito, K., 1959, Tokyo Astr. Obs. Report, 12, 243, in Japanese.

Saito, K. and Hata, S., 1964, Publ. Astr. Soc. Japan, 16, 240.

Yamanaka, D., 1983, private communication.

A BALLOON OBSERVATION OF THE THERMAL RADIATION FROM THE CIRCUMSOLAR DUST CLOUD IN THE 1983 TOTAL ECLIPSE

T.MAIHARA, K. MIZUTANI, N. HIROMOTO, H. TAKAMI and H. HASEGAWA

Department of Physics, Kyoto University,
Kitashirakawa, Kyoto 606, Japan

ABSTRACT. During the totality on June 11, 1983 in East Java, Indonesia, a near infrared photometric observation of the solar corona was made using a stratospheric balloon, which was successfully launched by a joint team of ISAS, Japan and LAPAN, Indonesia. The surface brightness distributions in four near infrared bands : at 1.25, 1.65, 2.25 and 2.8 μm, have been obtained in the outer coronal region. Noticeable excess emissions superposed on the strong coronal background emission have been recorded in the scan profiles at 1.25 and 1.65 μm, and less conspicuously at 2.25 μm as well, at about 4 $R_\odot$ from the sun. From the observed spectral and spatial characteristics, the excess emission component appears to originate in the thermal radiation from the circumsolar dust ring made of relatively large particles of about 100 μm in radius and with the olivine-like optical properties.

1. INTRODUCTION

One of the methods to study the properties of the interplanetary dusts is to measure the thermal radiation from them in the proximity of the sun. The merit of it is that the spectral feature would represent directly the optical properties reflecting the composition and size of dust particles. It is necessary to make such observations by avoiding the effect of the strong emission of the sun itself. Total eclipse enables this, and thus there have been a few observations, in which an excess emission at 2.2 μm has been detected (MacQueen 1968 ; Peterson 1969, 1971).

In the present observation we tried to measure the near infrared energy spectrum as well as the polarization of the particular emission feature formerly detected at about 4 $R_\odot$ from the sun. We utilized a stratospheric balloon on the occasion of totality on June 11, 1983. The successful ballooning was jointly performed by The Institute of Space and Astronautical Science, Japan and The Indonesian National Institute of Aeronautics and Space, which are described in Isobe et al. (1985) in this volume.

R. H. Giese and P. Lamy (eds.), Properties and Interactions of Interplanetary Dust, 55–58.

2. OBSERVATIONS

The instrument consisted of a 16 cm F/4 cassegrain telescope attached to a four-wavelength band simultaneous photometer, in which four 4-element array detectors of lead-sulfide and a filter/beam splitter system were installed. Effective wavelengths of each band are 1.25, 1.65, 2.25 and 2.8 μm respectively. At the 2.25 μm band, it is capable of measuring linear polarization by the additional optical components (a wire-mesh polarizer and a rotating half wave plate). Each detector subtends a field of view of about 5 arcmin projected in the sky.

Scanning were made, for about 3 minutes during the totality at the floating altitude of 30 km, between 6 $R_\odot$ on the west side of the sun and 14 $R_\odot$ on the east side. The scan paths crossed the ecliptic plane on the west side at ~ 4 $R_\odot$ where the actual data of surface brightness profiles were obtained, and continued slightly south of the ecliptic with an inclination of ~ 70 (cf. Fig.1 of Mizutani et al. 1984, paper I).

3. INFRARED POLARIZATION AND THE F CORONA

As Beard(1984) points out, measurement of infrared polarization of the corona is important to get a knowledge of the size of interplanetary dusts.

We have obtained 2.25 μm polarizations only in the inner region of the corona (Fig.1), and yet they are useful to discriminate between the K and F corona components. It is safely assumed that the polarization originates only in K corona ($P_K \approx 60\%$; see van de Hulst, 1950), while the F corona can be regarded to be non-polarized in this region (≲3 $R_\odot$). Hence the separation of the K corona component is

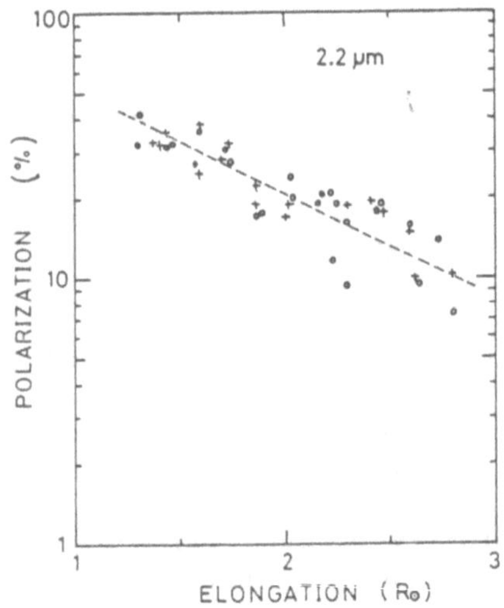

Fig.1 2.25 μm polarization data along the scan path

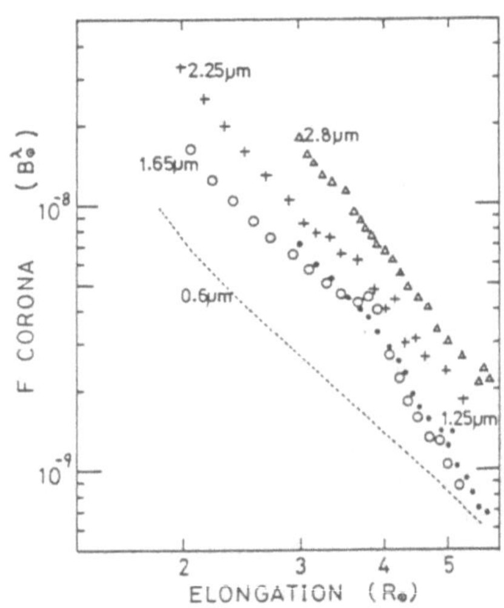

Fig.2 Derived F corona components on the west side of the sun (0.6 μm: Dürst 1982)

uniquely made. Since the scattering nature of the K corona has no wavelength dependence, it turns out that we have got the K corona brightness at each band. Thus we can extract the F corona profiles as shown in Fig.2.

The ordinate of Fig.2 is the intensity expressed by the unit of mean solar brightness : $B_{\odot}^{\lambda}$. It should be noted that near infrared light is considerably brighter than the 0.6 μm visual light, and besides, the brightness appears to become stronger at longer wavelengths in general.

This is an important information obtained for the first time by the present observation, because by analysing this we can learn the effective size range of dust particles as follows.

A spectrum of the F corona at $3R_{\odot}$, for instance, is plotted in Fig.3. Here we show calculated curves for a model F corona assuming the particle radius a = 1, 10, and 100 μm. In this calculation, we adopt the spatial distribution of $r^{-1.1}$, letting the density equal zero inside 3.8 $R_{\odot}$. From this figure it is obvious that the particles must be as large as 100 μm in radius or even larger.

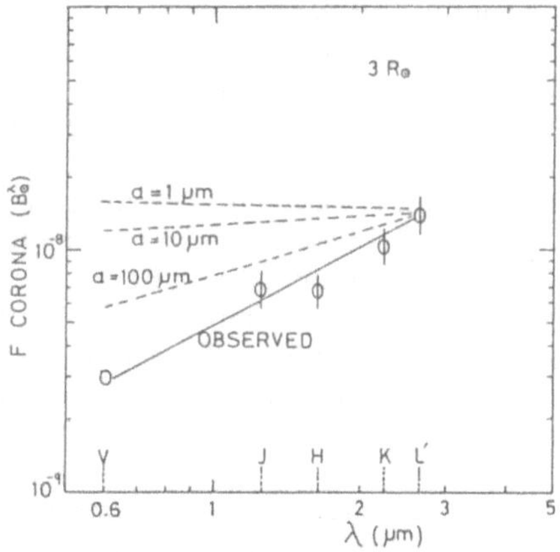

Fig.3 Spectrum of the F corona compared with calculations.

In the course of this calculation one can notice that, when we employ such large particles, the F corona light is largely contributed by dusts located relatively far from the sun. This fact is just consistent with the supposition that the F corona is virtually not polarized. It is also very important to note that the size spectrum in the region smaller than 100 μm in radius must be rather flat, i.e. the index of the power spectrum : $\gamma<2$, when expressed as $a^{-\gamma}$.

4. THERMAL EMISSION FROM THE INNERMOST DUST CLOUD

As reported in Paper I (Mizutani et al. 1984), some of the scan profiles of the surface brightness of the corona have demonstrated apparent excess emissions at several wavelength bands. Those features are also discerned in the F coronal profiles in Fig.2.

Is it possible to interprete it as the scattered solar radiation by a dust cloud in an annulus with density enhancement? Remember that the dust contributing to the F coronal emission is as large as 100 μm in radius. If we assume the particles responsible for the excess emission at $\sim$4 $R_{\odot}$ are in the same size range, it is difficult to expect such appreciable excess by scattering, because the phase functions of scattering in large ($\sim 90^{\circ}$) angles are so small that the density enhancement, if any, of a factor ($\lesssim$10) could be compensated.

Another aspect is the spectral characteristic of the excess component, which is presented in Fig.4 adding the data of previous observations. This spectral feature is also hard to understand as scattered light, but is easily ascribed to the thermal emission with a particular optical properties in the near infrared wavelengths. Thus we have tried to fit the spectrum to olivine particles with large size of, ~100 μm in Paper I. Absorption coefficients of olivine particles with various sizes have been calculated by Röser and Staude (1978).

Although we can get a consistent picture of the general interplanetary dust in terms of the claimed excess emission as well as the infrared F corona, there remains a problem that the derived temperature : 1300K is appreciably lower than the equilibrium temperature 1900K when olivine is assumed. In this connection we may have to presume even larger particles than 100 μm (in that case, the absorption in the 1.25 to 1.65 μm region becomes stronger, thereby the temperature taken here turning out to be higher). It is also possible to conceive a composite particle made up of olivine and other minerals with lower equilibrium temperature.

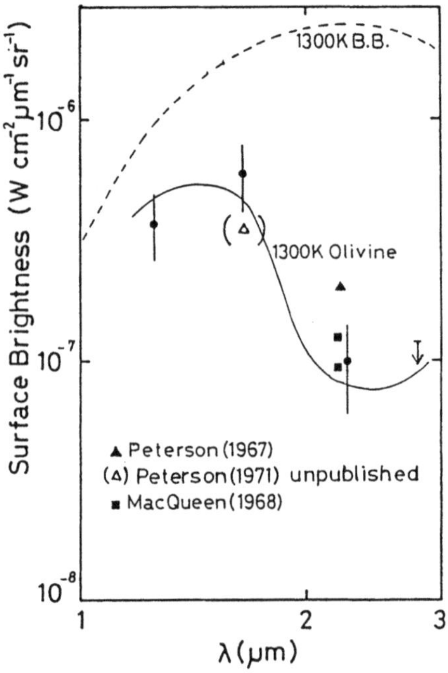

Fig.4 Emission sperctrum the excess component. Data of previous observations are also plotted.

ACKNOWLEDGEMENTS We would like to express our thanks to Dr. J. Nishimura, (ISAS), Dr. J. Soegijo (LAPAN), and their collaborators of both institutes for their great efforts throughout the beautiful balloon operation. We thank Dr. H. Tanabe who read this paper at the meeting on our behalf. Appreciations are also to Dr. S. Isobe for motivating the project including the present work.

REFERENCES

Beard, D. B. 1984, Astron. Astrophys., 132, 317
Dürst, J. 1982, Astron. Astrophys., 112, 241
Isobe, S., Tanabe, H., Hirayama, T., Koma, Y., Soegijo, J. and Baba, N. 1985, this volume
MacQueen, R. M. 1968, Astrophys. J., 154, 1059
Mizutani, M. Maihara, T., Hiromoto, N. and Takami, H. 1984, Nature, 312, P134 (referred to as paper I)
Peterson, A. W. 1969, Astrophys. J., 155, 1009
Peterson, A. W. 1971, Bull. Amer. Astron. Soc., 3, 500
Röser, S. and Staude, H. J. 1978, Astron. Astrophys., 67, 381
Van de Hulst, H. C. 1950, Bull. Astron. Inst. Netherlands, 11, 135

ON THE SOLAR DUST RING(S)

T. Mukai
Kanazawa Institute of Technology
Nonoichi, Ishikawa 921
Japan

ABSTRACT. Based on a mechanism to form the solar dust ring, it is proved that the observed peak in infrared F-corona cannot be explained by silicate type grain alone. Preliminary analysis on the recent infrared data of F-corona by Maihara et al.(1984) has suggested that the ring particle has different physical properties compared with the dust grains, which produce the background F-corona.

1. INTRODUCTION

In the F-corona, Peterson(1967) and MacQueen(1968) independently detected the peaks in infrared intensity for a wavelength λ of 2.2 μm over the range of 4-10 solar radii($R_\odot$) from the sun, and Lèna et al. (1974) confirmed the flux peak at 4 $R_\odot$ in the middle infrared at λ=10μm. Recently, Maihara et al.(1984) obtained the infrared brightness of F-corona at λ=1.25, 1.65, 2.25 and 2.8 μm by balloon observation at the total solar eclipse on June 10, 1983 in Indonesia. Their results have clearly shown the existence of the flux peak near 4 $R_\odot$.

Based on the dynamical behaviour of grains near the sun, Mukai et al.(1974), and Mukai and Yamamoto(1979) proposed a mechanism to form the solar dust ring and explained the observed flux peak as the combination of thermal and scattered light from such a solar dust ring consisting of grains suffering sublimation. Although their model has succeeded to construct the dust ring near the sublimation zone of grains, there still remains some uncertainty in identification of the material of ring particles. In this paper I shall reexamine whether silicate grains alone can produce the observed peak of infrared flux or not, and make a preliminary analysis of multiphotometric data by Maihara et al.(1984).

2. IS SILICATE A CARRIER OF INFRARED FLUX PEAK?

A complete dynamical analysis to form a solar dust ring has already been investigated in Mukai et al.(1974), and Mukai and Yamamoto(1979). I will therefore simply summarize the mechanism here. A radial velocity of the grain spiraling toward the sun under the Poynting-Robertson effect

R. H. Giese and P. Lamy (eds.), Properties and Interactions of Interplanetary Dust, 59–62.

suddenly decreases due to an increase of a relative radiation pressure on a grain. This is caused by a decrease of grain radius under sublimation. Consequently, a concentration of grains arises near the sublimation zone in the outer solar corona. The zodiacal cloud is displayed near the ecliptic plane and almost all of its grains fall toward the sun by the Poynting-Robertson effect, keeping its inclination of orbits. Then, such a concentration in number density of circumsolar grains would make a ring structure around the sun. It is called, therefore, the solar dust ring.

To produce a sharp peak of space density of grains, the temperature of grain should be almost independent of grain radius. In other words, when each grain with a different radius begins to sublime at nearly the same distance from the sun, the enhancement of spacial density of circumsolar grains becomes narrow and sharp. It was found in Mukai and Yamamoto(1979) that the ratio of the lifetime of grain with radius s, $t_s = s/|ds/dt|_{sub.}$ to a period of Kepler orbit, t_K becomes a good indicator of the position of the ring, where $|ds/dt|_{sub.}$ denotes a sublimation rate. Namely, the grain suffering sublimation keeps its solar distance from the sun against the Poynting-Robertson effect during $10^2 \sim 10^3$ orbital revolutions till it completly sublimes. This result agrees with the numerical calculations by Lamy(1976).

The shaded area in figure 1 denotes the region between $t_s/t_K = 10^2$ and 10^3. For silicate and silicate with impurity, I calculated a ratio of t_s to t_K as follows. That is, based on the Maxwell-Garnet expression the complex refractive index of heterogeneous silicate is derived(see Chýlek and Srivastava 1983). Then using the Mie theory and the energy balance equation for a grain, the temperature of the grain is obtained by the same treatment as shown in Mukai and Schwehm(1981). Finally, referring to the sublimation rate of silicate, a ratio of t_s/t_K is computed as functions of grain radius and solar distance. As shown in figure 1, a ratio of t_s to t_K varies with grain radius because the temperature of silicate strongly depends on grain radius(see Mukai and Schwehm 1981). This unfavorable feature in the formation of a sharp concentration of grains remains even when other materials such as impurities in the silicate matrix are included. Therefore, it can be said that silicate can hardly produce a sharp and narrow structure of ring near 4 $R_\odot$.

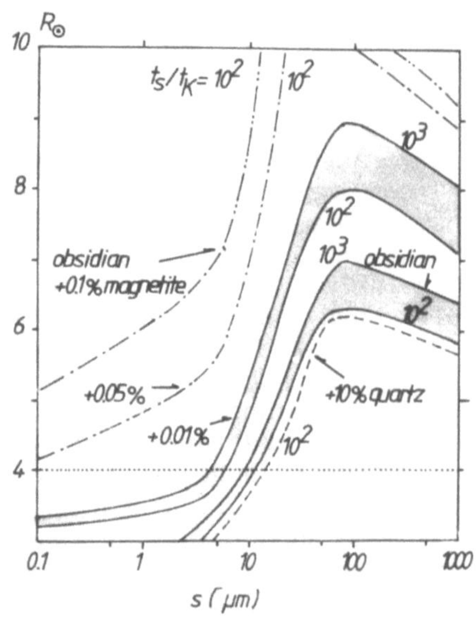

FIG.1. Computed ratio of t_s(a lifetime of a grain) to t_K(its Kepler period) for obsidian, and obsidian with impurity, where number of percentage means a volume fraction of impurity.

In addition, as shown in figure 2, thermal emission from silicate is about $10^2 \sim 10^3$ times smaller than the scattered radiance in the near infrared. Figure 2 comes from the model calculations for obsidian grains with 0.05% volume fraction of magnetite inclusions, where it is found that K-coronal radiance is roughly one tenth of scattered light by obsidian grains having a size distribution of s^{-4} between 0.01 μm and 1000 μm.

This result implies that in the near infrared, silicate cannot make a sharp peak of infrared flux because scattered light by ring particles is diluted by a large amount of light due to low angle scattering of grains far from the sun, and subsequently the flux peak becomes ambiguous.

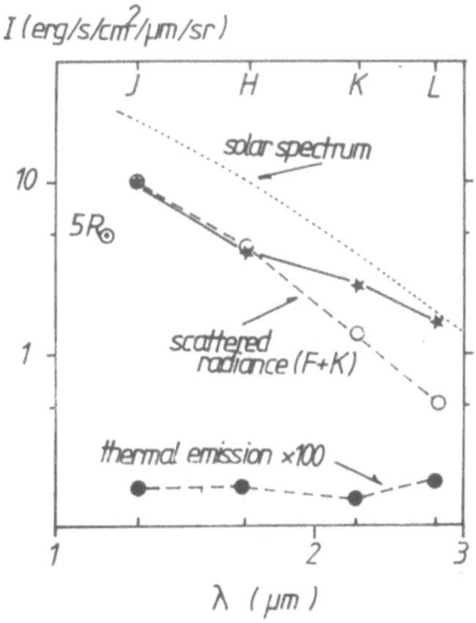

FIG.2. Energy spectrum from model calculations(circles) and observed results(stars).

3. PRELIMINARY ANALYSIS OF INFRARED F-CORONA OBTAINED BY BALLOON

Figure 3 shows the observed energy spectrum at each solar distance (Maihara et al. 1984). A similarity of these spectra to the solar energy spectrum(dotted curve), except at 4 $R_\odot$, strongly suggests that the observed flux in the near infrared outside the flux peak comes from the scattered light formed by grains extended between the sun and the earth.

Furthermore, a color ratio of J(1.25 μm) to H(1.65 μm) bands normalized by that of solar spectrum is applied to examine the observed results(see figure 4).

It is found that there is a weak 'dip' of color ratio near 4 $R_\odot$, which is mainly caused by the observed remarkable peak of flux in H band. Model calculation for the same silicate dust cloud(without dust ring structure) as that in figure 2, where a spatial distribution takes a $r^{-1.3}$-dependence, provides nearly constant value of $(J/H)/(J/H)_\odot \sim 1.4$ (solid line in figure 4).

Since outside the peak region the observed ratio does not vary with solar distance(see figure 4), it can be confirmed that the scattered light plays an important role in the coronal radiance.

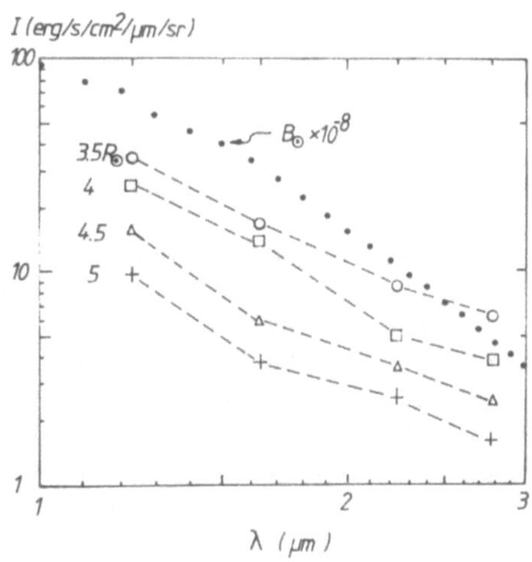

FIG.3. Observed energy spectrum of the infrared F-corona by Maihara et al.(1984).

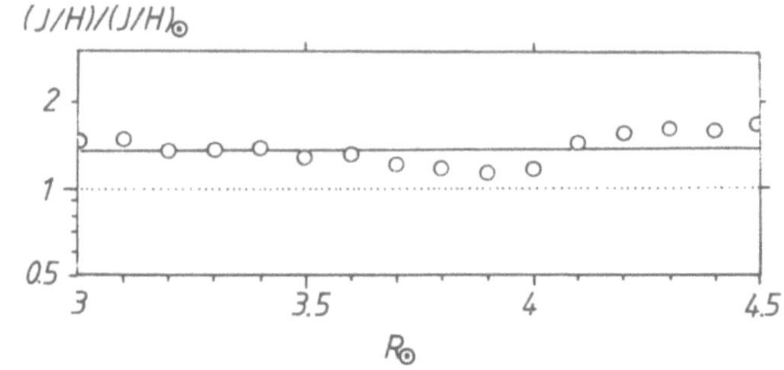

FIG. 4. Color ratio of J(1.25μm) to H(1.65μm) normalized by its solar value. Open circles show the observed results by Maihara et al.(1984). Solid line comes from the model calculations presented in the text.

This ratio approaches unity near the dip. There are alternative possibilities to interpret this feature, i.e. (a) if the scattered light produces this dip, ring particles might be larger and/or darker compared with background F-coronal particles, which seem to be silicate grains, and (b) if the thermal emission makes the dip, ring particles have the temperature of about 1800 K at 4 $R_\odot$.

The infrared observations reported by Maihara et al.(1984) have confirmed the existence of flux peak near 4 $R_\odot$. Preliminary analysis of their multiphotometric data considered here has suggested that the ring particles have a different nature compared with the grains existing outside the ring, which are perhaps silicate. Further measurements, especially polarimetric observations in the near infrared, and simultaneous observations at both near and middle infrared, are urgently needed.

I thank T. Maihara and K. Mizutani for communication of their infrared data of F-corona before publication.

REFERENCES

Chýlek,P. and Srivastava,V.:1983, Phys. Rev. B 27, p. 5098.
Lamy,P.L.:1976, Lecture Notes in Physics 48, p. 437.
Lèna,P., Viala,Y., Hall,D., and Soufflot,A.:1974, Astron. Astrophys. 103, p. 177.
MacQueen,R.M.:1968, Astrophys. J. 154, p. 1059.
Maihara,T., Mizutani,K, Hiromoto,H., Takami,H., and Hasegawa,H.: 1985, (this volume)
Mukai,T., Yamamoto,T., Hasegawa,H., Fujiwara,A., and Koike,C.:1974, Publ. Astron. Soc. Japan 26, p. 445.
Mukai,T. and Yamamoto,T.:1979, Publ. Astron. Soc. Japan 31, p. 585.
Mukai,T. and Schwehm,G.:1981, Astron. Astrophys. 95, p. 373.
Peterson,A.W.:1967, Astrophys. J. 148, p. L37.

Remarks by S.F.Singer(edited by T.Mukai);Liouville's theorem to describe the orbital evolution of an assembly of particles may lead us astray because the P.R.-effect is dissipative, and the evolution of the size distribution should be considered.

THE F-CORONA AND THE CIRCUM-SOLAR DUST EVIDENCES AND PROPERTIES [G. NIKOLSKY MEMORIAL LECTURE]

S. KOUTCHMY
Institut d'Astrophysique,
Paris
P.L. LAMY
Laboratoire d'Astronomie Spatiale,
Marseille

ABSTRACT. This review deals with the main properties of the F-corona. Analysis of its morphology and photometry allows to derive a new axisymmetric, non-spherical model. Polarization, color and infra-red properties are further considered. We suggest the existence of a variable "local" component surperimposed on a quasi-stationnary "far" component.

INTRODUCTION

Originally the Fraunhofer corona, the F-corona together with the K-corona forms the solar corona conspicuously seen during total solar eclipses. The dominating K-corona (more than 90 % of the coronal brightness) is produced by solar light scattered by free electrons; it is characterized by the absence of Fraunhofer lines of the solar spectrum which are completely smeared out by the large Doppler shifts of the high velocity electrons ($T \sim 2 \times 10^6 K$) of the corona. The F-corona is the component of the corona light which does exhibit solar Fraunhofer lines which enabled to recognize it as the inner Zodiacal Light caused by light scattering off interplanetary dust grains. In the polar regions, these spectral lines are observed as close as 1.1 $R_\odot$ (see Koutchmy and Magnan, 1973).

This lecture will review the main properties of the F-corona and present a new axisymmetric, non-spherical model. We shall further examine if those properties allow to delineate two components in the F-corona:

i) the "far" F-corona resulting from light diffracted (forward scattering) by interplanetary grains of the Zodiacal cloud which is essentially unpolarized, has approximately spherical symmetric and which dominates in the visible wavelengths

ii) the "local" F-corona resulting from light scattering by circum-solar dust grains which is probably polarized, highly non-spherical symmetric (equatorial enhancements) and which dominates in the infra-red wavelengths.

R. H. Giese and P. Lamy (eds.), Properties and Interactions of Interplanetary Dust, 63–74.

1. PHOTOMETRIC AND POLARIMETRIC ANALYSIS OF THE F-CORONA

1.1. Recent progress

The ground-based observations of Allen (1973) and Nikolsky (1975) and the aircraft (windowless) observations of Blackwell (1955) and Michard (1956) performed during the 1940-1970 period are now considered classical; they culminated in the synthesis of axially symmetric models of Van de Hulst (1953) and Saito (1970). Since 1970, considerable progress have been achieved on several fronts: instrumentation, technological means, methods, reduction procedure. Let us give a few examples:

- observations carried aboard high-altitude subsonic and supersonic aircrafts, balloons and rockets during total eclipses;
- coronographic observations carried aboard satellites (Skylab, Solwind, SMM,...) outside total eclipses;
- near-infrared imaging with IR emulsions and IR vidicons;
- infrared scannings (up to 12μm) and infra-red spectroscopy;
- the use of sophistical masks, such as neutral radial filters, avoiding over-exposition and therefore improving the spatial resolution;
- an improved understanding and modeling of the parasitic effects (coronal aureola, variation of the sky luminance) further checked by controlling the lunar background (Earthshine);
- two-dimensional photometry with fast microdensitometers and computers;
- sophisticated image processing such as numerical neutral filtering;
- improved absolute calibrations using stars in the field.

This, by no means, imply that all pre-1970 observations are superseded; several of them which benefited from exceptional conditions (such as the windowless aircraft observations which have never been repeated since) and which were carefully reduced are still extremely valuable as we shall see below.

1.2. Analysis of the flattening of the F-corona

Although the flattened aspect of both K and F coronae was perceived quite early, only the non-sphericity of the K-corona as demonstrated by large coronal streamers and transient events has been thoroughfully analyzed in connection with solar activity as defined by the Wolf number (sunspots). The ellipticity of the F-corona was always neglected leading to classical spherical models (e.g., Saito, 1970) although its natural extension, the Zodiacal Light, is known to be highly flattened. Quantitative equator / pole differences have been reported since 1975 (e.g. , Keller et al., 1975; Saito et al., 1977; Dürst, 1982). The analysis of the flattering of the F- corona faces two major problems:

i) the distortion introduced by the K-corona whose ellipticity varies further with time. In this context, it is interesting to point out that the equatorial F-corona is best observed during the period of <u>maximum</u> solar activity since the almost spherical K-corona tends to concentrate in the polar sectors (negative ellipticity). Conversely, the polar F-corona is best observed during the <u>minimum</u> of solar activity since the K-corona flattens in the equatorial direction and becomes quite faint in the polar regions (coronal holes).

ii) parasitic effects introduced by variations of the sky background (ground-based observations) or of the instrumental background (space coronographs). It is imperative that these backgrounds be very low but they must anyway be modelled and substracted for a correct analysis (see Michart and Sotirovsky, 1965; Koutchmy and Koutchmy, 1974; Dürst, 1982). The flattening index $\varepsilon_F = R_{eq}/R_{pol} - 1$ may now be studied as a function of R_{eq} in the following way: on a given isophote corrected for the sky background, R_{eq} and R_{pol} are measured; they are not however strictly recorded along the equatorial and polar directions but averaged over sectors of ±22.5° centered along those directions. The result for the 1983 eclipse (Koutchmy and Nitschelm, 1984) is shown in Fig. 1. The classical separation into three regions, $1<R<2\ R_\odot$ where the K-corona dominates, $R>4\ R_\odot$ where the F-corona dominates and $2<R<4\ R_\odot$ where both coronae are mixed is dramatically illustrated. One sees that starting from asymptotic sphericity as $R \to 1\ R_\odot$, the flattening index rapidly increases with R_{eq}. A set of carefully selected measurements on the basis of low sky backgrounds and high-quality photometry are regrouped in Fig. 2. Here, we explicitly assume that the F-corona is stationary and that the dispersion of the data points result only from errors of measurements. There is a good agreement with the result coming from photometric profiles (next section) and a smooth continuity with the ellipticity of the Zodiacal Light as determined by Dumont and Sanchez (1976). It should be kept in mind that this method is a first approximation since it allows to compare average equatorial and polar brightness and gives no information in the other directions.

1.3. Photometry of the F-corona

Once the luminance of K+F has been obtained as a function of solar distance after proper corrections for sky and instrumental backgrounds and the aureola effect, the separation between the K and F components requires further information either spectroscopic or polarimetric (e.g., Koutchmy and Magnant, 1973). We shall make use here of the polarization P_{K+F} which may be written:

$$P_{K+F} = \frac{K}{K+F} P_K + \frac{F}{K+F} P_K$$

Now $P_K(r)$ can be theoretically calculated assuming an homogeneous spherically symmetric model and can be shown to be rather insensitive to the adopted model and to rapidly converge toward the asymptotic value $P_K = 0.64$ at $5\ R_\odot$ (Saito, 1970). The classical condition $P_F=0$ valid up to $\sim 4\ R_\odot$ then allows to obtain an intensity profile of the F-corona. Beyond $4\ R_\odot$, the model can be extended on the basis of the variation of the density of electrons assuming it is a monotonically decreasing function of r. This is not too critical as K becomes rapidly quite faint. To the classical result of Blackwell et al. (1967) and to the classical models of Saito (1970) and Allen (1973) which are in fact averages of various measurements, we added the data of Koutchmy and Magnant (1973)

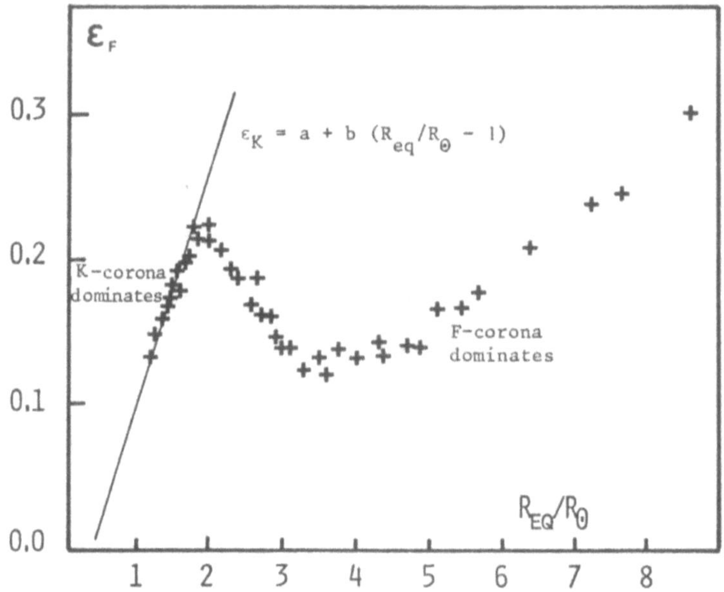

Fig. 1: The flattening index of ε_F the K-corona and of the inner F-corona versus equatorial heliocentric distance.

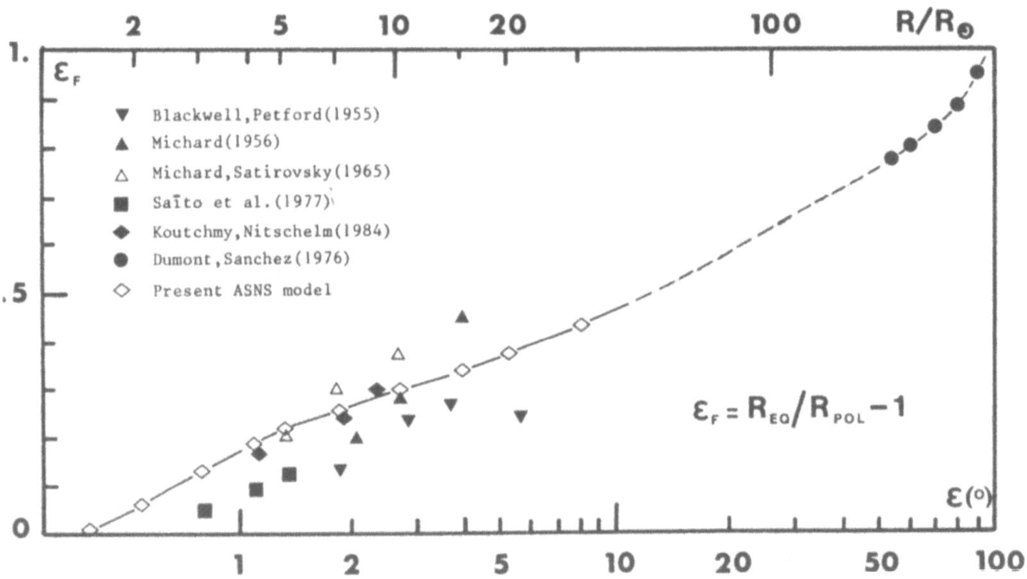

Fig. 2: The flattening index ε_F of the F-corona versus equatorial heliocentric distance in $R_\odot$ (top scale) or elongation in degrees (lower scale).

for the inner corona, of Koutchmy (1973) , of Keller et al. (1975), of Crifot-Magnant and Picat (1980) for a coronal hole and the most recent data of Dürst (1982) to build the axisymmetric non-spherical model ("ASNS") displayed in Fig. 3 valid in the wavelength interval 400-600 nm. Beyond 4 $R_\odot$, the intensity profiles are well represented by the power laws r^{-n} with n = 2.25 along the equator and n = 2.47 along the poles which smoothly connect to the inner Zodiacal Light (Leinert et al., 1976; Dumont and Sanchez, 1976). The ellipticity of the F-corona recalculated from these profiles is given in Fig. 2 and is in good agreement with the data points.

1.4. Polarization

The polarization of the F-corona is of crucial value for the understanding of the Zodiacal cloud as it can yield information on the relative importance of dust particles distributed along the line-of-sight and which diffract mainly non-polarized light and of circumsolar dust particles which scatter polarized light. Let us first consider the measurements of P_{K+F} which have been regrouped in Fig. 4 and which pertain to visible wavelengths. The situation is quite complex as P_{K+F} along the polar direction depends strongly on solar activity. We have therefore indicated this information. However, the reality of two diverging behaviors beyond 8 $R_\odot$ is conspiucous between windowless aircraft measurements of Blackwell (1955) and Michard (1956) later confirmed by the ground-based measurements of Michard and Sotirovsky (1965) and aircraft but through window measurements of Pepin (1970), Keller (1971) and Mutschecner et al. (1975). A possible "window" effect cannot be excluded and, as a matter of fact was noticed by one of the author during the supersonic aircraft (Concorde) observation of the 1973 eclipse. It must be further pointed out that the latter group of authors have attributed their high polarization to the K-corona (clouds of electrons) and not to the F-corona. We attempted to extract P_K using the theoretical result P_K $(r \gtrsim 5\ R_\odot)$ = 0.64 and obtained:

$R/R_\odot$	5	6	7	10	15
P_F	3×10^{-4}	0.007	0.016	0.021	0.020

2. COLORIMETRIC AND INFRA-RED ANALYSIS

Likewise the polarimetric data, the colorimetric and infra-red data offers the possibility to trace and characterize the circumsolar dust grains (so-called, "local" F-corona).

2.1. The color of the F-corona

The determination of the color of the F-corona is exceedingly difficult since it requires highly accurate photometry and calibrations in different spectral regions. To alleviate the calibration problem, it is assumed that the K-corona is solar-color (i.e., "white") so that the measure-

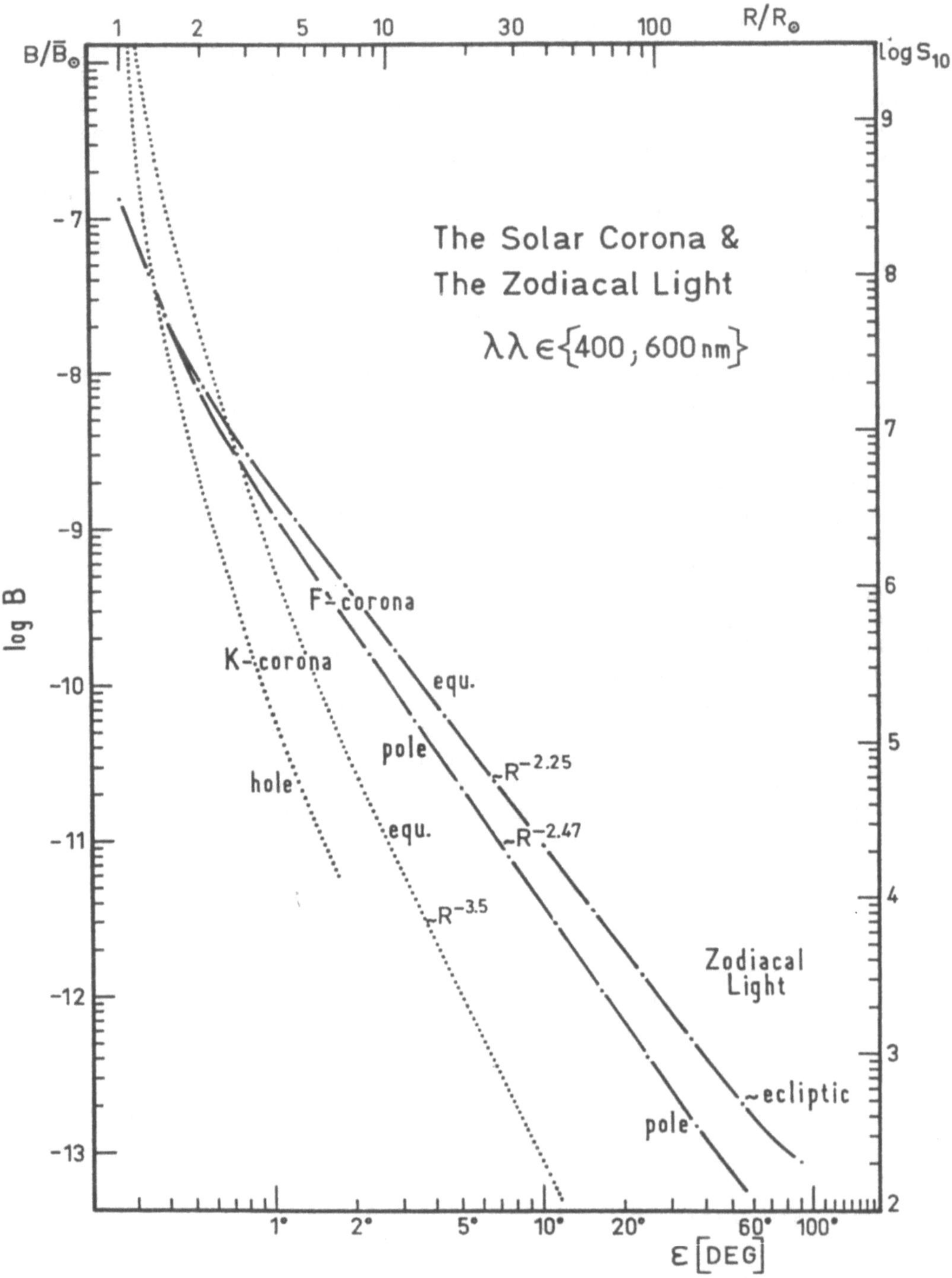

Fig. 3 : The axi-symmetric non-pherical model of the F-corona and its extension to the Zodiacal Light.

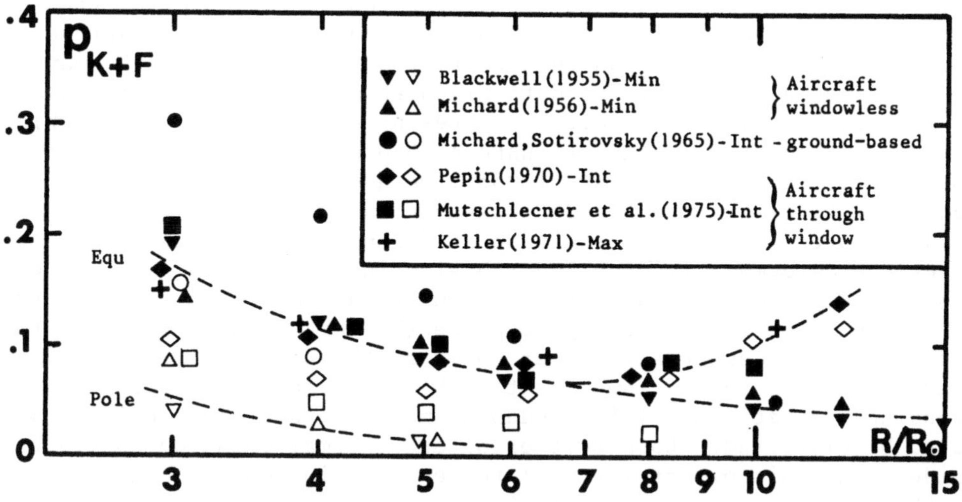

Fig. 4: The polarization of the total K+F corona versus heliocentric distance. The filled symbols pertain to the equatorial direction while the open symbols to the polar direction. The solar activity at the time of the observations is indicated by "Min", "Max" and "Int" for intermediate. The turnover beyond 8 $R_{\odot}$ may well reflect a window effect.

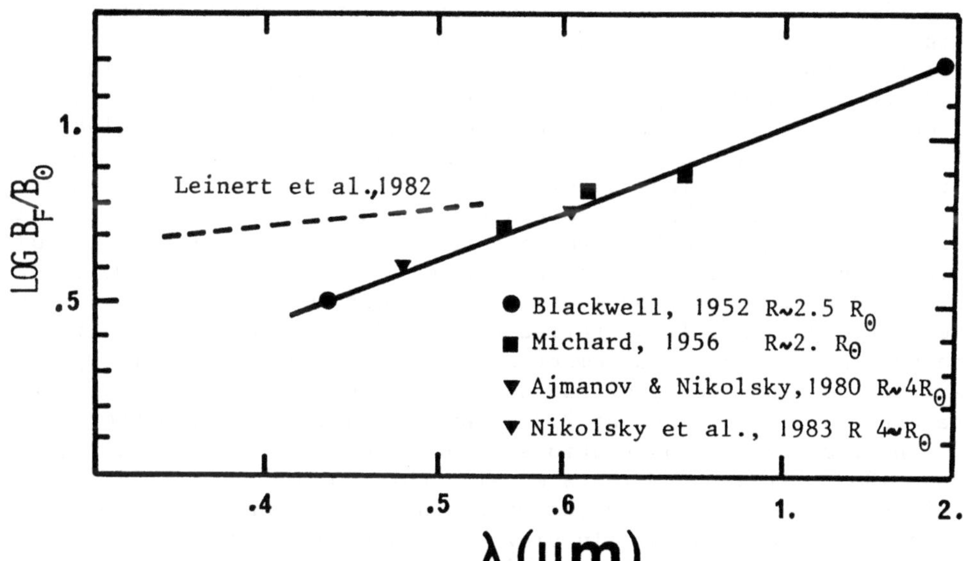

Fig. 5: The color index of the F-corona (see definition in the text) and of the inner Zodiacal Light (Leinert et al., 1982).

ments are in fact restricted to the comparison of the color of the F and K coronae. Following the method of Michard (1956), we present the variation of log $B_F/B_\odot$ on an arbitrary scale but versus the wavelength λ (instead of $1/\lambda$). The data given at different wavelengths (at least two) can, in general, be represented by straight lines which have been forced at the same ordinate at 640 nm. To the measurements of Blackwell (1952) and Michard (1956), we add the recent results of Ajmanov and Nikolsky (1980) and Nikolsky et al. (1983) in Fig. 5. It should be noted that they pertain to different locations in the F-corona. However the agreement is excellent considering further the variety of techniques employed and supports a strong reddening, much stronger that extrapolated from the Helios space probe (Leinert et al., 1982) in the limit $\varepsilon \to 0°$. The essentially solar color of the Zodiacal light at large appears to evolve in its inner part with a reddening increasing with decreasing elongations. This reddening may have different origins which need to be clarified. Both scattering and thermal emission should be considered as this effect may well reflect the increasing contribution from the latter process (see below).

2.2. The infra-red corona and thermal emission

The interest in infra-red observations of the F-corona was triggered by the theoretical prediction of Peterson (1963) of thermal emission bumps near 4 $R_\odot$. Near infra-red ground-based observations by Peterson (1967, 1969, 1971) and ground-based and balloon observations by Mac Queen (1968) have confirmed the prediction although an important quantitative disagreement is present (Fig. 6). Then the situation became confused following several negative reports. First, there was the observation at 10 μm carried out aboard the supersonic aircraft "Concorde" by Léna et al. (1974) whose result is plotted in Fig. 6. Then Mampaso et al. (1982) built a dedicated instrument using an externally occulting disk to perform infra-red observations outside solar eclipses benefiting from an improved detectivity. They reported no peak at all and further showed that their instrument could have detected a peak one order of magnitude de fainter than that observed by Peterson. An observation by Mutschlecner, Brownlee and Hall(quoted by Keller, 1981) performed at 0.7 and 2.2 μm at the February 1980 eclipse failed to record the inner part of the corona but indicate an enhancement at 9 $R_\odot$ and no other feature up to 20 $R_\odot$. At the same eclipse, Rao quoted by Zirker (1984) scanned the corona between 2 and 5.5 $R_\odot$ at 2.2 μm and did not detect the bump at 4 $R_\odot$. The most recent observations were performed with a balloon during the June 1983 eclipse by Maihara et al. (this volume): a preliminary analysis reveals an excess on the west side of the Sun only, especially strong at 1.25 and 1.65 μm. In view of this puzzling situation, we would like to make two comments. First, several authors have reported the existence of a (local?) component of the F-corona strongly reinforced in the equatorial regions (Mac Queen, 1968; Koutchmy, 1972) which appears to be distinctly different in its morphology from the classical flattening. Second, the alternance of positive and negative results may possibly result from the existence of transitory events. Nevertheless, it

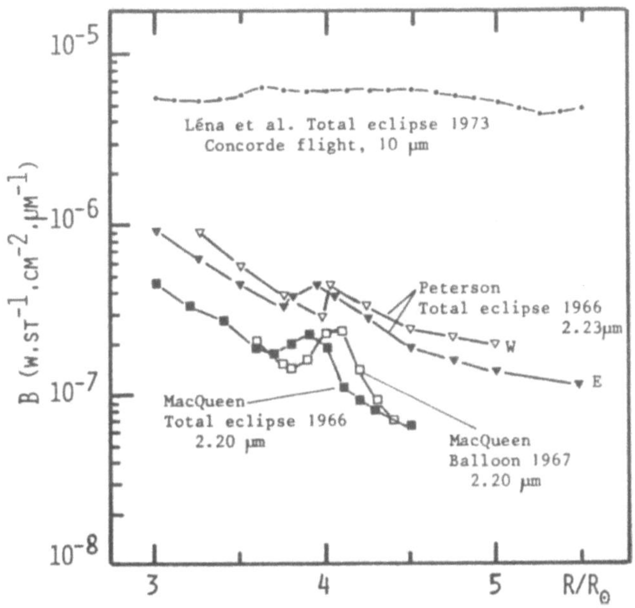

Fig. 6 : The infra-red brightness of the F-corona versus solar elongation showing the thermal feature in the near infra-red near 4 $R_\odot$.

is tempting to regroup all the positive resutls as we did in Fig. 7 with the data of Peterson, Mac Queen and Léna et al. (op. cit.) and of Mankin et al. (1970) for a distance of 4 $R_\odot$ to be compared with a solar color F-corona (white scattered light). In view of the large discrepancy between the above data, it is difficult to conclude although it appears that thermal emission dominates over scattered light beyond 3 μm. The high values observed in the 10 μm spectral domain could indicate the presence of silicates. Finally, it should be mentionned that there has been both predictions and theoretical interpretations of these results, in particular by Lamy (1974), Mukaï and Yamamoto (1979), Mukaï (this volume) and Grimshaw.

CONCLUSION

The properties of the F-corona in the optical domain appears now firmly established -with the possible exception of the polarization- allowing us to obtain a static model with the equatorial-polar asymmetry and having an high accuracy. The situation is far less satisfactory in the infra-red. Overall, it is difficult to assess the relative importance of the supposed two components "far" and "local" F-corona from the available observationnal data. Improved theoretical modeling is certainly required to clarify this question. Following classical observations, new externally occulted coronographs have revealed the permanent dynami-

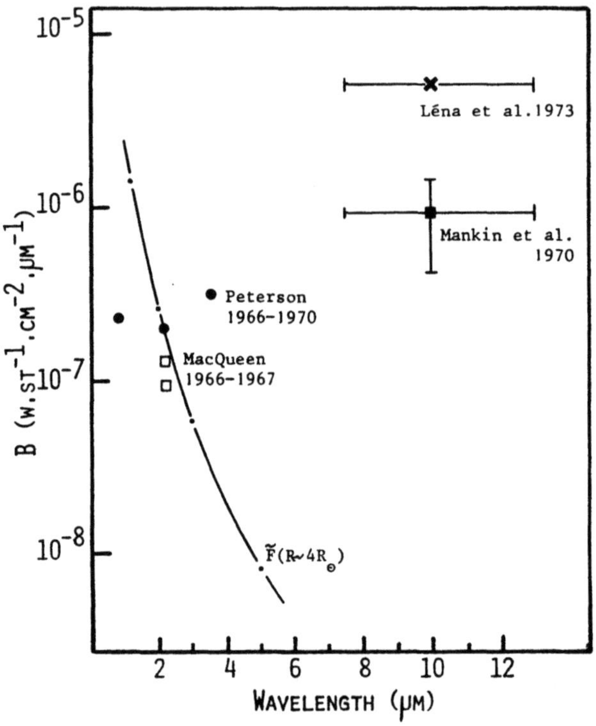

Fig. 7 : The infra-red spectrum of the F-corona near 4 $R_\odot$. The curve $\tilde{F}$ corresponds to a solar color corona.

cal state of the K-corona. Such a behaviour is not excluded for the F-corona as variations may be caused by
i) transitory equatorial enhancements;
ii) dust-free zones changing with the composition of the dust grains;
iii) debris of Sun-grazing comets as observed by Sheeley et al. (1982) and Michels et al. (1982).

In fact, the infra-red bumps at 4 $R_\odot$ may well have been transitory events.

If such ideas receive further support, then the F-corona should be viewed as a superposition of a quasi-stationnary "far" corona and a variable "local" corona, the variability resulting possibly from the interplanetary medium (comets) and from solar phenomenae (transient events, solar wind, solar cycle). Such exciting perspectives call, in our opinion, for infra-red imaging which should be done during solar total eclipses and for a long-life sophisticated space coronameter.

REFERENCES

Allen, C.W.: 1973, Astrophys. Quantities, The Athlone Press.
Ajmanov, A.K. and Nikolsky, G.M. : 1980, Solar Physics, 65, 171.
Blackwell, D.E. : 1952, MNRAS, 112, 652.
Blackwell, D.E. : 1955, MNRAS, 115, 629.
Blackwell, D.E. and Petford, A.D. : 1966, MNRAS, 131, 383 and 399.
Blackwell, D.E., Dewhirst, D.W. and Ingham, M.F. : 1967, Adv. Astron. and Astrophys. 5, 2.
Bohlin, J.D., Koomen, M.J., Tousey, R. : 1971, Solar Physics 21, 408.
Bohlin, J.D. : 1971, Solar Physics 18, 450-457.
Dürst, J. : 1982, Astron. Astrophys. 112, 241.
Howard, R.A. and Koomen, M.J. : 1975, IAU Coll. Nº 31 "Interplanetary dust and zodiacal light", Heidelberg, 66.
Keller, C.F. : 1971, Solar Phys. 21, 425.
Keller, C.F., Tabor, J.E. and Matuska, W. : 1975 BAAS, 7, 223.
Keller, C.F., and Liedenberg, D. : 1981, Los Alamos Science, LANL, 2,4.
Koutchmy, S. : 1972, Astron. Astrophys. 16, 103.
Koutchmy, S. : 1973, Probl. Kosm. Fiziki., 8, 28.
Koutchmy, S. and Magnan, F : 1973, Ap. J. 186, 671.
Lamy, Ph.: 1974, Astron. Astrophys. 35, 197.
Leinert, C. : 1975, Space Sci. Rev. 18, 281-339.
Leinert, C., Link, H., Pitz, E. and Giese, R.H.: 1976, Astron. Astrophys. 47, 221.
Leinert, C., Richter, I, Pitz, E, Hanner, M. : 1982, Astron. Astrophys. 110, 355.
Léna, P., Viala, Y, Hall, D. and Soufflot, A. : 1974, Astron. Astrophys. 37, 81.
Mac Queen, R.M. : 1968, Ap. J. 154, 1059.
Mac Queen, R.M., Ross, C.L. and Mattingly, T : 1973, Planet. Space Sci. 21, 2173.
Mampaso, A., Sanchez Magro, C. and Buitrago, J.: 1982, in "Sun and Planetary System", W. Fricke & G. Tedeki Eds, Reidel, p.257.
Mankin, W.G., Mac Queen, R.M. and Lee, R.H. : 1974, Astron. Astrophys., 31, 17.
Michard, R. : 1956, Ann. Astrophys. 19, 229.
Michard, R. and Sotirovsky, P. : 1965, Ann. Astrophys. 28, 96.
Michels, D.J., Sheeley, Jr. N.R., Howard, R.A., Koomen, M.J.: 1982, Science, 215, 1097.
Mukai, T. and Yamamoto, T. : 1979, P.A.S.J. 31, 585.
Mutschlecner, J.P., Keller, C.F. and Tabor, J.E. : 1975, 147th Meeting of the AAS.
Nikolsky, G.M. : 1975, The solar corona and the Interplanetary Medium, Ed. "Znanye", Moscow.
Nikolsky, G.M., Koutchmy, S. and Nesmjanovich, I.A. : 1983, Soln. Dannye, 4, 67.
Pepin, T.J. : 1970, Ap. J. 159, 1067.
Peterson, A.W. : 1963, Ap. J. 138, 1218.

Peterson, A.W. : 1967, Ap. J. Letters, 148, L 37.
Peterson, A.W. : 1969, Ap. J. 155, 1009.
Peterson, A.W. : 1971, B.A.A.S. 3, 500, 135th Meeting of the A.A.S.
Saito, K. : 1970, Ann. Tokyo Astron. Obs. vol.XIII, 93.
Saito, K., Poland, A.I. and Munro, R.H. : 1977, Solar Physics, 55, 121.
Sheeley, N.R., Howard, R.A., Koomen, M.J. and Michels, D.J. : 1982, Nature 300, 239.
Van de Hulst, H.C. : 1953, The Sun, ed. G. Kuiper, 262.
Weinberg, J.L. and Sparrow, J.G. : 1978 in "Cosmic dust", Ed. J.A.M. Mc Donnel (New York : Wiley-Interscience), 75.
Weiss-Wrana, K : 1983, Astron. & Astrophys. 126, 240.
Zirker, J.B. : 1984, Total Eclipses of the Sun, VNR Company Publ.

INTERPETATION OF F-CORONA RADIAL VELOCITY FIELD

L.I. Shestakova
Astrophysical Institute
Academy of Sciences
Alma-Ata 480068
USSR

ABSTRACT

The radial velocities of F-corona for total eclipse July 31, 1981 obtained at elongations (3-7) $R_\odot$ are interpretated assuming prograde motion in circular Keplerian orbits of circumsolar dust. In computations isotropic plus diffraction scattering functions are used. In supposition of silicate grains the best coincidence with observations is received for grains with radii about 0.4 μm with the presence of dust free zone (d.f.z.) around the Sun to heliocentric distances (6.5-14) $R_\odot$ and high grain concentration at the edge of this dust free region.

The presence of negative radial component of radial velocities can be explained by falling of the matter into the Sun under influence of Poynting-Robertson effect of solar wind for porous grains with radii less than 0.4 μm. The effect of radiation drag for these small grains is essentially less. Using grains with radii less than 0.4 μm the sizes of which decrease as approaching to the Sun one may explain the inner F-corona brightness.

R. H. Giese and P. Lamy (eds.), Properties and Interactions of Interplanetary Dust, 75.

INTERFEROMETRIC OBSERVATION OF THE F-CORONA RADIAL VELOCITIES FIELD BETWEEN 3 AND 7 $R_\odot$

P. V. Shcheglov
Sternberg Astronomical Institute
117234 Moscow
USSR

L. I. Shestakova, A. K. Ajmanov
Astrophysical Institute
480068 Alma-Ata
USSR

ABSTRACT. During the July 31, 1981 solar eclipse, F-corona interferograms near MgI λ 5184 Å were obtained using a Fabry-Perot etalon (FPE) with an FWHM of 0.5 Å (corresponding to 30 km/sec) and an image tube. Radial velocities V_r of the interplanetary dust (i.d.) were measured in different directions.

Both prograde and retrograde motions of i.d. in the ecliptic region is discovered. Most of velocity values do not exceed 50 km/sec. A negative velocity component appears after averaging all V_r for all directions. Its average increases to − 20 km/sec toward the Sun. Some ejections are observed. The strongest (+ 130 km/sec) is located at the north ecliptic pole at a distance of 6 to 7 $R_\odot$.

From the lack of unshifted Fraunhofer lines in the scattered sky light, we conclude that the sky brightness continuous component is predominant and its source is K-corona scattered light in the Earth's atmosphere.

1. INTRODUCTION

There are some difficulties in the problem of radial velocities V_r measurements during eclipses because of relatively low F-corona brightness at distances $R > 3\ R_\odot$. We know of only two such experiments: i) during the June 30, 1973 eclipse at a distance of 7.5 $R_\odot$ West of the Sun, the i.d. radial velocity (about + 50 km/sec) was measured using a Fabry-Perot spectrometer at MgI λ 5184 Å (Kozlov et al., 1974); ii) during the February 26, 1976 eclipse, the radial velocities at distances $R = 3.2\ R_\odot$ and $R = 4.3\ R_\odot$ West of the Sun were measured using the correlative radial velocity spectrometer (Beavers et al., 1980). The common feature of both velocity profiles is the existence of a maximum V_r = 50 km/sec. One can see that such measurements are

R. H. Giese and P. Lamy (eds.), Properties and Interactions of Interplanetary Dust, 77–80.

not numerous and that our work will alleviate this situation to a certain degree.

2. MATERIALS AND METHODS

The optical design of the instrument is described in details by Shcheglov et al. (1983). The joint expedition of Astrophysical Institute, Alma-Ata and Sternberg Astronomical Institute, Mocow for the observation of the July 31, 1981 total solar eclipse was lain near Shortandy of the Tselinograd region. The duration of totality was about 75 sec; interferograms of the corona and of the reference emission line of Ar λ 5187 Å were recorded on Kodak 103-aG emulsion. Only one photograph (53 sec) has normal photograph density and was used for the treatment.

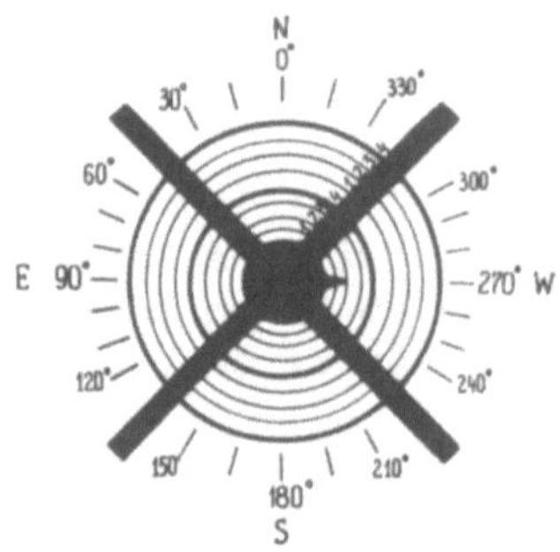

Figure 1. Representation of the 53 sec exposure interferogram. NS is the direction to the ecliptic poles. 1 - λ 5195 Å, 2 - λ 5192 Å, 3 - λ 5189 Å and 4 - λ 5184 Å MgI lines.

Fig. 1 represents the interferogram. It shows the central occulting disk which blocks the corona up to 2.5 $R_\odot$ and the 'cross' of the Ar lamp. Two interference orders are used for the treatment, each one has 4 rings formed by one or a few Fraunhofer lines. The definition of the center of the rings and the identification of the lines were made using the nonshifted lines of the day sky on a pre-eclipse photograph.

RMS deviation is 20 km/sec after the treatment using an automatic microdensitometer. The differential distortion errors relatively to Ar is small. The error due to the orbital motion of the Earth is less than 1 km/sec.

3. RESULTS

There is a velocity dispersion in the F-corona which exceeds the error bars at the change of the distance by value 0.5 $R_\odot$. The bifurcated lines both narrow and expansive are observed; mainly the bifurcation corresponds to the shift relatively main line to the negative direction.

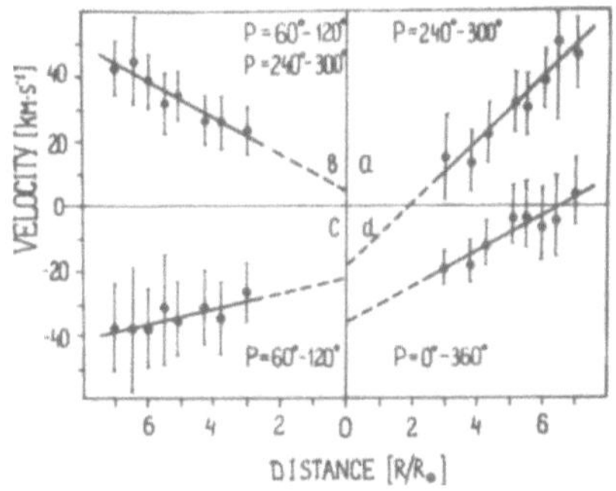

Figure 2. Measured and averaged radial velocities for different directions. a - West; b - West + (- East) - only direct Keplerian component; c - East; d - all components for all directions.

In the ecliptic region, the average velocity is negative on the east side of the Sun and positive on the west side. It proves the prograde Keplerian motion of matter. The average variation of the Keplerian velocity with the distance may be approximated by a linear dependence (Fig. 2b) V_r [km/sec] = (5.4 ± 0.7) $R/R_\odot$ + 5 ± 4.

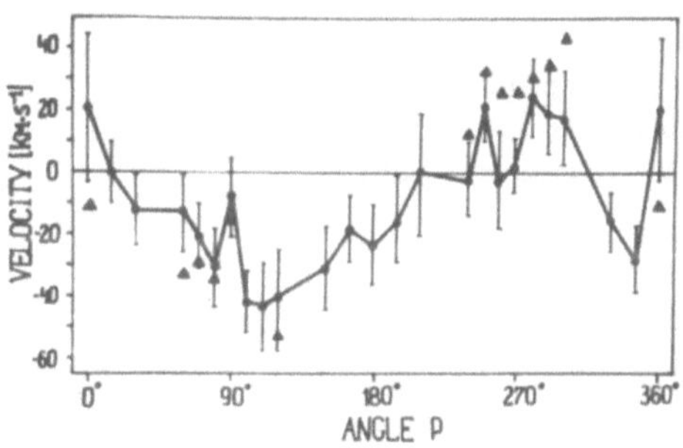

Figure 3. Solid line - V_r (P) averaged from 3 to 7 $R_\odot$; ▲ - retrograde motions and points with + 130 km/sec at P = 0° are excluded.

The absolute velocity averaged from 3 to 7 $R_\odot$ reaches a maximum near the ecliptic and a minimum near the poles (Fig. 3). There is a decrease of the velocity near the plane of ecliptic (P = 90 and P = 270°). The negative velocity component independent of the distance (Fig. 2d) shows the evidence of i.d. moving radially outward from the Sun. A local velocity peak of + 130 km/sec at the north ecliptic pole between 6 and 7 $R_\odot$ appears. On the negative, it is expressed in form of the distorsion of lines λ 5184 Å and λ 5189 Å. Up to a distance of 7 $R_{\odot}$ practically, there is no indication of a velocity component corresponding to nonshifted Fraunhofer lines in the scattered sky light.

4. DISCUSSION

The obtained velocity are values integrated over the line of sight. One can make no conclusion on the real dust velocity value at given distances from the Sun without a model (and assumptions) of the i.d. cloud.

Fig. 2a and 2c shows the prograde Keplerian motion. The different slopes for E and W is easily explained by the existence of a radial velocity component directed to the Sun and changing with distance (Fig. 2d). The asymmetric wide maximum of the prograde Keplerian velocities w.r.t. the abscissa axis may be explained by this cause too. The discovery of velocities in F-corona exceeding that the prograde Keplerian motion may be connected with the cometary dust scattered near the Sun, moving near the ecliptic in a retrograde direction. This comet was observed in July 20, 1981 (Sheely et al., 1982). The local decrease of the absolute velocitie for P = 90° and 270° (Fig. 3) may be also explained by the retrograde motion of cometary dust.

The absence of a "zero" velocity component corresponded to non-shifted lines in the eclipse sky is unexpected. It may be understood if one supposes that the sky brightness has a considerable continuous component. The scattering of the K-corona light in the Earth's atmosphere may be the main source of this component. According to our estimates, the Fraunhofer component of the eclipse sky brightness is weaker than the continuous component by at least one order of magnitude. This conclusion gives a new view on the possibility of spectral observations of the F-corona at large distances.

5. REFERENCES

Beavers,W.I., Eitter,I.I., Carr,P.H., Cook,B.C.: 1980, Astrophys. J., 238,349.

Kozlov,Ju.A., Novikov,S.B., Shcheglov,P.V.: 1974, Astron. Zirc. (USSR), No.808,1.

Shcheglov,P.V., Shestakova,L.I., Ajmanov,A.K.: 1983, Ann. Astrophys. Inst. of Acad. of Sci. of Kaz. SSR, 41,70.

Sheely,N.R., Howard,R.A., Koomen,M.I. and Michels,D.I.: 1982, Nature, 300,239.

The Motion of Interplanetary Dust Particles

I.R. East and N.K. Reay
Astronomy Group, Department of Physics, Blackett Laboratory, Imperial College, Prince Consort Road,
LONDON SW7 2BZ
ENGLAND

ABSTRACT. Radial velocity measurements on the solar MgI 5183.618Å Fraunhofer absorption line in the Zodiacal Light spectrum have been made with a microprocessor-controlled servo-stabilised Fabry-Perot interferometer. Observations were made at 5 and 10 degree intervals in the ecliptic plane between morning and evening elongations of 25 degrees. These new data are of much greater precision and coverage than any previously obtained. Obtained over a two year period, the observations show a consistent evening/morning asymmetry in the radial velocity curve with the Gegenschein receding from the Earth at 4 Km/sec. These data do not support the hypothesis that the majority of interplanetary dust grains are in hyperbolic trajectories.

1. THE INSTRUMENT

Because the Zodiacal Light is a very faint source it is necessary to use an interferometer of high luminosity. When only a narrow spectral region is to be observed the Fabry-Perot etalon offers a gain in luminosity over other types of instrument.

The MgI 5183.6Å Fraunhofer line was chosen because of it's depth, cleanness, freedom from blending and from contamination by airglow, geocoronal or meteoric emmission. It has been used with some success on previous occasions, (Hicks, May and Reay, 1974; James & Smeethe, 1970).

An etalon of diameter 80mm and Finesse ~24 with a plate separation of 87μm was employed giving a Free Spectral Range of 14.4Å.The etalon gap and paralellism were servo-stabilised using the technique of capacitance micrometry (Hicks, Reay and Scadden, 1974). Servo actuation is by piezo-electric transducers. An 80mm diameter 3 cavity interference filter was used as prefilter. Residual drift of the etalon plate spacing was less than λ/500 corresponding to a spectral drift of 0.03Å per hour. Periodic calibration scans of a cold Magnesium hollow cathode lamp ensured that systematic errors due to wavelength drift were always less than 0.01Å.

The photomultiplier employed was a RCA31043A-02 with a

R. H. Giese and P. Lamy (eds.), Properties and Interactions of Interplanetary Dust, 81–84.

gallium-arsenide photocathode which, when cooled to -70 degrees C, gave a dark count of between 1 and 4 Hz.

2. OBSERVATIONS AND DATA REDUCTION

The site chosen was the one used by Hicks et al (1974), Izana on Tenerife in the Canary Islands. Observing trips were made in March, July and September in 1981 and 1982 and also in March 1983. Two further trips in July and September 1983 were unfortunately both wiped out by bad weather. This still gave a total of more than 60 nights observing.

A spectral range of ~6Å centred on the MgI 5183.6Å absorption line was scanned at intervals of 0.25Å. Scans accumulated over several nights at elongation intervals of 5 or 10 degrees were co-added to improve signal to noise ratio. The summed scans were then cross-correlated with a gaussian profile of equal width (and similar shape) to that of the unshifted line when convolved with the instrumental profile. The peak of the cross-correlation function is then used to indicate the position and thus the shift of the line.

It is important to note that any small component of the Zodiacal Light originating from dust moving with velocities greater than about 50km/sec will not affect the position of the peak of the cross-correlation function as much as it would the first moment (or centre of gravity).

3. RESULTS

Figure 1 shows the sum total of the Doppler-shift results gathered over the whole observing program. The salient points to note are;

i) the red-shift of the Gegenschein,
ii) asymmetry between evening and morning,
iii) that the spread of the points is consistent with the calculated standard errors.

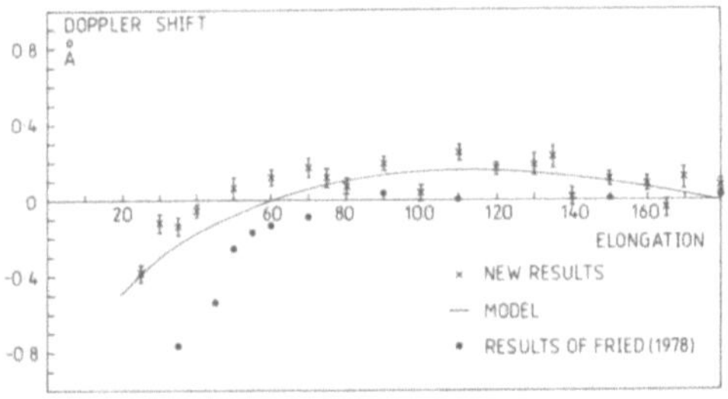

Figure 1 Measured Doppler-shift plotted as a function of elongation.

Note that the error at the Gegenschein is much less than elsewhere due to the deliberately applied priority to this point. Thus the red-shift of the Gegenschein is more significantly determined than the nature of the asymmetry.

Figure 2 shows the new results folded about the anti-solar point to ease comparison with the data from Fried (1978) (averaged for each elongation) and with a simple model in which the dust is in circular orbits, unaffected by radiation pressure, with a radial distribution of number density of the form;

$$n(r) = K/r^{1.3} \tag{1}$$

determined from Helios and Pioneer spacecraft, (Link et al., 1976). Diffraction effects are neglected, since these are only important for elongations below ~20 degrees, and isotropic scattering is assumed.

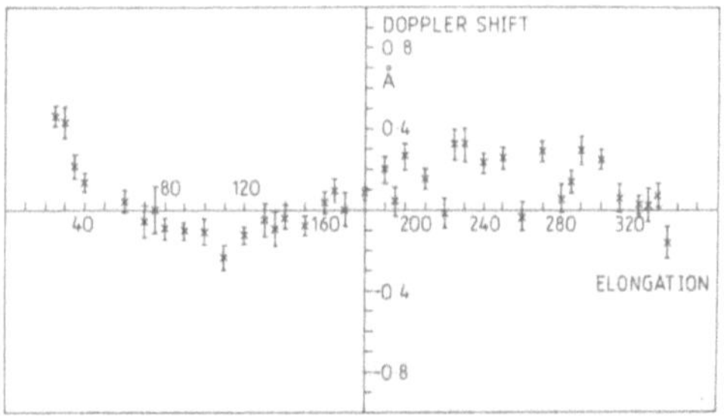

Figure 2. The new results folded about the antisolar point to facilitate comparison with the data of Fried and a simple model of circular Keplerian orbits.

This model is obviously not wholly inadequate. However it is inherently incapable of explaining the observed asymmetry or the recession of the Gegenschein. (The former is of course averaged out by the folding process).

4. INTERPRETATION OF DOPPLER-SHIFT DATA

4.1. Elongations less than 180 degrees

It may be shown that, in the expression for the observed Doppler-shift along a line of sight corresponding to a particular elongation, there exists a single factor χ which accounts for any deviations the dust may exhibit from circular Keplerian orbits, (East, 1984). It mayalso be shown also that χ factorises into three components, one each for the effect of radiation pressure, orbital eccentricity and orbital inclination provided radial invariance is assumed for optical properties, size distribution and distribution of orbital parameters. With the help of the Haug integral (Haug, 1958), these three factors are readily derived, (East, 1984).

Note that the effects of radiation pressure, orbital eccentricity and orbital inclination on Doppler-shifts measured in the ecliptic plane are indistinguishable. Also the measured values of χ as a function of ε may be used as a probe of radial invariance.

4.2 The Gegenschein

It may be shown (East, 1984) that to observe a red-shift in the Gegenschein implies directly the existence of a component within the Zodiacal dust cloud of particles in hyperbolic trajectory, i.e unbound to the solar system which are contributing to the observed Zodiacal Light.

Because the reduction technique employed is insensitive to any structure in the wings of the line the measured Doppler-shifts will be unaffected by a component with line of sight velocities near or in excess of the velocity of escape from the solar system at 1 A.U..

The fact that a shift in the Gegenschein has been observed can only be interpreted as implying that the particles originate at solar distances which are NOT << 1 A.U..

5. CONCLUSIONS

The weighted mean value of χ is between 0.85 +/-0.04 and 0.89 +/-0.04 depending only on choice of scattering function (observed VSF or isotropic) and radial distribution index (1.0 to 1.5) employed as the model.

There clearly exists a component of the Zodiacal dust cloud escaping from the solar system on hyperbolic trajectories which **does** contribute observably to the Zodiacal light. It is possible that this contribution is seasonally variable and that the observed recession of the Gegenschein is only a fairly coarse average value.

To be consistent with a red-shift being observed at all, the escaping particles must originate at solar distances which are **not** << 1 A.U.. This is consistent with conclusions reached by Whipple (1976) based on the model by Zook & Berg (1975) of submicron escaping particles as collisional debris.

ACKNOWLEDGEMENTS: The authors wish to thank the Instituto del Astrofisica de Canaries for use of site and facilities at the Observatorio de Izana, Tenerife.

One of us, (I.R. East), gratefully acknowledges receipt of an S.E.R.C. studentship during the greater part of this research.

The authors are indebted to Professor David Beard of the University of Kansas for many helpful and useful discussions.

REFERENCES

East I.R.: 1984, in preparation

East I.R. & Reay N.K.: 1984, submitted to *Astron. & Astrophys.* March 1984

Fried J.W.: 1978, *Astr. & Astrophys.*, **68**, 259

Haug U.: 1958, Z. *Astrophys.*, **44**, 71

Hicks T.R., May B., Reay N.K.: 1974, *Mon. Not. R. Astr. Soc.*, **166**, 439

Hicks T.R., Reay N.K., Scadden R.J.: 1974, *J. Phys. E. (Sci. Instr.)* **7**, 27

James J.F. Smeethe M.J.: 1970, *Nature*, **227**, 589

Link H., Leinert C., Pitz E., Salm N.: 1976, *Lect. Notes in Phys.*, **48**, 19

Whipple F.L.: 1976, *Lect. Notes in Phys.*, **48**, 403

Zook H., Berg O.E.: 1975, *Planet. Spa. Sci.*, **23**, 183

DOPPLER SHIFTS MEASUREMENTS OF THE ZODIACAL LIGHT AT THE PIC MIDI OBSERVATORY.

R. Robley[1], A. Bücher[1], S. Koutchmy[2], Ph. Lamy[3]

1) Observatoire du Pic-du-Midi, Toulouse.
2) Institut d'Astrophysique, Paris.
3) Laboratoire d'Astronomie Spatiale, Marseille.

Abstract: At the Pic-du-Midi Observatory, we have started a program to observe the spectrum of the zodiacal light in the region of the b absorption lines of MgI. We use a 1 meter spectrograph with a holographic grating followed by a focal reducer to obtain a 19 Å/mm dispersion over the micro-channel plate camera. Calibrated spectra obtained with an exposure time of 10 minutes systematically show earth atmospheric emission lines attributed to both NI and OH. The temporal variability of these emissions prevents an accurate determination of the line profiles. Our best processed measurements of Doppler shifts obtained during five nights do not allow to descriminate between circular and hyperbolic orbits of the interplanetary dust.

I - INTRODUCTION.

The first attempts to measure Doppler shift effects in the Hβ line of the spectrum of the zodiacal light using a Fabry-Perot etalon were reported by Ring et al. (1964), Reay and Ring (1968), and James and Smeethe (1970). Unfortunately the presence of a faint Hβ emission line due to the geocoronal emissions spoiled the measurements; the b-lines of MgI were subsequently favoured.

Recently Fried (1978) used a Griffin type spectrograph to produce the best results presently available. He found that, for elongations smaller than 60° the measured shifts imply velocities larger than those expected for Keplerian circular or elliptical orbits.

A point which seems to be well established is the prograde motions of the zodiacal cloud confirmed by all published measurements.

Since 1981, we have started Doppler shifts measurements of the zodiacal cloud at the Pic du Midi Observatory.

II - THE INSTRUMENT AND THE METHOD.

The observations are performed with a conventional Czerny-Turner type spectrograph of 1m focal length (entrance slit 1x80mm; holographic grating 160 x140 mm^2, 2400 gr/mm, 1st order).

The primary spectrum is re-imaged beyond a field lens using an image

R. H. Giese and P. Lamy (eds.), Properties and Interactions of Interplanetary Dust, 85–88.

reducer (effective aperture F/2) to produce a 19 Å mm^{-1} dispersion in the plane of the entrance window of the detector package. Several image tubes and photon counting system were tested, see Robley et al.(1982a). The best compromise was found with a 70mm photo-camera equiped with a zero distorsion proximity focused dual micro-channel plate with a S-20 photocathod. 103 aG film was used. The whole instrumental set-up is pointed directed to the morning or evening zodiacal cloud using an altazimutal mount in order to direct the 1x80mm^2 entrance slit accross the main axis of the cloud.
On each frame a calibration spectrum produced by a Mg hollow cathode lamp is recorded on both sides of the zodiacal light spectrum which extends over a 275 Å interval centered on the Mg I triplet. A sufficient signal to noise ratio is obtained with a 10 min exposure time when observing near the ecliptic plane. Several methods have been tested to measure the Doppler shifts, Robley et al. (1982a,1982b) and two of them have been retained:

1 - Optical measuring machine routinaly used for stellar radial velocities.
2 - Analog cross-correlations between calibration and zodiacal light spectra.

III - RESULTS AND DISCUSSIONS.

For ground-based observations several components are superposed along the line of sight, see Koutchmy et al. (1981), including a parasitic emission spectrum from the Earth atmosphere. Our observations systematically reveal a complexity and a variability of this component which seem to have been underestimated in the past.
The spectral internal observed is filled by the rotation-vibration spectrum of OH, especially the spectrum of the vibrational transition (9,2) corresponding to the 2 foundamental states $^2\Pi_{3/2}$ and $^2\Pi_{1/2}$. In addition, the well known emission of NI near 5200 Å is always present during the whole night, see figure 1 and seems to be distorded by an OH emission line.
All these night sky emissions appears to fluctuate in time over a scale of order of our exposure time that is 10 min. Consequently, the Fraunhofer spectrum of the zodiacal light is distorted, complicating the measurements of the shifts and leading to rather large error bars. Figure 2 shows an example of a spectrum used from a sequence obtained on the morning of Sept.8, 1983.
The whole set of our measurements is displayed in figure 3. The quiet large scatter of the data points is mainly due to the effects introduced by the OH emissions. It is the superiority of our technics to show simultaneously the F-spectrum of the zodiacal light together with the contaminating lines.As a consequence of this scatter it is not possible with the present data to infer the nature of the orbital motions of the interplanetary dust.

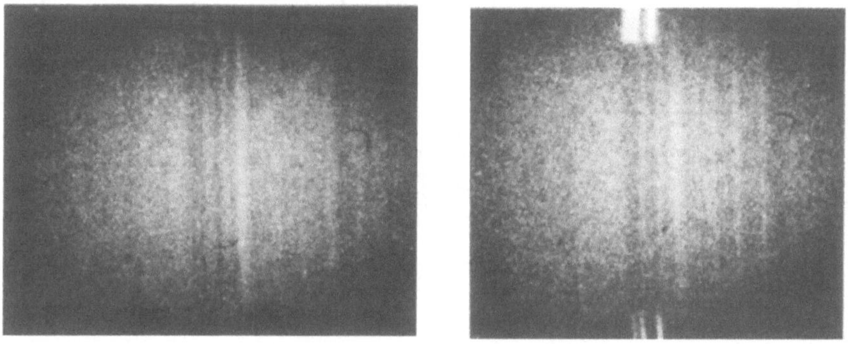

Figure 1 - Prints of the spectra N° 2 and 20 obtained during the night of September 3-4, 1983, in the direction of the north pole (h = 22°) showing the large variations of the OH emission lines near the MgI triplet (in absorption).

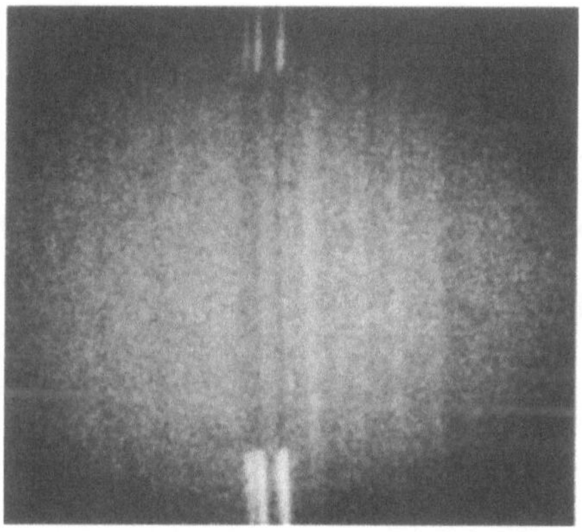

Figure 2 - Print of a typical spectrum obtained on the morning zodiacal light on Sept. 8, 1983 (East direction); the exposure time is 10 min.

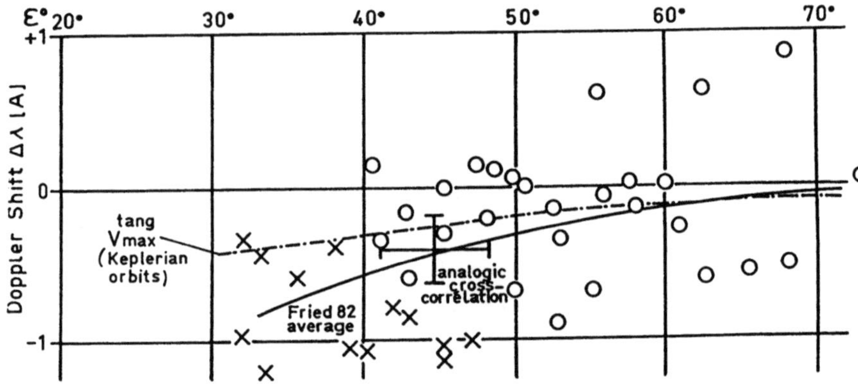

Figure 3 - Diagram displaying the whole set of Doppler shift measurements:O are for the morning observations (with reversed sign) and x for the evening ones. The result obtained using the cross-correlation method is shown with error bars. The heavy line is for the average results of Fried (1978) and the dashed line represents the theoritical result for Keplerian circular orbits.

REFERENCES.

Fried,J.W.: 1978, Astron. Astrophys. 68, 259.
James,J.F., Smeethe,M.J.: 1970, Nature 227, 588.
Koutchmy,S., Bücher,A., Robley,R.: 1981, Deuxième Coll. Nat. Telescope Spatial, Orsay, p. 39.
Reay,N.K., Ring,J.: 1968, Nature 219, 710.
Ring,J., Clarke,D., James,J.F., Daehler,M., Mack,J.E.: 1964, Nature, 202 167.
Robley,R., Bücher,A., Koutchmy,S.: 1982a, Sixth Eur. Reg. Meeting, Frike and Telek Eds, p. 255.
Robley,R., Bücher,A., Koutchmy,S., Lamy,P.L.: 1982b, Journées "Planétologie" INAG Ed., p. 205.

THE INTERPLANETARY MICROMETEOROID FLUX AND LUNAR PRIMARY AND SECONDARY MICROCRATERS

H. A. Zook[1], G. Lange[2], E. Grün[2], and H. Fechtig[2]
[1] NASA--Johnson Space Center, Houston, Texas, U.S.A.
[2] Max-Planck-Institut für Kernphysik, Postfach 10 39 80
6900 Heidelberg, F.R.G.

Abstract

We are gaining an increased awareness and understanding of Earth-orbiting space debris. Meteoroid experiments in near-Earth orbit must therefore now be able to differentiate between interplanetary meteoroids and space debris. Space debris impacts are not thought, however, to have significantly affected near-Earth meteoroid measurements carried out in the early 1960's. New experimental evidence also makes it appear very probable that most impact pits on lunar rocks with pit diameters smaller than 7 micrometers have been generated by lunar secondary ejecta impacts, and not by primary meteoroid impacts. In addition, ages determined from solar flare tracks in lunar rocks are not considered secure. Lunar crater production rates are more reliably deduced from meteoroid space experiments and not from solar flare track ages. When all of the above qualifications are taken into account, however, a rather satisfactorily self-consistent meteoroid flux versus mass distribution is obtained.

1. Introduction

Hanner et al.(1976) and Leinert et al.(1981) have determined broad features of the heliocentric distribution of meteoroids from zodiacal light observations between 0.3 and 3 AU. It is difficult, however, to derive much information about the size distribution of meteoroids from only the zodiacal light data obtained to date. Even the relatively easy question of whether or not sub-micron grains scatter a significant fraction of the observed zodiacal light is not similarly answered by everyone. Lamy and Perrin(1980), for example, suggest submicron grains could contribute significantly to scattered light, while Leinert et al.(1980) noted that the zodiacal light spectrum that they observe is reddened relative to the solar spectrum, rather than the reverse that would be expected if contributions by submicron grains were significant.

It is therefore necessary to make use of direct measurements such as the photographic and radar meteor data, the spacecraft in situ meteoroid sensing data, and the lunar impact crater data to obtain a relatively complete picture of the size distribution of meteoroids. Detailed orbital information is also better obtained through direct measurements. It is direct measurements that give rise to our knowledge of meteor streams, beta meteoroids(Berg and Grün,1973; Zook

R. H. Giese and P. Lamy (eds.), Properties and Interactions of Interplanetary Dust, 89–96.

and Berg,1975), and alpha meteoroids(Grün and Zook,1980).

Our primary purpose in this article is to point out that considerable care is required, however, to properly interpret either the spacecraft in situ data or the lunar crater data. Not all lunar craters are due to direct meteoroid impacts. (Note: in this paper, "crater" denotes central pit, if it exists, plus surrounding spall zone.) Analyses of the near-Earth data are similarly complicated by impacting Earth-orbiting debris.

2. In Situ Sensing of Meteoroids

Naumann(1966) carried out a most thorough analysis of meteoroid penetration data obtained with the Explorer XVI, Explorer XXIII, Pegasus I, Pegasus II, and Pegasus III Earth-orbiting satellites, as well as an analysis of the photographic and radar meteor data. From these data he derived that the slope of the log(number) versus log(mass) curve for meteoroids continuously decreased with decreasing meteoroid mass. Early examination of spacecraft windows for impact pitting (Zook et al.,1970; Cour-Palais,1974) confirmed the trend of the flux-mass curve established by Naumann; namely that the low level of observed window pitting was consistent only with a meteoroid flux-mass curve that "flattened" for meteoroid masses below about one milligram.

Hallgren and Hemenway (1976) and Nagel et al.(1976) noted, however, that a disturbing number of the impact craters found on the S-149 meteoroid impact experiment(flown during the 84 day Skylab IV mission in 1973) were coated only with aluminum in the pit interior. Clanton et al.(1980) also found, in a new analysis of impact craters found on the outer surface of the Skylab IV Command Module windows, that many of the smaller impact pits were coated with an aluminum-rich liner. They postulated that these craters resulted from impacts by aluminum oxide sperules that had been deposited in near-Earth space by ignition of solid rocket motors. Clanton et al. also found titanium in one pit. The titanium was postulated to result from a hypervelocity impact on the window of a flake of white thermal paint. It appeared that, for impact pits smaller than 20 micrometers in diameter, impacts by orbital debris contributed an important fraction of the total number of observed craters.

There is also an important debris population at larger particle sizes. Kessler and Cour-Palais(1978) noted that, at particle masses greater than a few hundred grams and at altitudes between 700 and 1200 km, the flux of derelict spacecraft and spacecraft fragments greatly exceeds the flux of meteoroids. More recently, Taff et al.(1984) used the two 0.8 meter telescopes associated with the M.I.T. Lincoln Laboratory Experimental Test System near Socorro, N.M. to search for centimeter-sized orbital debris. They detected several such objects for each hour of observation. According to Kessler (1984), the near-Earth orbital debris flux resulting from the Taff et al. study is at least one order of magnitude above the meteoroid flux for particle diameters in excess of one centimeter.

It is clear from the above evidence that it is absolutely necessary to be able to distinguish between orbital debris and meteoroids in order to determine an uncontaminated near-Earth meteoroid flux. Therefore it is proper to ask if the flux established by Naumann(1966) is a valid meteoroid flux. We believe that the flux determined by Naumann is, indeed, a valid flux for the following reasons:

1. The meteor data, with well determined trajectories, are clearly uncontaminated by orbital debris.

2. When corrected for Earth(or Moon) shielding, the Explorer XVI and Explorer XXIII 25 micrometer thick sensors were penetrated at a rate 2.3 times greater than were similar sensors on the five Lunar Orbiter spacecraft(Grew and Gurtler,1971). Such a near-Earth increase is entirely to be expected just due to the near-Earth gravitational focussing of meteoroids(see Zook,1975 for uncertainties in the expected gravitational increase). Thus the meteoroid penetration measurements taken in the early 1960's appear not to have significantly suffered from penetrations by space debris.

3. Brownlee et al.(1974) discovered two hypervelocity appearing impact pits on 800 square centimeters of aluminum foil that was exposed for 67 days during the Skylab IV mission. The two craters were 26 and 75 micrometers deep, respectively, and under chemical analysis, contained meteoritic material. As noted by Zook(1980), the near-Earth flux obtained from these two craters is 0.04 impacts/(square centimeter-year). This flux completely agrees with that obtained by Cour-Palais(1974) although the statistical weight of two impacts is not great. More important, the chemical analyses showed that, in this size range, impacts during 1973 were not largely caused by space debris.

3. The Lunar Microcrater Data

Most of the rocks returned from the Moon during the Apollo missions showed abundant evidence of meteoroid impact pitting. It was quickly understood that, in those cases where crater superposition was not a problem, the observed pit size distribution probably reflected the impacting meteoroid mass distribution. Because nearly all of the impact craters were hypervelocity-appearing, with glass-lined central pits, it was felt that very few of the observed pits were likely to have been caused by secondary lunar ejecta impacts. This feeling was strongly reinforced by the results of Schneider's (1975) laboratory experimental study on cratering due to secondary ejecta. Schneider created the high speed ejecta with a hypervelocity impact of a steel ball at normal incidence onto a glass plate. Before we further examine the question of secondary ejecta-caused craters, we first mention the differing impact crater size distributions obtained on lunar rocks by two different investigator groups.

Fechtig et al.(1974) compiled a variety of impact pit data on lunar rocks to obtain, using associated solar flare track ages, a composite meteoroid flux versus mass(or pit diameter) curve. Morrison and Zinner(1977) found that lunar rock 12054, collected during the Apollo 12 mission to the Moon, exhibited an unusually steeply increasing surface density of microcraters with decreasing pit diameter, for pit diameters below about 7 micrometers. They argued that this and other lunar rocks with similarly steep micron-to-submicron micropit distribution curves were the best candidates for the "true" lunar microcrater production curve. That is, they felt that these particular rock surfaces must not have been coated with the very thin layer of lunar dust that was often postulated to be causing a suppression of the creation of submicron pits on most lunar rocks--such suppression as would give rise to the "flatter" distribution observed by other groups, such as the Fechtig group. Both groups of investigators noted that the meteoroid fluxes that they derived were lower than the satellite-obtained meteoroid flux.

We now address the problem of the microcrater production rate. The microcrater production rate on a lunar rock is conventionally derived as follows: 1. The surface density of microcraters on the rock is measured; 2. The solar flare track density is measured versus depth into the same rock; 3. Using a track versus depth profile on a separate rock with an independently established exposure age as a standard, the solar flare track "exposure age" of the rock is obtained; 4. The crater production rate is then calculated by dividing the track age in item 3 into the crater surface density determined in item 1.

The solar flare track production rate is not, however, reliably known. The fact that different groups of investigators obtain contradictory results is one cause for caution. Zinner and Morrison(1976) report, for example, that there is a factor of 50 difference between the ratio they obtain for the surface microcrater density to the track density at 100 micrometer depth, and the corresponding ratio obtained by Hutcheon(1975). The reason for this difference is not yet understood. Furthermore, researchers such as Storzer et al.(1973) and Hutcheon et al.(1974) have obtained a very steep increase in solar flare track density as they proceeded from the interior of the lunar rock outward toward the surface, while Blanford et al.(1975) and Morrison and Zinner(1977) obtain a much smaller increase. That is, no "standard track density versus depth profile" has yet been agreed upon by workers in this field. Finally, Zook et al.(1977) and Zook(1980) argue that the solar flare track production rate could well have been higher about 20,000 years ago than is presently true. Such a time variation would give rise to excessive apparent surface exposure ages. Thus, one should use the direct in situ meteoroid sensing data, qualified as they are, rather than the track data to give the current impact pitting rate on lunar rocks.

We next wish to address the question of impact pitting by lunar secondary ejecta. Flavill et al.(1978) and Allison et al.(1981) reconsidered ejecta cratering, using both the Schneider(1975) data and new oblique angle impact data obtained with a Van de Graf accelerator(Flavill et al.,1978), but with a different primary mass distribution curve. They found that both their own data and the data generated by Schneider would suggest that many of the pits on lunar rocks between 1 and 10 micrometers in diameter were caused by impacting secondary ejecta.

Significant new experimental data have been recently obtained by Zook et al.(1984) using a light gas gun to propel glass and basalt projectiles at oblique angles onto sawed basalt rock surfaces. The impact velocities were typically in excess of 6 km/s and the ejecta were permitted to impact onto glass witness plates. These shots gave ejecta pitting rates that were two orders of magnitude higher than were obtained at normal incidence(by either Schneider, or Zook et al.). The resulting secondary impact pits look very similar to pits observed on lunar rocks, and the size distribution approaches the steepness observed by Morrison and Zinner(1977). Thus it now seems very probable that the lunar microcrater population for micron-sized pits(<7 micrometers in diameter) is usually dominated, depending on the particular lunar rock exposure geometry, by secondary ejecta cratering and not by primary interplanetary meteoroid cratering. The variations among different microcrater size distributions obtained from different lunar rocks may be due, not to variable lunar dust obscuration as often thought previously, but to variations in lunar rock surface exposure geometry relative to the primary and secondary cratering environment. Impact craters on lunar rocks with central pits larger than 7 micrometers are still believed to be directly caused by impacting meteoroids.

4. Conclusions

1. Earth-orbiting space debris generates impacts on Earth-orbiting sensors that can be mistaken for meteoroid impacts. It is not thought, however, that Earth-orbiting meteoroid experiments carried out in the early 1960's were significantly contaminated by debris impacts. Future studies will need to use either particle trajectory information or particle chemistry to separate meteoroids from man-made debris.

2. The lunar impact crater data is very probably dominated by secondary ejecta craters for impact pit diameters smaller than 7 microns. Also, the solar flare track ages established for lunar rocks are probably incorrect, and therefore should not be used to establish impact crater production rates.

3. The meteoroid mass distribution obtained from data on lunar impact pits larger than 7 microns closely fits the early satellite-obtained meteoroid data, as shown by Grün et al.(1984). Thus the lunar data, when properly interpreted, are of great value, and nicely complement the satellite and the zodiacal light data.

References

Allison, R.J. and McDonnell, J.A.M. (1981). "Secondary cratering effects on lunar microterrain: Implications for the micrometeoroid flux," Proc. Lunar Planet. Sci., 12B, 1703-1716.

Berg O.E. and Grün E. (1973) "Evidence of hyperbolic cosmic dust particles", In: Space Research VIII, Akademie Verlag, Berlin, 1047-1055.

Blanford, G.E., Fruland, R.M., and Morrison, D.A. (1975). "Long-term differential energy spectrum for solar-flare iron-group particles," Proc. Lunar Sci. Conf. 6th, 3557-3576.

Brownlee D.E., Tomandl D.A., Hodge P.W., and Hörz F. (1974) Elemental abundances in interplanetary dust. Nature 252, 667-669.

Clanton U.S., Zook H.A., and Schultz R.A. (1980) "Hypervelocity impacts on Skylab IV/Apollo windows". Proc. Lunar Planet. Sci. Conf. 11th, 2261-2273.

Cour-Palais B.G. (1974) "The current micrometeoroid flux at the Moon for masses <10^{-7} g from the Apollo window and Surveyor 3 TV camera results." Proc. Lunar Sci. Conf. 5th, 2451-2462.

Fechtig, H., Hartung, J.B., Nagel, K., Neukum, G. and Storzer, D. (1974). "Lunar microcrater studies, derived meteoroid fluxes, and comparison with satellite-borne experiments", Proc. Lunar Sci. Conf. 5th, 2463-2474.

Flavill, R.P., Allison, R.J., and McDonnell, J.A.M. (1978). "Primary, secondary and tertiary microcrater populations on lunar rocks: Effects of hypervelocity impact microejecta on primary population", Proc. Lunar Planet. Sci. Conf. 9th, p. 2539-2556.

Grew G.W. and Gurtler C.A. (1971) the Lunar Orbiter Meteoroid Experiments. NASA TN D-6266, 43 pages.

Grün, E. and Zook, H.A. (1980). "Dynamics of micrometeoroids in the inner solar system"; In: Solid Particles in the Solar System (eds. I. Halliday and B.A. McIntosh) D. Reidel, Dordrecht, 293-298.

Grün E., Zook H.A., Fechtig H., and Giese R.H. (1984) Collisional balance of the meteoritic complex. Submitted to Icarus.

Hallgren D.S. and Hemenway C.L. (1976) "Analysis of impact craters from the S-149 Skylab Experiment", Lecture Notes in Physics 48 (H. Elässer and H. Fechtig, Eds.) 270-274.

Hanner, M.S., Sparrow, J.G., Weinberg, J.L. and Beeson, D.E. (1976). "Pioneer 10 observations of zodiacal light brightness near the ecliptic: Changes with heliocentric distances", Lecture Notes in Physics, 48, 29-35.

Hartung J.B., Hauser E.E., Hörz F., Morrison D.A., Schonfeld E., Zook H.A., Mandeville J.-C., McDonnell J.A.M., Schaal R.B., and Zinner E. (1978) Lunar surface processes: Report of the 12054 consortium. Proc. Lunar Planet Sci. Conf. 9th, 2507-2537.

Hutcheon, I.D. (1975). "Micrometeorites and solar flare particles in and out of the ecliptic", J. Geophys. Res., 80. 4471-4483.

Hutcheon, I.D., MacDougall, D., and Price, P.B. (1974). "Improved determination of the long-term average Fe spectrum from 1 to 460 MeV/amu", Proc. Lunar Sci. Conf. 5th, 2561-2576.

Kessler D.J. and Cour-Palais B.G. (1978) Collision frequency of artificial satellites: The creation of a debris belt. Journal Geophys. Res., 83, No. A6, 2637-2646.

Kessler D.J. (1984) "Orbital debris issues," Advances in Space Research (In press).

Lamy, P.L. and Perrin, J.M. (1980). "Zodiacal light models with a biomdal population"; In: Solid Particles in the Solar System (I. Halliday and B.A. McIntosh, eds.) D. Reidel, Dordrecht, 75-80.

Leinert, C., Richter, I., Pitz, E., and Planck, B. (1981). "The zodiacal light from 1.0 to 0.3 AU", Astron. Astrophys., 103, 177-188.

Morrison, D.A., and Zinner, E. (1977). "12054 and 76215: New measurements of interplanetary dust and solar flare fluxes", Proc. Lunar Sci. Conf. 8th, 841-863.

Nagel K., Fechtig H., Schneider E., and Neukum G. (1976) "Micrometeorite impact craters on Skylab Experiment S-149", Lecture Notes in Physics 48, (H. Elsasser and H. Fechtig, Eds.) 271-278.

Nauman R.J. (1966). "The near earth meteoroid environment", NASA TN D-3717.

Schneider, E. (1975). "Impact ejecta exceeding lunar escape velocity", The Moon, 13, 173-184.

Storzer, D., Poupeau, G., and Kratschmer, W. (1973). "Track-exposure and formation ages of some lunar samples", Proc. Lunar Sci. Conf. 4th, 2363-2377.

Taff L.G., Beatty D.E., Yakutis A.J., and Randall P.M.S. (1984). "Low altitude, one centimeter, space debris search at Lincoln Laboratory's (M.I.T.) Experimental Test System", Advances in Space Research (in press).

Zinner, E., and Morrison, D.a. (1976). "Comment on 'Micrometeorites and solar flare particles in and out of the ecliptic', by I.D. Hutcheon", J. Geophys. Res., 81, 6364-6366.

Zook, H.A., Flaherty, R.E., and Kessler, D.J. (1970). "Meteoroid impacts on the Gemini windows", Planet. Space Sci., 18, 953-964.

Zook H. A. (1975) "The state of meteoritic material on the Moon", Proc. Lunar Sci. Conf. 6th,, 1653-1672.

Zook, H.A., and Berg, O.E. (1975). "A source for hyperbolic cosmic dust particles", Planet. Space Sci., 23. 183-203.

Zook, H.A., Hartung, J.B., and Storzer, D. (1977). "Solar flare activity: Evidence for large-scale changes in the past." Icarus, 32, 106-126.

Zook. H.A. (1980). "On lunar evidence for a possible large increase in solar flare activity $2x10^4$ years ago", Proc. Conf. Ancient Sun (eds. R.O. Pepin, J.A. Eddy and R. Merrill), 245-266.

Zook, H.A., Lange, G., Grün, E., and Fechtig, H. (1984). "Lunar primary and secondary microcraters and the micro-meteoroid flux", Lunar and Planet. Sci. XV, 965-966.

Discussion

R. Walker: I would like first to comment and then ask a question. 1. Microcraters are found not only on lunar rocks but also on crystals removed from the bottom of core tubes. In some cases we know (from isotopic measurements) that the crystals were exposed at the surface of the Moon at least $5x10^8$ to $1x10^9$ years ago. Thus the interplanetary dust has been a more or less constant component of the solar system for a good fraction of its total lifetime. 2. The contribution of secondary craters should be measurable by measuring crater populations on different parts of a large glass deposit. Glass on the very top of a rock should have the least secondaries, that on the side of a rock -- the most. Have you thought about this and is there an appropriate rock?

H. Zook: The answer is yes. Lunar rock 12054, returned during the Apollo 12 mission to the Moon, would be an excellent rock to use for this type of investigation. Its lunar orientation is well known, and its remarkably smooth glass surface faces in a variety of directions, including vertically where, presumably, secondary impacts should be rare or non-existent. Don Morrison (in Hartung, et al., 1978) did look at a chip from the top of this rock and found no craters smaller than 10 microns in diameter. A more complete study would be easy to do and should be carried out.

EXPLORER 46 METEOROID BUMPER EXPERIMENT:
EARTH ORBITAL DEBRIS INTERPRETATION

D. J. Kessler,
NASA, Johnson Space Center
Houston, Texas
U.S.A.

ABSTRACT.

The Meteoroid Bumper Experiment on Explorer 46 (launched 1972) was placed in Earth orbit to evaluate the effectiveness of using double-wall structures against meteoroids. This paper re-examines the data from this experiment. Certain sets of sensors were found to be penetrated much more frequently than other sets. The most plausible explanation is that nearly all of the penetrations were from an Earth orbiting population of particulates. In addition, because a large percentage of the penetrations occured soon after solid rocket motors were fired in space, the particulates are most likely 75 μm diameter aluminum oxide. Aluminum oxide particulates are a major exhaust product from solid rocket motors. The size of particulates from most current solid rocket motors is found to range between 0.1 μm to 20 μm. Modeling the orbits of particulates from these rockets predicts that measurements in Earth orbit of interplanetary dust in this size range are also likely to include Earth orbiting particulates from solid rocket motors.

R. H. Giese and P. Lamy (eds.), Properties and Interactions of Interplanetary Dust, 97.

ON THE EARTH'S DUST ENVELOPE

V.L.Barsukov and T.N.Nazarova
V.I.Vernadsky Institute of Geochemistry
and Analytical Chemistry, USSR Academy of
Sciences
117975 GSP Moscow V-334
USSR

ABSTRACT. The data of studying meteoric particles on Elektron-1 and Elektron-3 space vechicles were processed in a new way which made it possible to get the first idea of the Earth's dust envelope structure. Meteoric matter is distributed in it not unformly, but as aggregations of particles moving around the Earth in more or less stable orbits at certain distances from the Earth's surface.

The planets are the gravitation centers. They posess dust envelopes with different densities, structures and length.

To reveal the Earth dust envelope structure one must know the position of the particle in circumterrestrial space at the moment of its recording by the space vehicle.

To get the idea of the structure of the Earth's dust envelope we made an attempt to process once more in a different than early way the data obtained in 1964 from the space vehicles Elektron-1 and Elektron-3 /1/.

Elektron-1 was launched on January 30, 1964. Its orbital parameters were: perigee 406 km, apogee 7,100 km, inclination 61°, revolution period 169 minutes. Elektron-3 was launched on July 11, 1964 (perigee 405 km, apogee 7,040 km, inclination 60.87° and period 168 minutes).

Information from Elektron-1 over the period from January 31, 1964 to March 5, 1964 (6th to 302nd orbit) and from Elektron-3 from July 11, 1964 to August 3, 1964 (1st to 206th orbit) was processed once again in a different way.

Within these periods meteoric particles were recorded by Elektron-1 in 33 orbits during 41 two-minute information accumulation cycles and by Elektron-3 in eight orbits during eight cycles.

All cycles, during which the recording of the im-

R. H. Giese and P. Lamy (eds.), Properties and Interactions of Interplanetary Dust, 99–104.

pacts of meteoric particles took place, were put to their positions in the orbit and plotted on the graph (Fig.1)

As evident, the greater part of them was grouped at definite distances from the Earth which were close to some of the distances predicted for search by Dr.Yu.K. Gulak on the basis of his theory of the distribution of solid interplanetary matter in space and near gravitational centres /2/.

It should be noted that of eight cycles during which meteoric particles were recorded from Elektron-3 which operated about four months after Elektron-1 seven cycles fell on the same distances from the Earth as during the experiment from Elektron-1.

Meteoric particles at the same distances from the Earth were recorded on the ascending or descending parts of the orbits which followed each other in a row or with intervals.

This picture can take place if meteoric particles move around the Earth in more or less stable orbits lying at certain distances from its centre. In this case meteoric particles are the Earth's satellites and form its dust envelope.

The character of the recordings gives grounds to believe that meteoric matter along its orbits is distributed not uniformly, but as individual moving aggregations. The orbits of the particles cross each other when the particles are recorded at the same distances from the Earth on the ascending and descending parts of the satellite's orbit.

To verify the conclusions we made on the basis of the experiments from Elektron-1 and Elektron-3 we used the published data of the meteoric experiment by Dr. H. Fechting, Dr. E.Grün, and Dr. G.Morfill from the HEOS-2 satellite /3/. This experiment yielded a great amount of information. The data on the mass, the impact velocity and the direction of the movement of meteoric particles in near-Earth space were obtained. Many particles had the low impact velocity.

We made an attempt to process the HEOS-2 data in a way similar to the processing of the data obtained from Elektron-1 and Elektron-3.

In the orbit of HEOS-2 the position of all recorded particles were marked. The authors of the experiment classified the registered particles according to time profile under three categories - groups, swarms and random particles.

Fig.2 shows one distances from the Earth of all particles recorded in the perigee area from about 10,000 km to 60,000 km.

The graph clearly shows the regularity in the position of particle registration sites. A considerable amount of them lies at the same distances from the Earth within about 250 km, including the particles recorded on the ascending and descending parts of HEOS-2 orbits and and for the particles moving from different directions.

The characteristics of the meteoric particles, obtained from the experiment, and the specific features of their distribution in near-Earth space show that most of them are Earth satellites and move around it in more or less stable orbits lying at definite distances from its surface.

The character of the distribution of meteoric particles - Earth satellites - is similar to the distribution observed from Elektron-1 and Elektron-3: the particles at the same distances from the Earth were observed from HEOS-2 irregularly. The time intervals between them had different durations, which indicate their not uniform distribution along their orbits.

Thus, the results of the meteoric experiments from Elektron-1, Elektron-3 and HEOS-2 are in agreement. They indicate the presence of the Earth's dust envelope of the above-mentioned structure.

CONCLUSIONS

The processing of the data on meteoric particles from Elektron-1 and Elektron-3 by a different method made it possible to get the first qualitative idea of the structure of the Earth's dust envelope.

Meteoric matter is distributed in the Earth's dust envelope not uniformly, but as individual aggregations, which move around the Earth in more or less stable orbits lying at certain distances from the Earth's surface and which have different angles of the inclination to the plane of its equator.

The data of the meteoric experiment by Dr. H.Fechtig, Dr. E.Grün, and Dr. G.Morfill from the HEOS-2 satellite which were processed by us in a way similar to the meteoric experiments from Elektron-1 and Elektron-3 have proved to be in agreement with each other.

ACKNOWLEDGMENTS

Authors are grateful to Prof. T.M.Eneyev for his participation in the discussion of our work, valuable advise and remarks.

REFERENCES

1. Barsukov V.L., Nazarova T.N.: 1983, in Astronom.Vestnik, N 4, "Nauka" Moskow, p.238.
2. Gulak Yu.K.: 1972 Isvestia VUZor, Physics USSR, 8.
3. Fechtig H., Grün E. and Morfill G.: Max-Plank Institut für kernphysic, MPIH-1978 v 8.

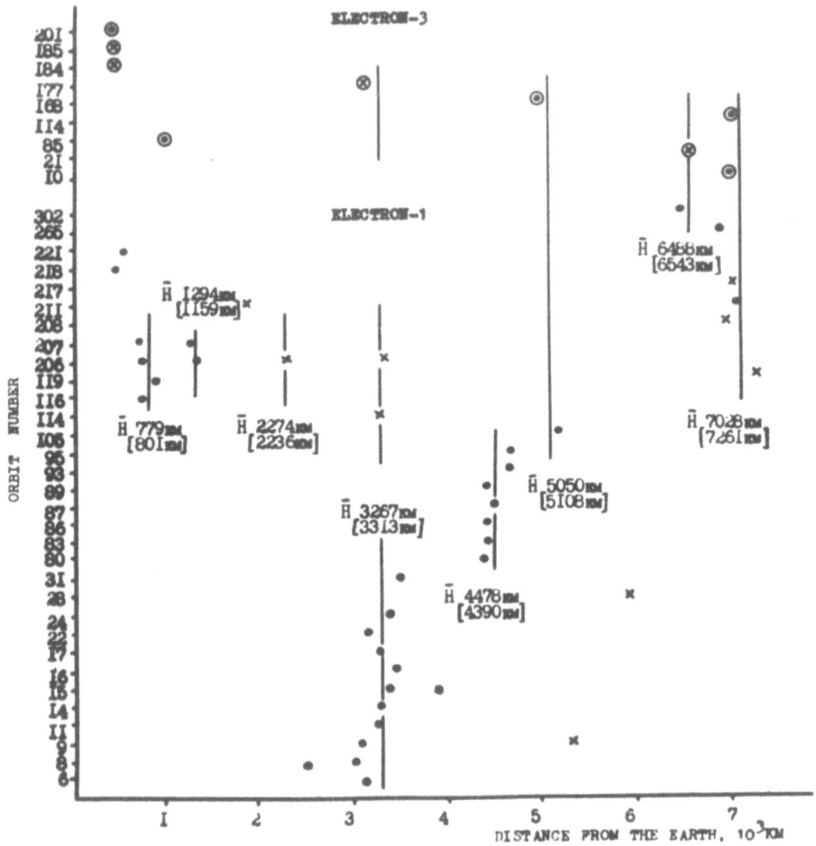

Fig.1 The distances of information accumulation cycles from the Earth's surface during which impacts of meteoric particles were recorded.
• -ascending part of the cicuit of Elektron-1;
x -descending part of the circuit of Elektron-1;
⊙ -ascending part of the circuit of Elektron-3;
⊗ -descending part of the circuit of Elektron-3;
[...km]-distance of the meteoric particles aggregation predicted for search by Yu.K.Gulak.

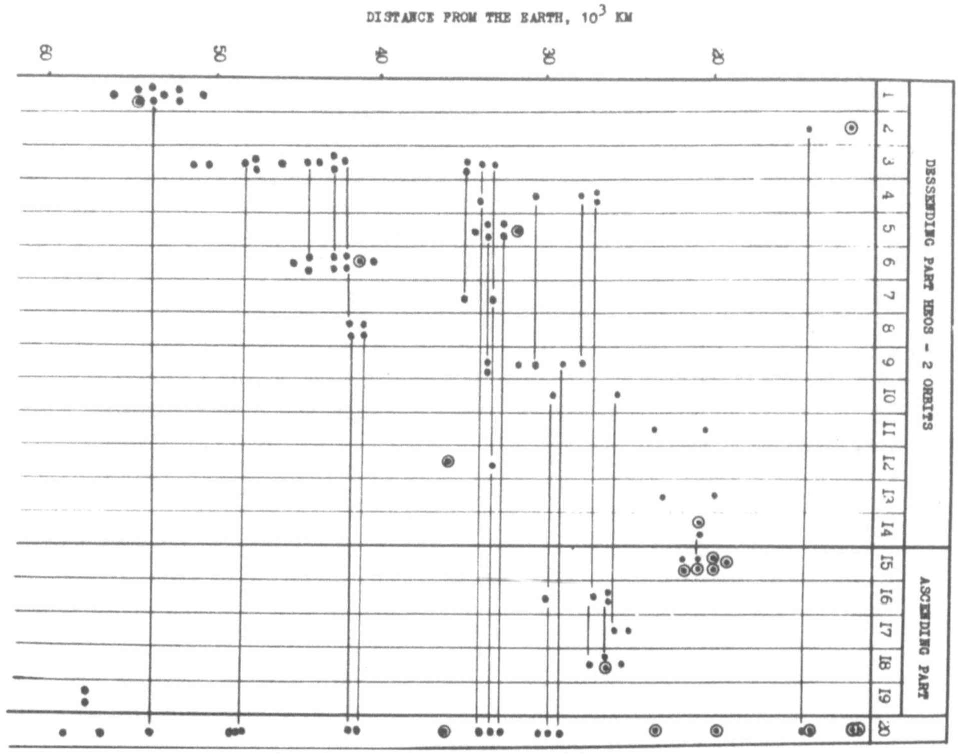

Fig.2 The distances from the Earth of all particles recorded in the perigee area from 10,000 km to 60,000 km. Groups: 2,10-13: swarms:1,3-9, 14-19, random particles-20. The particles whose impact velocities are large for being regarded as Earth satellites are marked by the sign

Discussion.
Dr. Kessler' question:
Is it possible that Elektron-1 and Elektron-3 detected the needles from "Ford Space Needle" experiments performed several time in the early 1960's?

Dr. Fechtig' remark:
There is a possibility that HEOS-2 detector registrated the particles of solid rocket motors near the Earth ($< 10R_{earth}$)

Author' answer.
It can not be asserted that a single needle ("Ford Space Needle Experiments) and the particle of solid rocket motors could not be registered by Elektron-1, Elektron-3 and HEOS-2 meteoric detectors, though this possibility is small.
But if meteoric detectors of those satellites would register on essential quantity of man-made debris, their spatial location is not ordered.
Calculation has shown that millions of years and sufficient value of space density of man-made debris are necessary for formation of debris structure in space.
It seems to us that swarms registered by HEOS-2 are the structural elements of the Earth dust envelope.
They were registered not only at distances less than 60,000 km, but also at distances about 200000 km from the Earth.

ORBITS OF INTERPLANETARY DUST PARTICLES INSIDE 1 AU AS OBSERVED BY HELIOS

E. Grün, H. Fechtig and J. Kissel
Max-Planck-Institut für Kernphysik
Postfach 10 39 80
D - 6900 Heidelberg
F.R.G.

ABSTRACT. The in situ dust detectors on board the Helios spaceprobes detected impacts of micrometeoroids between 0.3 AU and 1 AU distance from the sun. Among the measured quantities were mass, impact speed and flight direction of the dust particles. Radial variations of the flux, azimuthal distribution and impact speed are discussed. Significantly different results were obtained for the "apex" particles observed by both ecliptic and south sensors and for the "eccentric" particles detected only by the south sensor. The radial spatial density variation of the "eccentric" particles is compatible with that derived for zodiacal light particles. Whereas the spatial density of "apex" particles peaks at 0.5 to 0.6 AU. From the measured quantities probability distributions of orbital elements were derived for the observed micrometeoroids. About 60% of the observed micrometeoroids ("apex" particles) are on low energy orbits ($a \lesssim 0.6$ AU). About 30% of the particles ("eccentric" particles) had high energy orbits ($a \gtrsim 0.9$ AU). More than 10% of the observed particles show high probabilities for travelling on hyperbolic orbits.

1. INTRODUCTION

The Helios spacecrafts orbit the sun between 0.3 AU and 1 AU. We report on data from the micrometeoroid detector taken during the first 10 orbit revolutions of Helios 1. Preliminary data from Helios 2 support the findings. 235 impacts were recorded in the 5 years time interval. 83 meteoroid impacts were observed by the ecliptic sensor and 152 by the south sensor. Detailed descriptions of the instrument and the data were published by Grün et al. (1980) and Grün (1981). Most of the particles detected by the ecliptic sensor had masses of $10^{-13} \lesssim m \lesssim 10^{-10}$g. The particles observed by the south sensor had masses $10^{-15} \lesssim m \lesssim 10^{-9}$g. This difference has been attributed by Grün et al. (1980) to two different particle populations, one ("apex" particles) made out of material of normal density ($\rho \gtrsim 1$ g/cm³) - only this population of micrometeoroids could be observed by the ecliptic sensor - and another population ("eccentric" particles) of low densities $\rho < 1$ g/cm³ which was, in addition, observed by the south sensor. In this paper we concentrate on the

R. H. Giese and P. Lamy (eds.), Properties and Interactions of Interplanetary Dust, 105–111.

spatial and orbital distributions of micrometeoroids observed by Helios.

2. FLUX DISTRIBUTION

The radial distribution of the dust flux could only be derived after a long integration time. Fig. 1 shows the result for both sensors. Since

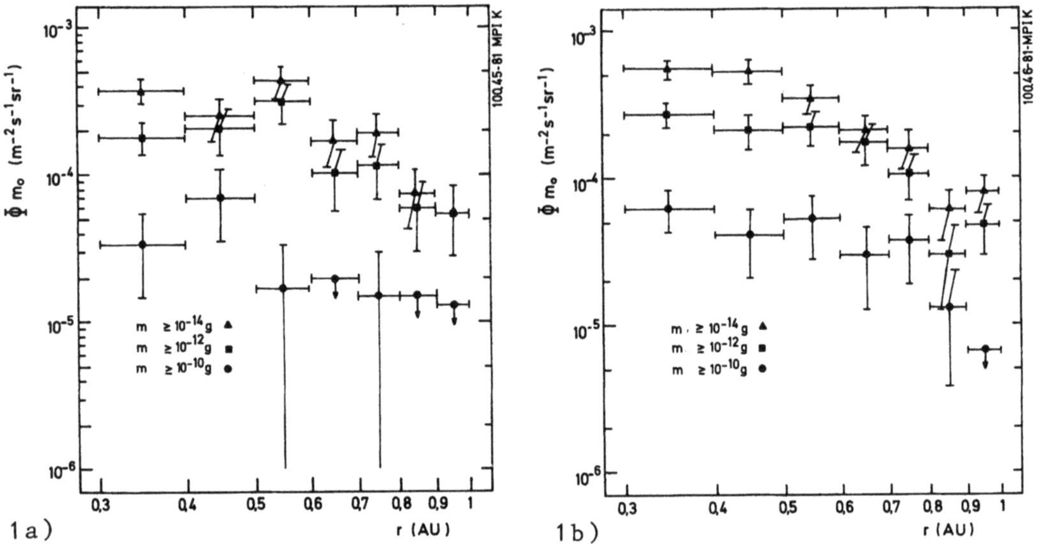

Fig. 1: Radial variation of the micrometeoroid flux observed by Helios. The fluxes were averaged over a 0.1 AU interval. The statistical errors correspond to the particle numbers observed. Upper limits are shown where no particles were detected. The fluxes given for $m \geq 10^{-12}$ g and 10^{-14} g are lower limits since the instrument did not record particles impacting at speeds $v \lesssim 5$ km/s and 20 km/s, respectively.
a) ecliptic sensor b) south sensor

Helios is a spinning spacecraft the fluxes shown are averaged over all directions in the ecliptic plane. Flux values have been calculated from the number of impacts detected during the corresponding time intervals and by taking into account the sensitive areas of 54.5 cm^2 and 66.5 cm^2 and effective solid angles of 1.04 sr and 1.39 sr for the ecliptic and south sensor, respectively. Both sensors measured fluxes increased by a factor between 3 and 10 going inward from 1 to 0.3 AU. However, there is a significant difference in the radial variation of the fluxes measured by both sensors. Besides a somewhat reduced flux in the radial interval 0.8 to 0.9 AU ($m \geq 10^{-12}$ and 10^{-14} g) compared to both adjacent intervals the flux onto the south sensor is steadily increasing towards the sun. This is compatible with the steady increase in zodiacal light intensity reported by Leinert et al. (1981). In contrast to that, the flux recorded by the ecliptic sensor has its maximum ($m \geq 10^{-12}$ and 10^{-14} g) in the radius interval between 0.5 and 0.6 AU and stays constant or even decreases further in.

3. AZIMUTHAL DISTRIBUTION

During one spin revolution of the Helios spacecraft each sensor scans a full circle in azimuth. Fig. 2 shows the azimuthal distribution of impacts on both sensors in various distance intervals. Outside 0.6 AU

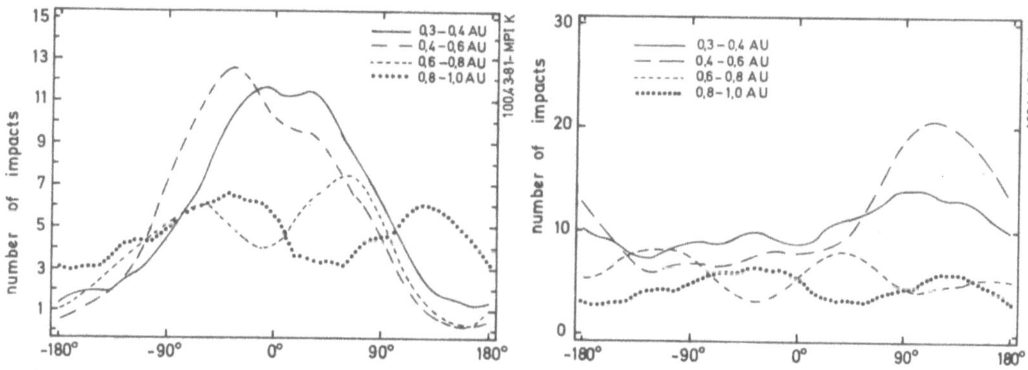

2a) 2b)

Fig. 2: Azimuthal distribution of impacts. An individual impact is represented by the angular probability distribution centered on the actual sensor pointing direction. Apex direction ≈ 0°, solar direction = 90°. Azimuthal distributions are shown for 4 different radius intervals.
a) ecliptic sensor b) south sensor

the azimuthal distributions are generally flat with perhaps a slight enhancement around the apex direction (0°). Inside 0.6 AU the azimuthal distribution from ecliptic sensor is strongly peaked towards the apex direction. Similar distributions have been observed by the micrometeoroid experiments on board the Pioneer 8 and 9 spacecrafts (Berg and Gerloff, 1971) for big particles and the HEOS-2 satellite (Hoffmann et al., 1975). A possible source of these "apex" particles has been discussed by Grün and Zook (1980). A quite different azimuthal distribution has been detected by the south sensor inside 0.6 AU. The number of impacts from the apex direction is smaller than that observed by the ecliptic sensor. This indicates that the average inclination of "apex" particles is smaller than 30°. Most of the impacts were detected when the south sensor was pointing approximately in the solar direction (90°). An even stronger increase of the dust flux from the solar direction has been observed by the Pioneer 8 and 9 experiments for small particles (Berg and Grün, 1973). Zook and Berg (1975) interpreted the "solar" particles to be small fragment particles (β-meteoroids) leaving the solar system on hyperbolic orbits. The number of β-meteoroids observed by Helios is much smaller than that detected by the Pioneer spacecrafts.

4. IMPACT SPEED DISTRIBUTION

The speed of micrometeoroids impacting the Helios detectors has been determined. There is a bias for small particles ($m \lesssim 10^{-12}$g) towards high speeds and for big particles ($m \gtrsim 10^{-12}$g) towards low speeds. Small particles with low speeds are not detectable and big particles with high speeds cause saturation of the measurement and no speed value can be derived. In the mass range $10^{-13}\text{g} \leq m \leq 10^{-10}$g no bias exists for intermediate impact speeds 10 km/s $\leq v \leq$ 20 km/s. Fig. 3a shows the radial speed distribution for particles observed by the ecliptic sensor. The average impact speed rises from 7 km/s in the interval 0.9 to 1 AU to 23 km/s at 0.3 to 0.4 AU.

With the south sensor the highest average impact speed (∿ 19 km/s) in the mass range $10^{-13} \leq m \leq 10^{-10}$g were observed between 0.8 and 1 AU, the lowest average speed (∿ 8 km/s) at 0.6 to 0.8 AU and further in a steady increase towards 16 km/s at 0.3 to 0.4 AU had been measured. In Fig. 3b however we show another subset of the speeds measured with

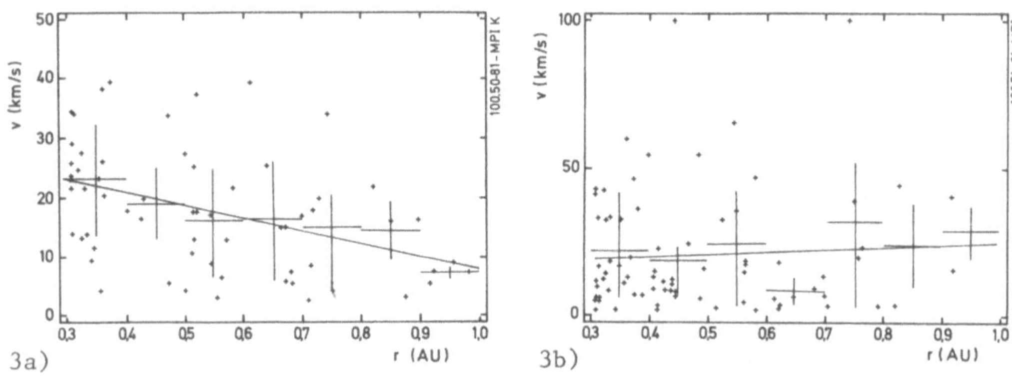

Fig. 3: Radial variation of impact speeds. Small dots are values for individual particles. Crosses are average values over 0.1 AU intervals with 68% of the measurements lying within the error bars. Also a linear fit through all measurements is shown.
a) particle with masses $10^{-13} \leq m \leq 10^{-10}$g recorded by the ecliptic sensor;
b) "eccentric" particles observed by the south sensor.

the south sensor. Particles on "eccentric" orbits have been identified by Grün et al. (1980) to be those which are only recorded by the south sensor and not by the ecliptic sensor. These particles have the highest average impact speeds (∿ 30 km/s) at 1 AU and slightly lower (∿ 20 km/s) at 0.3 AU. Such a speed variation together with the flux variation observed by the south sensor is compatible with the spatial density variation determined by Leinert et al. (1981) according to $n \sim r^{-1.3}$. The "apex" particles observed by the ecliptic sensor show a quite different radial spatial density profile. Their spatial density has its maximum at 0.5 to 0.6 AU and decreases both inwards and outwards.

5. ORBITAL DISTRIBUTION

From the measured impact speed, the actual pointing direction of the sensor and the position and velocity of the Helios space probe, the heliocentric velocity of micrometeoroids can be determined. Because of the large field-of-view of the sensors and the large uncertainty of the impact speed measurement only probability distributions for the direction and the amount of the velocity vector are obtained. These distributions are transformed into probability distributions of orbital elements for each individual particle, taking into account its radiation pressure constant β. For a more detailed description of the method and its biases see Schmidt and Grün (1979,1980). Fig. 4 shows the pro-

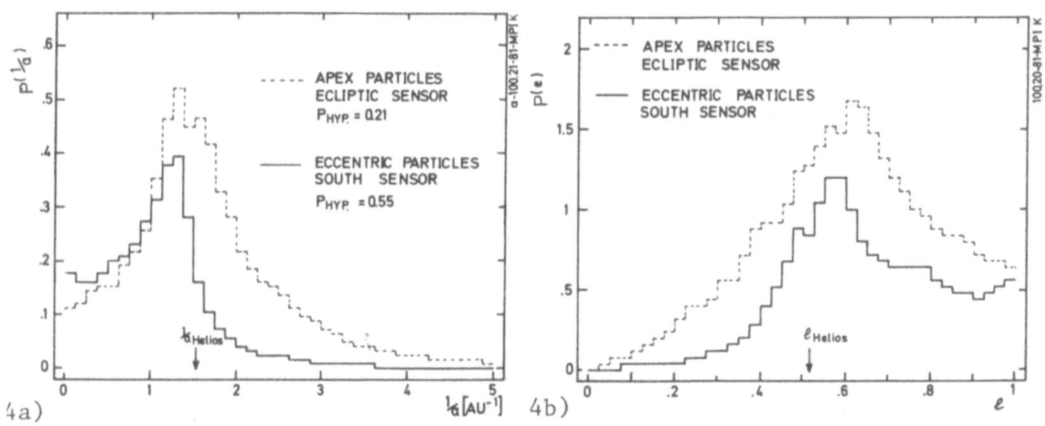

Fig. 4: Probability distribution of orbital elements for "apex" particles and "eccentric" particles. The total probabilities are normalized to 1. The strong enhancement of the probability at the values of the orbital elements of Helios is a bias involved with this method.

a) distribution of the reciprocal semi-major axes. The probability for hyperbolic orbits p_{HYP} is indicated;
b) distribution of eccentricities.

bability distributions of orbital elements for the two different populations of micrometeoroids observed by Helios: "apex"-particles and "eccentric" particles. Because of the biases involved with this method only relative results will be discussed here.

The "apex" particles show a much higher probability for bound orbits than for hyperbolic orbits (p_{HYP} = 0.21). Contrary to that, "eccentric" particles have maximum probability for hyperbolic orbits (p_{HYP} = 0.55). However, the absolute value of this probability will be too high due to a methodical bias. The average reciprocal semi-major axes (Fig. 4a) of "apex" particles is larger than that of Helios (1/a Helios = 1.54 AU^{-1}) in contrast to that of "eccentric" particles which is smaller. "Apex" particles which constitute about 60% of all observed micrometeoroids are on low energy orbits ($\bar{a} \lesssim 0.6$ AU), whereas 30% of them ("eccentric" particles) are on high energy orbits ($\bar{a} \gtrsim 0.9$ AU). At least 10% of the observed particles show high probabilities for travelling on hyperbolic orbits.

The probability distributions of the eccentricities are displayed in Fig. 4b. Both distributions have maxima and means above the value of Helios (e_{Helios} = 0.52). This shows that likewise the "apex" particles move on high eccentric orbits with an average eccentricity of $\bar{e}$ = 0.6

REFERENCES

Berg O.E. and Gerloff U. (1971) "More than two years of micrometeorite data from two Pioneer satellites", Space Res. **XI**, 225.

Berg O.E. and Grün E. (1973) "Evidence of hyperbolic cosmic dust particles", Space Res. **XIII**, 1047.

Grün E. (1981) "Physikalische und chemische Eigenschaften des interplanetaren Staubes - Messungen des Mikrometeoritenexperiments auf Helios", Bundesministerium für Forschung und Technologie, Forschungsbericht W 81-034.

Grün E., Pailer N., Fechtig H. and Kissel J. (1980) "Orbital and physical characteristics of micrometeoroids in the inner solar system as observed by Helios 1", Planet. Space Sci., **28**, 333.

Grün E. and Zook H.A. (1980) "Dynamics of micrometeoroids in the inner solar system", in: 'Solid Particles in the Solar System' (Eds. I. Halliday and B.A. McIntosh), Reidel, Dordrecht, 293.

Hoffmann H.J., Fechtig H., Grün E. and Kissel J. (1975) "Temporal fluctuations and anisotropy of the micrometeoroid flux in the Earth-Moon system", Planet. Space Sci. **23**, 985.

Leinert C., Richter I., Pitz E. and Planck B. (1981) "The zodiacal light from 1.0 to 0.3 AU", Astron.Astrophys. **103**, 177.

Schmidt K.D. and Grün E. (1979) "The distribution of orbital elements of interplanetary dust in the inner solar system as detected by the Helios spaceprobe", Space Res. **XIX**, 439.

Schmidt K.D. and Grün E. (1980) "Orbital elements of micrometeoroids detected by the Helios 1 spaceprobe in the inner solar system", in: 'Solid Particles in the Solar System' (Eds. I. Halliday and B.A. McIntosh), Reidel, Dordrecht, 321.

Zook H.A. and Berg O.E. (1975) "A source for hyperbolic cosmic dust particles", Planet. Space Sci. **23**, 183.

CALIBRATION OF THE GALILEO/ISPM DUST DETECTORS WITH IRON PARTICLES

J.R. Göller, E. Grün
Max-Planck-Institut für Kernphysik
P.O. Box 10 39 80
D-6900 Heidelberg
W-Germany

ABSTRACT. The impact-ionization-detectors for the Galileo-/ISPM-Missions were calibrated using about 7000 iron particles with speeds between 1 km/s and 67 km/s and masses between 10^{-15} g and 2 x 10^{-10} g. These calibrations showed that the impact charge (Q) was proportional to the particle mass (m) at a fixed velocity. The speed dependent charge yield (Q/m) showed significant deviations (up to one magnitude) from a power law.

There are three independent methods (risetimes of electron and ion pulses, amplitude ratio between channeltron and ion pulse) for measuring particle speed and two possibilities (charge yields of electron and ion pulses) for mass measurement. This calibration showed that the speed can be determined within a factor of 1.2 and the mass within a factor of 2.0.

1. SCIENTIFIC OBJECTIVES

The Galileo dust detector shall measure the distribution and the dynamical properties (speed, mass, flight direction) of micrometeorites in Jupiter's vicinity, especially their interaction with the Jovian natural satellites and its magnetosphere.

The ISPM dust detector will determine the three-dimensional distribution of the zodiacal cloud and offer means to distinguish between the different models of its shape.

Both missions are able to do measurements in the asteroid belt and to detect dust of cometary origin.

2. THE SENSOR

The sensor is an impact-ionization-detector based on the HEOS 2-detector (Hoffmann et al., 1975). Its main improvements are:

- the increase of the target area (1000 cm², which is about 10 times the area of HEOS 2)
- the use of entrance grids to measure the electrical charge of a dust

R. H. Giese and P. Lamy (eds.), Properties and Interactions of Interplanetary Dust, 113–115.

particle entering the sensor
- the introduction of a channeltron, which measures part of the impact-ions. This helps to identify reliably impacts and to determine the speed of a particle.

The measured quantities are
- the amplitudes of ion, electron and channeltron pulses
- the risetimes of ion and electron pulses
- the primary charge of a particle.

3. CALIBRATION AND RESULTS

The calibrations were done at the Heidelberg dust accelerator using about 7000 iron particles with speeds between 1 km/s and 67 km/s and masses between 10^{-15} g and 2×10^{-10} g. Even at high velocities (> 20 km/s) enough particles were obtained to get good statistics for the data evaluation. Four units (both flight units, the Galileo flight spare unit and the ISPM engineering model) were calibrated.

It was found that for fixed velocities the impact charge was proportional to the particle mass, as shown in Figure 1 for 5 different speeds.

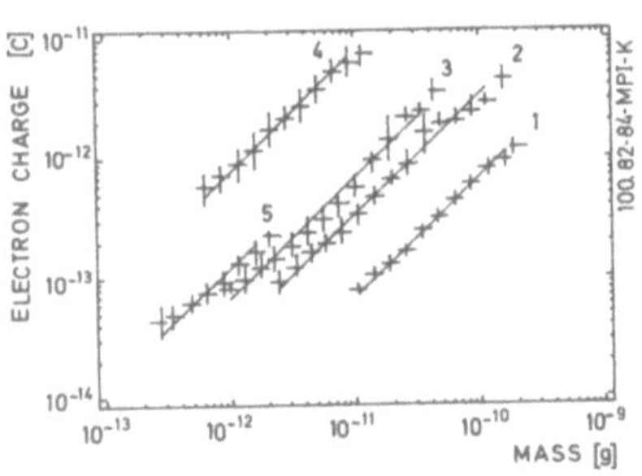

Speed ranges:
1: 2.2 - 2.5 km/s
2: 3.8 - 4.3 km/s
3: 5.8 - 6.3 km/s
4: 7.8 - 8.5 km/s
5: 10.0 - 11.0 km/s

Figure 1. Electron charge versus particle mass. For optical reasons curve 4 shows the charge multiplied by 10. The plotted lines show proportionality.

Therefore the charge yield (impact charge/particle mass, Q/m) could be plotted as a function of velocity (Figure 2). The differences in yield between the units are smaller than the deviations of the curves. So all the measurements could be combined.

In earlier experiments (Hoffmann et al., 1975, Grün, 1981) the charge yield was found to be a power law of the velocity. This is not the case for the Galileo/ISPM detectors. The charge yield shows a deficiency up to one magnitude in the region between 4 km/s and 20 km/s.

This may be caused by energy dissipation in melting processes of target and/or projectile. Studies of Eichhorn (1976) on the impact flash showed similar deviations, depending on the materials.

In addition there is a geometrical factor. The DIDSY-detector (Grün, 1984) shows similar but not as strong deviations. It has a mainly plane target, in contrast to the curved configuration of the

Galileo/ISPM detector.

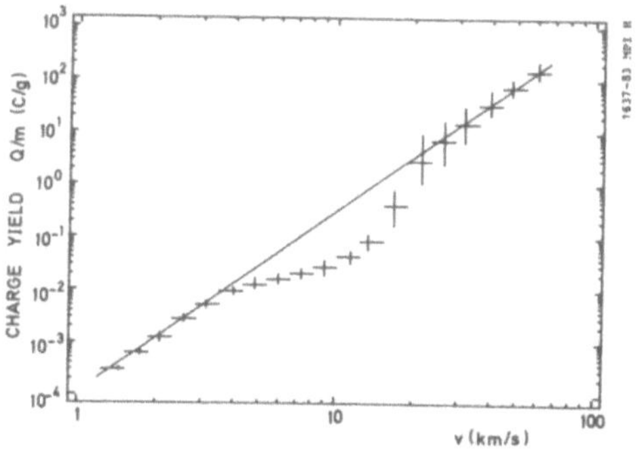

Figure 2. Charge yield Q/m (ions) plotted against velocity. In the region between 4 km/s and 20 km/s Q/m shows a deviation from a power law, fitted at low and high velocities.

The speed of a particle can be determined by three methods: the rise-times of electron and ion pulses and the amplitude ratio of channeltron and ion signals. The accuracy (1 σ) for each method is a factor of about 1.7. Using all the three possibilities the deviation is a factor of 1.2.

The particle mass can be determined from the speed (calculated as above) and the charge yields of impact ions and electrons. For one possibility the accuracy factor is ∿2.2, using both ∿2.0.

These factors are of course only valid for iron particles. Calibrations of earlier detectors showed that the calibration curves also hold for other projectile materials, but with somewhat reduced accuracy.

The overall accuracy of the speed and mass determination with the Galileo/ISPM detectors will be better than that of the Helios experiment which was characterized by factors 2 and 10, respectively.

REFERENCES

Eichhorn, G. (1976). 'Analysis of the hypervelocity impact process from impact flash measurements', Planet. Space Sci. 24, 771.

Grün, E. (1981). 'Physikalische und chemische Eigenschaften des interplanetaren Staubes - Messungen des Mikrometeoritenexperimentes auf Helios', Forschungsbericht BMFT-FB-W 81-034.

Grün, E., Busoletti, E., Minafra, A., Kuczera, H., McDonnell, J.A.M. (1984). 'First calibration measurements with the dust impact detector DIDSY-IPM', Adv. Space Res., in press.

Hoffmann, H.-J., Fechtig, H., Grün, E., Kissel, J. (1975). "First results of the micrometeoroid experiment S215 on HEOS 2 Satellite", Planet. Space Sci. 23, 215.

THE LDEF INTERPLANETARY DUST EXPERIMENT

S.F. Singer
George Mason University, Fairfax, VA 22030
J.E. Stanley
University of Virginia, Charlottesville, VA 22903
P. Kassel
NASA Langley Research Center, Hampton, VA 23665

SUMMARY. The Long Duration Exposure Facility was launched for the first time on April 6, 1984 by the NASA space shuttle Challenger. An array of solid-state detectors record the arrival time and approximate direction of an impacting particle. Two levels of detector sensitivity provide an indication of particle energy and mass. The active area is nearly 1 m^2. We therefore expect count rates of about 30 per day over an exposure time of about a year. The orbit of the particle cannot be obtained, except statistically. We know the orientation of the detector, the position of LDEF in earth orbit, and the position of the earth in relation to other celestial bodies. From this information we can extract essential orbit information.

An earlier flight on the Explorer 46 satellite gave first evidence of the existence of submicron-sized particles, mostly associated with fresh meteor streams. We hope to obtain more precise data and estimate lifetimes in interplanetary space.

To study the fate and origin of IP (interplanetary) dust, we measure various kinds of time variations. Among the most interesting is the secular variation, i.e., the flux in various meteor streams, as a function of the passage of a comet.

One of the challenging problems will be to distinguish IP dust from man-made space debris. The separation will depend on knowing something of the orbits of debris. Probably more important will be the chemical analysis of the particles. The two types of information are in a sense complementary.

INTRODUCTION

Impact and penetration experiments are different in character from optical observations of IP dust (such as zodiacal light - Z.L.) and from collection experiments. Impacts give the time of arrival and often the direction of the particle's motion, in addition to its momentum or kinetic energy. A great variety of interesting characteristics can in principle be derived from impact measurements: particle orbits, concentration, size distribution; further analysis must combine

R. H. Giese and P. Lamy (eds.), Properties and Interactions of Interplanetary Dust, 117–120.

impact results with those of optical, collection, and other experiments.* (For a fuller discussion, see Hawaii Zodiacal Light Symp. (1) and Ref. 5). Now, IR emissions and crater statistics (on spacecraft and lunar samples) provide additional methods of observation.

For example, the analysis of the Pioneer penetration experiment (2) has allowed us to say something about the orbits, concentration, size distribution, and optical albedo of particles beyond the earth's orbit (3). Especially useful was the analysis of Pioneer data obtained near Jupiter (4).

EXPLORER 46 RESULTS

But the analysis of the Explorer 46 data, first reported at the Ottawa IAU Colloquium (7) and amplified at a conference at Cornell University (8), has given the following new results:

(i) The existence of submicron-sized particles ($\sim 0.1\mu$ and of mass $\sim 10^{-16}$ gm) was deduced from the fact that the higher-sensitivity impact detectors showed a higher counting rate. We further deduce that radiation pressure is not an important perturbing force (7, 8).

(ii) While "old" meteor streams (Halley's Comet) showed no increase, recently refreshed streams (Encke's Comet) showed the largest increases. From this result we deduce:

(a) The submicron particles are injected directly by a comet and not produced in a secondary manner by the meteor streams. The calculated injected mass of $\sim$ 5000 tons appears reasonable (7, 8).

(b) The "lifetime" of submicron particles (in the meteor stream orbit) is about three years. We have deduced that electromagnetic perturbing forces could explain the observed lifetime (7, 8), assuming that the particles had acquired a (reasonable) potential of about three volts.

PROSPECTIVE LDEF RESULTS

The Explorer 46 data were quite limited. However, the LDEF experiment should provide much higher counting rates, direction of arrival, and also much better time resolution. This in turn should allow us to say something about:

(i) particles in hyperbolic orbits (β- meteroids from the sun's direction) (9),

(ii) particle swarms (possibly lunar ejecta (10),

(iii) morning-to-evening asymmetry (1),

(iv) effects of the earth's orbit eccentricity (1),

(v) presence of interstellar dust.

---

* It should be noted here that impact (or penetration) detectors do not measure the dust particle concentration directly, nor the flux. They measure the flux above a certain threshold (which depends on mass and relative velocity of the particle). (See e.g., Ref. 1, or 5, 6).

Following the completion and analysis of the first LDEF experiment, it would be most desirable to repeat it during and following the passage of Halley's Comet. In addition, one would like to carry out the experiment at different orbital altitudes above the base of the exosphere. Many of the interesting phenomena depend on altitude in a unique manner (see 1, 5, 6).

SPACE DEBRIS

The question has been raised to what extent the Explorer 46 data (and LDEF) are contaminated by space debris (produced, e.g., by rocket firings). The impact detector cannot, of course, distinguish the source of a single impact; but statistical analysis can separate space debris from IP dust.

(i) Explorer 46 observed a high rate of impact for a few hours just after being put into orbit.

(ii) The increases during meteor streams clearly demonstrate the predominant presence of IP dust.

(iii) The lifetime of space debris would be limited because of the density of the residual atmosphere. Therefore, one should see a decay of debris impact following, say, a rocket firing and, of course, a time association.**

The LDEF data, because of their much higher quality (count rate, timing, and direction) should allow a more detailed analysis of this phenomenon.

** We are aware of the suggestion that submicron particles be artificially injected into orbits around the earth (Ian Strong, Los Alamos, this volume). Such an experiment might provide a controlled means of studying the space debris problem.

REFERENCES

1. Singer, S.F. and Bandermann, L.W. (1967). Nature and origin of zodiacal dust. In The Zodiacal Light and the Interplanetary Medium (J.L. Weinberg, Ed.). NASA SP-150.

2. Humes, D.H. (1976). The Jovian meteoroid environment. In Jupiter (T. Gehrels, Ed.), pp. 1052-1067. Univ. or Arizona Press, Tucson.

3. Stanley, J.E., Singer, S.F., and Alvarez, J.M. (1979). Interplanetary dust between 1 & 5 AU. Icarus 37, 457-466.

4. Singer, S.F., and Stanley, J.E. (1976). Interplanetary dust particles near Jupiter. Icarus 27, 197-205.

5. Bandermann, L.W., and Singer, S.F. (1969). Interplanetary dust measurements near the Earth. Rev. Geophys. 7, 759-797.

6. Singer, S.F. (1961). Nature 192, 321-325.

7. Singer, S.F. and Stanley, J.E. In Solid Particles in the Solar System (I. Halliday and B.A. McIntosh, eds.) Reidel, Boston 1980.

8. Singer, S.F. and Stanley, J.E. Comets produce submicron particles in the solar system. In Cosmology & Astrophysics (Y. Terzian and E.M. Bilson, eds.), Cornell University Press, Ithaca, (1982).

MEASUREMENTS OF THE ELEMENTAL AND ISOTOPIC COMPOSITION OF INTERPLANETARY DUST COLLECTED ON LDEF

H. Fechtig[1], F. Hörz[2], E. Igenbergs[3], E. Jessberger[1], H. Kuczera[3], G. Lange[1], N. Pailer[4], S. Sutton[5], P. Swan[5], R. Walker[5] and E. Zinner[5]
[1]Max-Planck-Institut für Kernphysik, D-6900 Heidelberg, FGR
[2]NASA/Johnson Space Center, Houston, TX 77058 USA
[3]Technische Universität München, D-8000 München, FRG
[4]Dornier System GmbH, D-7790 Friedrichshafen, FGR
[5]McDonnell Center for the Space Sciences
Washington University, St. Louis, MO 63130 USA

ABSTRACT: A passive interplanetary dust collection experiment, currently in orbit aboard LDEF (Long Duration Exposure Facility), is described. The collectors, germanium target plates covered by metallized Mylar foils, are designed for secondary ion mass spectrometry (SIMS) measurements of the elemental and isotopic compositions of residues resulting from micrometeoroid (> 10^{-10} grams) impacts. Impact simulation experiments have demonstrated the validity of the collection concept. Quantitative elemental analyses are complicated by the non-uniform distribution of projectile-derived elements.

1. INTRODUCTION

The Long Duration Exposure Facility (LDEF), deployed by Space Shuttle (STS-13) in April, 1983, and scheduled for retrieval in February, 1984, carries 57 science and technology experiments [1]. One such experiment, described here, is designed to collect residues from impacting interplanetary dust particles for laboratory measurement of elemental and isotopic compositions [2]. In addition, the experiment will provide information on the flux and size distribution of micrometeoroids.

Isotopic measurements are the prime experimental objective. Such measurements, performed on interplanetary dust particles collected in the stratosphere, have shown already that interplanetary dust is primordial in the sense that some isotopic ratios differ from the average solar system values [3]. Orbital dust collections are complementary to stratospheric collections because different particle selection effects exist. Orbital collections also provide an opportunity to collect interstellar particles which are not expected to survive atmospheric entry because of their high velocities.

R. H. Giese and P. Lamy (eds.), Properties and Interactions of Interplanetary Dust, 121–126.

2. EXPERIMENTAL DESIGN

A total of 240 capture cells are deployed covering an area of 2 m^2 distributed equally on the trailing and leading surfaces of the gravitationally stabilized LDEF structure. Each of the 240 cells (figure 1) consists of a pure, polished Ge target plate and a 2.5 μm thick Mylar foil separated by 0.2 mm. The cover foil is coated with 100Å of Au/Pd on the space-facing surface, to prevent erosion of the plastic by the residual atmosphere at LDEF orbital altitudes [4], and 800Å of Ta on the Ge-facing surface, to facilitate chemical microanalyses. The choice of the materials (Mylar, Ge, Ta and Au/Pd) was based on considerations of chemical purity, mechanical and thermal properties, and economy. At impact velocities ~10 km/sec, particles break-up, melt, and/or vaporize during penetration of the foil and subsequent impact on the Ge. The resulting residues/impact ejecta are deposited on the Ge and the underside of the foil. Approximately 100 particles greater than 10 μm in size are expected to be captured by these collectors during LDEF's 10 months in orbit.

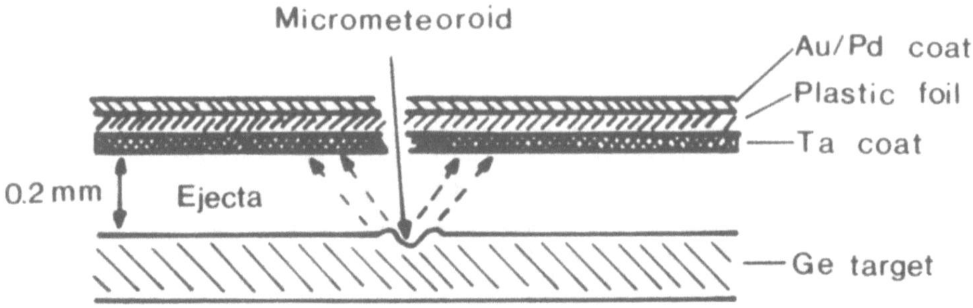

Figure 1

Collection principle for micrometeoroid collectors aboard LDEF.

The collected particle residues will be analyzed by a variety of techniques to determine the isotopic, elemental and physical characteristics of interplanetary dust. Emphasis will be placed on the measurement of isotopic ratios of selected elements (C, Mg, Si, S, Ca, Fe, Ni) by secondary ion mass spectrometry (SIMS). Elemental abundances will also be obtained by SIMS. Morphological studies of the impact and deposition regions will be performed in an SEM.

3. INITIAL RESULTS OF IMPACT SIMULATION EXPERIMENTS

Impact simulation experiments are being pursued in preparation for the return of the collectors. They are aimed at understanding the relationships between projectile characteristics (size, velocity, density and chemical composition) and impact structure morphology and they are also required to enable us to perform quantitative isotopic and elemental analyses of thin residues on the collector substrates. Another approach will be implantation of Ge substrates and Ta films with known doses of different ions ("marker ion technique" [5]).

Impact simulations are being conducted on 2.5 cm diameter collectors, similar in design to those on LDEF, bombarded in a plasma gun located at the Technical University, Munich. The Munich facility is capable of accelerating individual dust particles 50–100 μm in size to known velocities on the order of 10 km/sec.

SIMS analyses in a modified CAMECA IMS 3F ion microprobe are obtained by bombarding the specimen surface with O_2^+ ions to obtain positively charged secondary ions (for elements such as Na, Mg, Si, Fe) and with Cs^+ for negative secondaries (C, S). For elemental analysis, a the 3-dimensional distribution of projectile atoms is determined from depth profiles taken at regular intervals across the specimen surface. Isotopic measurements are made at high mass resolution to eliminate molecular interferences.

Test samples bombarded with glass (Na-rich and Corning lunar analog) and pyroxene particles 100 μm in size are currently being studied. After bombardment, the Mylar foil typically contains several large penetration holes (~100 μm) and hundreds of smaller holes (down to a few μm) each of which is associated with an impact region on the Ge. The small holes are produced by small particles which are secondary fragments broken from larger particles during acceleration in the plasma gun.

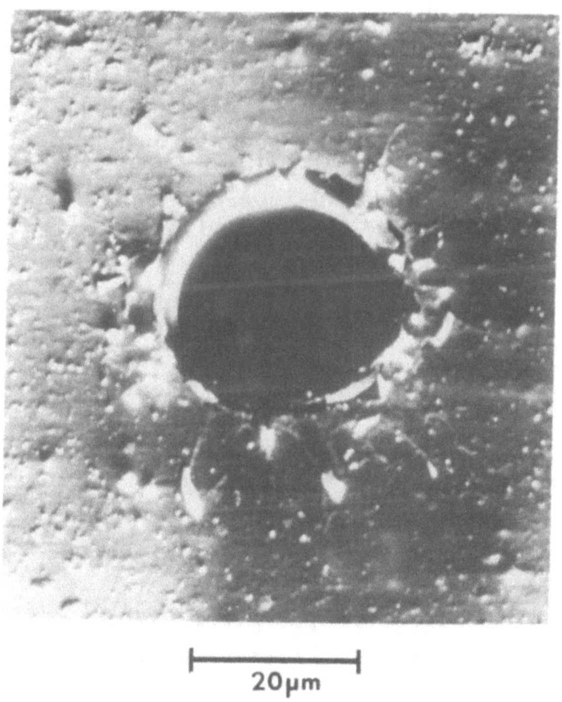

Figure 2:

SEM micrograph of penetration hole produced in tantalum coated Mylar by a glass particle accelerated by the Munich Technical University plasma gun. Tantalum, germanium and projectile fragments/residue surround the 20 μm diameter hole.

3.1. Location of Impacts

Deposition halos on the plastic foil and Ge associated with penetration holes as small as several µm (figure 2) can easily be located with a low-powered, binocular microscope under oblique illumination. The impact structures on the Ge are particularly easy to identify because of their characteristic, concentric ring morphology (described below).

3.2. Secondary Ion Yields

Secondary ions from projectile residues can readily be measured with the ion probe for the major elements of interest (e.g., Na, Mg, Al, Si, K, Ca, and Fe). A typical traverse across a residue region on the Mylar foil produced by pyroxene is shown in figure 3. The ion signals are normalized to the $^{181}Ta^+$ signal from the substrate. The scans shown are obtained by integrating signals over an intermediate (~30–100Å) depth range from the depth profiles. The maximum $^{40}Ca^+$ signal was about 100 times higher than the background away from the impact feature. Similar scans are measured on the corresponding Ge structures.

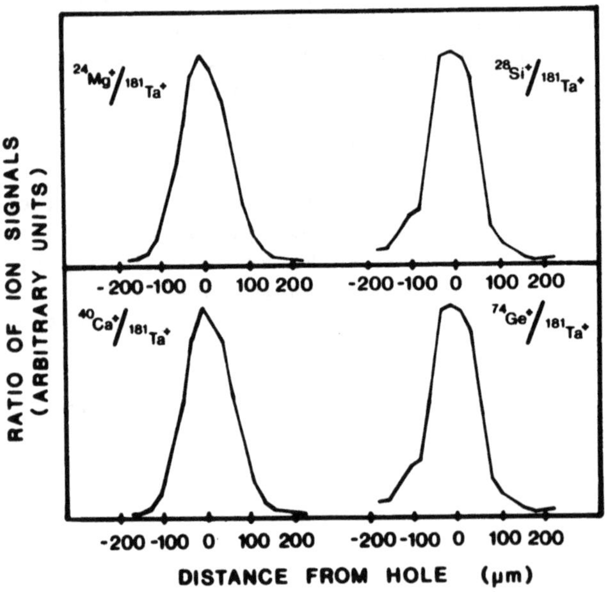

Figure 3:

Normalized plots of ionprobe signal from Mylar in the vicinity of projectile residue versus position for a depth interval of ~30–100Å and four different isotopes, ^{24}Mg, ^{28}Si, ^{40}Ca and ^{74}Ge. Regions containing projectile-derived residue are easily identified by well-defined signal maxima.

3.3. Spatial Distributions of Residues

Spatial distributions of the projectile residues vary considerably from element to element. Highly volatile elements (Na, K) tend to be distributed over a larger area around the impact location than lower volatility elements (Mg, Si, Ca, Fe). Consequently, elemental ratios vary with distance from the penetration hole.

3.4. Impact Structure Morphologies

Mylar: SEM observations of regions on the underside of the Mylar foil adjacent to penetration holes reveal abundant splashes and fragments. EDX analysis in these areas show only Ge indicating that Ge excavated by the impact dominates over the projectile elements.

Germanium: Impact regions on the Ge exhibit a variety of morphologies. Those corresponding to large holes (~100 µm) possess features typical of hypervelocity impacts, i.e. deep craters, extended spalled regions and ejecta-bearing rims. For holes less than about 30 µm in size the impact features consist of complex cell structures (figure 4). These structures are remarkably similar to that observed by McDonnell et al. [6] for an STS-3 impact on Kapton after penetration of a 5µm Al foil. This structural similarity indicates that these terrestrial shots simulate extraterrestrial events reasonably well. The small Ge impact structures

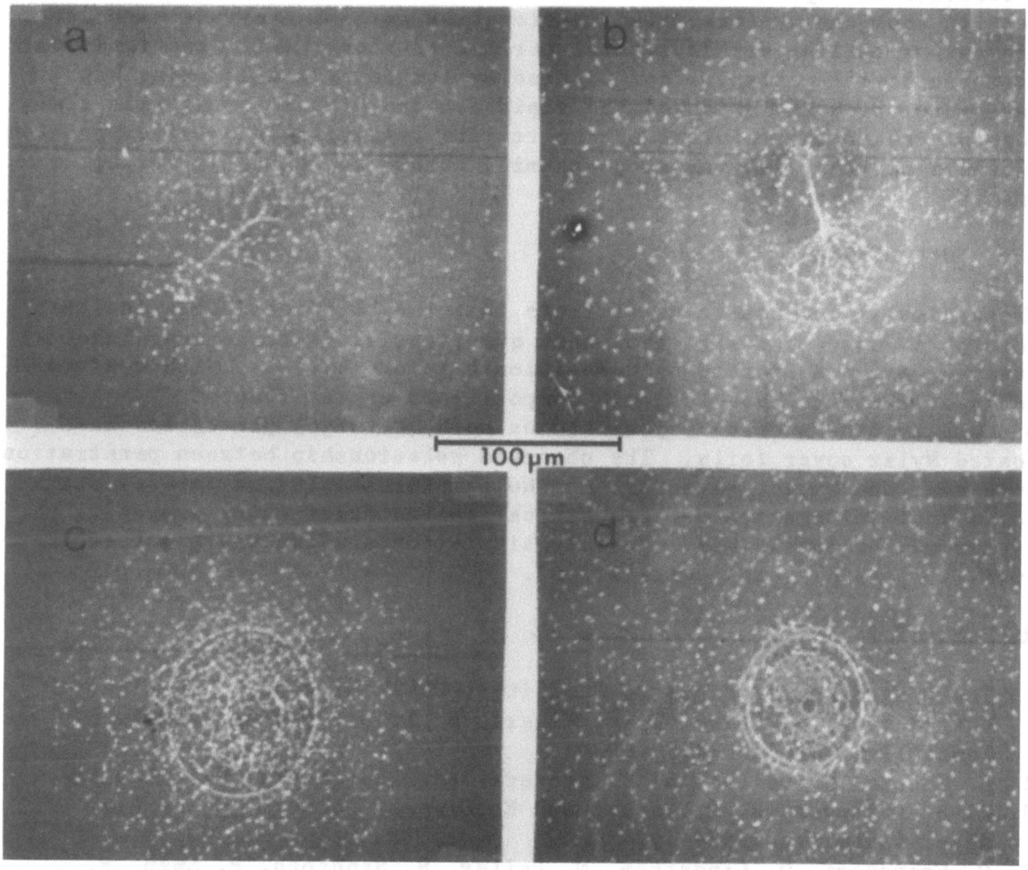

Figure 4:

Germanium impact structures for 4 projectiles with different penetration hole size, 11.3 (a), 12.0 (b), 13.3 (c) and 20.0 (d) microns.

lack substantial relief, as would be expected for impact craters, and contain considerable amounts of projectile material. These observations show that small particles break up during penetration of the Mylar and impact the Ge as liquid and/or vapor in a dispersed jet of roughly conical shape. The morphology of the Ge structures relates systematically to penetration hole diameter (figure 4). Smallest holes (<10 μm) are associated with clusters of "etch pits" while holes ~ 20 μm in size correlate with well-developed cell structures terminated at the perimeter by well-defined rings. Still larger holes (~30 μm) relate to regions showing evidence of substantial excavation in addition to cells and rings.

3.5 Isotopic Measurements

Magnesium and silicon isotopic measurements have been made at high mass resolution on impact residues from the simulation experiments [2]. The results show the typical mass fractionation of SIMS analysis with isotopic ratios falling within several per mil of linear mass fractionation. Isotopic measurements are complicated by low ion signals during high mass resolution analysis and the rapid decline of the signal with time (depth) due to the thinness of the residue coating. This latter problem can be avoided by lateral step-scanning across the deposits during analysis.

4. CONCLUSIONS

These simulation results demonstrate that the LDEF collection/analysis concept is sound but also show that quantitative isotopic and elemental analyses are complicated by deposition heterogeneities. SIMS measurements of elemental and isotopic compositions of intercepted micrometeoroids should be possible on both Ge target plates and Ta-coated Mylar cover foils. The observed relationship between penetration hole size and impact structure morphology for simulation impacts suggests that information on the physical characteristics of impacting particles may be obtainable. Further simulation studies are required, however, to quantify such correlations.

REFERENCES

1. The Long Duration Exposure Facility (LDEF): Mission 1 Experiments, edited by L. G. Clark, W. H. Kinard, D. J. Carter, Jr. and J. L. Jones, Jr., NASA SP-473 (1984).
2. E. Zinner, N. Pailer and H. Kuczera, Adv. Space Res. **2**, 251 (1983).
3. K. D. McKeegan, R. M. Walker and E. Zinner, Geochim. Cosmochim. Acta, (submitted 1984).
4. P. Fraundorf, D. Lindstrom, N. Pailer, S. Sandford, P. Swan, R. Walker and E. Zinner, "AIAA Shuttle Environment Meeting," 131 (1983).
5. E. Zinner and R. M. Walker, Proc. 6th Lunar Sci. Conf., 3601 (1975).
6. J.A.M. McDonnell, W.C. Carey and D.G. Dixon, Nature **309**, 237 (1984).

PROSPECTS FOR COSMIC DUST EXPERIMENTS ON THE PLANNED REFLIGHT OF LDEF

R.M. Walker and E. Zinner
McDonnell Center for the Space Sciences
Washington University
St. Louis, Missouri 63130
U.S.A.

ABSTRACT

The Long Duration Exposure Facility (LDEF-I), which contains a number of cosmic dust experiments, is due to be launched in the spring of 1984 and recovered about a year later. Current plans call for re-fitting the LDEF spacecraft with a large area of plastic nuclear track detectors and relaunching (LDEF-II) for a flight that will last about 2 years. The main purpose of the mission is to extend primary cosmic ray abundance measurements to the actinide region. A meeting was held at Washington University in December 1983 to discuss the problems and prospects for cosmic dust experiments on LDEF-II. Most participants were drawn from the LDEF-I community of investigators. The meeting resulted in a report which treated the scientific rationale for LDEF-II dust experiments, discussed various implementation options, and concluded with a set of summary recommendations. We discussed this report and summarized the status of LDEF-II as of this meeting. It is important to note that the report serves equally well as a basis for discussion of dust experiments on future space stations.

R. H. Giese and P. Lamy (eds.), Properties and Interactions of Interplanetary Dust, 127.

ACOUSTIC PENETRATION AND IMPACT DETECTOR FOR MICROMETEOROID AND SPACE DEBRIS APPLICATION

H. Kuczera, H. Iglseder,
U. Weishaupt, E. Igenbergs
Lehrstuhl für Raumfahrttechnik
Technische Universität München
F.R.G.

ABSTRACT

An active detector is described which recently has been developed for laboratory impact measurements and which already has been proposed for cosmic dust flight experiments. The detector consists of a very thin penetration foil and an impact plate. By means of piezo detectors the elastic-wave propagation times between the impact location and the piezo detectors can be measured in the foil and in the target. The evaluation of these signals gives the time of the event, the impact location, the projectile velocity, the flight path direction and at least the order of magnitude of mass of the impacting particle. This type of sensor will meet all requirements for particle detection in the diameter range from a few microns (dust particles) up to approx. a few millimeters (space debris) at all velocities. Small-sized projectiles can be detected by sensitive amplifiers. Low-energy electronics including data storage can be used for triangulation and impact time calculation. First experimental results which have been achieved in the Munich Accelerator Facilities will be presented.

To be published in Adv. Space Res.

R. H. Giese and P. Lamy (eds.), Properties and Interactions of Interplanetary Dust, 129.

AN EMPIRICAL PENETRATION EQUATION FOR THIN METALLIC FILMS USED IN CAPTURE CELL TECHNIQUES

W.C. Carey, J.A.M. McDonnell and D.G. Dixon
Space Sciences Laboratory
University of Kent at Canterbury
Canterbury, Kent CT2 7NT, United Kingdom

ABSTRACT. The problem in the determination of impact particle parameters from hypervelocity impact damage on a foil or multilayer structure is one which deserves close attention in the era of ready access to recovered space impact events. In the MFE experiment (McDonnell et al 1984), we chose metallic foils, and this paper addresses the decoding of perforations in metallic foils, primarily of aluminium of typical dimensions 5 µm (hard temper rolled). Data is examined to establish for the first time a penetration relationship which extends over:

1. marginal and thin sheet perforation;
2. velocities varying from some 2 to 69 kms^{-1};
3. particle densities varying from some 1 g cm^{-3} (ices?) to 7.8 g cm^{-3} (iron?)

1. INTRODUCTION

The mathematical description of hypervelocity impacts onto solid material commonly takes the form of empirical equations derived from experimental data. Such equations are typically of two types:

i) crater size equations for both thin foils and semi-infinite targets;

ii) marginal or perforation limit equations applied to impact on thin foils.

Another approach is to model the actual physical processes which take place in hypervelocity impact using a three-phase equation of state and a numerical computer code. This latter method has been successfully used to simulate impacts on the Giotto meteoroid shield (Arnadeau et al 1984) but requires accurate data, profound understanding of the physical phenomena and extensive computing power. The empirical method, then, continues to be used although the equations cannot always be reliably extrapolated beyond the experimental limits which are usually less than 20 kms^{-1} for micron dimensioned particles in a Van der Graaff

R. H. Giese and P. Lamy (eds.), Properties and Interactions of Interplanetary Dust, 131–136.

dust accelerator and less than 10 kms^{-1} for millimetre-sized particles in a light gas gun.

A list of commonly-encountered crater size equations is shown in Table I and plotted against reliable experimental data for iron projectiles impacting aluminium foil (Figure 1). Most noticeable is the turn over in the experimental data particle diameter d decreases, preceding a perforation limit. These phenomena are not represented in any of the equations which have only limited regions of coincidence with the data.

Table I

Three empirical crater-size equations

Nysmith and Denardo	$\frac{D}{d} = 0.88\rho_m^{\frac{1}{2}}(\frac{f}{d})^{0.45}\ V^{\frac{1}{2}}$	(1)
Sawle	$\frac{D}{d} = 1.3.2\ (\frac{\rho_m}{\rho_f}.\frac{v}{c})^{0.2}\ (\frac{f}{d})^{2/3}$	(2)
Maiden	$\frac{D}{d} = 0.45\ (\frac{f}{d})^{2/3}\ V + 0.9$	(3)

A list of commonly-encountered crater-size equations relating final hole size D to target and incident particle parameters.

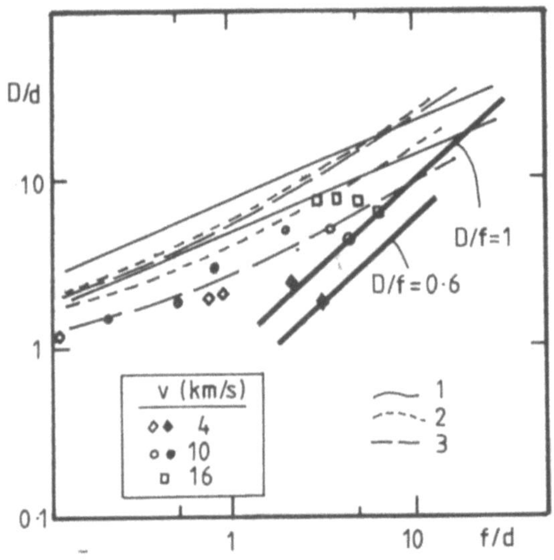

Figure 1. Experimental data points for iron projectiles impacting aluminium foil at 4, 10 and 16 km/s are shown. The plotted lines are given by the equations of Maiden et al, Sawle and Nysmith and Denardo for velocities of 4 and 10 kms^{-1}, scaled to iron on aluminium where appropriate.

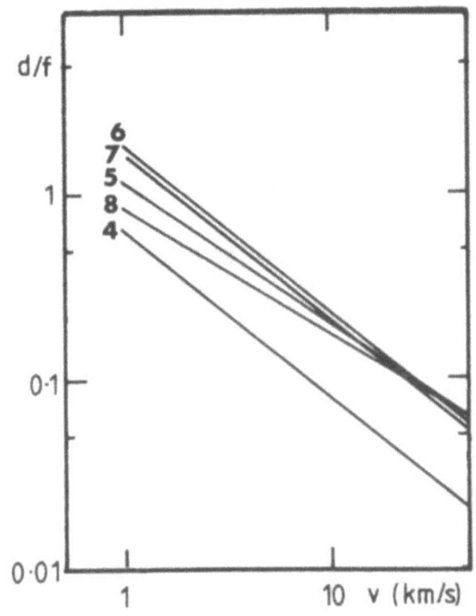

Figure 2. Perforation limit equations plotted for iron projectiles impacting 5 μm aluminium foil.

2. EMPIRICAL EQUATION DERIVATION

It is therefore necessary to derive an equation which describes the regime near the perforation limit in which the perforation diameter, D, starts to decrease. There is already an equation (McDonnell 1979) which describes the perforation limit of iron particles on aluminium foil and this limit falls on the D/f = 1 curve shown in Figure 1, for high velocities.
For velocities of 4 kms^{-1} and below, the perforation limit lies on the D/f = 0.6 line, implying that the morphology of lower-velocity marginal perforations is different from that at higher velocities. A list of perforation limit equations is shown in Table II and plotted for iron on aluminium for 5 micron diameter particles in Figure 2.

Table II

Marginal or Perforation-Limit Equations

Nauman	$f/d = d^{0.056} V^{0.875} \rho^{0.52}$	(4)
McDonnell	$f = 0.79\, V^{0.763}\, d$	(5)
Pailer and Grün	$f = \dfrac{1}{\varepsilon^{0.006} \rho_f^{0.05}}\, m^{0.4} \rho_m^{0.33} (V\cos\alpha)^{0.88}$	(6)
Fish and Summers	$f/d = 0.57 (\frac{1}{\varepsilon})^{1/8} (\frac{\rho_m}{\rho_f})^{\frac{1}{2}}\, V^{7/8}\, d^{1/18}$	(7)
Cour-Palais	$f = 3.56 \times 10^{-4} \rho^{0.148}\, m^{0.352}\, V^{0.67}$	(8)

A list of perforation limit equations relating foil thickness f to target and incident particle parameters.

List of symbols

v = velocity		d = projectile diameter	
D = perforation diameter		D_t = perforation diameter at top of foil	
f = foil thickness		D_1 = perforation diameter at lower edge of foil	
ρ_f = density of foil		ρ_m = density of projectile	
ε = foil ductility		α = impact angle	

Note: Units are not consistent throughout all the above equations.

As an important constituent of the cosmic dust population is likely to be low density material rather than iron, so a perforation limit for low velocity (ρ = 1) particles on aluminium has been calculated. Arnadeau et al (1984) have published results for a low density particle impacting an aluminium meteor shield at 68 kms^{-1} near marginality. Results of experiments using iron projectiles on gold foil (McDonnell, unpublished) may also be examined by assuming these are equivalent to unit-density particles impacting aluminium foil because the projectile-target density ratios are very similar (Fish and Summers 1965). The limits are found to lie on D/f = 2 for low velocities but are expected to extend (as for iron projectiles) to D/f = 1 for higher velocities where the perforation diameters are relatively large. The limit equation for low density (ρ = 1) particles impacting aluminium foil is thus:

$$f/d = 0.46\ V^{0.64} \tag{9}$$

where v is in kms^{-1}. This is plotted in Figure 3 with other perforation-limit equations scaled (when necessary) for low density particles.

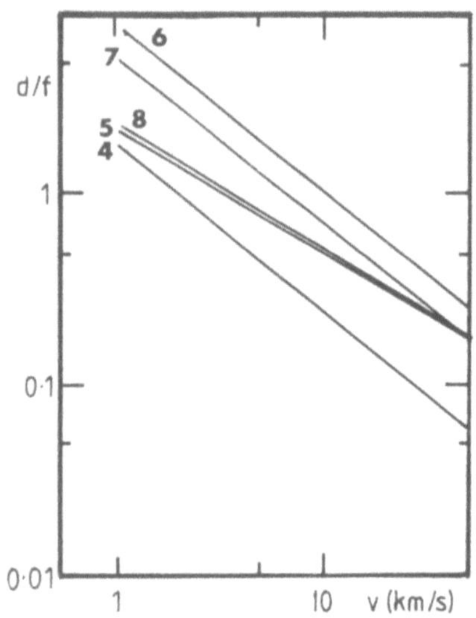

Figure 3. Perforation limit equations plotted for particles with ρ_m = 1 impacting alumium foil.

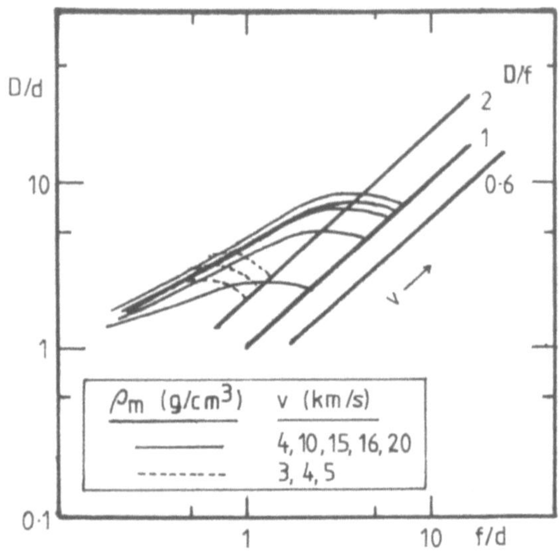

Figure 4. The empirical crater-size equation for iron on aluminium plotted against experimental data. Data for low density particles is also shown.

Taking the D/f = 1 as the marginal perforation limit for iron on aluminium and using the data points shown in Figure 1, a crater size equation has been derived for velocities of 4 to 20 kms^{-1}. The equation takes the form:

$$\frac{D}{d} = 1 + 1.5(f/d)\ V^{0.3} \left[\frac{1}{1 + (f/d)^2\ V^{-n}}\right] \tag{10}$$

where, for $2 < V < 20$, where

$$n = 1.02 - 4 \exp(-0.9 V^{0.9}) - 0.003(20 - V) \quad (11)$$

The term outside the square brackets is fitted to the data points where D is increasing and is modified by the term inside the brackets. The factor n fits the curve through the perforation limits. This equation is shown plotted in Figure 4, with the experimental data for iron on aluminium to which it is fitted and the data for low density ($\rho_m = 1$) particles on aluminium for comparison.

For Fe-Al at $V > 20$ kms^{-1} we have:

$$\frac{D}{d} = 1 + 1.5(f/d)\, V^{0.3}. \left[\frac{1}{1 + (f/d)^2 . V^{-1.02}}\right] \quad (12)$$

3. CONCLUSIONS

We have developed a single equation covering the marginal and very thin foil perforation properties which fits the available data for iron particles impacting aluminium foil at velocities up to 20 kms^{-1}. This situation undoubtedly is not the most common of impact events in space, and hence we have commenced the scaling of this to lower density projectile data, e.g. cometary particles of density ~ 1 g cm^{-3}. A marginal perforation limit is derived which will be extended to cover non-marginal perforations in the future.

4. REFERENCES

Arnadeau, F., Winkelmuller, G., and de Rouvray, A. 'The Giotto Micrometeoroid Impact and Structural Analyses', Synthesis Report, Engineering System International, ESA/ESTEC contract no. 4637/81/NL/AB-CCN07, April 1984.

Cour-Palais, B.G. 'The Comet Halley Micrometeoroid Hazard Workshop' ESA SP-153, p. 85, October 1979.

Fish, R.H. and Summers, J.L., 'The effect of material properties on threshold perforation', Proc. Hypervelocity Impact Symposium 7th, Vol. VI, 1965.

McDonnell, J.A.M., and Alexander, W.M., 'Aspects of Multilayer Target Penetration: (1) Physical Processes', COSPAR 1983.

McDonnell, J.A.M., Carey, W.C. and Dixon, D.G., 'Cosmic dust collection by the capture cell technique on the Shuttle 'Pathfinder' mission', , Nature, Vol. 309, No. 5965, pp. 237-240, 1984.

McDonnell, J.A.M., 'Factors affecting the choice of foils for penetration experiment in space', Space Research X, pp. 314-325, 1970.

Maiden, C.J., Gehring, J.W. and McMillan, A.R., 'Investigation of fundamental mechanism of damage to thin targets by hypervelocity projectiles', NASA TR-63-225, 1963.

Nauman, R.J., NASA TN D 3717, 1966.

Nysmith, C.R. and Denardo, B.P., 'Experimental investigation of the momentum transfer associated with impact into thin aluminium targets', NASA TN D 5492, 1969.

Pailer N. and Grün, E., 'The penetration limit of thin films', Planet. Space Sci. 28, pp. 321-331, 1980.

Sawle, D.R., 'Hypervelocity impact in thin sheets and semi-infinite targets at 15 kms^{-1}', LRL, UCB, AIAA Hypervelocity Impact Conf. Cincinatti, 1960.

SOME REMARKS ON THE DENSITY OF INTERPLANETARY DUST GRAINS

P.L. LAMY
Laboratoire d'Astronomie Spatiale
Marseille

ABSTRACT. The relevance of the bulk density as a physical parameter characterizing interplanetary dust grains is discussed. The various measurements which lead to a determination of this parameter are reviewed. The specific case of the collected interplanetary dust grains is considered.

The bulk density of interplanetary dust grains has been and is still a matter of controversy. This quantity cannot, in general, be directly measured; it is used to relate the mass and the size of a grain. This duality stems from physics itself as there are interactions sensitive to the mass (e.g., gravitational forces) while others are sensitive to the size or the cross-section (e.g., light scattering, radiation pressure, gas and plasma interactions). The measuring technics of the grains reflect this duality as, for instance, impact sensors are generally sensitive to the kinetic energy and thus to the mass, while optical sensors are sensitive to the cross-section. One sees that the density is not strictly speaking the relevant parameter, but what is needed is a relationship between mass and average cross-section.

1. SOURCES OF INFORMATION ON THE DENSITY

1.1. Lunar microcraters

The ratio P/D_c of the depth to the crater diameter is known to be a good indicator of the density of the impacting particle. After conflicting results, the situation was clarified by Le Sergeant and Lamy (1980) who based their analysis on 284 craters. They concluded that low-density grains ($\lesssim$ 1g/cm^3) are not detected, that grains in the diameter range 1 - 500 µm have densities of the order of 3g/cm^3 typical of silicates and that, below 1 µm, two components are present, one with the above density of 3g/cm^3 and the other with a higher density possibly indicating a metallic composition. It has been argued (e.g., Hanner, 1980; Fechtig, 1982) that the data of Nagel et al. (1976) indicates a "low-density" group ($\sim$ 1g/cm^3-. This is not correct as discussed by

R. H. Giese and P. Lamy (eds.), Properties and Interactions of Interplanetary Dust, 137–140.

Le Sergeant and Lamy (1980) as the ratio P/D_c of 0.4 for this group (or $D_c/P = 2.5$) implies a density of 2 to 2.5 g/cm^3. In fact, the calibrations of Vedder and Mandeville (1974) indicated for polystyrene projectiles ($\rho = 1$ g/cm^3) $P/D_c \sim 0.2$, a value significantly lower than that pertaining to the "low-density" group of Nagel et al.

With the exception of collected grains for which all physical parameters can in principle be measured, the P/D_c criterion yield a straightforward relationship between mass/density and cross-section/size, while being free of assumptions. Therefore, the above result may be regarded with an high level of confidence. However, crater formation by low-density aggregats has never been studied and it may be argued that either they form craters too shallow to be detected or they rapidly coalesce at impact and thus behave as smaller normal-density projectiles.

1.2. Meteors

Relating the quantities observed when a meteor penetrates into the Earth atmosphere to the original mass of the meteor and its density is far from being straightforward and in fact, requires many assumptions. The complexity of the physics which may not be accounted for by the classical equations (Hughes, 1978) as well as the uncertainty on various parameters probably preclude obtaining more than an order of magnitude for the density of the meteors.

For radio-meteors in the mass range 10^{-6} - 10^{-2} g, Verniani (1973) found a mean density of 0.8 g/cm^3. For photographic meteors whose masses are larger than $\sim 10^{-2}$ g, even lower densities (0.3 g/cm^3) were reported (e.g., Millman, 1976). Ceplecha (1967) and Benyuch (1974) have, on the contrary, obtained density in the range 1.4 - 4 g/cm^3.

The result of Verniani (1973) may certainly be questioned on the basis of the lunar microcraters data which overlap the lower end of his mass interval and which definitely do not support a density of 0.8 g/cm^3.

The whole question certainly needs to be revised. The physical description of meteor penetration has recently been improved by Pecina and Ceplecha (1984). They have not yet reported results for the density of meteors but Ceplecha (1983) already pointed out the importance of using "instant" atmospheric density profiles in comparison with standard atmospheres. Such progress should hopefully lead to a better estimate of the density of meteors in the near future.

1.2. The Helios data

The Helios sensors were not specifically designed to measure the density of impacting grains. However, the ecliptic sensor detected approximately twice less events than the south sensor although is was identical to the former except for a thin protective film against solar wind particles. Because of their orbital characteristics, these surplus events should have been detected by the ecliptic sensor. This question has received considerable attention and Grün et al. (1980) found that the front film has a discriminating effect w.r.t. the density of impacting

grains and concluded that 30 % of interplanetary dust particles have densities below 1 g/cm^3. This analysis calls however for several remarks.
i) The relevant calibrations performed by Pailer and Grün (1980) cover only the following ranges: 1.25 - 7.85 g/cm^3 for the projectile density, 10^{-13} - 10^{-10} g for the mass and 1.4 - 13 km/sec for the velocity. The range of velocities is probably too low, but more important, no experimantal results are avaible for $\rho < 1.25$ g/cm^3; henceforth, the interpretation of the Helios data is based on extrapolations.

The fact that the maximum densities are systematically less than 1 g/cm^3, down to 10^{-2} g/cm^3 (not to mention 10^{-5}) looks highly suspicious.
ii) It is clear from the study of Carey et al. (this volume) that, in the case of the perforation limit for particles with $\rho = 1$ g/cm^3 impacting aluminium foil, the Pailer and Grün (1980) equation gives the highest d/f or, in other terms, systematically favors low densities.
iii) After correcting for the difference in sensitivity between the ecliptic and the south sensor, the excess of impacts reduces to ~ 1.4; a simple inspection of the experimental results of Pailer and Grün (1980) shows that this reduction of sensitivity may well be accounted for by particles having densities between 1.25 and 7.8 g/cm^3.

The above remarks suggest that the results of Helios for the density of interplanetary grains as reported by Grün (1980) must be considered with some precautions.

1.4. Collected interplanetary dust grains (IDG)

Collected IDG represent potentially the best source of information for the density. Although no direct measurements have been reported, estimates are possible based on composition and physical appearance and Brownlee (1978) indicated a range of 1 to 3 g/cm^3. The possibility of low-density concerns, however, only the small fraction ($\sim$ 6 %) of chondritic aggregates - which are definitely not fluffy - the remaining 94 % having the bulk density of the minerals composing them. This is also the situation for meteorites whose density lies in the range 2.2 g/cm^3 (Cl) to 7.85 g/cm^3 (Iron).

2. CAN LOW-DENSITY PARTICLES REALLY EXIST ?

An absolute constaint on the density of IDG comes from their composition. The bulk density of solid, non-volatile materials present in the solar system is larger than ~ 2 g/cm^3 (ices are left out of the present discussion since icy grains cannot be present at 1 AU). With this condition, how far can the apparent density be reduced? I made a very simple investigation by scaling a 5 µm diameter chondritic aggregate of 0.4 µm "elementary" grains to a 25 mm diameter sphere made of pyrex spheres (density 2.5 g/cm^3) of 2 mm diameter and considered two extreme cases for which the cross-sections as well as the masses were perfectly defined so that the apparent densities could be calculed:
i) One outer monlayer of pyrex spheres yield an density of 0.63 g/cm^3 while the fraction of void amount to 75 %;

ii) Complete filling of the 25 mm sphere by the pyrex spheres yield a density of 1.65 g/cm^3, the fraction of void being 34 %.

None of the collected chondritic aggregates looks like the first extreme case and Brownlee (1978) was correct when he indicated a density larger than 1 g/cm^3. One could still further reduces the density by removing some of the pyrex spheres from the outer monolayer while retaining approximately the cross-section but not by much. The conclusion is therefore quite clear: an aggregate having a fraction of void as large as 34 % still has a density larger than 1 g/cm^3; only under extreme conditions with much larger fraction of void does the density becomes less than 1 g/cm^3 but it never becomes much less than 1 g/cm^3.

As the most direct analysis (lunar microcraters) and physical evidences lead to rejecting low densities for interplanetary dust grains while they are only supported by indirect methods (meteors) or extrapolations (Helios), it seems quite reasonable to accept this rejection.

Finally, the above simulation suggests a possible method to determine the density of the collected interplanetary dust grains whose shape is not too complicated. It consists in measuring the mass and the average of several cross-sections (from scanning electron microscope photographs). Then a mean size can be defined and consequently, a mean density can be determined.

REFERENCES

Benyuch, V.V.: 1974, Astron. Vestnik 8, p. 96
Brownlee, D.E.: 1978, Cosmic Dust (Wiley) p. 295
Ceplecha, Z.: 1967, NASA SP-135, p. 35
Ceplecha, Z.: 1983, Asteroids Comets Meteors (Eds. C.I. Lagerkvist and H. Rickman), p. 435
Fechtig, H.: 1982, Comets (Ed. L. Wilkening), p. 370
Grün, E., Pailer, N., Fechtig, H., Kissel, J.: 1980 Planet. Space Sci. 28, p. 333
Hanner, M.S.: 1980, Icarus 43, p. 373
Hughes, D.W.: 1978, Cosmic Dust (Wiley), p. 123
Le Sergeant d'Hendecourt, L.B., Lamy, P.L.: 1980 Icarus 43, p. 350
Millman, P.M.: 1976, Lect. Notes Phys. 48, p. 359
Nagel, K., Neukum, G., Fechtig, H., Gentner, W.: 1976, Earth Planet. Sci. Lett. 30, p. 234
Pailer, N., Grün, E.: 1980, Planet. Space Sci 28, p. 321
Pecina, P., Ceplecha, Z.: 1984, Bull. Astron. Inst. Czechosl. 35
Vedder, J.F., Mandeville, J.C.: 1974
Verniani, F.: 1973, J. Geophys. Res. 78, p.8429

A CORONASCANNER FOR A SUN ORIENTED SPACE MISSION

S. KOUTCHMY
Institut d'Astrophysique, Paris

P.L. LAMY
Laboratoire d'Astronomie Spatiale, Marseille

The quantitative study of the solar corona between 3 and 20 $R_{\odot}$ is still a challenging task. Inherent difficulties are the proximity of the Sun and the still bright inner corona (< 3 $R_{\odot}$), the low brightness of the outer corona and the low contrast of possible structures. A spinning, externally-occulted coronograph having its field-of-view limited to a sector of the corona presents several advantages over conventional coronographs for this study such as higher rejection of stray light and azimuthal uniformity over the whole field-of-view. Such an instrument could address the following aspects of the F-corona:

- photopolarimetry and colorimetry
- ellipticity, axis of symmetry
- temporal/spatial variations
- equatorial enhancements (rings)
- local component.

R. H. Giese and P. Lamy (eds.), Properties and Interactions of Interplanetary Dust, 141.

COLLECTION OF COSMIC DUST: PAST AND FUTURE

D.E. Brownlee
Department of Astronomy
University of Washington
Seattle, Washington 98195, U.S.A.

The collection of cosmic dust began in the middle of the last century with the recovery of spheres from the ocean floor and from Greenland ice (Murray and Renard 1891). The occurence of native metal in some of the deep sea spheres was the first clue that they were extraterrestrial. The deep sea spheres were described as "chondres" and their origin was attributed to the atmospheric melting of meteors. Polar ice and the ocean floor sediments often contain only minor amounts of magnetic terrestrial particles > 100μm and in these sites it is possible to collect rather large magnetic extraterrestrial particles that fell in historic times. In the intervening century an extensive series of particle searches were carried out in nearly all likely types of terrestrial collection sites. These included glaciers, islands, beaches, deserts, lakes, rooftops, rainwater, and all levels of the atmosphere up to low Earth orbit. Most of these efforts were not successful in collecting particles that were later proven to be extraterrestrial. In addition to the earlier deep sea and polar work, successful recoveries were made from a beach sand, a desert, and the stratosphere. All of these efforts are described in an excellent review by Hodge (1981).

Dust particles whose extraterrestrial origin can be positively proven are routinely collected at the present time both from sediments and from the stratosphere. The analysis of these samples gradually has taken on new significance both as a source of information about the physical properties of cosmic dust and as a source of cosmochemical information about primitive solar system materials. The sequence of events that led to the current "modern era" of dust collection and analysis began with the pioneering efforts of Curtis L. Hemenway, to whom this talk is dedicated. Dr. Hemenway, starting in the late 1950s, began an extensive series of collection efforts that revolutionized cosmic dust collection and catalyzed a major expansion of the cosmic dust field. His first dramatic new experiment was the Venus Flytrap rocket collector he built along with Bob Soberman (Hemenway and Soberman 1962). Many rocket collectors followed that he flew into the mesophere and into noctilucent clouds. In space he pioneered collections by inspecting Mercury capsule parts and he built

R. H. Giese and P. Lamy (eds.), Properties and Interactions of Interplanetary Dust, 143–147.

recoverable clamshell-like collectors exposed on Gemini missions (Hemenway et al. 1967). The first micrometeoroid impact craters recovered from space were obtained by these experiments which were, in addition, the first experiments directly retrieved from space by an astronaut during extravehicular activity (EVA). One collector was launched on an Agena and later picked up by Gemini. He also flew collectors on Skylab. An airlock he designed played a crucial role in saving the Skylab mission when it was used for erection on an emergency sunshade to cool the spacecraft. Ironically, the original problem was caused by a mechanical failure of a micrometeoroid shield during launch. On the space collectors and also on many of the rocket experiments Curt graciously invited investigators to build their own surfaces and fly them in his experiments. This established the tradition of joint investigators now used in LDEF and also provided a mechanism for many new and unestablished investigators (including the author) to get started in the then rapidly expanding und very exiting field.

Hemenway also conducted an extensive set of balloon collections. His balloon top settling plate collector, "Sesame" was the first experiment flown at the U.S. national scientific balloon flight facility at Palestine, Texas (Hemenway et al. 1967b). After many Sesame flights, he then developed the "Magellan"collector - a large funnel collector that was flown in the southern hemisphere below a superpressure balloon that circled the globe (Hallgren and Hemenway 1976). This was the pioneering effort for recovering a large payload flown on a long duration, superpressure balloon.

All of Hemenway's experiments were daring advances forward that utilized the rapidly advancing technology of space flight, ballooning and laboratory technique. These collections radically advanced the state of the art, even though only the craters on his orbital experiments can be confidently proven to be extraterrestrial. Most of the collections were controversial and did not collect particles that could be proven to be extraterrestrial on chemical or isotopic grounds. There were two major problems that plagued collections made in the 1960s. One was that it was widely, but not unanimously, believed that the flux was so high that particles could be collected with sounding rockets and settling plate collectors mounted on balloons. We now know that the flux of 10μm particles is only $1m^{-2}day^{-1}$ and these techniques are not practical even if contamination problems could be totally eliminated. The only successful airborne particle collections have been made by balloon and aircraft collections that cleanly sample huge volumes of air. These devices collect one 10μm particle per 1000 ft^3 of air, a rate of one per hour. The second problem with earlier collections was the difficulty of analysis of collected samples. The scanning electron microscope (SEM) became commercially available in the 1960s, and Hemenway was quick to adapt this revolutionary new surface imaging technique to cosmic dust particles. Unfortunately, the commercial availability of solid state X-ray analyzers for the SEM did not come until several years later. This device provided, for the firs time, the capability for doing nondestructive chemical analysis of microparticles in seconds. Prior to the availability of this device, elemental analysis on particles could only be done with X-ray crystal spectrometers which are used on microprobes and were available on a few electron microscopes. The long integration times requird for these analyses caused great difficulty for analysis of small particles. An example is the analysis of high atomic number elements in "stardust" particles collected on Hemenway's rocket experiments. Spectrum scans for these very small particles took so long that the particles became obliterated by beam contamination (diffusion pump oil). This obliteration made it difficult to verify the analysis results by a second analysis, or even to confirm that the beam was always exactly on the sample.

With the technology available prior to about 1970 it was difficult to identify collected particles as either terrestrial or extraterrestrial. In the majority of cases it was simply assumed that the recovered samples were extraterrestrial by virtue of their morphology or location of collection. Unfortunately, these two criteria are really only useful in deep ocean sediments and even then they are by no means infallible. In the past 15 years new information on extraterrestrial particles and new and fairly common analytical tools have made the identification of at least the majority of cosmic dust particles a fairly straightforward task. Fortunately, the majority of cosmic dust particles larger than 5μm have chondritic elemental compositions. Within a factor of 2-3 typical abundances match type 1 carbonaceous chondrites for all elements. No known terrestrial particle matches chondritic composition for more than a few elements. As an operational criterion, when a particle is analyzed in an electron microprobe or SEM, if it has chondritic abundances for C, Na, Mg, Al, Si, S, Ca, Cr, Mn, Fe and Ni, it is almost certainly extraterrestrial. Particles that melt during atmospheric entry to form spheres differ from the chondritic pattern in that they are usually depleted in volatile elements such as Na, S and C and siderophile elements such as Ni, Co and Cr (a siderophile under reducing conditions). Many cosmic dust particles, of course, do not have chondritic compositions and establishing an extraterrestrial origin for these is more difficult. Typical large deviations from a chondritic pattern occur when a particle's composition is dominated by a single mineral. Regardless of its composition, if a collected particle was a small particle in space, then its extraterrestrial nature can be absolutely verified by the detection of implanted solar wind (Hudson et al. 1981), solar flare tracks (Bradley et al. 1984) or the existence of ^{26}Al, ^{10}Be, or ^{53}Mn produced by cosmic ray bombardement (Nishiizumi 1983; Raisbeck et al. 1983). The composition of a particle can be determined in seconds but the other techniques for determining its origin are complex and cannot be routinely used on a large number of samples. The track and rare gas tests can be done on 10μm particles but only if they were not strongly heated during atmosphere entry. The cosmogenic isotope effects require particles of 300μm in diameter or larger.

Particle collections experiments are currently being conducted in the stratosphere, polar ice, ocean sediments and earth orbit. The most pristine material will always come from the stratosphere, but particles will generally be smaller than 50μm because of the flux problem. Important goals of future atmopsheric collections will be to collect particles from meteor showers or fireballs and to collect rare large particles (~50μm) that have suffered minimal alteration by atmosphereic heating. Air sampling collectors become increasingly inefficient for larger particles and perhaps the best approach for > 50μm particles is the Hemenway Magellan collector, which flies a large funnel (settling plate) for a period of months. A few percent of 50μm particles enter the atmosphere at very low incidence angle and suffer only minor heating. This process actually allows a much smaller fraction of millimeter-size particles to survive without melting and thus become legal "micrometeorites". A major goal of future sediment collections will be to collect these giant micrometeorites. These particles are unmelted, but are not nearly as well preserved as the atmospheric dust because of weathering processes at the Earth's surface. Their size, however, makes them very important for a variety of analytical techniques that are best done in large samples. Future efforts on the sediment particles should also include detailed studies to detect temporal changes in the flux, size or composition of particles that have entered the atmosphere over the 2×10^8 -year record of Earth history preserved in deep sea sediment cores.

The collection area that will undoubtedly see the greatest change in coming years is space. Although it is probably impossible to collect particles in space in a truly nondestructive manner, it should be possible to collect analyzable material in a nonselective fashion in regards to physical properties and, most importantly, it should be possible to collect particles with known orbital parameters. Some of the stratospheric particles are highly porous fragile objects. It is possible that there are even more fragile particle types that do not survive atmospheric entry and could only be collected in space. While it does not appear practical to capture such fragile particles intact, it should be possible to capture solid debris resulting from the impact into a capture cell or the collection surface. Chemical analysis of the fragments can provide an excellent means of characterization of the original particle.

One of the most important aspects of space collections is to collect samples from a known source. The straightforward approach, of course, is take the collector directly to the source, such as a cometary coma, and then return it to the Earth. Such missions are currently under study for low cost sample returns with flyby velocities of 10 kms^{-1}. For Earth orbiting experiments "matching" a single particle with a known source requires precise measurement of its orbital parameters by electronic techniques. Because of orbital evolution it cannot be expected that many collected particles could be reliably matched with a known source but some can. This is particularly true for large particles whose collisional lifetimes are much smaller than their Poynting-Robertson evolution times. With a suitably large collector it appears possible to collect a significant number of 10μm particles from known comets and even contemporary interstellar grains in transit through the solar system.

The current work on collected particles has provided a detailed chemical and structural information on what are probably typical interplanetary dust particles in the 5μm-50μm size range. Most of these particles are black aggregates of small grains but there is actually wide diversity among the collected samples. Particles range from solid mineral chunks such als FeS to fragile porous aggregates composed of millions of tiny but distinct mineral grains. Future work will provide increasingly detailed information about the source materials of interplanetary dust. Work on dated sediments has provided initial data on the time history of the Earth's accretion of dust and has indicated that future efforts may provide powerful constraints on the temporal variations in the meteoroid complex. Recent work on tracks in stratospheric micrometeorites and cosmogenic isotopes in deep sea spheres has provided the first real measurement of particle lifetimes that can be used to constrain models of the dynamical evolution of interplanetary dust.

REFERENCES

Bradley, J.P., Brownlee, D.E. and Fraundorf, P., 1984: Science, **226**, 1432-1434.
Hallgren, D.S. and Hemenway, C.L., 1976: 'Magellan Collectors of Large Cosmic Dust Particles' in Interplanetary Dust and Zodiacal Light, Lecture Notes in Physics,**48**, eds. H. Elsässer and H. Fechtig, Springer Verlag, New York.
Hemenway, C.L. and Soberman, R.K., 1962: Astron. J., **67**, 256-266.
Hemenway, C.L., Hallgren, D.S. and Kerridge, 1967a: Space Res., **8**, North Holland, 521-535.
Hemenway, C.L., Hallgren, D.S. and Coon, R.E., 1967b: Space Research, **7**, 1423-1431.
Hodge, P.W., 1981: Interplanetary Dust, Gordon and Breach, New York, 280 pp.
Hudson, B., Flynn, G.J., Fraundorf, P., Hohenberg, C.M. and Shirck, J., 1981: Science, **211**, 383-386.
Murray, J. and Renard, A.F., 1891: Rep. Sci. Results Voyage H.M.S. Challenger 3, Neill and Co., Edinburgh.
Nishiizumi, K., 1983: Earth Planet. Sci. Lett., **63**, 223-228.
Raisbeck, G.M., Yiou, F., Klein, J., Middleton, R., Yamakoshi, Y. et al.: 1983, Lunar Planet. Sci. **XIV**, 622-623.

COMPLEMENTARY LABORATORY MEASUREMENTS OF INDIVIDUAL INTERPLANETARY DUST PARTICLES

A. Fahey, K. D. McKeegan, S. A. Sandford,
R. M. Walker, B. Wopenka and E. Zinner
McDonnell Center for the Space Sciences
Washington University
St. Louis, MO 63130 USA

ABSTRACT. Complementary analysis techniques including electron microscopy (SEM/EDX and TEM), molecular spectroscopy (FTIR and Raman), and secondary ion mass spectrometry (SIMS), are used to study individual dust particles collected in the stratosphere. Large deuterium enrichments and solar flare tracks show that most particles in the "chondritic" class are interplanetary dust particles (IDPs). Infrared transmission spectra of most IDPs fall into three major classes (layer-lattice silicates, pyroxenes and olivines). TEM and Raman measurements confirm this classification. The IR spectra show certain similarities to spectra observed in comets and protostars. In particular the 6.8 µm features observed in protostars and IDPs may have a common origin. Large D excesses are observed in IDPs of the first two IR classes. The correlation of D/H ratios with the C concentration indicates a carbonaceous carrier of the excess D. The D enrichments and IR spectra provide links to interstellar molecular cloud material.

1. INTRODUCTION

"Chondritic" dust particles collected in the stratosphere [1] consist primarily of interplanetary dust material. The extraterrestrial origin is shown by solar-type gases [2-3], large D excesses [4,5] and solar flare tracks [6]. In spite of the small mass (1-50 ng), it is possible to perform a number of microanalytical measurements on a given dust grain. Table 1 gives a summary of analyses performed on nine different IDPs. Arbitrary names are given to individual particles.

2. MEASUREMENT TECHNIQUES

Energy dispersive x-ray analysis (EDX) is used to select particles of the "chondritic subset" [7]. These are then examined by other techniques.

Infrared transmission spectra from 2.5 to 25 µm are measured by

R. H. Giese and P. Lamy (eds.), Properties and Interactions of Interplanetary Dust, 149–155.

Table 1. A summary of the analysis techniques applied to the study of each stratospheric dust particle.

Particle name	Infrared Class[6]	Isotopic Measurements H	C	Mg	Si	Other Analyses Done EDX	TEM	Raman
Mosquito		X	X	X	X	X		X
Skywalker	layer-	X	X	X	X	X	X[12]	X
Calrissian	lattice	X	X			X	X[14]	X
Lea	silicates	X				X		X
Low-Ca						X	X[13]	
Spray-2	pyroxene	X		X	X	X	X[14]	X
Essex		X				X	X[14]	X
Attila	olivine	X				X	X[17]	X
Jedai						X	X[14]	X

FTIR spectrometry [8]. These measurements show that the majority of IDP spectra fall into three different classes termed "layer-lattice silicates", "pyroxenes" and "olivines" based on similarities to terrestrial mineral spectra [9]. Figure 1 shows representative spectra of the three IR classes. While no individual spectral type matches the emission spectrum of Comet Kohoutek, a reasonable match is obtained by adding roughly equal amounts of "layer-lattice silicate" and "pyroxene" IDP spectra [8,9]. The lack of agreement in the position and width of the 10 μm silicate feature of the "olivine" class with the Comet Kohoutek spectrum shows that olivine cannot be a major component of that comet's dust. Besides the 10 μm silicate feature and 3.0 and 6.0 μm hydration bands, a prominent band at 6.8 μm is seen in the spectra of the "layer-lattice silicate" group. This band is strongest in Calrissian, where an accompanying band at 11.4 μm suggests that these features are due to a carbonate mineral. The 6.8 μm feature is seen in the spectra of some protostars [10] where it might have a similar origin.

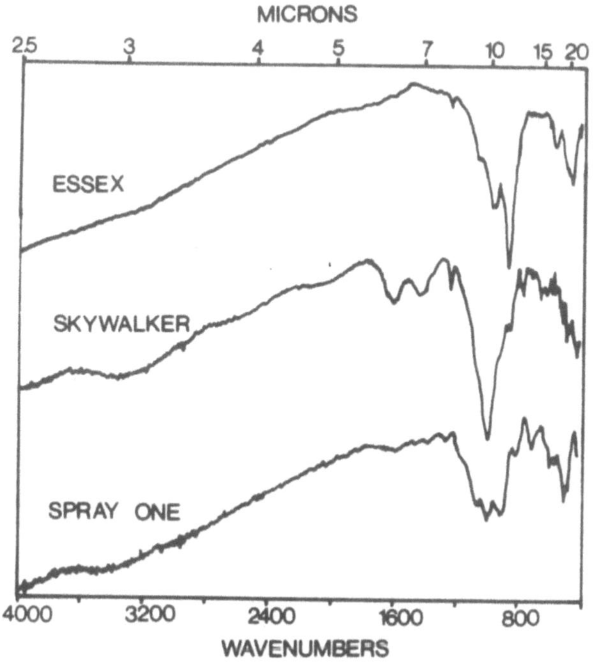

Figure 1

Infrared transmission spectra of three IDPs representative of the "olivine" class (Essex), "layer-lattice silicate" class (Skywalker) and "pyroxene" class (Spray One).

High resolution transmission electron microscopy including electron diffraction measurements provide information on the mineralogy and morphology of the IDPs. Previous TEM observations have shown that IDPs contain highly unequilibrated material which is distinct from that seen in meteorites [11-13]. The recent discovery of solar flare tracks in IDPs [6] provides a novel way to prove the extraterrestrial origin of stratospheric dust particles. All the TEM observations on the IDPs listed in Table 1 were made at Arizona State University [14-16,19].

Some information on the chemical state of C in the IDPs can be provided by the laser Raman microprobe [17,18]. Such information is of special interest since the D enrichments correlate with C concentration. Fig. 2 shows Raman spectra obtained from Calrissian and Attila. The two lines seen in the spectra are due to graphite. The relative strength, widths, and exact positions of the two peaks are known to vary with the degree of crystallinity of the graphite. In IDPs a constant pattern is seen in different fragments of a given particle but differences are observed between particles.

SIMS isotopic measurements of several elements (H, C, Mg, Si in this study) can be made in individual fragments of a single IDP [5]. D/H measurements show large D excesses in most of the IDPs measured so far (Fig. 3). The D excess is distributed heterogeneously on a size scale of several µm and is correlated with the C/O, N/O, and S/O ratios, indicating a carbonaceous, probably organic, carrier for the excess D. In contrast, Mg and Si measured in three particles exhibit isotopically normal values. The C isotopic composition is homogeneous in a given IDP, but there seem to be slight differences between different particles.

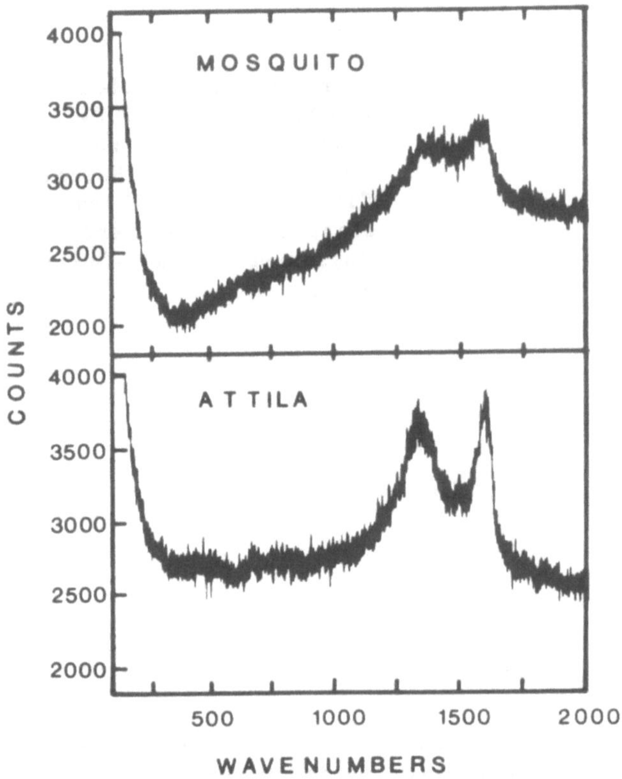

Figure 2:

Raman spectra of two IDPs. The two lines are associated with graphite.

Figure 3:

Range of δD values measured in seven IDPs. Terrestrial actinolite was used as a standard defining the δD scale.

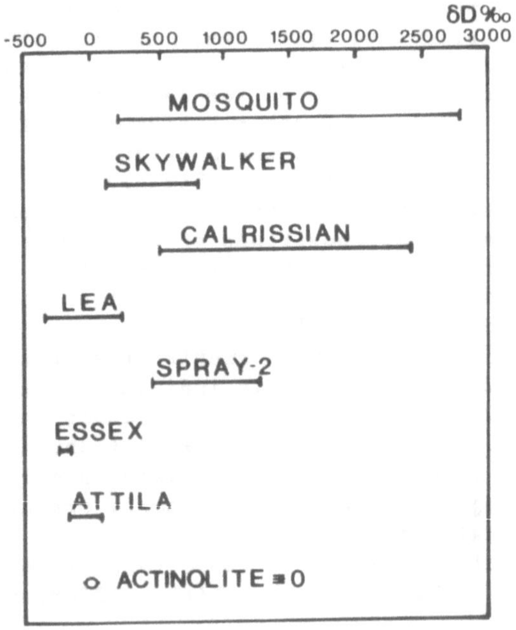

3. COMPARISON OF RESULTS OBTAINED BY DIFFERENT TECHNIQUES.

IR spectroscopy is the simplest method found so far for classifying particles of the chondritic subset into different groups. We next discuss the relationships that exist between an IDP's IR classification and its properties measured by other analytical methods.

3.1 IR and TEM

Two particles in the "layer-lattice silicate" class, Skywalker and Low-Ca, have been studied in the TEM by Tomeoka and Buseck [14,15]. In both cases an abundance of poorly crystalline phyllo-silicates was observed. Both are distinct from meteorites and there are considerable differences in the detailed mineralogy of the two particles. Solar flare tracks were seen in Skywalker (Tomeoka, private communication) proving its extraterrestrial origin. In the "olivine" class, Jedai, Essex and Attila, have been examined by Christoffersen and Buseck [19]. Jedai and Essex are primarily composed of fosterite crystals of irregular and equidimensional shape. These two IDPs also show differences in detail: the Fe content of the olivines in Jedai is typically 5-8% whereas it is 15-20% in Essex. Attila also contains abundant olivines, in addition to FeS, and a form of Fe-Zn sulfide not seen previously in IDPs. Tracks are abundant in the olivine crystals of Attila, thus providing proof that some particles of the "olivine" class are extraterrestrial. The only particle of the "pyroxene" class examined in the TEM, Spray-2, contains both small pyroxene and olivine crystallites imbedded in a low-Z matrix. A pyroxene line is also seen in the Raman spectrum of this particle.

In summary, TEM observations confirm that the IR classifications reflect the major silicate phases present in different particles. However, particles whose IR spectra are nearly identical can show substantial differences in their detailed mineralogical structures when viewed at high magnification.

3.2 IR and SIMS

Figure 3 shows the range of δD values measured in different fragments of seven IDPs. Two of the three particles from the "layer-lattice silicate" class, Skywalker and Calrissian, show large D excesses. D excesses in Lea are smaller, but still outside the limits of terrestrial materials, proving the extraterrestrial origin of all three particles. The extraterrestrial origin of Calrissian is especially significant since the strong 6.8 μm feature in the IR spectrum of this particle, likely due to a carbonate, is similar to features observed from protostars [10]. Spray-2 of the "pyroxene" class also exhibits large D excesses. The two particles of the "olivine" class, Essex and Attila, show no excess D. While Essex is slightly depleted in D, Attila's hydrogen isotopic composition is terrestrial within the measurement errors. Nevertheless, the existence of abundant solar flare tracks in Attila proves that it is extraterrestrial. Thus there are interplanetary dust particles with a terrestrial hydrogen isotopic composition.

Large deuterium enrichments seen in certain fractions of primitive meteorites have been interpreted as showing the presence of "interstellar" material [20]. Similar arguments suggest that IDPs also contain primordial molecular cloud material.

4. CONCLUSIONS

Large D excesses and the presence of solar flare tracks confirm that many particles in the chondritic subset of the stratospheric collection are interplanetary dust particles. The magnitude of the D excesses and their association with a carbonaceous phase suggests that, like certain primitive meteorites, IDPs contain material formed in interstellar molecular clouds [20]. The similarity of the IR spectra of "layer-lattice silicate" particles that exhibit large D excesses with those of protostars surrounded by dust shells provides an additional link between IDPs and interstellar molecular clouds. The 10 µm feature in the emmission spectrum of the Comet Kohoutek can be fitted by a mixture of the spectra of the "layer-lattice silicate" and "pyroxene" types, consistent with, although not proving, a cometary source for these IDPs.

REFERENCES

[1] Brownlee D.E., Pilachowski L., Olszewski E., and Hodge P.W. (1980) In Solid Particles in the Solar System, (eds. I. Halliday and B.A. McIntosh), D. Reidel.

[2] Rajan, R. S., Brownlee, D.E., Tomandl, D., Hodge, P.W., Farrar IV, H., and Britten, R. A. (1977) Nature 267, 133.

[3] Hudson B., Flynn G. J., Fraundorf P., Hohenberg C.M., and Shirck J. (1981) Science 211, 383-386.

[4] Zinner E., McKeegan K.D., and Walker R.M. (1983) Nature 305, 119.

[5] McKeegan K. D., Walker R. M. and Zinner E., Geochim. Cosmochim. Acta, submitted.

[6] Bradley J.P. and Brownlee D.E. (1984) 47th Meteoritical Soc. Conf., U-1.

[7] Fraundorf P., McKeegan K.D., Sandford S.A., Swan P., and Walker R.M. (1982) J. Geophys. Res. 83, A403.

[8] Sandford S.A. and Walker R.M. (1984) Astrophys. J., submitted.

[9] Sandford S.A., Fraundorf P., Patel R., and Walker R.M. (1982) Meteoritics 17, 276.

[10] Soifer V. T., Puetter R. C., Russell R. W., Willner S. P., Harvey P. M. and Geillett F.C. (1979) Ap.J. 232, L53.

[11] Fraundorf P. (1981) Geochim. Cosmochim. Acta 45, 915.

[12] Christoffersen R. and Buseck P.R. (1983) Science 222, 1327.

[13] Bradley J.P., Brownlee D.E., and Veblen D.R. (1983) Nature 301, 473.

[14] Tomeoka K. and Buseck P.R. (1984) 47th Meteoritical Soc. Conf., S-1.

[15] Tomeoka K. and Buseck P.R. (1984) Earth Planet. Sci. Lett. 69, 243.

[16] Bradley J. P., Christofferson R., Tomeoka K., personal communication.

[17] Fraundorf P., Patel R.I., Walker R.M., Freeman JJ. and Adar F. (1982) Lunar Planet. Science XIII, 231.
[18] Wopenka B. and Sandford S. A. (1984) 47th Meteoritical Soc. Conf., S-5.
[19] Christoffersen R. and Buseck P. R., to be published.
[20] Yang J. and Epstein S. (1983) Geochim. Cosmochim. Acta 47, 2199.

PROTON INDUCED X-RAY EMISSION (PIXE) ANALYSIS OF METEORITIC MICROSAMPLES

R. Wallenwein, H. Blank, E.K. Jessberger, K. Traxel
Max-Planck-Institut für Kernphysik
Postfach 10 39 80
6900 Heidelberg
W. Germany

ABSTRACT. Interplanetary dust particles (IDPs) collected in the stratosphere are thought to be mostly of cometary origin. They survived heating during their deceleration in the earth's atmosphere (1). Because of their small size (< 50 μm) and mass (< 10^{-7}g) they are difficult to analyse. Special preparation and examination methods have been developed for their investigations (2). We set out to study the trace-element-composition of these particles using the Heidelberg proton microprobe.

At Heidelberg a proton microprobe has been constructed with a beam spot of ≥ 2 μm x 3 μm and has been successfully used for the quantitative analysis of trace elements in minute (10 μm^3) inclusions in lunar materials and meteorites (3). We reported on exploratory studies of the applicability of this microprobe for the trace element analysis of IDPs. First test measurements were made on thick meteoritic samples. These samples consist of material from the meteorites Hallingeberg and Allegan, which was ground to nominally less than 1 μm grain size and subsequently compacted with high pressure. The reason for choosing that material was that we expected chemical uniformity on a very small scale. We took as standard Lunar Analog Glas, containing 13 elements of different concentrations. The agreement between the PIXE results and the bulk analysis of the meteorites is quite satisfactory within an error rate of 10%.

For further comparing tests we used individual particles of the carbonaceous chondrite Murchison. The particles were mounted on a thin (200-300 Å) carbon foil held by a nickel grid. We analysed a total of ten particles with a typical diameter of about 20 μm and determined their elemental composition. We found discrepancies to the values from bulk Murchison. However, we want to mention that the bulk analysis done by Jarosewich was performed with 40g of the Murchison meteorite.

The next step will be to use PIXE for minor element analysis and also to analyse real IDPs. We set out to produce some thin standards with our ion implantation equipment for further measurements.

ACKNOWLEDGEMENTS: We thank K. Fredriksson for the Hallingeberg and

R. H. Giese and P. Lamy (eds.), Properties and Interactions of Interplanetary Dust, 157–158.

Allegan pellets, E. Zinner for the preparation of the Murchison samples, G. Lange and D. Maas for advice and help.

REFERENCES

(1) - F.L. Whipple, Proc.Nat.Acad.Sci. USA 37, 19, 1951.

(2) - D.E. Brownlee, Microparticle studies by sampling techniques, Cosmic Dust, p. 295-336, Wiley, New York, 1978.

- G.J. Flynn, P. Fraundorf et al., Chemical and structural studies of "Brownlee" particles, Proc.Lunar Planet.Sci.Conf. 9th, p. 1187-1208, 1978.
- P. Fraundorf, Microcharacterization of interplanetary dust collected in the earth's stratosphere, Dissertation Washington University, St. Louis, Missouri, 1980.
- P. Fraundorf, Interplanetary dust in the TEM, Geochimica et Cosmochimica Acta, Vol. 45, p. 915-943, 1981.

(3) - The Heidelberg Proton Microprobe, Nucl.Sci.Appl., Vol. 1, p. 33-55, 1980.

- H. Blank, Spurenelemente in koexistierenden opaken Oxyden lunarer Gesteine gemessen mit einer Protonenmikrosonde, Dissertation University of Heidelberg, 1982.

PHYSICAL PROPERTIES OF SUBMICRONIC CARBONACEOUS PARTICLES CANDIDATE AS COSMIC DUST

A. Borghesi, E. Bussoletti and L. Colangeli
Gruppo Astrofisico, Physics Department
University of Lecce
73100 Lecce
Italy

ABSTRACT. We present here a compendium of our laboratory measurements of the physical properties of submicronic particles made by some carbonaceous materials candidate as cosmic dust: amorphous carbon, silicon carbide. Comparison with data obtained by other authors is presented and discussed in view of astrophysical applications.

1. INTRODUCTION

Solid particles produce the absorption and scattering of starlight representing one of the main agents which control transfer of radiation in the interstellar medium.

In spite of a large amount of direct observations, the composition, size distribution, structure and morphology of grains are still under debate.

Laboratory work on this subject has been developed since three years in our Department with the aim of performing an accurate analysis of the physical properties of different materials as possible candidates for cosmic dust.

At present we have investigated carbonaceous materials such as various types of amorphous carbon and silicon carbide, expected to be formed around carbon stars (Borghesi et al., 1983a,b) and in supernovae explosions (Greenberg, 1978).

2. EXPERIMENTAL

A detailed description of the experimental procedures has been reported elsewhere (Borghesi et al., 1983a,b); we summarize here the main items.

2.1. Amorphous carbon.

Two different production methods have been used for this material:
AC, by striking an arc between two amorphous carbon electrodes in a controlled Ar atmosphere at different ambient pressures;

R. H. Giese and P. Lamy (eds.), Properties and Interactions of Interplanetary Dust, 159–162.

BE-XY, by burning benzene and xylene in air at room pressure.

2.2. Silicon carbide.
Different samples of commercial mixture of α-SiC polytypes have been processed in this case. The raw material has dimensions much larger than 1 μm so that the particles were initially ground and sedimented in acetone to select appropriate size ranges for the grains.

The tendency of grains to clump together was prevented by appropriate ultrasonic treatment.

High magnification TEM analysis enables to determine particles shapes, their dimensions and size distributions. The morphological properties of grains are reported in Table I.

Table I. Morphological properties of the samples.

material	elongation	grain shape
AC	80 Å	spheroidal
BE-XY	300 Å	spheroidal
SiC	4000 Å	irregular

SEM analysis is in progress to study in detail the surface morphology of the particles in order to clarify the open question about the degree of "porosity" of the dust which was pointed out by Abadi and Wickramasinghe, 1976, Abadi et al., 1976, and Blanco et al., 1980.

Diffraction methods are used to determine the crystallographic structure of the grains. AC, BE and XY samples appear amorphous, while SiC samples appear essentially exaedric.

Transmission analysis is performed on double-beam Perkin-Elmer spectrometers - Mods. 330 and 683 - to cover the wavelength range 2000 Å - 40 μm.

3. RESULTS

Figure 1 reports the Q_{ext}/a values versus λ determined in our observations for amorphous carbon in the range 2000 Å - 40 μm for BE, XY and AC samples.

Absolute values for BE and XY samples agree within the experimental errors; while for AC samples they appear systematically lower; however the spectral slope is the same for the three kind of particles, i.e. $Q_{ext}/a \propto \lambda^{-1}$.

Our data are in good agreement with previous observations performed by Koike et al., 1980, on amorphous carbon obtained by means of the same production methods.

It is interesting to note that our spectra show a bump falling around 2500 Å while graphite measured by Huffman, 1975, presents a bump at around 2300 Å. Present data do not allow to individuate which, between the two materials, is the most likely to produce the observed interstellar feature. However, theoretical papers presented by Czyzak and Santiago, 1973, and Seki and Hasegawa, 1981, seem to indicate that some sort of amorphous carbon may be more likely.

Figure 2 reports the average of several measurements of the Q_{ext}/a for α-SiC in the range 25 - 40 μm.

The characteristic absorption band at 11.4 μm is clearly appearing, as well as the weak phonon resonances falling at 10.6 and 12.7 μm. The continuum, out of the band, presents a $\lambda^{-1.3}$ trend.

Our results are in good agreement with those presented by Dorschner et al., 1977, and Friedemann et al., 1981. A comparison with the observations of β-SiC performed by Stephens, 1980, shows that, in this case, the main band is splitted in two peaks while the central one is absent. This behaviour may be interpreted as due to the different crystallographic composition (polytypes for α-SiC, cubic for β-SiC).

By comparing astronomical observations with α-SiC and β-SiC bands the former appears to be the most indicated to simulate the cosmic spectra.

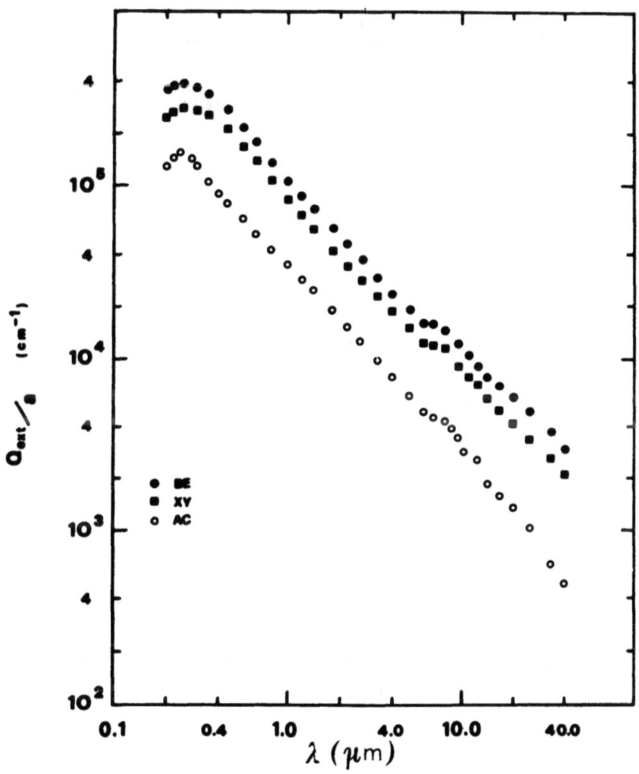

Figure 1. The extinction efficiency of BE, XY, AC samples (mean values).

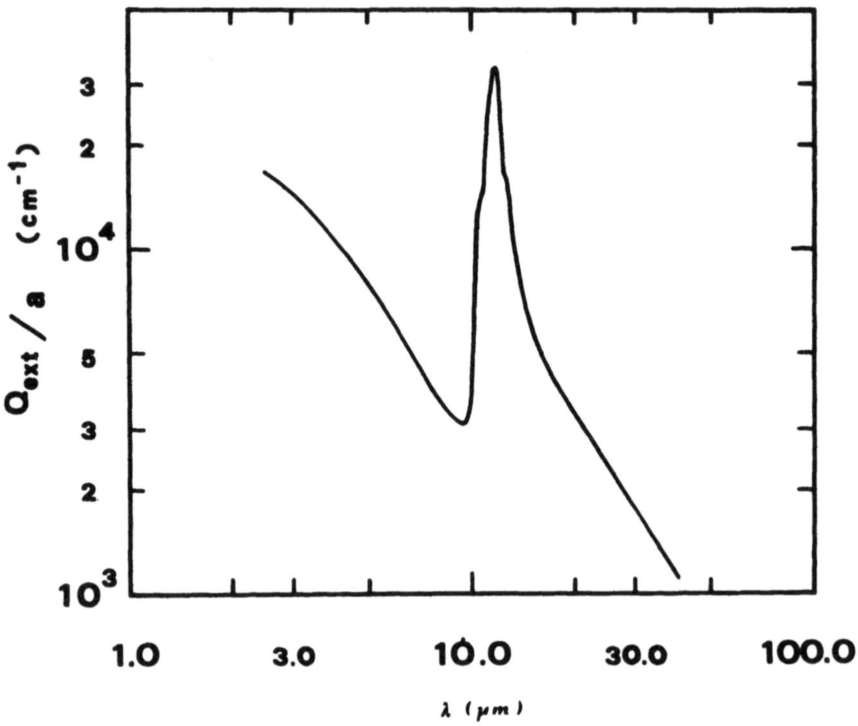

Figure 2. The extinction efficiency of α-SiC samples (mean values).

References.

H. Abadi, P. Joshi, S. Ramadurai and N.C. Wickramasinghe; Nature, 263, 214, (1976).

H. Abadi and N.C. Wickramasinghe; Astrophys. Space Sci., 39, L31 (1976).

A. Blanco, A. Borghesi, E. Bussoletti and S. Fonti; Astrophys. Space Sci., 68, 505 (1980).

A. Borghesi, E. Bussoletti, L. Colangeli, A. Minafra and F. Rubini; Infrared Phys., 23, 85 (1983a).

A. Borghesi, E. Bussoletti, L. Colangeli and C. De Blasi; Infrared Phys., 23, 321 (1983b).

S.J. Czyzak and J.J. Santiago; Astrophys. Space Sci., 23, 443 (1973).

J. Dorschner, C. Friedemann and J. Gurtler; Astron. Nachr., 298, 279 (1977).

C. Friedemann, J. Gurtler, R. Schmidt and J. Dorschner; Astrophys. Space Sci., 79, 405 (1981).

J.M. Greenberg; "Cosmic dust" (ed. J.A.M. McDonnell) p.187, Wiley, New York (1978).

D.R. Huffman; Astrophys. Space Sci., 34, 175 (1975).

C. Koike, H. Hasegawa and A. Manabe; Astrophys. Space Sci., 67, 495 (1980).

J. Seki and H. Hasegawa; Progr. Theoretical Phys., 66, 903 (1981).

J.R. Stephens; Astrophys. J., 237, 450 (1980).

A DISCOVER OF IRON SPHERULES WITH GOLDEN Ni-Fe CORE AND SILICATE SPHERULES WITH GOLDEN Ni-Fe INCLUSION IN DEEP-SEA SEDIMENTS

Peng Hanchang*, Zhuang Shijie**, Liu Zhenkun***, and Yu Zhong****

* First Institute of the National Bureau of Oceanography, Qingdao, China

** Institute of Geology Ministry of Metallurgical Industry, Guilin, China

*** Institute of HUNAN Metallurgical Industry Changsha, China

**** Scientific Research Academy for Petroleum Exploration and Development, Beijing, China

ABSTRACT: Recently, we found iron spherules with golden Ni-Fe core and silicate spherules with golden Ni-Fe inclusion in deep-sea sediments. This finding is of significance for further understanding of the origin of the deep-sea spherules.

1. INTRODUCTION

Since microscopic magnetic spherules in the deep-sea sediments were found by Marray and Renard (1891), many scientists have performed the overall studies of them, and divided them into three types: the iron spherules (type I), the silicate spherules (type S) and the glassy spherules (type G).

For the iron spherules, because of discoveries of Ni-Fe cores, whose chemical composition is similar to the iron meteorite, ones have an identical viewpoint on their extraterrestrial origin. But the shapes, sizes, colors, shines, surface structures of cores, and their place in the iron spherules have not been reported. For the silicate spherules, although many evidences have been discovered to prove that the are extraterrestrial, many people have been doubting their genuine origin.

Recently, we found many iron spherules with golden Ni-Fe core and a few silicate spherules with golden Ni-Fe inclusion in the deep-sea sediments taken from the North Pacific Ocean, and observed and analysed the cores and inclusions in details by using a microscope and a scanning electron microscope. Their chemical compositions were measured by using electron probing. The results show that these deep-sea spherules are different from material formed on the Earth, and may be cosmic material flying off the surfaces of meteoroids as they entered the atmosphere of the Earth.

R. H. Giese and P. Lamy (eds.), Properties and Interactions of Interplanetary Dust, 163–168.

2. MICROSCOPIC CHARACTERISTICS

Under the stereoscopic microscope magnifying 50-60 times, the iron spherules can clearly be seen as black or brownish-black globular, oval, button, teardrop or other irregular shape. Their grain-size was general 100-200 μm. Most of them have strong magnetic properties.

The golden Ni-Fe cores were found in broken button-shaped iron spherules. The cores have metallic luster, globular. Their sizes are 1/4 - 1/8 of whole spherule. Because their hardness was clearly larger than that of their crust, they often fall down. All of the Ni-Fe cores are near the button-chaped plane or pit. Results of electron probing analysis indicate that Ni content of button-shaped iron spherules reaches 2.77% to 52.29%.

The silicate spherules are dark-green or greyish-white, globular, oval or irregular particles. There are some black spots with different sizes on the surface of some of them, and there are some circular pits on the spots, which seem to be the trace produced by airflow. Under the scanning electron microscope, octahedral magnetite cramed in the circular pits on the iron spots were found.

The golden Ni-Fe inclusions were found in broken Fe-Si mixing silicate spherule. The inclusions are globular, and have metallic luster. Their sizes are about 1/4 of whole spherule. Because their hardness is clearly larger than that of the crust, they will be dislocated when moving them lightly. Results of electron probing analysis indicate that the crust is olivine.

3. STRUCTURE AND CHEMICAL COMPOSITION

3.1 Scanning electron microscope analysis

To observe the iron spherules with golden Ni-Fe core and silicate spherules with golden Ni-Fe inclusion, a scanning electron microscope was used, by which it can clearly be seen that Ni-Fe core is near the button-shaped plane. The surface of the core is rough, and some of them also have a cavity. The cores are easily dislocated because of crack between the cores and their crust. In x-ray NiKa plane distribution pictures, nickel element of core concentrates highly (Fig.1). Results of electron probing analysis indicate that Ni content reaches 79.1% (Fig.1 A and C).

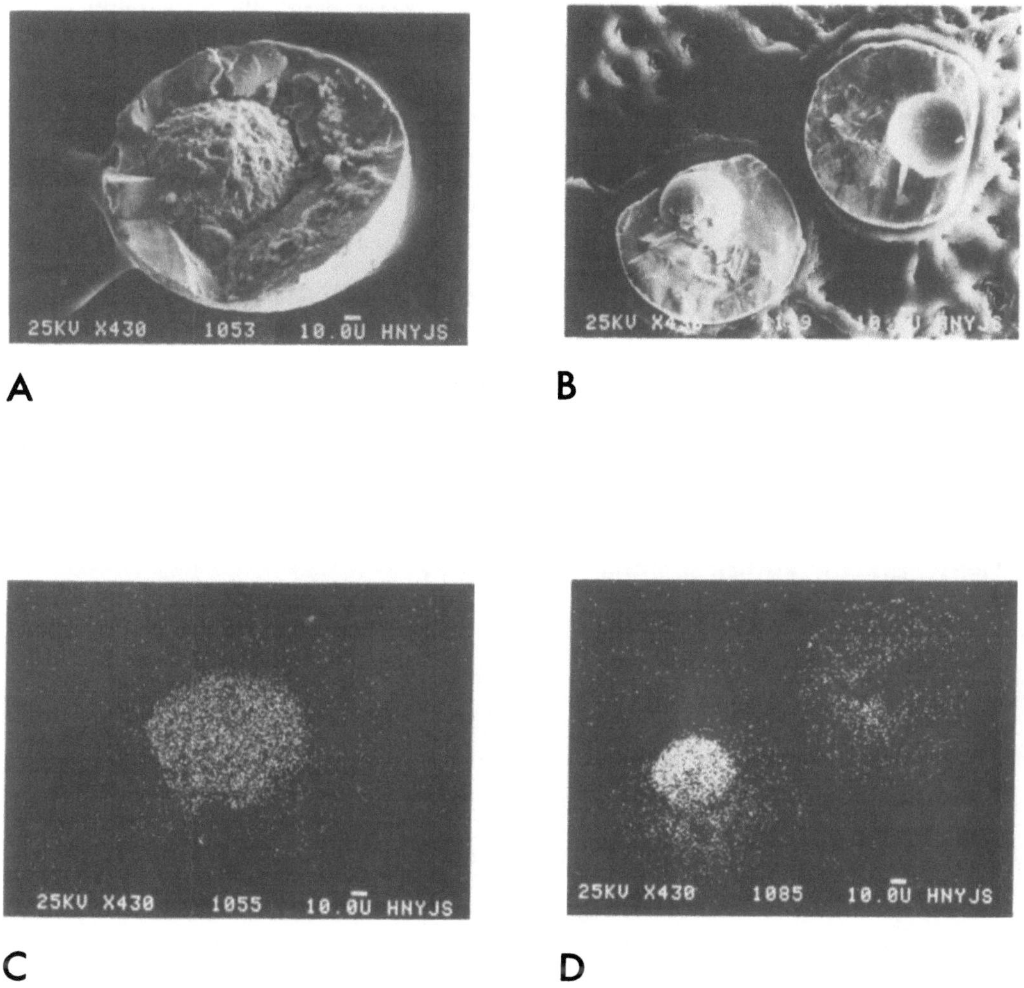

Figure 1. SEM Photographs of the iron spherules. A and B are broken iron spherules with the Ni-Fe core; C and D are respectively NiKa plane distribution pictures of A and B.

The Ni-Fe inclusions usually are near the sides of the spherules. They are easily dislocated because their hardness is larger than that of the crusts. The Ni-Fe inclusion is globular, and there are some parallel grooves on their surfaces. The crusts are olivine. X-ray NiKa plane analysis indicates that nickel elements are highly concentrated in the Ni-Fe inclusions (Fig.2).

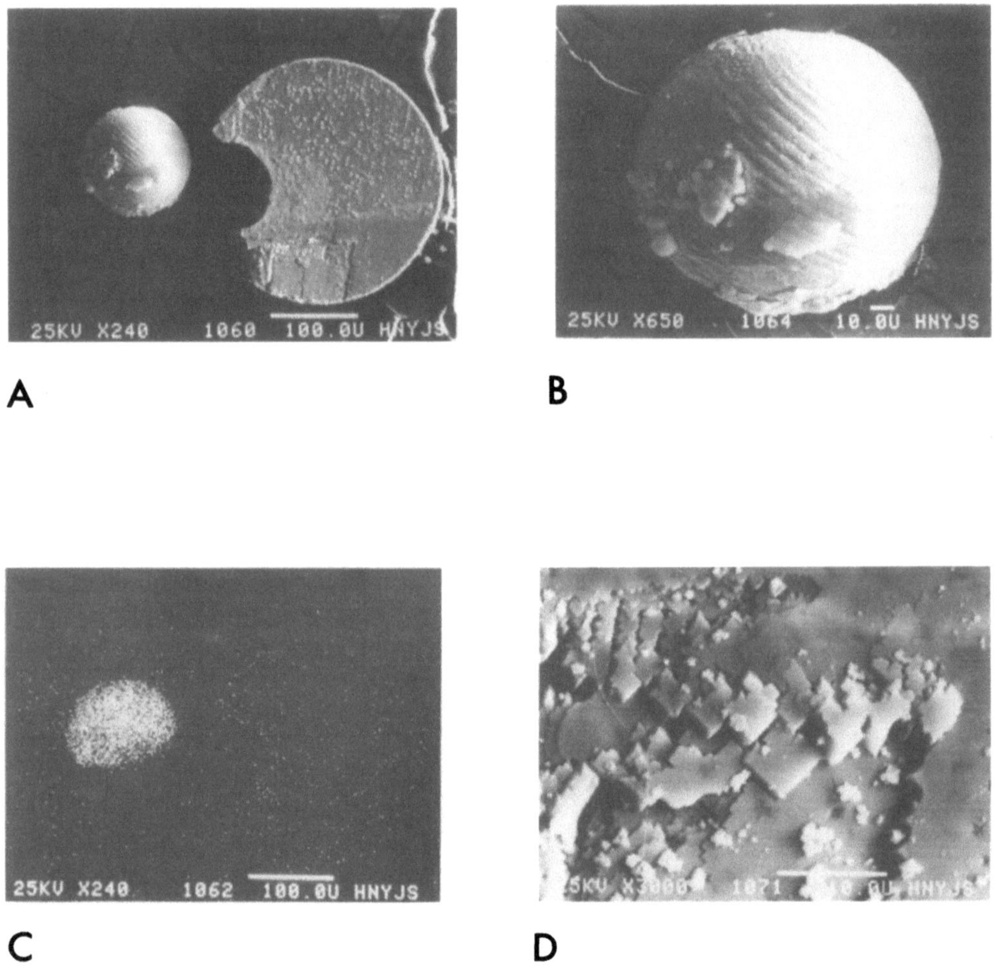

Figure 2. SEM Photographs of the silicate spherules with Ni-Fe inclusion. A and B are the Ni-Fe inclusions falling down from the crust; C is NiKa plane distribution pictures of the Ni-Fe inclusions; D is olivine in the crust.

3.2 Electron probing analysis

The chemical compositions were analysed by using an electron probe. The analysis results are shown in Table I.

TABLE I

Results of electron probing analysis of iron sperules and silicate spherules with golden Ni-Fe cores (or inclusions)

	Oxides								
	MgO	Al_2O_3	SiO_2	CaO	Ti_1O_2	Na_2O	K_2O	MnO	FeO
M_{2-5}									
Crust	tra.	tra.	tra.	0.02	tra.	tra.	tra.	0.51	94.56
Core	tra.	tra.	tra.	0.16	tra.	tra.	tra.	0.14	-
M_{12-21}									
Crust	-	-	-	-	-	-	-	-	-
(I)Core	-	-	-	-	-	-	-	-	-
M_{20}									
Crust	-	-	-	-	-	-	-	-	-
Core	-	-	-	-	-	-	-	-	-
M_{2-39}									
(S)Crust	20.31	tra.	32.45	0.18	tra.	tra.	tra.	tra.	47.39
Inclu.	-	-	-	-	-	-	-	-	-

	Metals						Remarks
	Fe	Co	Ni	Cu	Au	S	
M_{2-5}							
Crust	-	0.21	2.17	tra.	tra.	-	
Core	19.96	1.05	79.10	tra.	tra.	-	Fig. 1 - A.C
M_{12-21}							
Crust	72.69	0.21	0.43	-	-	-	
(I)Core	46.26	1.89	52.29	-	-	-	
M_{20}							
Crust	68.02	0.43	1.01	-	-	-	
Core	70.12	0.55	28.07	-	-	-	
M_{2-39}							
(S)Crust	-	tra.	tra.	-	-	-	
Inclu.	35.76	0.65	33.29	-	-	31.16	

Note: (I) Iron spherurles; (S) Silicate spherules;
"-" Content is Zero; tra. trace.

As seen from Table I, the chemical compositions of the crusts and cores in the iron spherules with golden Ni-Fe cores are completely different. The major chemical component of the crusts is Fe(or FeO), also including a small amount of Ni, Co, etc.; the major chemical component of the core is Ni, minor component being Fe, and a small amount of Co, etc..

The major chemical components of crusts in the silicate spherules with golden Ni-Fe inclusion are FeO, SiO , MgO and others, which are the chemical components of the olivine; the major chemical components of the includion are Fe, Ni and S, and a small amount of Co, which are the chemical components of the Pentlandite.

4. DISCUSSION

The above analysis results indicate the following facts:

1) The shapes, sizes, colors, lusters, surface structures and existing places of the core contained in the iron spherules are first observed. That has an important significance for further exploring and proving the origin and formation mechanism of deep-sea iron spherules.
2) Most of the iron spherules with the button-shaped structure contain golden Ni-Fe core, whose Ni content is very high. Therefore, their characteristics can be used to identify them among the polluting spherules.
3) The discovery of the silicate spherules in the Fe-Si mixing silicate spherules and the golden inclusions, whose features are identical with the Ni-Fe core in the iron spherules, further proves that the deep-sea silicate spherules are extraterrestrial material. It is highly possible that they are products flying out of the surfaces of the iron-stone meteorites. Because of incomplete disjunction of material, they are different both in structure and in composition.

REFERENCES

Murray, J. and Renard, A.F., 1891, Report on the scientific results of the H.M.S. "Challenger" during the years 1873-76, Deep-sea Deposits, 327.

Brownlee, D.E. et al., 1981, Extraterrestrial components, in: The Sea, **7**, p.733.

Blanchard, M.B. et al., 1980, Earth and Planetary Science Letters, **46**, p. 178.

Yamakoshi, K., 1981, Analyses of Large-sized Spherules and Core-Nucleus Spheres from Marine Sediments, 14th Lunar and Planetary Symposium.

Peng, H.C. et al., 1981, A Preliminary Study on Deep-sea cosmic dust, Kexue Tongbao, **26**, p. 906.

$^{26}Al/^{10}Be$ IN DEEP SEA SPHERULES AS EVIDENCE OF COMETARY ORIGIN

G.M. Raisbeck* and F. Yiou*, J. Klein** and
R. Middleton**, D. Brownlee***
* Laboratoire Rene Bernas, C.S.N.S.M.,
BP N°1, 91406 Orsay, FRANCE
** University of Pennsylvania, Philadelphia, PA 19104,U.S.A.
*** University of Washington, Seattle, Wa 98195, U.S.A.

ABSTRACT. Using accelerator mass spectrometry the cosmogenic isotopes ^{26}Al (half-life 716,000 years) and ^{10}Be (1.5 x 10^6 years) have been measured in deep sea magnetic spherules. Because ^{26}Al can be abundantly produced by relatively low energy solar flare particles, while ^{10}Be is mainly produced by higher energy galactic cosmic rays, the ratio $^{26}Al/^{10}Be$ of a body irradiated in space increases as the size of the body decreases. Our measured ratios of $^{26}Al/^{10}Be$ in the spherules are much larger than found in meteorites, and indicate irradiation in interplanetary space of a parent body of $\lesssim 1$ cm diameter. Since most bodies of this size entering the atmosphere are associated with cometary orbits, our results strongly suggest that these spherules represent cometary debris. Our results also suggest an irradiation time in space of the order of 10^6 years for the parent bodies.

1. INTRODUCTION

The extraterrestrial origin of deep sea magnetic spherules has been suspected ever since their discovery, almost a hundred years ago (Murray and Renard, 1891). Modern studies have strongly supported this point of view (Brownlee, 1981), but the exact nature of the parent material has remained uncertain. There is considerable evidence to suggest that the spherules are ablation products of larger bodies which have impacted with the atmosphere (Brownlee, 1981). A minority view is that the spherules exist already in interplanetary space as melted objects from asteroidal collisions (Parkin et al., 1983).

If the spherules are ablation products, the size of the initial objects is of critical importance. Most meteoroids entering the earth's atmosphere weigh $\lesssim 1$ mg, have cometary type orbits, and presumably represent cometary debris. On the other hand, it is generally believed that at least a significant fraction of the larger size objects giving rise to meteorites come from asteroids (see references in Brownlee, 1981).

R. H. Giese and P. Lamy (eds.), Properties and Interactions of Interplanetary Dust, 169–174.

On the basis of ^{53}Mn data, Nishiizumi (1983) and Nishiizumi and Arnold (1982) have argued that spherule parant bodies are meteorite size objects. The development of the technique of accelerator mass spectrometry now makes possible the measurement of other cosmogenic isotopes in these spherules. We earlier reported measurements of ^{10}Be (half-life 1.5 million years) and ^{26}Al (716,000 years) in three groups of chondritic type spherules (Raisbeck et al., 1983). We noted that the $^{26}Al/^{10}Be$ ratios suggested $\lesssim$ 1 cm size objects as the parent bodies for at least some of these spherules. We also mentioned that similar measurements in individual spherules were highly desirable to test this hypothesis, as well as to give additional information on parent size and irradiation times. We report here our first results in a program to measure ^{10}Be and ^{26}Al in individual spherules.

2. EXPERIMENTAL PROCEDURE AND RESULTS

The chondritic spherules considered here were magnetically scavenged from the ocean floor using the "cosmic muck rake", at the same (KK2) or nearby (KK1) location as those reported as sample B-1 in Raisbeck et al. (1983). Their sizes and weights are given in Table I. After weighing, the spherules were dissolved, and an aliquot taken for ^{27}Al measurements by flameless atomic absorption spectrometry. Carrier solutions of ^{9}Be (0.3 or 0.5 mg) and ^{27}Al (0.5 mg) were then added, and the two elements separated on a Dowex 50 x 8 ion exchange column. After further purification, the samples were transformed into BeO and Al_2O_3 for measurement on the accelerator. The procedures for ^{10}Be and ^{26}Al measurements on the University of Pennsylvania tandem accelerator are described in Klein et al. (1982) and Middleton et al. (1983), respectively. Because of the order of magnitude less material used, the uncertainties in the measurements are larger than in our earlier work. The uncertainty on the ^{26}Al measurements is mainly due to the statistics of the number of atoms detected. The ^{10}Be measurements, on the other hand, were limited by interference from a ^{10}B induced background. Because of this interference we were able to obtain only upper limits to ^{10}Be concentrations for two of the spherules.

The results are given in Table I. For convenience we also give the earlier results from Raisbeck et al. (1983) for the group of spherules (B-1) recovered by the same technique, and at the same place, as the spherules studied here.

TABLE I

^{26}Al and ^{10}Be in deep sea magnetic spherules

Sample	Diameter (microns)	Weight (μg)	^{27}Al (%)	^{26}Al (10^9 atoms/g)	^{10}Be (10^9 atoms/g)	$^{26}Al/^{10}Be$	Irradiation age (My)
B-1[a]	430-550	1780	1.12 ±.11	46 ± 7	5.6 ± 0.8	8.2 ± 1.7	0.9
KK1-1	465	140	0.83 ±.08	61 ± 9	10.7 ± 3.6	5.7 ± 2.1	2.1
KK1-2	445	133	0.61 ±.06	58 ± 10	2.7 ± 1.6	21 ± 13	0.4
KK1-3	445	118	0.47 ±.05	25 ± 5	< 1.5	> 17	< 0.2
KK1-4	485	152	0.88 ±.09	31 ± 7	5.0 ± 2.0	6.1 ± 3.4	0.8
KK2-3	620	316	1.08 ±.11	33 ± 6	< 1.0	> 25	< 0.2
Chondritic Meteorites (saturated)			0.87	20-35[b]	20-22[c]	1 - 1.5	
Chondritic (c2) cosmic dust (saturated)				189[d]	14[d]	~ 14	

a From Raisbeck et al. (1983)
b From Heymann and Anders (1967);
c From Moniot et al. (1982) and Sarafin et al. (1984)
d From Reedy (1984)

3. DISCUSSION

The results for the individual spherules in Table I tend to support our earlier interpretation. We include in Table I the saturation values found for ^{10}Be and ^{26}Al in chondritic meteorites, and calculated values for very fine chondritic dust irradiated to saturation in interplanetary space at 1 AU. For irradiation times short compared to the ^{26}Al half-life, the $^{26}Al/^{10}Be$ ratio would be about twice these values. In Fig. 1 we show a curve for the $^{26}Al/^{10}Be$ ratio as a function of depth from the surface of a body of chondritic composition irradiated in space. This curve has been estimated in the following way. The concentrations at zero depth have been taken from the calculations of Reedy (1984) for chondritic dust. It can be noted that the corrected value of Michel et al. (1982) gives a similar result for ^{26}Al. The values at large depths are taken from observed meteorite results. The shape of the curve between these points has been estimated from calculated ^{26}Al depth profiles in lunar soils and rocks (Reedy et al., 1984, and private communication) and chondritic meteorites (Michel et al., 1982) assuming ^{10}Be remains relatively constant. It is realized that there are significant uncertainties in such a procedure. Among other things, the calculations are quite sensitive to the assumed solar flare spectrum, chemistry and size of the parent body, and erosion effects. However, such calculations have given reasonable fits to observed ^{26}Al profiles, particularly in lunar samples (Reedy et al., 1983) and we believe Fig.1 should be sufficiently accurate for the purpose of the discussion that follows.

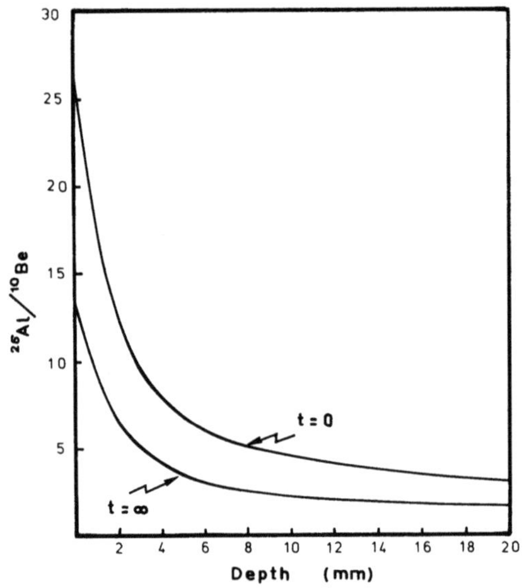

Figure 1. Calculated $^{26}Al/^{10}Be$ as a function of depth (assuming a density of 3 g/cm³) for chondritic matter irradiated in interplanetary space at 1 AU for a time very short (t = 0) or very long (t =∞) compared to ^{26}Al half-life (716,000 years). See text for details.

Although the uncertainties are large, it can be seen from Table I that the $^{26}Al/^{10}Be$ ratios of all the spherules are considerably larger than found in meteorite size objects. In fact, in two of the spherules (KK1-1, KK1-2), the ^{26}Al concentrations themselves are much larger than ever found in chondritic type meteorites. This supports our earlier interpretation that a significant fraction of the ^{26}Al in the spherules has been formed by solar flare particles. From Fig.1 we can see that the $^{26}Al/^{10}Be$ ratios in the spherules are consistent with irradiation in parent bodies having sizes of $\lesssim$ 1 cm. Since most bodies of this size entering the atmosphere are associated with cometary like orbits, it seems reasonable to conclude that the spherules represent cometary debris.

As can be seen in Table I, the concentration of ^{10}Be in meteoritic and dust size chondritic matter is estimated to vary only modestly. It is for this reason that we felt justified in assuming an approximately constant ^{10}Be concentration in calculating $^{26}Al/^{10}Be$ for Fig.1. Using this same assumption, and for irradiation times of the order of the ^{10}Be half-life or shorter, it is possible to estimate the irradiation times (T_i) from the observed ^{10}Be concentrations using the expression:

$$T_i = -\frac{1}{\lambda} \ln (1 - R)$$

where λ = the ^{10}Be decay constant (4.62×10^{-7} years) and R is the ratio of the ^{10}Be concentration to the saturation value (which we take as being $(17 \pm 3) \times 10^9$ atoms/g). The irradiation times calculated in this way are given in Table I. Although these estimates are probably only good to a factor of 2 or so, the implied exposure ages are considerably shorter than found for most meteorites (Reedy et al., 1983).

It must be emphasized that the above irradiation times assume no loss of ^{10}Be from the spherules, either during their entry into the atmosphere, or during the time they spent in contact with sea water. Similarly, the conclusions based on the $^{26}Al/^{10}Be$ ratios assume no preferential loss of either of the isotopes during the same stages. The ^{27}Al concentrations in the spherules, are in reasonably good agreement with concentrations in carbonaceous chondrite meteorites, suggesting that any loss of this element must be quite modest. Unfortunately a similar test for ^9Be is very difficult because of its very low abundance. It should, however, be possible to test for this possibility in the future by studying similar spherules recently recovered from polar ice (M. Maurette, private communication). These spherules are much less severely etched than those found in sea sediments, and the volatile element contents of some of them suggest a much milder heating in the atmosphere.

Acknowledgements: We thank J. Lestringuez for help with sample preparation, and R. Reedy for making calculations relative to this work.

4. REFERENCES

Brownlee, D.E., 1981, in The Sea, **7**, J. Wiley and Sons.
Heymann, D. and Anders, E., 1967, Geochim. Cosmochim. Acta, **31**, 1793.
Klein, J., Middleton, R. and Tang, H.-Q., 1982, Nucl.Inst. and Meth., **193**, 601.
Michel, R., Brinkmann, G. and Stuck, R., 1982, Earth Planet. Sci. Lett., **59**, 33; Erratum, **64**, 174.
Middleton, R., Klein, J., Raisbeck, G.M. and Yiou, F., 1983, Nucl.Inst.and Meth., 218, 430.
Moniot, R.K., et al., 1982, Lunar and Planet. Sci., **13**, 536.
Murray, S. and Renard, A.F., 1891, Rept. Sci. Results Voyage H.M.S. Challenger, **3**, Neill and Co., Edinburgh.
Nishiizumi, K., 1983, Earth Planet. Sci. Lett. **63**, 223.
Nishiizumi, K. and Arnold, J.R., 1982, Lunar Planet. Sci., **13**, 594.
Nogami, K. and Yamakoshi, K., 1984, Geochem. J. (in press).
Parkin, D.W., Sullivan, R.L. and Bull, R.K., 1983, Geophys. J. R.Astron. Soc., **75**, 473.
Raisbeck, G.M., Yiou, F., Klein, J., Middleton, R., Yamakoshi, K. and Brownlee, D.E., 1983, Lunar and Planet. Sci., **14**, 622.
Reedy, R.C., Arnold, J.R. and Lal, D., 1983, Science, **219**, 127.

Reedy, R.C., 1983, Meteoritics, **18**, 383.
Sarafin, R. et al., 1984, Nucl. Inst. and Meth., **B5**, 411.

Walker: To clarify the question of the distinction between ablation - which would remove the outer skin with a high $^{26}Al/^{10}Be$ compared to the interior - and irradiation in space as small objects: what is the max. $^{26}Al/^{10}Be$ seen in surfaces of lunar rocks? In short, what is the dividing line in terms of $^{26}Al/^{10}Be$?
Answer: Although there exist detailed ^{26}Al profiles, I am not aware of any experiments that have measured $^{26}Al/^{10}Be$ in the very surface layers of lunar rocks. Even if they were done, however, a direct comparison with the spherules would not be possible because the lunar material has a higher concentration of ^{27}Al, which is an important progenitor for production of ^{26}Al from solar flares. However, our interpretation of the $^{26}Al/^{10}Be$ results implying irradiation in small bodies is not based on the high value alone. The preablation surface of a chondritic meteorite should indeed have a high $^{26}Al/^{10}Be$ ratio. However most chondritic meteorites are observed to be saturated in ^{10}Be, and therefore presumably their surface layers would be also, contrary to what is observed in the spherules.

Lamy: Your conclusion for parent bodies $\lesssim$ 1 cm may not necassarily imply a cometary origin. The sporadic background, which is only partly connected to comets, is far superior to the contribution by streams associated to comets.
Answer: Our suggestion that small parent objects ($\lesssim$ 1 cm) imply a likely cometary origin is based simply on the fact that most bodies of this size entering the atmosphere (> 95% according D.W. Hughes, private communication at this meeting) are estimated to have comet like orbits, even if only a much smaller fraction are associated with specific streams.

CHEMICAL AND ISOTOPIC COMPOSITIONS OF REFRACTORY ELEMENTS IN DEEP SEA SPHERULES

Kazuo Yamakoshi and Koichi Honma
Institute for Cosmic Ray Research
University of Tokyo
Tanashi Tokyo 188
Japan

ABSTRACT. In order to study the inhomogeneity of the solar materials at the initial state, the enrichment factors and isotopic anomalies of refractory elements in deep sea spherules relative to the average solar abundance were examined using instrumental neutron activation analysis. Fairly high enrichment factors and isotopic anomalies of refractory elements were observed in several spherules, however, after the ablation process of the spherules during the atmospheric heatings they were not inconsistent with those of normal values in the solar system.

1. INTRODUCTION

During the developing time of the solar system, it is verified, the peak temperatures were inhomogeneous throughout the solar system and also the dynamical mixing of the condensed materials were imperfect. Especially the chemical and isotopic compositions of the elements, which have considerably high condensation temperatures, in the most primitive fractions in carbonaceous chondrites and also in the interplanetary dust are thought to be good indicators for the studies on the pre-history of the solar system.

2. ENRICHMENT FACTORS OF SIDEROPHILE ELEMENTS

In Fig.1, we compared the contents of refractory siderophile elements in deep sea spherules and metal grains from carbonaceous chondrites. The data of stony spherules determined by Ganapathy et al. (1978) are also plotted. The enrichment factors are defined as the ratio of the elemental compositions to those of type 1 carbonaceous chondrites.

According to the condensation theory in equilibrium, the condensed materials in the solar nebula have the same chemical compositions with those of the solar abundance.

R. H. Giese and P. Lamy (eds.), Properties and Interactions of Interplanetary Dust, 175–178.

Grossman and Ganapathy (1976) have given the mean enrichment factors as 18.6 for refractory siderophile and rare earth elements in metal grains from Allende meteorite. However, Palme and Wlotzka (1976) and Palme et al. (1982) gave extremely high enrichment factors of refractory elements in metal grains from carbonaceous chondrites. These very high enrichment factors could be hardly explained by the condensation theory in equilibrium, so they proposed a model of the alloy-condensation of the refractory elements in the solar nebula.

However, we have not yet found such high factors in deep sea spherules. These factors are coincided by chance with those given by Grossman and Ganapathy. However, these factors of the spherules can be explained sufficiently by melting and concentration processes of micrometeorites during the atmospheric heatings.

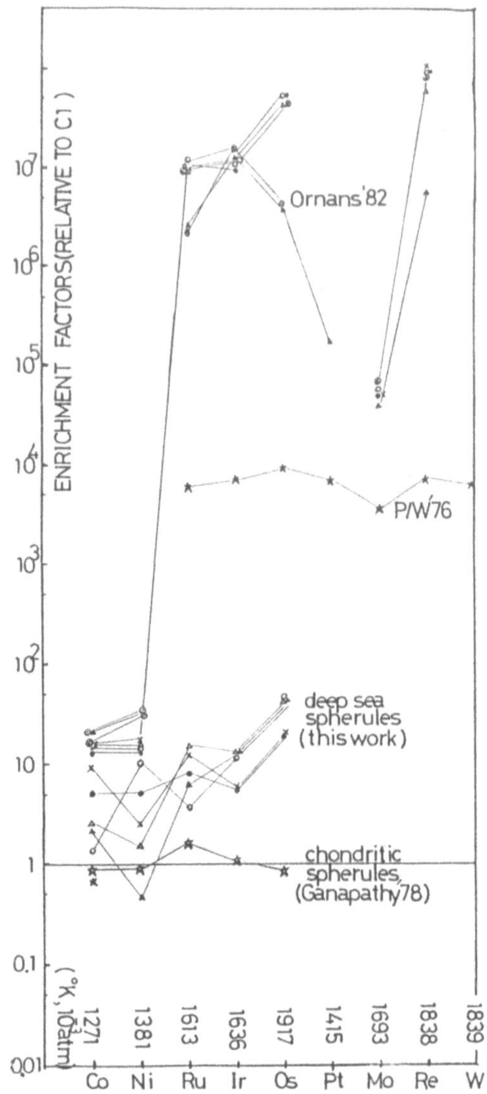

Figure 1. Enrichment factors of refractory elements in deep sea spherules and metal grains from carbonaceous chondrites.

3. (Ru/Os) vs. (Ir/Os) Plot.

In Fig.2, one trial was performed. The correlation between (Ru/Os) and (Ir/Os) is examined to obtain effective clues to understand some kinds of grouping of the data. The data of these metal grains in carbonaceous chondrites determined by Chicago group fall on the trajectory (a). And the end members are the bulk compositions of type 1 carbonaceous chondrites. However, the deep sea spherules fall on another curve (b), whose end members are ordinary iron meteorites, ex. Canyon Diablo. In addition, a strange family of the data from Ornans carbonaceous chondrite gathers on the line (c). This trial is indeed suggestive, however at the present time, no decisive conclusion can be drawn.

4. ISOTOPIC COMPOSITIONS OF OSMIUM AND IRIDIUM

Osmium and Iridium are very suitable elements for studies on the pre-solar and extra-solar materials. In this work, Os-184, Os-190 and Os-192 and also Ir-191 and Ir-193 were measured by instrumental neutron activation analysis.

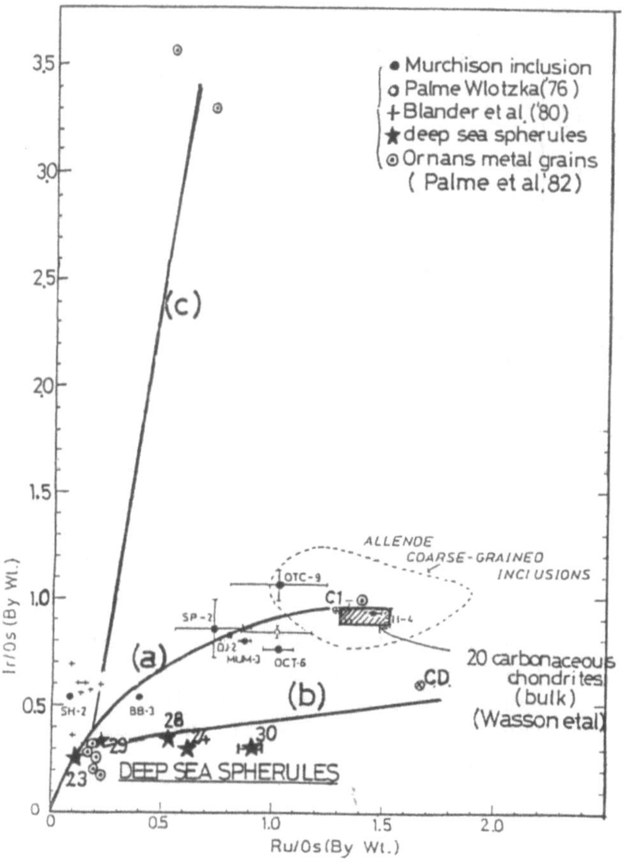

Figure 2. (Ru/Os) vs. (Ir/Os) plots of the deep sea spherules and also metal grains in many carbonaceous chondrites. (Details; see in the text.)

Os-184 is produced through p-process only and Os-192 is induced through r-process only in super-novae explosions. The other nuclides are produced both through r- and s-process.

According to Seeger et al. (1965), the ratio of the nuclides produced through r- and s-process is 7 for Os-190, 14 for Ir-191 and 22.8 for Ir-194, respectively. However, the half-lives of Os-193 (induced from Os-192) and Ir-194 (induced from Ir-193) are so short, that precise measurements of these nuclides could not yet succeeded. In Fig.3, the ratios of Os-185 and Os-191 activities in deep sea metallic spherules and also in Canyon Diablo are shown. And we measured Os reagent samples in many times and we have the reference values shown also in Fig.3.

Almost all spherules are not inconsistent with the reference values, if we take $\pm 2\ \sigma$ as the statistical errors, these data suggest not so strange results.

5. INTERIM CONCLUSION

Many experimental data obtained in metal grains from carbonaceous chondrites suggest the inhomogeneity of the condensed materials in the solar nebula, however, from the deep sea spherules we could obtain not so decisive evidence in chemical and isotopic compositions.

The authors are grateful to Mr. K. Matsumoto, the Metal Mining Agency of Japan, for the deep sea sediments used in this work.

REFERENCES

Chicago group; Kawabe, I.: 1982, private communication.

Ganapathy, R., Brownlee, D. E. and Hodge, P. W.: 1978, Science 201, 1119 - 1121.

Grossman, L. and Ganapathy, R.: 1976, Geochim. Cosmochim. Acta 40, 331 - 344.

Kallemeyn, G. W. and Wasson, J. T.: 1981, Geochim. Cosmochim. Acta 45, 1217 - 1230.

Palme, H. and Wlotzka, F.: 1976, Earth Planet. Sci. Lett. 33, 45 - 60.

Palme, H. and Wlotzka, F., Nagel, K. and El Goresy, A.: 1982, Earth Planet. Sci. Lett. 61, 1 - 12.

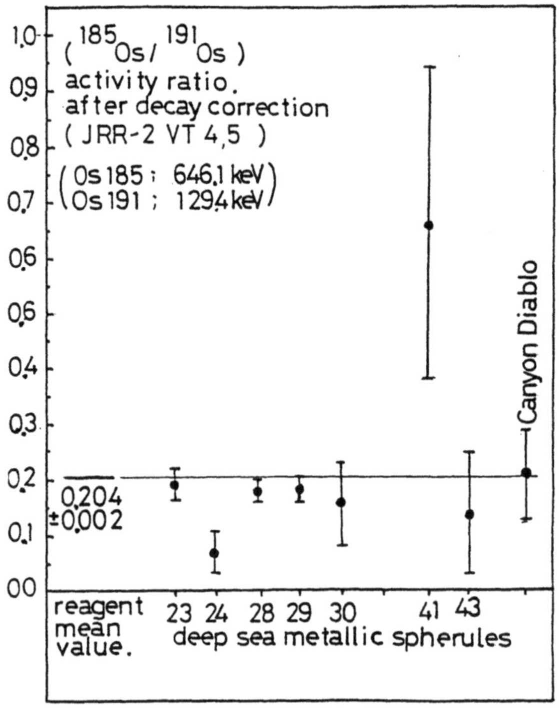

Figure 3. The activity ratio of (Os-185/Os-191) in deep sea spherules and Canyon Diablo. The reference value of the reagent is also shown.

COSMIC RAY EXPOSURE AGE DETERMINATIONS OF COSMIC SPHERULES FROM MARINE SEDIMENTS

Kazuo Yamakoshi
Institute for Cosmic Ray Research
University of Tokyo
Tanashi Tokyo 188
Japan

Abstract. The cosmic ray exposure ages of deep sea metallic spherules were determined by various methods; low level countings (Ni-59), neutron activation analysis (Mn-53), high energy accelerator mass spectrometry (Be-10,Al-26) and mass spectrometry (K isotopes). The exposure ages of 0.3 - 50 Ma were obtained. According to Poynting-Robertson effect, the starting points (supplying sources) are located at inner region of the orbit of Saturn.

1. INTRODUCTION

This is a summary paper of the cosmic ray exposure age determinations of cosmic, metallic spherules from marine sediments. This work is based upon the data obtained preliminary by various methods [1] - [4].

Until now, the identification of the deep sea spherules as extraterrestrial has been studied on the basis of chemical and petrological data. An independent and more definitive characterization would be the presence of cosmic ray induced (cosmogenic) nuclides in the spherules. In addition to the identification, these cosmogenic nuclides carry also some potential information on the irradiation history of the cosmic spherules.

2. THE CALCULATION FORMULA OF THE COSMIC RAY EXPOSURE AGES

The cosmic ray exposure age (T) is calculated as follows;
(a) The amounts of the accumulated, cosmogenic stable nuclides (including extremely long-lived radionuclides; ex.K-40) (S) is devided by the production rate (R), so we have

$$T = S/R.$$

(b) The activity ratio (A) of the two long-lived radionuclides induced from the same target material (ex; iron) by

R. H. Giese and P. Lamy (eds.), Properties and Interactions of Interplanetary Dust, 179–181.

cosmic rays is obtained in the following formula;

$$A = \frac{\lambda_1 N_1}{\lambda_2 N_2} = \frac{g_1[\,1-\exp(-\lambda_1 T)]}{g_2[\,1-\exp(-\lambda_2 T)]} ,$$

where g = the production rate of the nuclide at 1 A.U. from the sun, λ = the mean life of the nuclide. So the common age (T) can be calculated. Here, we have $g = n\sigma f$, where n = the target atom numbers, σ = the production cross section and f = the cosmic ray intensity. In the case of the solar cosmic rays, the flux is dependent on the radial distance from the sun. However, we can cancell out the radial factors from the ratio of (g_1/g_2). Therefore, the cosmic ray exposure age (T) is free from the origin, motion, source position and trajectories of the dust (micro-meteorites) in the solar system. In the case of the high energy Galactic cosmic rays, the cosmic ray intensity is approximately constant except the closest region of the sun. Therefore, the ratio of (g_1/g_2) is also independent of the dust motion and orbits in the interplanetary space.

3. THE OBTAINED RESULTS

From [1] to [3], the data are obtained from the determinations of two cosmogenic radionuclides and the last one is obtained by isotopic anomalies determined with mass spectrometry.

TABLE I

reference	spherules size (μm)	nuclides	method	exposure age (Ma)	starting position
[1]	total sediments	Ni-59 Mn-53	LLD & (n,γ)	0.30	2.6 AU (s=24 um)
[2]	metallic 180 - 330	Be-10 Al-26	AMS	3.9	2.0 AU (s=260 um)
[3]	metallic 180 - 260	Ni-59 Mn-53	LLD & (n,γ)	0.01	~1.0 AU
[4]	metallic	K-39	MASS		
	360	K-40	SPECTRO-	21 + 15	3.4 AU
	340	K-41	METRY	3.4 +2.9	1.7 AU
	300			19 + 8	3.5 AU
	300			52 + 23	5.7 AU

[LLD] = low level radiation countings, [AMS] = accelerator mass spectrometry.

In the first experiment [1] we have measured Ni-59 and Mn-53 in the nickel and manganese fractions extracted chemically from large volume of deep sea sediments. Therefore, we have not seen any meteoritic materials in that sediments. We obtained a "mean" value of the cosmic ray exposure age as 0.3 Ma. In this experiment we have not discussed the size, chemical compositions,colour,density and other properties of the spherules, so to said, we have skipped the recognition of cosmic materials in the sediments.

In the second work, the nuclides were determined by the tandem Van de Graaf, University of Pennsylvania[2].

In the third experiment [3], the cosmic ray exposure age of the metallic spherules in several size groups have been studied in order to obtain the information of the dust motion in the interplanetary space due to Poynting-Robertson effect. However, manganese nuclides induced in iron parent bodies by nuclear reaction is liable to evaporate away from the parent bodies by atmospheric heatings [5]. Therefore, we could obtains apparently such a short age.

In the fourth work, the cosmic ray exposure ages were determined through the excess amounts of K-40 induced by spallogenic reaction with Galactic cosmic rays using a high sensitive mass spectrometer [4].

4. CONCLUSION

According to Poynting-Robertson effect, the cosmic ray exposure age (T) is written with the formula;

$$T = 700 \cdot ad(r^2 - 1),$$

where a = the size of the dust, d = the density of the dust and r = the starting point of the dust (AU),from which the dust is removed and irradiated by the cosmic rays.
Of coarse, the size of the spherules is not the same of the pre-atmospheric, original size of the dust. Therefore, we take the spherule-size as the lower limit, so that the cosmic ray exposure time is obtained as the upper limit.

The data used in this work suggest, that almost all spherules have come from the inner region of the orbit of Saturn.

REFERENCES

[1] Yanagita, S.,Yamakoshi, K. and Imamura, M.:1979, Proc.Intern.Conf.Cosmic Rays (Kyoto) OG-12-20.
[2] Raisbeck, G.M., Yiou, F., Klein, J., Middleton, R. and Yamakoshi, K.:1982, Proc.45th Meteoritical Soc.Meeting.
[3] Yamakoshi, K.,Ohashi, H. and Imamura, M.: 1982, Proc. 5th Intern.Conf.Geochronology (Nikko) pp 401-402.
[4] Kobayashi, K., Ohki, M. and Ohashi, H.: 1982, ibid. pp 186-187.
[5] Nogami, K. and Yamakoshi, K.;1984, Geochem.J. in press.

COSMIC SPHERULES, ASTEROID COLLISIONS AND THE SOLAR CONSTANT

D.W. Parkin
School of Physics
University of Bath
U.K.

ABSTRACT

In 1980 (Phil. Trans. Roy. Soc. 297, 495) and in 1983 (Geophys. Jour. Roy. Astr. Soc. 75, 473) it was proposed that cosmic spherules are formed as splash ejecta in asteroidal collisions. The Poynting-Robertson effect size-sorts the spherules as they spiral to Earth in circular orbits. The time taken in My is given by

$$\Delta t = 3.5 \times 10^{-7} D\delta \quad (a^2 - 1)$$

where a (in AU) is the collision distance from the Sun, D (in µm) is the diameter of the spherules and δ (in kg/m^{-3}) its density. In the 1983 paper, we tested the hypothesis by plotting the size of all the iron spherules (≥ 43 µm diam.) against time for two dated N. Pacific cores. Sloping lines, giving a ~ 2.1 AU, could be discerned; but their existence could be disputed - see figure 6.

In current work, all the spherules from core V21-65 have been cracked open to find δ. Out of 220 spherules, 65 have rusted metal globules and these ($D\delta$) points give a confused plot, because of inaccurate δ-values. However, 66 spherules have their metal globules well preserved. Since globule diameter and %Ni can be accurately measured, δ can be calculated. These ($D\delta$) points show two sloping lines with some clarity, giving a ~ 3.4 AU; also, there are vague lines, giving a-values well within the main asteroid belt.

To establish the hypothesis, wide cores are needed, giving an abundance of undamaged spherules. To obtain accurate a-values, the Robertson formula requires correction because of the spherule's "hummocky" oxide surface. Also, any waviness in the sloping lines could be a measure of a variable solar constant.

A new kind of iron spherule was found because of the cracking; 13 enclose opaque beads of true glass (rich in Fe, Ni and Si, no Mg) instead of metal globules. Crushed bits are deep ruby-red and nonmagnetic.

R. H. Giese and P. Lamy (eds.), Properties and Interactions of Interplanetary Dust, 183.

OBSERVATION OF HIGH $^3He/^4He$ RATIO IN OCEAN SEDIMENTS — EVIDENCE FOR EXTRATERRESTRIAL MATERIAL CONTAMINATION

M. Ozima, M. Takayanagi, S. Zashu and S. Amari
Geophysical Institute
University of Tokyo
Tokyo 113
Japan

Abstract. Occurrence of anomalously high $^3He/^4He$ ratio ($> 5 \times 10^{-5}$) is rather common in the Pacific deep sea sediment. The $^3He/^4He$ ratio correlates well with the ^3He content, whereas the ^3He content shows approximate inverse relation with the sedimentation rate. Input of stratospheric dust of about 1 ppm or less which corresponds to about 2000 ton/y fallout rate can explain the above observation.

1. Introduction

We have measured 39 sediments from 12 different sites, 10 sites from the western to central Pacific and 2 sites from the Atlantic Ocean. We found $^3He/^4He$ ratios higher than 5×10^{-5} for 6 sites, well above the values generally observed in common terrestrial materials, in accordance with previous works (1, 2).

All the sediments we studied are piston-cored samples. While two Atlantic sediments (DSDP samples) are of Cretaceous age (3, 4), all the Pacific samples are of Quaternary or at most of late Tertiary age (5, 6). Petrographic descriptions and other data of the samples are given in literatures (3-6). The detailed experimental procedures for the gas extraction and the mass spectrometric analysis were described elsewhere (7).

2. $^3He/^4He$ ratios in sediments

Although because of the small number of the data, it is difficult to draw a general conclusion on the distribution of $^3He/^4He$ ratio in ocean sediments, there appears no clear correlation between the isotopic ratio and geography. However, it is important to note that high $^3He/^4He$ value ($> 5 \times 10^{-5}$) is rather common in the Pacific sediments. Except for diamonds (8), the highest $^3He/^4He$ ever reported for terrestrial materials is 5.2×10^{-5} for Hawaiian olivine xenoliths (9) and the values for common terrestrial rocks such as granites are much less than 10^{-5} (10). ^3He production in the earth is practically negligible

R. H. Giese and P. Lamy (eds.), Properties and Interactions of Interplanetary Dust, 185–189.

(11). It is also difficult to attribute the high $^3He/^4He$ ratio to some artificial nuclear fallout or debris, since samples were taken at more than two different levels in a core considerably below the sea water bottom and in several cases the deeper level showed the higher $^3He/^4He$ ratio. Since extraterrestrial materials have $^3He/^4He$ ratio generally higher than 10^{-4} with high helium content ($^4He \geq 10^{-5}$ cm^{3}STP/g)(10), it is reasonable to assume contamination of some extraterrestrial materials in the sediments to account for the high $^3He/^4He$ ratio ($> 5 \times 10^{-5}$). This assumption is supported by the well-defined correlation between the ^3He content and $^3He/^4He$ ratio shown in Fig. 1. The correlation can be explained as mixing of some materials with high ^3He content and $^3He/^4He$ ratio with sediment which has generally low He content (12).

3. Cosmic dust in sediments

If the global fallout of cosmic dust is responsible for the high $^3He/^4He$ ratio as well as for the high ^3He content, we expect that there is inverse relation between the concentration of the cosmic dust and the sedimentation rate, since in high sedimenting region, the fallout cosmic dust must be diluted by terrestrial materials. In Figure 2, we plot ^3He content against the sedimentation rate which is taken from literatures (3-6). Two Atlantic samples are quite isolated from Pacific samples. This may be due to helium-loss from the Atlantic samples, since they are older by a factor of a hundred than the Pacific sediments. Ignoring the Atlantic samples, there still appears an approximately inverse relation between the ^3He content and sedimentation rate. If we assume uniform fallout rate of cosmic dust and also the initial ^3He content in sediment is negligible, ^3He concentration in the sediment which can be taken as a measure of the amount of the cosmic dust contamination can be expressed as a function of sedimentation rate (r) and flux of the cosmic dust fallout (f),

$$(^3He)_{sed} = \frac{\alpha f}{\rho r} \qquad (1)$$

where α and ρ denote the concentration of ^3He in the extraterrestrial material and density of the sediment. For the ^3He concentration in the cosmic dust, we take the value measured for stratospheric cosmic dust ($\sim$ 10-μm-diameter); i.e., $\alpha \simeq 4 \times 10^{-5}$ cm^{3}STP/g (see the next paragraph). Putting $\alpha = 4 \times 10^{-5}$ cm^{3}STP/g and $\rho = 2$ g/cm^3 in Eq.(1), we can construct a family of curves corresponding to different values of f. In Figure 2, we show a curve corresponding to $f = 4 \times 10^{-10}$ g/cm^2·y, which approximates the data. The value f ($= 4 \times 10^{-10}$ g/cm^2·y) corresponds to a cosmic dust fallout rate of about 2000 tons/y. The value may be favorably compared with the estimates made on satellite, radio and visual observations of about a few hundred tons per year for the particle size below 10 μm in diameter (13). Cosmic dust which has a larger than 10-μm-diameter may not be important in the helium inventory in sediments, because the smaller ones occupy only a small fraction of the total mass of the cosmic dust, while the larger one is likely to loose helium due to atmospheric impact heating. We believe that the approximate inverse relation but not random relation between the ^3He content and the sedimentation rate is significant and the relation is attributable to the global cosmic dust fallout.

Assuming that the cosmic dust is responsible to the high $^3He/^4He$ ratio in Figure 1 we also show a mixing curve between the cosmic dust and the sediment. For the cosmic dust, we take $^4He = 0.1$ cm^{3}STP/g (14) and assumed solar wind implanted helium, that is, $^3He/^4He = 4 \times 10^{-4}$. For the uncontaminated sediments, we assumed a typical crustal value 10^{-8} for $^3He/^4He$ ratio (11) and $^4He = 3 \times 10^{-7}$ cm^{3}STP/g (12). From Figure 1, it is evident that less than 1 ppm contamination of the cosmic dust can account for the observed high $^3He/^4He$ as well as the high ^3He content.

Microscopic examination did not show any evidence of extraterrestrial materials. Neutron activation analyses showed slightly higher Ni (200 - 600 ppm) and Co contents (100 - 200 ppm), but Cr (30 - 50 ppm) is slightly lower than that observed in common crustal rocks. We could not detect Ir. Here, it should be emphasized that owing to extremely small amount of ^3He in terrestrial samples ($\lesssim 10^{-12}$ cm^{3}STP/g)(10) and to enormous enrichment of helium in cosmic dust ($\sim 10^{-1}$ cm^{3}STP/g)(14), even slight contamination of the latter still results in conspicuous anomaly of ^3He content and in $^3He/^4He$ ratio. Since no other element has such enormous contrast in the concentration between the cosmic dust and terrestrial materials in addition to their marked depletion in the earth, it is not surprising if cosmic dust contamination effect is only seen in ^3He content and in $^3He/^4He$ isotopic ratio. Finally, the ubiquitous occurrence of cosmic dust in deep ocean sediment would find important application in oceanography. If the fallout rate is estimated independently, ^3He content in ocean sediments would yield continuous record of the sedimentation rate, or vice versa.

References

(1) Merrihue, C., Annals N.Y. Acad. Sci, 119, Art. 1, 351-367 (1964).
(2) Krylov, A.Ya., Mamyrin, B.A., Silin, Yu.I. and Khabarin, L.V., Geochemistry International, 10, 202-205 (1973).
(3) Initial Report of the Deep Sea Drilling Project, 43, 19-321 (1979).
(4) Ibid, 43, 323-391 (1979).
(5) Kobayashi, K., Tonouchi, S., Furuta, T. and Watanabe, M., Bull. Ocean Res. Inst. Univ. of Tokyo, No. 13, 1-148 (1980).
(6) Preliminary Report of the Hakuho Maru Cruise KH80-3, p209 (1981).
(7) Ozima, M., Zashu, S. and Nitoh, O., Geochim. Cosmochim. Acta, 47, 2217-2224 (1983).
(8) Ozima, M. and Zashu, S., Science, 219, 1067-1068 (1983).
(9) Kaneoka, I. and Takaoka, N., Earth Planet. Sci. Lett., 39, 382-386 (1978).
(10) Ozima, M. and Podosek, F.A., Noble gas geochemistry, p367, Cambridge Univ. Press (1983).
(11) Tolstikhin, I.N., in Terrestrial Rare Gases, (Ed. Alexander, E.C.Jr and Ozima, M.) 33-62 (Japan Science Soc. Press., 1978).
(12) Posodek, F.A., Honda, M. and Ozima, M., Geochim. Cosmochim. Acta,

44, 1875-1884 (1980).
(13) Hughes, D.W., in Cosmic dust (Ed. J.A.M. McDonnell), 123-185 (Wiley, 1978).
(14) Rajan, R.S., Brownlee, D.E., Tomandl, D., Hodge, R.W., Farrar, H.IV. and Britten, R.A., Nature, 267, 133-135 (1977).

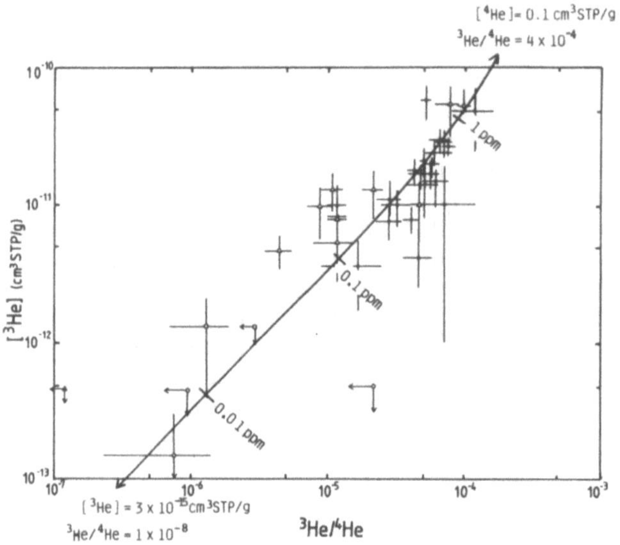

Figure 1. ^3He content in sediments is plotted against $^3He/^4He$ ratio. The curve indicates a mixing curve between He in the cosmic dust (^4He = 0.1 cm^{3}STP/g, $^3He/^4He = 4 \times 10^{-4}$) and in the sediments ($^3He = 3 \times 10^{-15}$ cm^{3}STP/g, $^3He/^4He = 10^{-8}$). Numerical figures shown on the curve indicate the degree of the contamination (in ppm) of the cosmic dust in sediments.

Δ : KH-80-3	o : KH68-4	• : KH75-3-5
□ : KH-75-3-11	◇ : DSDP	

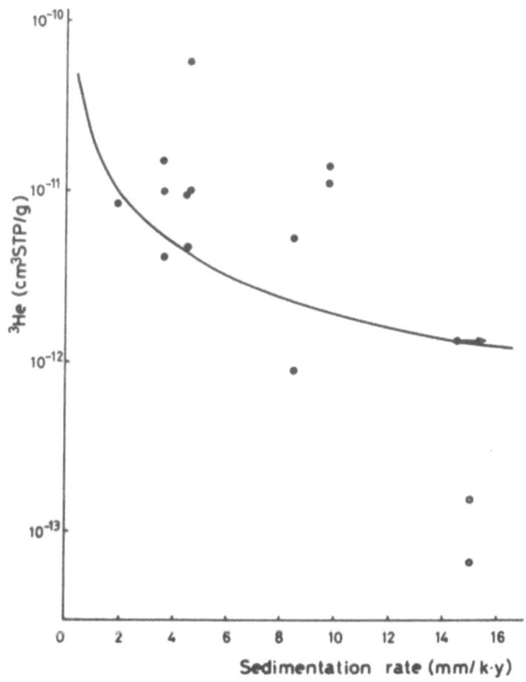

Figure 2. ^3He content is plotted against sedimentation rate. Two hollow circles indicate the Atlantic DSDP samples which are of Cretaceous age. The Pacific samples (solid circles) show approximate inverse relation between ^3He content and sedimentation rate. The curve is constructed by assuming uniform fallout rate (2000 tons/y) of cosmic dust (^4He = 0.1 cm^3STP/g, $^3He/^4He = 4 \times 10^{-4}$).

COMETARY DUST AND THE PERIODICITY OF RAINFALL SINGULARITIES

Z. Kviz
School of Physics
University of New South Wales
P.O. Box 1 Kensington 2033
Australia

ABSTRACT. Periodicity in rainfall singularities in the U.S.A. corresponds to the periods of two comets whose orbits are sufficiently close to Earth's orbit. These effects are connected with the position of a stratospheric jet stream over the territory of the U.S.A. This investigation indicates that some processes, so far not properly understood, could be responsible for the influence of cometary dust on the triggering of atmospheric precipitation.

More than thirty years ago Bowen (1953) unfolded a hypothesis on the influence of meteoric dust on precipitation. Thirty days after the maximum activity of a meteor stream the amount of precipitation seemed to be generally higher in many regions in the world than on other calendar days. According to Bowen's original hypothesis the meteoric dust acts as a rain trigger introducing freezing nuclei into the clouds in the sense of "natural cloud seeding". A clear connection between meteor streams and a singularity in precipitation was found by Bowen (1956) for the meteor stream of γ-Draconids (Giacobinids) on 9th October and the rainfall peak on 9th November. This rainfall peak shows the same periodicity as the orbital period of the meteor stream and the comet Giacobini-Zinner, namely 6.6 years.

In order to check the possible influence of other comets without known meteor streams, the present author selected the comet Grigg-Skjellerup with orbital period 5 years as the most suitable. The last third of the month of May was tested in the 120-year series of Prague rainfall data (1840-1960). The date of close approach between the orbits of the Earth and of the comet is 26th April and thus the rainfall peak should be expected on 26th May if the dust is concentrated in the cometary orbit and if the seeding effect or any other trigger mechanism exists. Such a peak with the periodicity of 5 years was found in the Prague rainfall records by the autocorrelation and power spectra method (Kviz 1964) - see Figure 1. A tentative conclusion was drawn that the invisible and fine cometary dust may actually act as a rain trigger supplying more nucleating material than visible meteors.

In 1969 the author was given an opportunity to study such a

R. H. Giese and P. Lamy (eds.), Properties and Interactions of Interplanetary Dust, 191–196.

periodicity of rainfall peaks for Australian rainfall records during a 3-year Fellowship at the CSIRO, Division of Radiophysics, Sydney, Australia. Only two comets were studied, Giacobini-Zinner and Grigg-Skjellerup. These are the only comets with suitable orbital periods and with a sufficiently close approach to the Earth's orbit. Many localities with longer rainfall records (more than 70 years) were studied by the method of auto-correlation and power spectrum for Australia and a few other countries as well (New Zealand, U.S.S.R., France, Rhodesia).

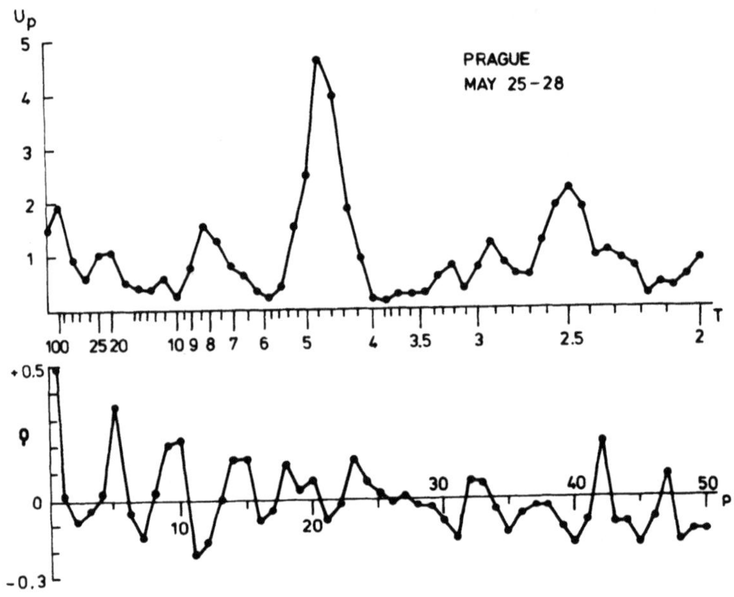

Figure 1. Autocorrelation of Prague rainfall (120 years) occurring between 24th-28th May - lower curve. Upper curve represents the power spectrum and indicates clearly the peak around 5 years.

Many localities revealed the expected periodicity, but not all of them. The geographical grouping of the localities with positive correlation was obvious but no clear clue could be found to explain this geographical grouping.

Unfortunately, localities in Australia with rainfall records long enough for autocorrelation treatment are concentrated along the coast, especially the eastern coast, and a full pattern of geographical dependence of the effect could not be revealed. The only suitable area on the globe, large enough and with evenly distributed meteorological stations and with long uninterrupted series of rainfall records, is the United States of America. However, the autocorrelation method did not reveal any clear clue even for the network of US meteorological stations.

As the autocorrelation and power spectrum method gives only the period and not the phase, another method was tried. The orbital period of a comet was divided into several equal sections and the amount of rainfall which occurred in each individual section of the period was added together. This method was tested for two localities with long series of records - Prague, Czechoslovakia and Sydney, Australia. The series were split into two parts in order to test the persistence of the effect. Graphs in Figure 2 show the result.

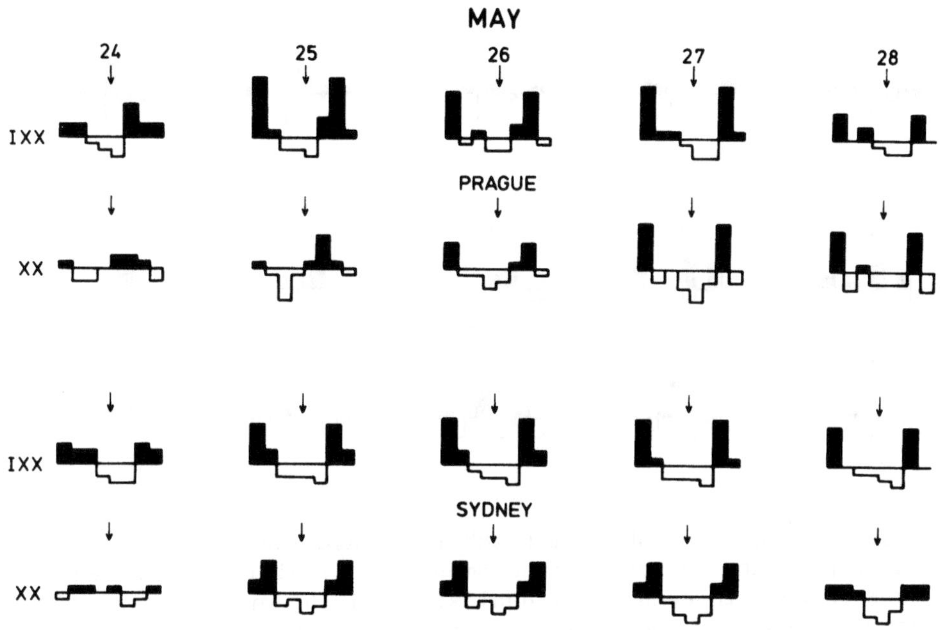

Figure 2a. Test of the "phase method" for Prague and Sydney. Each histogram represents the normalized amount of precipitation between 24th-28th May,in each individual sixth of the orbital period of the respective comet. The closest approach of the comet is indicated by an arrow. Data is separated for 19th and 20th century.

This was considered as an indication that for some localities, and for the season of the year considered, the rainfall occurs when the comet is closer to the Earth and for the others when the comet is further away from the Earth. Using this phase method for some US meteorological stations it has been possible to find some pattern in the geographical grouping of localities with high rainfall when a comet is close to the Earth. It seems that such localities lie under the jet stream - see Figure 3.

These results indicate that there may exist some, so far not understood, processes in the atmosphere, responsible for the influence of cometary dust on the rainfall predominantly under the jet stream. Perhaps by studying this influence on other continental areas of the globe we could gain information on the average behaviour of jet streams in the past.

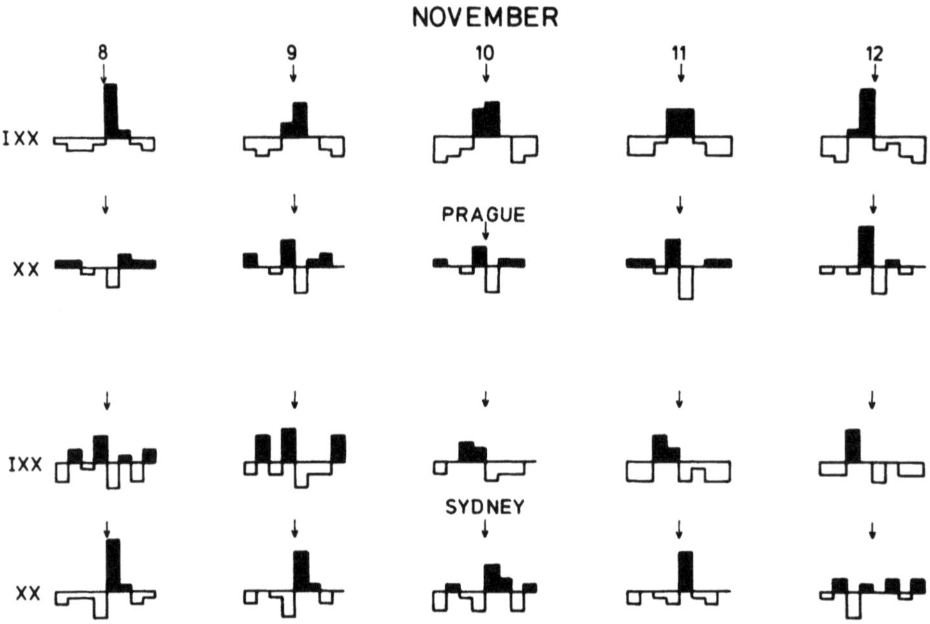

Figure 2b. Test of the "phase method" for Prague and Sydney. Each histogram represents the normalized amount of precipitation between 8th-12th November, in each individual sixth of the orbital period of the respective comet. The closest approach of the comet is indicated by an arrow. Data is separated for 19th and 20th century.

Results presented in this paper are obviously the first step based on the investigation of 45 US meteorological stations and the research should be extended to a larger number of stations and also for other countries, especially these with a dense network of meteorological stations, such as India. Perhaps the countries close to the Himalayas may yield the best results because of the relatively stable position of the subtropical jet stream in that region.

The investigation of comets and other bodies of our planetary system proved that they bear quite a high amount of water ice. This fact has not been known at the time of publication of Bowen's hypothesis. The water itself, passing through low temperature layers of the stratosphere, perhaps may provide the best freezing nuclei to trigger the rainfall. This question should be the topic of future research concerning the meteor-rainfall correlation.

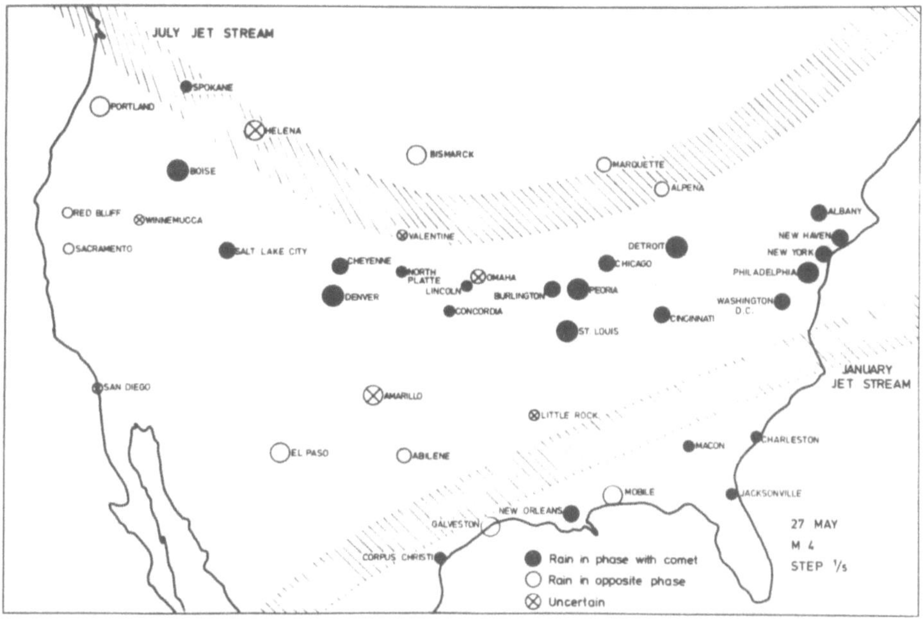

Figure 3a. Distribution of the tested meteorological stations in USA for 24th-28th May with the rain "in phase" with the comet Grigg-Skjellerup, stations with rain "out of phase" and stations with no clear results.

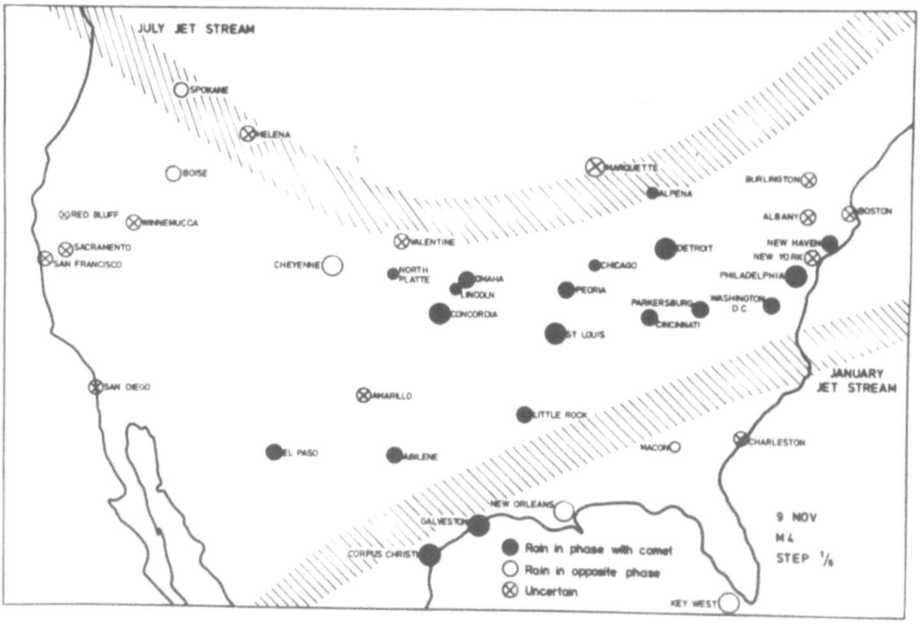

Figure 3b. As for 3a but for 9th-12th November and for the comet Giacobini-Zinner.

References:

Bowen E.G. 1953, Austr. J. Phys. 6, 490.
Bowen E.G. 1956, Tellus 8, 394.
Kviz Z. 1965, Meteorol Zpravy (Meteorol. Bulletin - in Czech) 18, 177.

LABORATORY INVESTIGATIONS AND OPTICAL PROPERTIES OF GRAINS

R.H. Zerull
Ruhr-Universität Bochum
Bereich Extraterrestrische Physik
P.O. Box 10 21 48
4630 Bochum 1, F.R.G.

ABSTRACT. A survey on different techniques of scattering measurements is given. Typical results are presented. Efficiency and limits of theoretical methods available are shortly summarized. Special emphasis is given to the demands on future laboratory experiments needed for the interpretation of polarization and colour effects produced by interplanetary and cometary dust.

1. INTRODUCTION

Light scattering problems can be roughly divided into two groups: In the case of the "D i r e c t S c a t t e r i n g P r o b l e m" the physical properties of the particles are known and their light scattering properties are wanted. A typical example for the application of the findings is the radiation transfer in the earth's atmosphere. But such results are also necessary for the solution of the "I n v e r s e S c a t t e r i n g P r o b l e m". In this case the access to the particles in question is difficult and thus the light scattering signature is a very important if not the only source of information concerning their physical properties. Typical examples for this case are interstellar, interplanetary, and cometary dust and dust in planetary atmospheres. According to the broad field of applications numerous publications deal with the problem to derive scattering functions theoretically or experimentally.

2. FORMALISM

Generally a scattering process may be described proceeding from a linear transformation (see van de Hulst, 1957).

$$\begin{pmatrix} I_{s1} \\ I_{s2} \\ U_s \\ V_s \end{pmatrix} = \frac{\lambda^2}{4\pi^2 r^2} \begin{pmatrix} A_{11} & A_{12} & A_{13} & A_{14} \\ A_{21} & A_{22} & A_{23} & A_{24} \\ A_{31} & A_{32} & A_{33} & A_{34} \\ A_{41} & A_{42} & A_{43} & A_{44} \end{pmatrix} \cdot \begin{pmatrix} I_{o1} \\ I_{o2} \\ U_o \\ V_o \end{pmatrix}$$

R. H. Giese and P. Lamy (eds.), Properties and Interactions of Interplanetary Dust, 197–206.

Index 0 characterizes the Stokes vector of the incident radiation, index s that one of the scattered radiation. λ is the wavelength; r is the distance between scattering particle and observer. The indices 1 and 2 characterize the direction of the electric vector perpendicular and parallel to the scattering plane, respectively. The significance of all elements of the Mueller-matrix has been investigated by Perry et al., 1978 and Bottiger et al., 1980. Their measurements indicate, that for most applications the elements A_{11}, A_{12}, A_{21}, and A_{22} are dominant. They determine I_{s1} and I_{s2} of the Stokes-vector of the scattered radiation if the incident radiation is unpolarized or either polarized perpendicular or parallel to the scattering plane. It should be mentioned, however, that for special applications the evaluation of elliptical polarization can be very powerful. In this case other matrix elements also play a significant role, for instance A_{34} for the investigation of biological cells (Bickel and Stafford, 1980).

The elements $A_{12} \equiv A_{21}$ (they are identical if certain symmetry conditions are satisfied, see van de Hulst, 1957) characterize the cross-polarizing effect of particles; they are zero for single scattering spheres, but can reach the same order of magnitude than A_{11} and A_{22}in the case of large irregular transparent particles.

Proceeding from the matrix elements mentioned above, the total scattering function $i = A_{11} + A_{22} + 2\ A_{12}$ and the degree of linear polarization

$$P = \frac{A_{11} - A_{22}}{i}$$

can be defined. These quantitaties illustrate the scattering properties in the examples given at the end of this contribution.

3. METHODS TO DETERMINE SCATTERING PROPERTIES

3.1 Theoretical Methods and their Limits

Despite continuous improvement of computers the theoretical methods to determine scattering functions are still restricted concerning shape, refractive index, size and/or orientation. Exact solutions, comparable to Mie-theory for spheres, exist for infinite cylinders (Rayleigh, 1881; Lind and Greenberg, 1966) and for spheroids (Asano and Yamamoto, 1975, Schaefer, 1980). Other methods are restricted to certain sizes (particles very small compared to the wavelength: Rayleigh, 1881), special refractive indices (Debye, 1915) or only minor deviations from spherical shape ("Point Matching": Oguchi, 1973; Bates, 1975; Yeh and Mei, 1980. "Perturbation Method" : Yeh, 1964; Erma, 1969). The "T-Matrix Method" (Waterman, 1971; Barber and Yeh, 1975) as well as the theory of Purcell and Pennypacker (1973) reach their limits, if larger particles and different orientations are to be considered, due to the immense demand on computing time. Other methods are not generally applicable because they are based one simplifying assumptions (Gustafson, 1980; Wolff, 1975, 1981; Mukai, 1980; Lumme and Bowell, 1981; Schiffer and Thielheim, 1982, 1984 this volume;

Chiappetta, 1980; Perrin and Lamy, this volume). The same holds for some semiempirical methods (e.g. Chylek et al., 1976; Pollack and Cuzzi, 1980; Giese et al., 1978). Nevertheless, the agreement between observed scattering features and the results obtained by the theoretical methods mentioned above can be surprisingly good. But because of the restrictions also mentioned laboratory measurements are still necessary. An important application of them is to determinate the range of validity of theoretical methods.

3.2 Experimental Methods

Depending on the specific application different experimental methods can be applied in order to obtain scattering results. Fig. 1 gives a schematic overview.

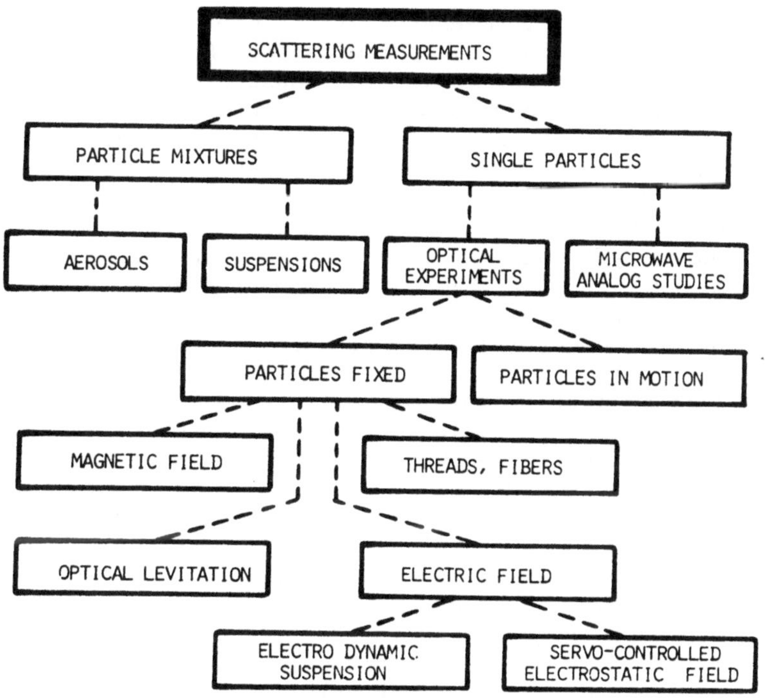

Figure 1. Experimental Methods for Scattering Measurements.

The measurements can be roughly divided into two groups, considering single particles or ensembles of particles, respectively. The latter case is optimized for rapid overall in-situ analysis of particles, where details of single particles are of minor interest. Typical examples are the measurements of Holland and Gagne (1970) and Perry et al. (1978).

The experimental effort becomes by far higher for single particle measurements. This holds in particular for microwave analog studies as initiated by Greenberg et al., 1961 and Giese and Siedentopf, 1962. They take advantage of the fact, that the electromagnetic laws are valid over a wide range of wavelengths, i.e. 4 orders of magnitude between the optical and the microwave region. Analog conditions in both regions are provided for the same ratio of particle size to wavelength and by choice of materials with the same refractive index. Scattering results for mono- and polydisperse mixtures must be composed of data for different orientations and sizes, which can become quite laborious. The advantage of microwave analog studies is, that orientation, size, shape, structure, and refractive index of the particles can be varied systematically and exactly. Microwave Laboratories for applications concerning cosmic dust exist at the "Space Astronomy Laboratory", Gainesville, Florida and at "Bereich extraterrestrische Physik", Ruhr-Universität Bochum. At Gainesville the priority has been given to measurements of dielectric regular nonspherical particles (cylinders, spheroids: see Schuerman et al., 1981), including extinction and radiation pressure efficiency. At Bochum,highly irregular structures of more or less absorbing particles have been a special point of interest (Zerull et al., 1980).

For the most part measuremednts of single particles are carried out directly in the optical region. The measurement and evaluation of the light flashes of moving particles is a typical technique for in-situ experiments; systematic studies, however, should be performed having the particles fixed. In this case the measurement of the physical and chemical properties (microscope, SEM, mass spectrometer etc.) and conclusions concerning their connection with observed scattering features become possible. Fixing of particles can be obtained using threads (for example spider threads: Saunders, 1980). In order to obtain reliable results the application of this principle should be restricted to particles which are sufficiently large compared with the thread diameter.

The other suspension methods work without contacting the particle. Adequate to balance gravity are radiation pressure and magnetic or electric forces. Restrictions concerning the particle material must be observed in the case of magnetic (Beams, 1954) and optical levitation (Ashkin and Dziedzic, 1971, 1980). The latter requires low absorption to avoid evaporation of the particles in the strong laser beam. The advantage of optical levitation is, that radiation pressure automatically centers the particle within the laser beam.

Charged particles can be suspended by electric forces. A servo-controlled "Millikan-field" (Electrostatic suspension: Wyatt and Phillips, 1972; Bottiger et al., 1980) can be used for this purpose as well as a superposition of AC- and DC-fields ("Electrodyamic suspension": Wuerker et al., 1959; Straubel, 1960; Weiss-Wrana, 1983, Weiss-Wrana et al., this volume).

4. RESULTS AND FUTURE NEEDS

The extent of this contribution does not allow to give a complete overview of all laboratory measurements available. Instead of that it may, however, be tried to give a rough idea about the relations beween observed scattering features and physical properties of the particles, illustrated by a few examples.

4.1 Relations between Scattering Features and Particle Properties

Table I illustrates the connection between particle properties (size, material, outer shape and inner structure) and typical features of measured quantities (intensity, linear polarization, colour, cross polarisation).

TABLE I

Connection between Particle Properties and Scattering Features.

QUANTITY MEASURED	TYPICAL FEATURES	CONNECTION WITH PARTICLE PROPERTIES
INTENSITY	ENHANCEMENT OF FORWARD SCATTERING	SIZE
	ENHANCEMENT OF BACKSCATTERING	SHAPE MATERIAL
LINEAR POLARIZATION	LOCATION AND MAGNITUDE OF POSITIVE MAXIMUM LOCATION OF NEUTRAL POINT MAGNITUDE OF NEGATIVE BRANCH	MATERIAL SHAPE STRUCTURE SIZE
COLOUR EFFECTS	CHANGE OF FEATURES ABOVE WITH COLOUR	MATERIAL STRUCTURE SIZE
CROSS POLARIZATION	MAGNITUDE OF CROSS-POLARIZING ELEMENTS OF MUELLER-MATRIX	MATERIAL SHAPE SIZE

Concerning intensity and linear polarization, the information available is quite comprehensive, regarding laboratory measurements and observations, as well. Some typical examples of laboratory measurements are shown in the next section.

All features observed in the scattering pattern can vary with wavelength. The following reasons may be responsible for such colour effects:

- The index of refraction varies with wavelength.
- The particle size is in the range, where the scattering efficiency strongly depends on the ratio of size and wavelength.
- Microstructure of the particles, i.e. single constituents may be in the size range mentioned above, and/or the distance between the constituents, relative to the wavelength, varies.

The reproduction of the correct colour dependence is a powerful tool to check the validity of different models of the zodiacal light cloud and cometary dust, as well. Unfortunately some basic requirements for full use of this criterion are still lacking such as a comprehensive knowledge and understanding of the colour effects produced by different types of particles. Promising experiments to solve this problem, however, are on the way (see Weiss-Wrana et al., this volume). Available observations are also somewhat ambiguous, in particular concerning the colour dependence of polarization. Comparing different results for comets the situation is almost confusing (Michalsky, 1981; Kneissel et al., 1983; Doose et al., 1974; Kiselev et al., 1978; see also Weiss-Wrana et al., this volume). But there are also different statements concerning the colour dependence of the polarization of the Zodiacal Light (Weinberg and Sparrow, 1978; Leinert et al., 1981). Space experiments in the near future will probably help to remove ambiguities and allow, together with appropriate laboratory results, to take full advantage of the significance of colour effects.

Another scattering effect sensitive to particle properties is cross-polarization. It can easily be measured in the laboratory illuminating the particles with a linearly polarized beam. If the incident light is unpolarized, as in the case of the sun, cross polarization is still present but cannot be separately measured. This reduces the value of that criterion at the moment. It is mentioned nevertheless as it may gain importance by chance of future space experiments, for instance during a comet rendezvous mission, where dust particles can be illuminated by artificial light sources.

4.2 Typical Scattering Measurements

This contribution may be concluded by presenting the results of 6 typical scattering measurements in order to illustrate the effects of size, shape, material, and structure of particles, as generally mentioned in the previous section. In each figure results of microwave analog studies (left hand side, Zerull et al. 1977, 1980) are contrasted to laser scattering experiments (right hand side, Weiss, 1981; Weiss-Wrana, 1983). All results are compared to Mie-calculations or Fresnel-reflection curves of equivalent spheres; the base of comparison are the volume (microwave measurements) or the mean geometric cross section (laser results), respectively. Referred to the optical region, the microwave particles simulate particles of several microns, whereas the particles measured with the laser equipment are approximately one order of magnitude larger.

Fig. 2 illustrates effects of increasing deviation from spherical shape. Rough spheres (roughness ~ 1/10 wavelength) do not scatter significantly different from smooth spheres (left diagram). Backscattering enhancement, in particular, is fully preserved. This is obviously not the case for the deformated glass sphere (right diagram).

Fig. 3 illustrates, again for dielectric material, how stronger deviations from spherical shape affect the scattering diagram. Basically the same effects occur as already observed in the case of a deformated glass sphere: Less polarization, increased intensity at medium scattering angles, no enhanced backscattering.

Fig. 4 illustrates the typical scattering properties of highly irregular absorbing particles. Lower polarization with the maximum closer to 90° and intensity continuously increasing toward backscattering are the dominant differences with respect to spheres.

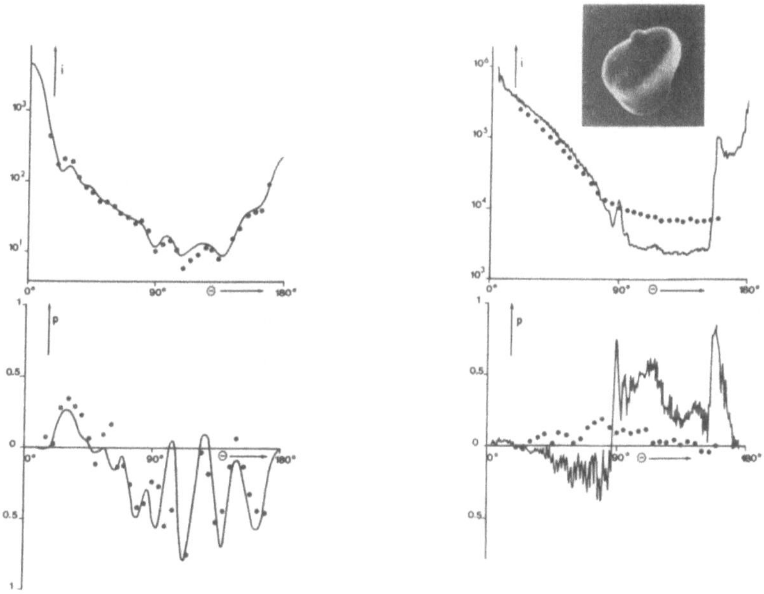

Figure 2. ····· Rough Spheres ——— Smooth Spheres $m = 1.57-0.006i$ $5.3 \leq \alpha \leq 14$

····· Deformed Glass Sphere ——— Ideal Sphere (Mie) $m = 1.51$ $\alpha = 320$

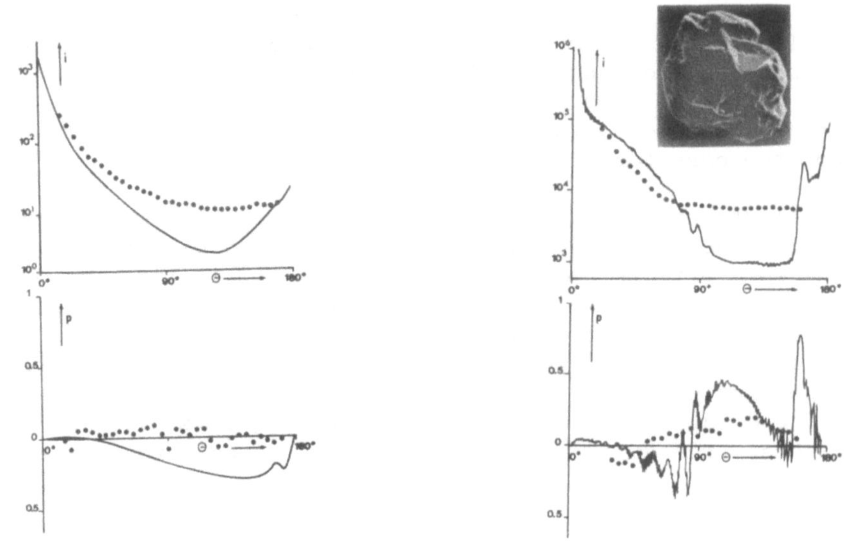

Figure 3. ····· Nonspherical Particles
——— Spheres (Mie)

$m = \begin{cases} 1.5 \; -0.005i \\ 1.57 \; -0.006i \quad 1.9 \leq \alpha \leq 17.8 \\ 1.70 \; -0.015i \end{cases}$

····· Quartz Particle
——— Sphere (Mie)
$m = 1.54 \qquad \alpha = 150$

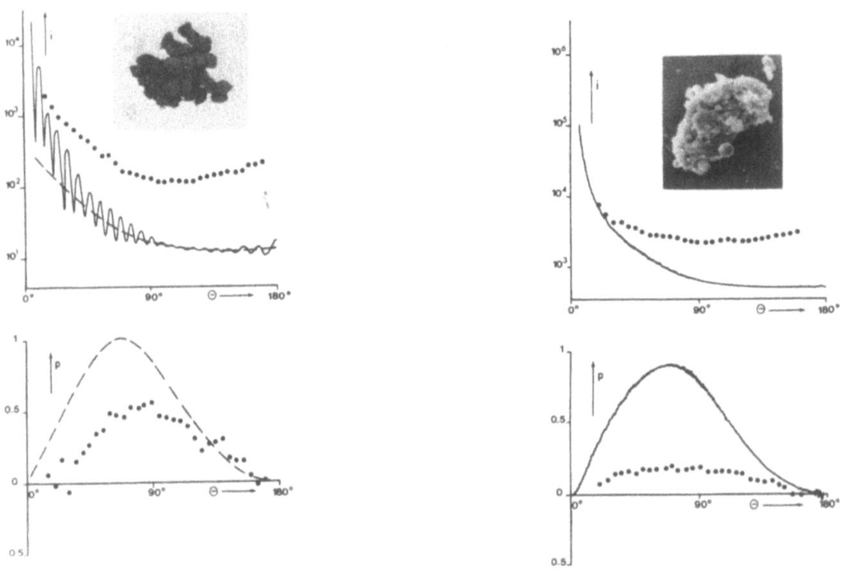

Figure 4. ····· Fluffy Particle
——— Sphere (Mie)
----- Sphere (Fresnel Refl.)

$m = 1.45 - 0.05i \qquad \alpha = 27$

····· Flying Ash Particle
——— Sphere (Mie)
$m = 1.55 - 0.01i \qquad \alpha = 190$

5. REFERENCES

Ansano, S., Yamamoto, G., 1975, Appl. Opt., **14**, p.29.
Ashkin, A., Dziedzic, J.M., 1971, Appl. Phys.Lett., **19**, p.283.
Ashkin, A., Dziedzic, J.M., 1980, in: Light Scattering by Irregularly Shaped Particles (D. Schuerman, ed.), Plenum Press, New York, p.233.
Bates, R.H.T., 1975, IEEE Trans. MTT, **23**, p.605.
Barber, P., Yeh, C., 1975, Appl. Opt., **14**, p. 2864.
Beams, J.W., 1954, Science, **120**, p. 619.
Bickel, W.S., Stafford, M.E., 1980, in: Light Scattering by Irregularly Shaped Particles (D.Schuerman, ed.), Plenum Press, New York, p. 299.
Bottiger, J.R., Fry, E.S., Thompson, R.C., 1980, in: Light Scattering by Irregularly Shaped Particles (D.Schuerman, ed.), Plenum Press, New York, p.283.
Chiappetta, P., 1980, Astron. Astrophys., **83**, p. 348.
Chylek, P., Grams, G.W., Pinnick, R.G., 1976, Science, **193**, p. 480.
Debye, P., 1915, Ann. Physik, **46**, p. 809.
Doose, L.R., Coffeen, D.L., 1974, in: Planets, Stars and Nebulae Studied with Photopolarimetry (T. Gehrels, ed.), Univ. of Arizona Press, p.818.
Erma, V.A., 1969, Phys. Rev., **179**, p.1238.
Giese, R.H., Siedentopf, H., 1962, Zeitschr. f. Naturforschung, **17a**, p. 817.
Giese, R.H., Weiß, K., Zerull, R.H., Ono, T., 1978, Astron. Astrophys., **65**, p.265.
Greenberg, J.M., Pedersen, N.E., Pedersen, J.C., 1961, J. Appl. Phys., **32**, p.223.
Gustafson, B.A., 1980, Reports from the Observatory Lund, **17** (Ph.D-Thesis).
Holland, A.C., Gagne, G., 1970, Appl. Opt., **9**, p. 1113.
Hulst, H.C., van de, 1957, Light Scattering by Small Particles, John Wiley, New York.
Kiselev, N.N., Chernova, G.P., 1978, Sov. Astron., **22**, p. 607.
Kneissel, B., Schwehm, G.H., Leinert, C., 1983, in: Proceedings of the International Conference on Cometary Exploration Vol.II (T.I. Gombosi, ed.), Budapest, Hungary, p.177.
Leinert, C., Richter, I., Pitz, E., and Planck, B., 1981, Astron. Astrophys., **103**, p. 177.
Lind, A.C., Greenberg, J.M., 1966, J. Appl. Phys., **37**, p.3195.
Lumme, K., Bowell, E., 1981, Astron. J., **86**, p.1694.
Michalsky, J.J., 1981, Icarus, **47**, p.388.
Mukai, S., Mukai, T., Giese, R.H., 1980, in: Light Scattering by Irregularly Shaped Particles (D.Schuerman, ed.), Plenum Press, New York, p.219.
Oguchi, T., 1973, Radio Sci., **8**, p. 31.
Perry, R.J., Hunt, A.J., Huffman, D.R., 1978, Appl. Opt., **17**, p. 2700.
Pollack, J.B., Cuzzi, J.N., 1980, in: Light Scattering by Irregularly Shaped Particles (D. Schuerman (ed.), Plenum Press, New York, p.113.

Purcell, E.M., Pennypacker, C.R., 1973, Astrophys. J., **186**, p.705.
Rayleigh, Lord, 1881, Phil. Mag., **12**, p.81.
Saunders, M.J., 1980, in: Light Scattering by Irregularly Shaped Particles (D. Schuerman ed.), Plenum Press, New York, p.237.
Schaefer, R.W., 1980, Ph.D. Thesis, State Univ. New York at Albany.
Schiffer, R., Thielheim, K.O., 1982, J. Appl. Phys., **53**, p.2825.
Schuerman, D.W., Wang, R.T., Gustafson, B.A.S., Schaefer, R.W., 1981, Appl. Opt., **20**, p. 4039.
Straubel, H., 1960, Acta Physica Austriaca, **8**, p. 265.
Thompson, R.C., Bottiger, J.R., Fry, S., 1980, Appl. Opt., **19**, p. 1323.
Watermann, P.C., 1971, Phys. Rev., **D3**, p. 825.
Weinberg, J.L. and Sparrow, J.H., 1978, in: Cosmic Dust (ed. J.A.M. Mc Donnell), Wiley-Interscience, p.75
Weiss, K., 1981, Laserstreuexperimente an Einzelteilchen zur Interpretation des optischen Verhaltens interplanetaren Staubes, Diss. Ruhr-Universität Bochum, Forschungsbericht BMFT-FB-W 81-047.
Weiss-Wrana, K., 1983, Astron. Astrophys., **126**, p.240.
Wolff, M., 1975, Appl. Opt., **14**, p.1395.
Wolff, M., 1981, Appl. Opt., **20**, p.2493.
Wuerker, R.F., Shelton, H., Langmuir, J., 1959, J. Appl. Phys., **30**, p.342.
Wyatt, P.J., Phillips, D.T., 1972, J. of Colloid and Interface Sci., **39**, p.125.
Yeh, C., 1964, Phys. Rev., **135**, A p. 1193.
Yeh, C., Mei, K.K., 1980, in: Light Scattering by Irregularly Shaped Particles (D. Schuerman ed.), Plenum Press, New York, p.201.
Zerull, R., 1973, Mikrowellenanalogieexperimente zur Lichtstreuung an Staubpartikeln, Diss. Ruhr-Universität Bochum, Forschungsbericht BMFT-FB-W 73-18.
Zerull, R.H., Giese, R.H., 1974, in: Planets Stars and Nebulae Studied with Photopolarimetry (T.Gehrels ed.), University of Arizona Press, Arizona, p. 901.
Zerull, R.H., Giese, R.H., Weiß, K., 1977, Appl. Opt., **16**, p. 777.
Zerull, R.H., Giese, R.H., Schwill, S., Weiß, K., 1980, in: Light Scattering by Irregularly Shaped Particles, (D. Schuerman ed.), Plenum Press, New York, p.273.

6. DISCUSSION

Koutchmy: I wonder if there exist not any attempts to compute theoretically and also to measure the V-Stokes parameter (circular polarization) and if not, why?

Answer: The elements of the Mueller-matrix relevant for the determination of the parameter V of the Stokes vector of the scattered radiation have been measured by Bottiger et al., 1980, Thompson et al., 1980, Perry et al., 1978, and Bickel and Stafford, 1980.

REMOTE SENSING OF THE ZODIACAL CLOUD ALONG SECANTS TO EARTH'S ORBIT

René Dumont
Observatoire de Bordeaux
B.P. 21, F-33270 Floirac, France

Anny-Chantal Levasseur-Regourd
Service d'Aéronomie du C.N.R.S.
B.P. 3, F-91370 Verrières-Le-Buisson, France

ABSTRACT. Partial but significant localization along the line of sight (l.o.s.) of the contributions to the integrated zodiacal brightness can be achieved, without recourse to rash physical assumptions, if we focus on the two intersections between the l.o.s. and the terrestrial orbit. Then, there exists on the l.o.s. two other points or "nodes", the contributions of which can be retrieved with outstandingly low uncertainty. The photometric results at these nodes (heliocentric change of scattering efficiency, backscattering phase function etc.) are highly reminiscent of previously ascertained features, and they encourage extensions to the polarimetric, the spectrometric and the radiometric cases. The salient results are, respectively : a positive radial gradient of local polarization degree ; orbital velocities in excess over the keplerian ones, at least outside the earth's orbit ; the derivation of local temperatures from IRAS data.

1. METHOD OF THE SECANT. RETRIEVAL OF THE SCATTERING COEFFICIENT

In zodiacal light studies, the mere knowledge of optical quantities integrated along the l.o.s. is dramatically insufficient, and efforts have to be made to localize the information.

Several previous attempts have been published (Dumont, 1973 ; Dumont and Sánchez, 1975 ; Leinert *et al.*, 1976 ; Schuerman, 1979 ; Mujica *et al.*, 1980 ; etc.), the first two of which introducing a rigorous inversion of the brightness integral if the l.o.s. lies in the ecliptic plane. The properties derived from this inversion (mean volume scattering function and polarization curve ; radial gradient of density $\nu \simeq 1.2$) have received such further support that it could hardly have been a wrong approach. However, these works, like many others, had the disadvantage of an excessive model-dependence, especially because of the assumption of a homogeneous zodiacal cloud, which has recently become more questionable. Pioneer 10 photometric data (Schuerman, 1980) are in conflict with it. The radial change of (integrated) polarization degree observed, in constant helioecliptic directions, by the two Helios

R. H. Giese and P. Lamy (eds.), Properties and Interactions of Interplanetary Dust, 207–213.

probes, also invalidates the homogeneity (Leinert *et al.*, 1981).

Our new approach (Dumont and Pelletanne, 1981 ; Pelletanne, 1982 ; Dumont, 1983 ; Dumont and Levasseur-Regourd, 1984) avoids major physical assumptions. It is valid in the ecliptic plane, since the l.o.s. must be secants to the terrestrial orbit (Fig. 1). Calling Z the zodiacal

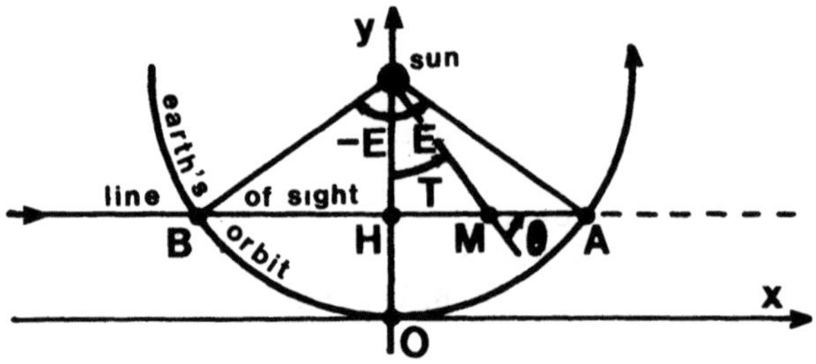

Fig. 1. Method of the secant : Geometry.

brightness observed, and assuming things to remain unchanged leftward of B while the earth moved from B to A, the subtraction Z(A) - Z(B) provides the brightness contribution by the chord BA. If B tends towards A, or the secant towards the tangent, an infinitesimal section of the l.o.s. is isolated and its contribution is derived. Per unit-length, this contribution was known (Dumont, 1973) to be equal to the derivative $-(dZ/d\varepsilon)_{\varepsilon=90°}$ where ε is the elongation.

For a slight departure between the secant and the tangent, a crude approximation to the contribution by the mid-point H is the linear interpolation between B and A. Such an approximation becomes rapidly improper as y increases, and also if we attempt to retrieve the contributions of sections anterior to B. To the observed values of the integral at B and at A, can be added a third constraint, which is merely the fact that no contribution can be given by a section at the infinity. Similarly, were the l.o.s. complete not truncated at A, the contribution would also be zero by a remote section at ∞ on its rightward extension.

In order to remove these infinite values of the abscissa, let us change the variable x into the angle T (= "scattering pointer"). The values of T at the true observing locations are ± E (= "elongation pointer"). Then, the expressions of the brightness integral for a virtual observer at M are, equivalently :

$$\text{(linear frame)} \qquad Z = S \int_{-\infty}^{x} r^{-2} \mathcal{D} \, dx$$

$$\text{(angular frame)} \qquad Z = \frac{S}{m} \int_{-\pi/2}^{T} \mathcal{D} \, dT$$

where S is the solar intensity, $r = \overline{\odot M}$, $m = \overline{\odot H}$. $\mathcal{D}$ is a proportionality coefficient (L^{-1}) representing the efficiency of the elemental volume at M to scatter the sunlight in the x-direction. $\mathcal{D}$ is the "Directional scattering coefficient" or the "Directional scattering cross-section of the unit-volume". One may recognize in $\mathcal{D}$ the classical n.σ product of the number density times the phase function (Giese, 1968), but the separation would require extra assumptions.

Then, any curve $\mathscr{D}(T)$ joining the point (-90°, 0) to the point (+90°, 0) and fulfilling the observed integrals at -E and at +E will be a theoretical solution, provided that it never takes negative values - since any section does contribute or does not to the integrals, but it can subtract nothing in such an optically thin medium.

If, in addition, we assume these curves to be smooth (otherwise, how could the changes of brightness be smooth as the observing direction or location varies ?) the point is that they have to constrict in two kinds of foci which we call the "nodes" (Fig. 2). A mathematical treatment with various 5-parameters forms, i.e. one degree of freedom over and above the 4 constraints, shows these nodes to be remarkably insensitive, both in T and in $\mathscr{D}$, to the mathematical model. Trigonometric series have been chosen, but if they are changed into polynomials, hyperbolic series or others, the curves and especially the nodes are practically invariant, shifting by less than 1° in abscissa and less than 1 % in ordinate.

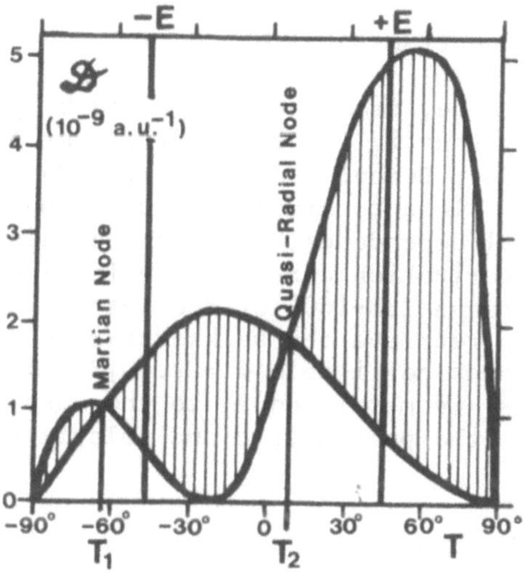

Fig. 2. Variation vs. the angle T (i.e. along the l.o.s.) of the Directional Scattering Coefficient $\mathscr{D}$ (a.u.$^{-1}$) with its uncertainty. Occurrence of the two "Nodes". (Example with E = 45°.)

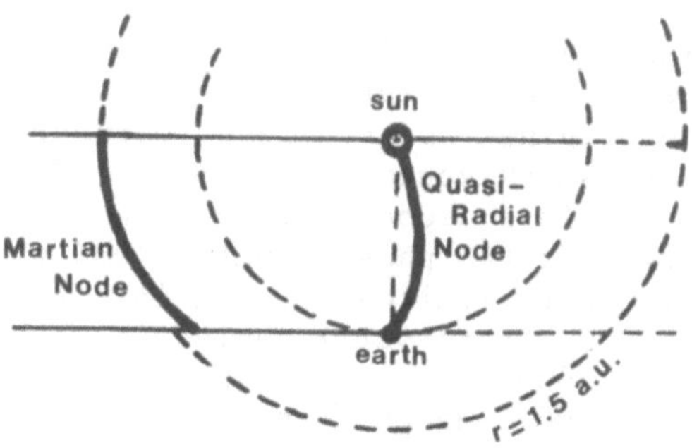

Fig. 3. Loci of the two nodes in the ecliptic plane as the secant (i.e. E or m) varies. The node T_1 is quasi-concentric at 1.5 a.u. ; the node T_2 is quasi-radial.

The abscissae T_1, T_2 of the nodes therefore correspond to regions of the l.o.s. where the uncertainty on the local contribution, or on the scattering coefficient, is considerably smaller than elsewhere. As E or m varies, the loci of the nodes in the ecliptic plane appear on Fig. 3. It happens that the node T_1 is at practically a constant heliocentric distance 1.5 a.u. and we called it the "Martian Node". The node T_2 weakly departs from the mid-point of the chord, and it can be named the "Quasi-Radial Node".

2. PHOTOMETRIC AND POLARIMETRIC RESULTS

The local value of the Directional scattering coefficient $\mathcal{D}$ at each node can be expressed as a function of the brightnesses Z(E), Z(-E) observed at the two intersections l.o.s.-orbit :

$$\mathcal{D}_i = \frac{1}{S} [\mu_i \; Z(E) + \nu_i \; Z(-E)]$$

The coefficients μ_i and ν_i are dependent of the node (i = 1 or 2), of the secant (i.e. of E or m), but almost not dependent of the mathematical model.

Using this formula separately for the perpendicular and for the parallel components of Z, gives access to the local polarization degree $\mathcal{P}$ at each node. The observed brightness and polarization for 30° < ε < 150° which is the classical elongation range of zodiacal studies (the gegenschein will be treated on its own) will allow to retrieve the local values of $\mathcal{D}$ and of $\mathcal{P}$ along the two nodes, i.e. from 0.5 to 1.0 and at 1.5 a.u. heliocentric distance.

2.1. Martian Node : The backscattering functions at 1.5 a.u.

Figure 4 shows $\mathcal{D}$ and $\mathcal{P}$ at the Martian node, from the integrated brighnesses Z and polarizations P compiled by Fechtig *et al.* (1981). Most of the original observations (Weinberg, 1964 ; Dumont and Sánchez, 1975 ; Levasseur-Regourd and Dumont, 1980) lead to similar results. Due to the fair deconvolution from any radial dependence, these curves also represent the phase function and the polarization curve at the level of Mars, both limited in principle to the range 138-160° of the scattering angle θ. It is still easier, however, to retrieve the backscattering range 160-180°, because $\mu_1 Z(E)$ then becomes negligible w.r.t. $\nu_1 Z(-E)$: $\mathcal{D}_1$ becomes proportional to Z(-E) and $\mathcal{P}_1$ is merely equal to the observed polarization, P(-E). In other words, the photometric and polarimetric

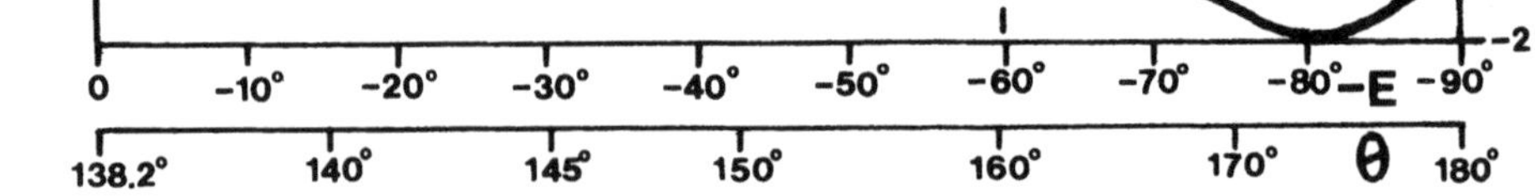

Fig. 4. The Directional Scattering Coefficient $\mathcal{D}$ (a.u.$^{-1}$) and the local Polarization Degree $\mathcal{P}$ at 1.5 a.u. retrieved for all secants (0 ⩽ m ⩽ 1 a.u.). In the right part, one can asymptotically recognize the "profiles" of the gegenschein at 1 a.u.

profiles of the gegenschein, as seen from 1 a.u., directly display the backscatter parts of the phase function and of the polarization curve at 1.5 a.u. - taking notice, however, of the ratio asymptotically equal to 3/2 between the scales for the elongations and for the scattering angles.

2.2. Quasi-Radial Node : The heliocentric change of $\mathcal{D}$ and of $\mathcal{P}$

Figure 5 shows $\mathcal{D}$ and $\mathcal{P}$ at the second node, from the same observational sources. Since this node remains not far from the radius $\overline{\odot 0}$, these curves are fair approaches towards the radial dependences of $\mathcal{D}$ and of $\mathcal{P}$,

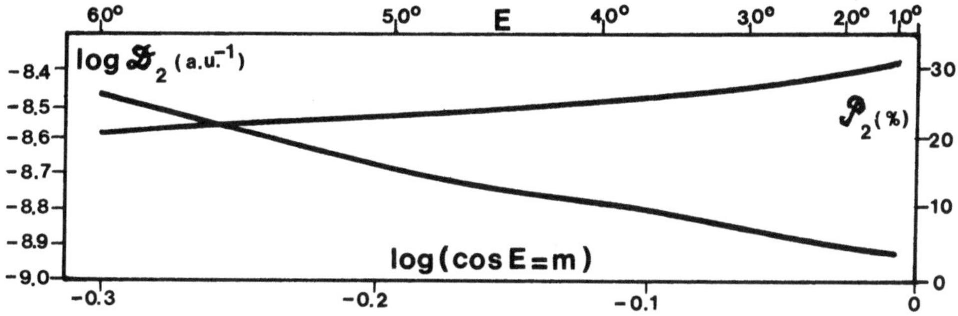

Fig. 5. The Directional Scattering Coefficient $\mathcal{D}$ (a.u.$^{-1}$) and the local Polarization Degree $\mathcal{P}$ (for $\theta \simeq 90°$) retrieved for $0.5 \leqslant$ m(or r)$\leqslant$ 1 a.u. Were the cloud homogeneous, the slope of $\mathcal{D}$ in log scales would be that of the "space density", $-\nu$. We find $\mathcal{P}$ radial-dependent (with the sign observed by Helios).

both at right scattering angle, therefore with a fair deconvolution from the angular effect. In logarithmic scales, the curve $\mathcal{D}_2$ has a mean slope $-\nu = \log\mathcal{D}_2/\log m$ equal to -1.44, slightly different but reminiscent of the values of ν representing the "space density" slope as long as the homogeneity was assumed.

The curve $\mathcal{P}_2$ shows that the local polarization degree drops from ~ 0.3 to ~ 0.2 between 1 and 0.5 a.u. This also is highly reminiscent of a previous result, which is the heliocentric decrease of (l.o.s.-averaged) polarization seen from the Helios probes as they went towards the sun. This radial gradient of local polarization implies the dust complex to be heliocentric dependent, invalidating once more the converse assumption.

3. KINEMATICAL RESULTS : LOCALIZED VELOCITIES FROM DOPPLERSHIFTS

The method can be extended to Doppler-Fizeau Spectrometry, leading to localized information about the kinematics of the dust. The integration of the elemental brightness is replaced by the integration of the product (elemental brightness • x-component of the local mean velocity vector).

The Quasi-Radial Node has the advantage that the x-component is the tangential component. It can be retrieved between 0.5 and 1 a.u. Left

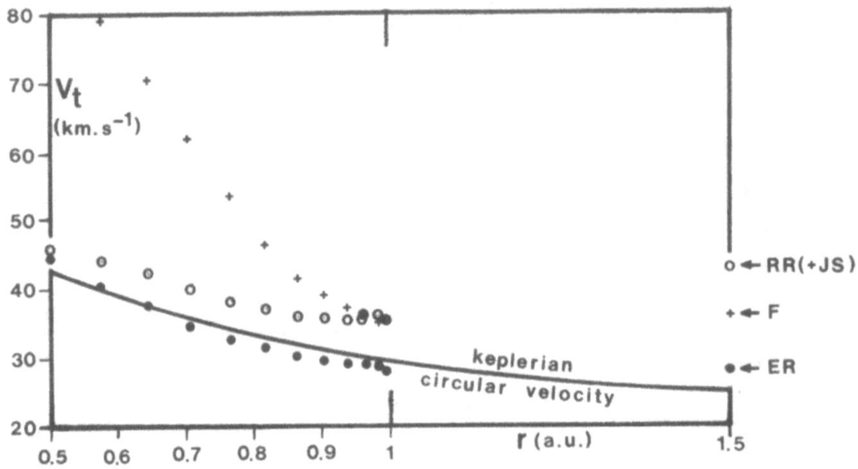

Fig. 6. The tangential component of the local volume-orbital velocity retrieved from Dopplershifts (r = 0.5 to 1 and 1.5 a.u.).

half of Fig. 6 shows the results from, and the discrepancies between, the observational sources for the Dopplershifts (Reay and Ring, 1968, extended towards small elongations by James and Smeethe, 1970, (o) ; Fried, 1978, (+) ; East and Reay, 1984 and this volume, (•)). For the latter observations, provisionally restricted to the more reliable morning light + the gegenschein, the tangential component differs little from the keplerian circular velocity.

At the Martian Node (on right edge of Fig. 6) the first two sources both lead to important excess (× 1.7 and × 1.5) over the keplerian velocity 24.3 km.s^{-1}. East and Reay's new data are tentatively interpreted by a tangential component 28 km.s^{-1}, i.e. hardly superior to the keplerian, plus a radial escape component of the order of 2 km.s^{-1}, which is reported by the authors to be seen, free of obliquity, as a "receding gegenschein", and which seems able to partly account for their evening-morning asymmetry.

4. RADIOMETRIC RESULTS : LOCAL TEMPERATURES FROM INTEGRATED EMISSION MEASUREMENTS BY IRAS

Finally we have attempted to extend the "method of the secant" to infrared radiometry. From the integrated emission-brightnesses $I_\lambda(B)$, $I_\lambda(A)$ at a wavelength λ where the scattered light is negligible w.r.t. the thermal emission, can be retrieved the local emissivities $\mathcal{E}_\lambda$ at each node. If two wavelengths are observed, the temperature of the dust is accessible, assuming that it radiates like a black or grey body : The ratio of the local emissivities is a function of the temperature only.

The recent observations by IRAS of the diffuse infrared background are described by Hauser *et al.*, 1984, with a preliminary separation into Zodiacal Emission and Galactic components. The wavelengths 12 and 25 μm are favourable, as giving the highest signals, with lowest galactic contamination. Also rather favourable are some of the directions observed, which can provide us with 3 secants at 5, 10 and 15° elongation-pointer. For the 3 secants and the 2 nodes the provisional temperatures are :

E	Martian Node (∿ 1.5 a.u.)	Quasi-Radial Node (∿ 1 a.u.)
5°	217 K	258 K
10°	215 K	262 K
15°	212 K	269 K

Taking account of the exact positions of the nodes and of the observational uncertainties, we are led to 260 ± 25 K at 1 a.u. and to 212 ± 15 K at 1.5 a.u. These localized and weakly model-dependent temperatures can be compared to those given by IRAS team, which are averaged on the l.o.s., i.e. admittedly too hot for remote sections and too cold for nearby ones (due to the non-uniqueness of their solution, even this average appears to be uncertain by almost 100 K). In addition, a gain of precision on the temperatures means that information becomes available about the albedo of the dust, which we tentatively infer to be smaller at Mars than at the Earth (Levasseur-Regourd and Dumont, 1984).

REFERENCES

Dumont, R. 1973, *Planet. Space Sci.* **21**, 2149.
Dumont, R. 1983, *Planet. Space Sci.* **31**, 1381.
Dumont, R., Levasseur-Regourd, A.C. 1984, *Planet. Space Sci.*, in press.
Dumont, R., Pelletanne, B. 1981, *C.R. Acad. Sci. Paris* **293 II**, 377.
Dumont, R., Sánchez, F. 1975, *Astron. Astrophys.* **38**, 405.
East, I.R., Reay, N.K. 1984, *Astron. Astrophys.*, in press.
Fechtig, H. *et al.* 1981, *Landolt-Börnstein* Neue Serie **VI 2 a**, 228.
Fried, J.W. 1978, *Astron. Astrophys.* **68**, 259.
Giese, R.H. 1968, *ESRO Scientific Note* **41**.
Hauser, M.G. *et al.* 1984, *Astrophys. J.* **278** L-15.
James, J.F., Smeethe, M.J. 1970, *Nature* **227**, 589.
Leinert, C. *et al.* 1976, *Astron. Astrophys.* **47**, 221.
Leinert, C. *et al.* 1981, *Astron. Astrophys.* **103**, 177.
Levasseur-Regourd, A.C., Dumont, R. 1980, *Astron. Astrophys.* **84**, 277.
Levasseur-Regourd, A.C., Dumont, R. 1984, *C.R. Acad. Sci. P.*, in press.
Mujica, A. *et al.* 1980, *Planet. Space Sci.* **28**, 657.
Pelletanne, B. 1982, Thesis, Univ. Bordeaux.
Reay, N.K., Ring, J. 1968, *Nature* **219**, 710.
Schuerman, D.W. 1979, *Planet. Space Sci.* **27**, 551.
Schuerman, D.W. 1980, *I.A.U. Symp.* **90**, 71.
Weinberg, J.L. 1964, *Ann. Astrophys.* **27**, 718.

A METHOD FOR DERIVING THE MEAN VOLUME SCATTERING PHASE FUNCTION FOR ZODIACAL DUST

S. S. Hong
Space Astronomy Laboratory
University of Florida
1810 NW 6th Street
Gainesville, Florida 32609
USA

ABSTRACT. A linear combination of 3 Henyey-Greenstein phase functions is substituted for the mean volume scattering phase function in the zodiacal light brightness integral. Results of the integral are then compared with the observed brightness to form residuals. Minimization of the residuals provides us with the best combination of Henyey-Greenstein functions for the scattering phase function of zodiacal dust particles.

1. INTRODUCTION

Most theoretical methods for deriving an empirical scattering function for interplanetary dust particles require taking derivatives of the observed zodiacal light brightness. Observational errors are easily amplified in the process of differentiation and likely to distort the scattering function derived therefrom. As an alternative, we take an integral approach: A utilization of the versatile Henyey-Greenstein (HG) function as a parameterized trial scattering function will enable us to employ the method of non-linear least squares in determining the mean volume scattering phase function for zodiacal dust particles.

2. BRIGHTNESS INTEGRAL WITH THE HENYEY-GREENSTEIN FUNCTION

A linear combination of three HG functions

$$\phi\ (\Theta;\ g_k, w_k\ ) = \sum_{k=1}^{3} \frac{w_k}{4\pi} \frac{(1 - g_k^2)}{(1 + g_k^2 - 2g_k \cos\ \Theta\ )^{3/2}} \tag{1}$$

having different asymmetry factors, g_k, adequately describes such general features as the forward peak, isotropic middle, and backward enhancement in a particle scattering pattern. Here, w_k represents relative weights of the three components, and Θ denotes scattering angle. With the usual power-law relation, $n(r) = n_o(r_o/r)^{\nu}$, for the heliocentric density distribution of particles, the zodiacal light

R. H. Giese and P. Lamy (eds.), Properties and Interactions of Interplanetary Dust, 215–218.

brightness $Z(\varepsilon)$ in the ecliptic is given by

$$Z(\varepsilon) = \frac{\zeta\,\overline{\sigma}}{\sin^{\nu+1}\varepsilon} \int_{\varepsilon}^{\pi} \Phi(\Theta) \sin^{\nu}\Theta \, d\Theta \quad , \tag{2}$$

where $\zeta = F_o n_o r_o$ with F_o being the solar flux at $r = r_o$; $\Phi(\Theta)$ is the mean volume scattering phase function of interplanetary particles and $\overline{\sigma}\Phi(\Theta)$ becomes the differential scattering cross-section. For an arbitrary ν, one may rely on numerical means to evaluate the brightness integral. However, for $\nu=1$, the integral with HG scattering phase function has an analytical solution. Additionally, one can generalize the result of the special case $\nu=1$ to cases other than unity.

For $\nu=1$, substitution of eqn (1) into eqn (2) for $\Phi(\Theta)$ gives

$$Z(\varepsilon) = \frac{\zeta_1\overline{\sigma}_1}{4\pi} \sum_{k=1}^{3} w_k \, H(\varepsilon;\, g_k) \quad , \tag{3}$$

where

$$H(\varepsilon;\, g_k) = \frac{1}{\sin^2\varepsilon} \, \frac{1 - g_k}{g_k} \left[\frac{1 + g_k}{(1 + g_k^2 - 2g_k\cos\varepsilon)^{\frac{1}{2}}} - 1 \right] \tag{4}$$

for non-zero asymmetry factors and $H(\varepsilon;0) = 1/(1 - \cos\varepsilon)$ for zero asymmetry factor; $H(\pi;g_k) = \frac{1}{2}(1 - g_k)/(1 + g_k)^2$ at the anti-solar point. Notice that subscript "1" is attached to ζ and $\overline{\sigma}$ to remind us that $\nu=1$.

Normalizing both theoretical $Z(\varepsilon)$ and observed $Z_{obs}(\varepsilon)$ brightness to their corresponding values at elongation 90° [$z(\varepsilon) \equiv Z(\varepsilon)/Z(\pi/2)$ and $z_{obs}(\varepsilon) \equiv Z_{obs}(\varepsilon)/Z_{obs}(\pi/2)$], we form the reduced chi-square as

$$\chi_r^2 = \frac{1}{imx - p} \sum_{i=1}^{imx} \eta^2(\varepsilon_i) \left[z(\varepsilon_i) - z_{obs}(\varepsilon_i) \right]^2 \quad , \tag{5}$$

where imx is the number of observational values used in the analysis and p is the number of parameters used in the parametric representation of the unknown scattering function. To make the importance of each observed brightness nearly equal, we introduce the statistical weighting function $\eta^2(\varepsilon_i) = 1/z_{obs}(\varepsilon_i)$. Minimization of the reduced chi-square with respect to the HG parameters determines the best parameters for the mean volume scattering phase function $\Phi_1(\Theta)$ of the zodiacal dust, the subscript "1" again serving as a reminder of $\nu = 1$.

Since the equation of direct differential inversion

$$\Phi(\Theta = \varepsilon) = \frac{1}{\zeta\,\overline{\sigma}} \left[-(\nu+1)\cos\varepsilon \, Z(\varepsilon) - \sin\varepsilon \frac{\partial}{\partial\varepsilon} Z(\varepsilon) \right] \tag{6}$$

holds for an arbitrary power-law exponent (Dumont, 1976), our special case result gives

$$\Phi_1(\Theta = \varepsilon) = \frac{1}{\zeta_1\overline{\sigma}_1} \left[-2\cos\varepsilon \, Z(\varepsilon) - \sin\varepsilon \frac{\partial}{\partial\varepsilon} Z(\varepsilon) \right] . \tag{7}$$

Therefore we can eliminate the differential term from eqn (6) to obtain for arbitrary ν :

$$\Phi(\Theta = \varepsilon) = \frac{\zeta_1\bar{\sigma}_1}{\bar{\sigma}\zeta}\left[\Phi_1(\varepsilon) - (\nu - 1)\cos\varepsilon\,\frac{Z(\varepsilon)}{\zeta_1\bar{\sigma}_1}\right] \quad . \qquad (8)$$

Once ν is fixed, all the quantities inside the square brackets become known; hence the functional form of $\Phi(\Theta)$ is determined. Since generalization by eqn (8) does not involve any differentiation, the influence of observational errors on the derived scattering function becomes minimal.

3. THE MEAN VOLUME SCATTERING PHASE FUNCTION

Non-linear least squares analysis given to combined ground and space observations (Dumont and Sánchez, 1975; Leinert, Link, Pitz and Giese, 1976) provides us with the solution in Table I. Although w_3 is as small as 0.005, the third HG component plays an increasingly important role towards large scattering angles. For example, 28% of $\Phi_1(\pi)$ comes from this component, while only 7% comes from the more heavily weighted first component. Observed brightening of the zodiacal light towards the anti-solar point requires this strongly backward throwing (g_3 = -0.813) HG function in the mean volume scattering function of the zodiacal dust. The resulting scattering phase function is depicted in Fig 1 by a dashed line. In the same figure, the zodiacal light brightness (solid line) resulting from eqn (3) with the parameter values from Table I is compared with the observations (dots) at 23(=imx) selected elongations.

Table I. Henyey-Greenstein Parameters for $\Phi_1(\Theta)$

Asymmetry factors g_k	0.698	-0.203	-0.813
Relative weights w_k	0.665	0.330	0.005
$\zeta_1\bar{\sigma}_1$ in $S_{10}\,(V)_\odot \cdot sr$		4.61×10^3	

The uniqueness of the solution given in Table I can be assessed by examining whether or not the equal chi-square contours have an isolated minimum in the HG parameter planes. We first kept one asymmetry factor fixed, then at every possible grid point of the remaining two asymmetry factors, we found the local-minimum reduced chi-square by varying the relative weights. As an example, we displayed in Fig 2 such contours on the (g_1 ,g_2) plane with g_3 and w_3 kept at their best values. To each contour level chi-square values are given in units of 10^{-4} ; a few percent random error in the observation corresponds to roughly 20 in this unit. The contours in Fig 2 clearly bear out a well-defined minimum, and the same was found from the contours on both (g_2 , g_3) and (g_3 ,g_1) planes.

$\Phi_1(\Theta)$ was generalized by eqn (8), and the results are displayed in Fig 3. At low scattering angles the derived scattering function is sensitive to the power-law exponent; for $\nu > 1$, it reverses its rise towards the forward direction. As observations below elongation 15° were excluded from the analysis, the reality of such reversal and its relation to the density very close to sun needs additional investigation.

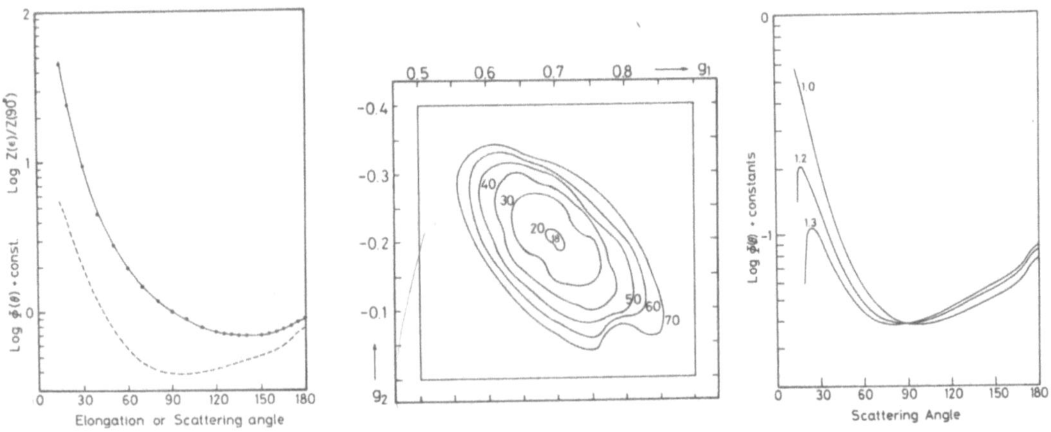

Fig 1: Comparison of observational (dots) and computed (solid line) brightness of zodiacal light using the dashed line HG phase function.
Fig 2: Projection of minimum chi-square surface on the (g_1, g_2) plane.
Fig 3: Generalization of $\Phi_1(\Theta)$ to cases ν slightly larger than unity.

4. CONCLUDING REMARKS

Versatility of the HG function in portraying widely varying scattering patterns and its rendering of an analytical solution to the brightness integral enables us to establish the uniqueness of the derived solution. The overall scattering characteristics found from our integral approach are in general agreement with previous investigations. What is unique about this study is that characteristics of the mean volume scattering phase function for the zodiacal dust are quantified by specifying numerical values for the HG parameters. Finally, it should be pointed out that an extension of this method to polarization amounts to only changes in notation, because the brightness integral holds for both components of the polarized brightness.

ACKNOWLEDGEMENT: This work was supported by the U.S. Air Force Office of Scientific Research.

REFERENCES

Dumont, R., Sánchez, F.: 1975, Astron. Astrophys. 38, 405.
Dumont, R.: 1976, in Interplanetary Dust and Zodiacal Light, Lecture Notes in Physics 48, ed. by H. Elsässer and H. Fechtig, p. 115.
Leinert, C., Link, H., Pitz, E., Giese, R. H.: 1976, Astron, Astrophys. 47, 221.

MICROWAVE AND LASER FACILITIES TO DETERMINE SCATTERING AND COLOUR SIGNATURES RELATED TO THE PHYSICAL PROPERTIES OF DUST PARTICLES

K. Weiss-Wrana, R. H. Giese, R. H. Zerull
Ruhr-Universität Bochum
Bereich Extraterrestrische Physik
P.O. Box 10 21 48
4630 Bochum 1, F.R.G.

ABSTRACT. At Bochum there exist three experimental setups for measurements of the scattering properties of dust particles. The microwave analog experiment is the best choice for systematic studies of particles in the size range between 0.5 and 10 times the wavelength (correspondig to about 0.25 to 5 µm in the optical region). The multicolour laser experiment providing electrodynamic suspension of single dust particles is optimized for studies of particles in the size range between 20 and 150 µm. Another laser experiment is designed for rapid analysis of particle mixtures in view of in-situ applications. The measuring principles of all instruments are shortly summarized.

1. MICROWAVE ANALOG EXPERIMENT

The Bochum microwave analog experiment operates at a frequency of 35 GHz, corresponding to a wavelength of 8.57 mm. The range of scattering angles is between 15° and 170°. The size range of scattering particles is between 0.5 and 10 wavelengths. The lower limit is due to the sensitivity of the receiver and error signals of the suspension mechanism, whereas the upper limit is determined by observing the far field condition. These limits correspond in the optical range ($\lambda = 0.5$ µm) to particle sizes between 0.25 and 5 µm. For further details of the equipment see Zerull, 1976. For investigations of larger particles our laser facilities have to be used.

2. MULTICOLOUR LIGHT SCATTERING EXPERIMENT

With the multicolour light scattering experiments (Fig. 1) the scattering properties of single particles of the size between 20 µm and 150 µm can be measured.

We extended our former light scattering experiment (Weiss, 1981; Weiss-Wrana, 1983) to the investigation of colour effects. A Crypton

R. H. Giese and P. Lamy (eds.), Properties and Interactions of Interplanetary Dust, 219–221.

Ion laser emits blue (λ = 476 nm), green (λ = 531 nm), yellow (λ = 568 nm) and red (λ = 676 nm) light. The direction of the linearly polarized light can be turned 90° by applying the half wave voltage to a pockels cell. The charged solid particle is suspended by using the electrodynamic balance which is a superposition of a Millikan-field acting against the gravitational force and a nonuniform alternating field which has a focusing effect (Weiss-Wrana 1983). The charged particle is kept within a small region in space over a long period of time.

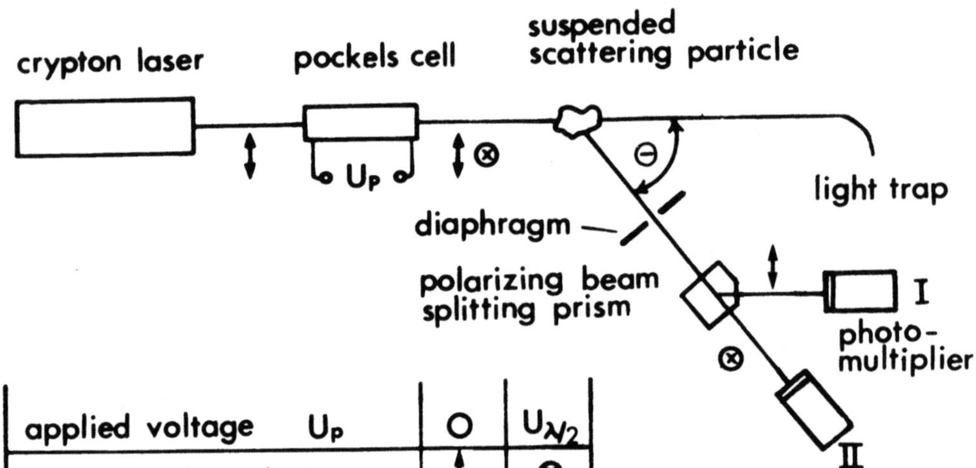

applied voltage U_P	0	$U_{\lambda/2}$
incident light, polarization	↕	⊗
scattered light PM I	I_{11}	I_{21}
PM II	I_{12}	I_{22}

Figure 1. Schematic diagram of the multicolour light scattering experiment.

The optical device for detecting the scattered intensity is mounted on an arm which may be rotated through an angle from 15° to 165° by a stepping motor. The scattered intensity is analyzed in the following way: A polarizing beam splitter prism (Foster prism) splits the two orthogonaly polarized components of the scattered light. Photomultiplier I and II detect simultaneously the intensity with the direction of polarization like the illuminating light and the intensity with the direction of polarization rotated by 90°, the so called "cross polarization". The polarization of the illuminating light changes periodially from perpendicularly to parallel polarized with respect to the scattering plane.

For the investigation of colour effects the measurement of the intensity of the four matrix elements can be repeated for four wavelengths. The absorption coefficient of the material mainly depends on the wavelength. Furthermore in the case that surface elements or building stones of agglomerates are in the same order as the wavelength effects of diffraction are expected, which result in colour dependence of both, intensity and especially of linear polarization. This will be proved by experiments in the near future.

3. PHOTOPOLARIMETER FOR RAPID PARTICLE ANALYSIS

For rapid analysis of particle mixtures a special photopolarimeter was built up, restricted to measurements at only 3 scattering angles, as illustrated in Fig. 2.

One sensor at 5° measures the near forward scattered intensity, which is important for size determination. The sensors at 90° measure intensity, linear polarization, and cross polarization in order to discriminate different types of particles. For the same reason a third sensor at 175° is provided which measures the degree of enhancement of backscattering. For details see Zerull, 1982.

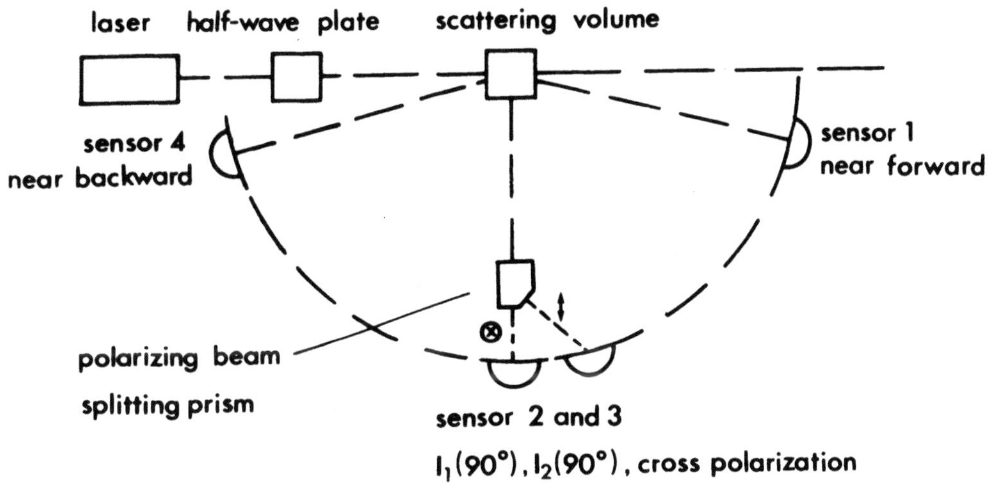

Figure 2. Schematic diagram of the particle analyzing photopolarimeter.

4. REFERENCES

Weiss, K., 1981. BMFT-FB-W81.

Weiss-Wrana, K., 1983, Astron. Astrophys., **126**, p.240.

Zerull, R.H., 1976, Beiträge zur Physik der Atmosphäre, **49**, p.168.

Zerull, R.H., 1982, in: Atmospheric Aerosols, their Formation, Optical Properties and Effects (ed. A. Deepak), Spectrum Press, Hampton, Virgina, p.167.

EXPERIMENTAL RESULTS ON SCATTERING BY IRREGULAR AND METEORITIC DUST PARTICLES RELATED TO PHOTOPOLARIMETRY OF ZODIACAL LIGHT AND COMETS

K. Weiss-Wrana, R. H. Giese, R. H. Zerull
Ruhr-Universität Bochum
Bereich Extraterrestrische Physik
P.O. Box 10 21 48
4630 Bochum 1, F.R.G.

ABSTRACT. The investigations of light scattering by larger meteoritic and terrestrial single grains (size range 20 μm to 120 μm) demonstrate that the scattering properties of irregularly shaped dark opaque particles with very rough surfaces resemble the characteristic features of the empirical scattering function as derived from measurements of the zodiacal light. Purely transparent or translucent irregularly shaped particles show a quite different scattering behaviour. Furthermore irregular and multicomponent fluffy particles in the size range of a few microns were modelled by microwave analog measurements in order to explain positive and negative polarization of the light scattered by cometary dust grains.

1. INTERPLANETARY DUST

Some properties of the interplanetary dust particles are comparatively well known such as the size distribution of larger (≳ 5 μm) grains near 1 AU, the spatial distribution in the ecliptic, and the brightness of the zodiacal light as a function of elongation. The main contribution of the zodiacal light is produced by particles between 10 and 100 μm in size. Many of the interplanetary particles collected by aircraft in the stratosphere are non-spherical porous opaque aggregates. Each reasonable zodiacal light model has to take into account these properties.

We could investigate the scattering pattern of real cosmic materials, samples of the meteorites Allende and Murchison. Meteorite Murchison is very similar to the extraterrestrial carbonaceous chondritic materials (Brownlee, 1976). Meteorite Allende shows white inclusions imbedded in a fine-grained dark matrix.

Fig. 1a illustrates the scattering function i and the degree of linear polarization P for a dark grummy particle of the matrix of Meteorite Allende (28 μm in size). Contrary to the behaviour of absorbing spheres, the irregularly shaped particle shows an increase in intensity in the backscattering range. The degree of linear polarization reaches a maximum of about 40% at a scattering angle

R. H. Giese and P. Lamy (eds.), Properties and Interactions of Interplanetary Dust, 223–226.

Θ ≃ 80°. These are the typical scattering patterns of terrestrial highly irregularly shaped absorbing particles investigated by light scattering experiments (Weiß-Wrana, 1983) and by the microwave analoge technique (Zerull et al., 1980).

In the case of a less absorptive particle of the inclusion (diameter = 50 μm) the values of the positive degree of polarization are significantly smaller without the distinct maximum of polarization, see Fig.1b. Internal scattering processes and internal refraction are more probable for slightly absorbing agglomerates than for strongly absorbing ones. The contribution of the neutral component is greater, which significantly decreases the degree of polarization. This agrees with the scattering properties of terrestrial less absorptive fly-ash particle (Weiss-Wrana, 1983).

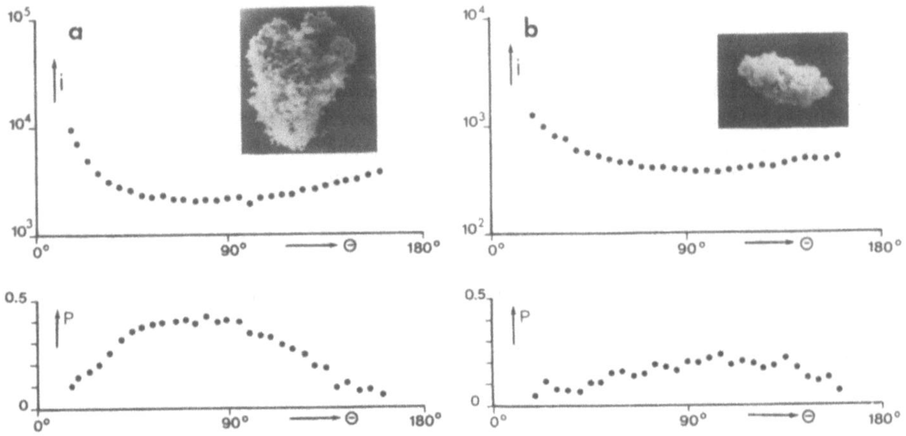

Figure 1. Total intensity i and degree of linear polarization P of meteoritic samples. a) 28 μm particle of the dark matrix of meteorite Allende. b) 50 μm particles of the white inclusion of meteorite Allende.

Fig. 2 illustrates the general shape of the empirical scattering function σ(Θ) and the corresponding degree of polarization P(Θ) derived from the brightness of the zodiacl light by inversion of the integration along the line of sight (Dumont and Sanchez, 1975; Leinert, 1978): The intensity is clearly enhanced toward backscattering. The degree of linear polarization reaches a maximum of 35% at a scattering angle Θ ≃ 85°. Our light scattering experiments show that irregularly shaped opaque particles can resemble the scattering function and the corresponding polarization. Large purely transparent and translucent irregularly shaped particles like quartz and andesite cannot be the dominating component of the interplanetary particles producing the zodiacal light (for details see Weiss, 1981 and Weiss-Wrana, 1983).

Using the experimental data of 7 different terrestrial and 3 meteoritic particles covering the size range of 20 to 120 μm and taking into account the size distribution (Giese and Grün, 1976) and the composition of the collected interplanetary particles (Brownlee et al., 1976) the scattering behaviour of a polydisperse mixture was calculated (Weiss-Wrana, 1983).

As demonstrated in Fig. 2 the scattering patterns of the particle mixture resembles the characteristic features of the zodiacal light quite well.

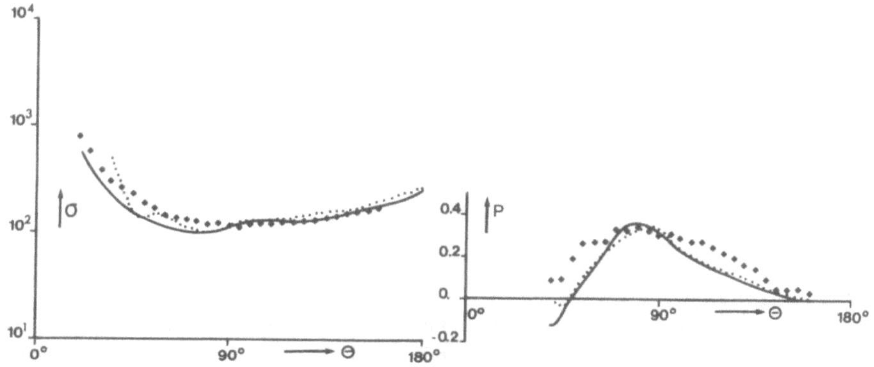

Figure 2. Zodiacal light model: Empirical scattering function σ and degree of polarization P of the interplanetary dust compared with experimental data (light scattering) of a particle mixture of 10 irregularly formed particles (rhombs) in the size range between 20 μm and 120 μm (Weiss-Wrana, 1983). The dotted line corresponds to the data of Dumont and Sanchez (1975), the solid ones to Leinert's data (1978, 1983).

2. COMETARY DUST

Recent photopolarimetric measurement of Comet West give some impressions about the variation of the brightness and linear polarization of sunlight scattered by dust in the cometary coma over a considerable region of scattering angles. Fig. 3 shows the intensity measured by Ney and Merrill (1976) and the degree of polarization as averaged by Zerull and Giese (1983) over the measured quantities of different observers (Merrill, 1976; Isobe 1978; Michalsky, 1981; Kiselev, 1981). The significant features not only of comet West but also of other ones are: The intensity is clearly enhanced toward backscattering. The degree of linear polarization reaches a positive maximum of 25-30% at a scattering angle $\Theta \simeq 90°$ and has a neutral point at $\Theta \simeq 160°$ followed by a branch of slightly negative polarization to about -5%.

Assuming that the main distribution of cometary grains has a size of just a few microns these scattering features can be explained by our microwave analog experiments (Zerull and Giese, 1983). As demonstrated in Fig. 3 a polydisperse mixture of five small two-components fluffy particles containing dielectric and absorbing constituents show the same scattering characteristic. Purely irregularly nonabsorbing compact and fluffy particles cannot resemble the optical properties in the visible, they show rather isotropic scattering for larger scattering angles and no pronounced polarization (Zerull et al., 1980).

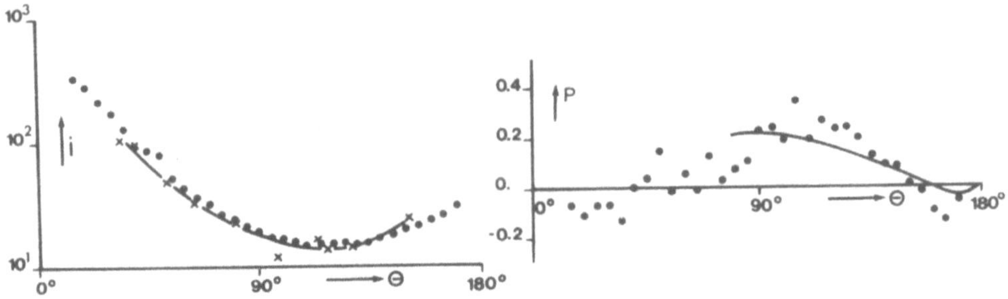

Figure 3. Cometary dust model for comet West: Observed intensity (solid line with crosses, Ney and Merill, 1976) and degree of polarization (solid line, as averaged by Zerull and Giese, 1983) compared with experimental data (microwave analog experiment) of a polydisperse mixture of two-component fluffy particles (index of refraction m = 1.5 and m = 1.65 - 0.25i) corresponding to 2 to 4 μm in size in the visible (dots).

3. REFERENCES

Brownlee, D.E., Tomandl, D.A., Hodge, P.W.: 1976, in Lecture Notes in Physics, **48** (eds. H. Elsässer and H. Fechtig), Springer Verlag, Heidelberg, p. 279.

Dumont, R., Sanchez, F.: 1975, Astron. Astrophys., **38**, 405.

Giese, R.H., Grün, E.: 1976, in Lecture Notes in Physics, **48** (eds. H. Elsässer and H. Fechtig), Springer Verlag, Heidelberg, p.135.

Isobe, S., Saito, K., Tomita, K. and Maehara, H., 1978, Publ. Astron. Soc. Japan, **30**, 687.

Kiselev, N.N., Chernova, G.P., 1981, Icarus, **48**, 473.

Leinert, C., 1978, 1982, private communication.

Ney, E.P., Merrill, K.M., 1976, Science, **194**, 1051.

Michalsky, J.J., 1981, Icarus, **47**.

Weiss, K., 1981, BMFT-FB-W81.

Weiss-Wrana, K., 1983, Astron. Astrophys., **126**, 240-250.

Zerull, R.H., Giese, R.H., Schwill, S., Weiss, K., 1980, in Light Scattering by Irregularly Shaped Particles (ed. Schuerman, D.W.), Plenum Press New York and London, p. 273.

Zerull, R.H., Giese, R.H., 1983, in Cometary Exploration II, Proceedings of the International Conference in Cometary Exploration, Hungarian Academy of Science, Budapest, p.143.

LABORATORY RESULTS ON POLARIZATION PROPERTIES OF ELONGATED PARTICLES AND COMPARISONS TO DUST IN THE TAIL OF COMET IKEYA-SEKI (1965 VIII)

BO Å. S. GUSTAFSON
Space Astronomy Laboratory, University of Florida
1810 NW 6th Street
Gainesville, Florida 32609
U.S.A.

ABSTRACT. Data obtained through microwave analog measurements are used to compare optical polarization characteristics of silicate particles ranging in shape from spheroids and cylinders to chains of spheres. It is confirmed that the dependence of the degree of polarization on scattering angle is a powerful indicator of particle shape. The reversal and steep gradient in the degree of polarization found by Weinberg and Beeson (1976a,b) in the dust tail of comet Ikeya-Seki (1965 VIII) is suggestive of laboratory measurements corresponding to 1.6 μm long, 0.4 μm in diameter silicate cylinders.

1. INTRODUCTION

Krishna Swamy (1978) showed that the run of the degree of polarization with scattering angle observed in the tail of comet Ikeya-Seki (1965 VIII) could result from size segregation among spherical particles. Segregation along the tail is expected due to faster diffusion away from the tail for low mass particles. But spherical particles are probably rare in a disintegration event. Cubes or other rough shapes may be a better approximation. Sekanina and Farrell (1980) found independent evidence for elongated grains in the tail of comet West (1976 VI), and suggested that chain like aggregates may have been common. Schuerman et al. (1981) and Zerull (1985) concluded that there are marked deviations from Mie polarization patterns for the case of non-spheres. The degrees of polarization produced by chains of five spheres, similar sized circular cylinders and spheroids are compared to illustrate the effect of detailed shape including sharp edges. In this paper, we concentrate on the degree of polarization. This quantity, a ratio of intensities, is independent of the (unknown) number of particles along the line of sight. In addition, the angular dependence of the degree of polarization is found to be a powerful indicator of particle parameters (see also Schuerman et al., 1981 and Zerull, 1985).

R. H. Giese and P. Lamy (eds.), Properties and Interactions of Interplanetary Dust, 227–230.

2. OBSERVED DEGREE OF POLARIZATION

Comet observations covering several scattering angles involves more than one set of particles. Either the dust is observed at substantially different times or is located at different distances from the comet's nucleus (unless observations are made simultaneously from more than one point in space). Weinberg and Beeson's (1976a,b) multicolor observations of brightness and polarization cover scattering angles θ from 116 to 136 degrees, with detailed results given at 5300 Å along the tail axis (Figure 1). Krishna Swamy (1978) estimated that the dust responsible for these data range from 0.8 to 1.6 AU from the cometary nucleus.

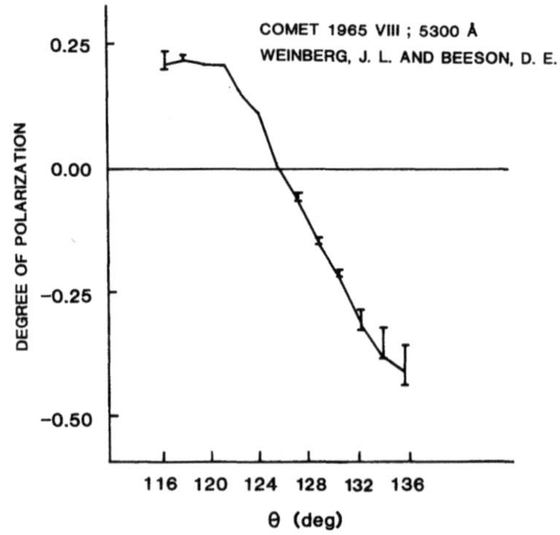

Figure 1. Angular dependence of the degree of polarization in the tail of comet Ikeya-Seki.

3. SCATTERING RESULTS

All scattering data are averaged over a uniform distribution of orientations to simulate scattering by an ensemble of randomly oriented particles. To illustrate the effects of particle shape, particles of equal surface area are compared. For the cylinders and chains of spheres, scattering data were obtained experimentally using the microwave analog

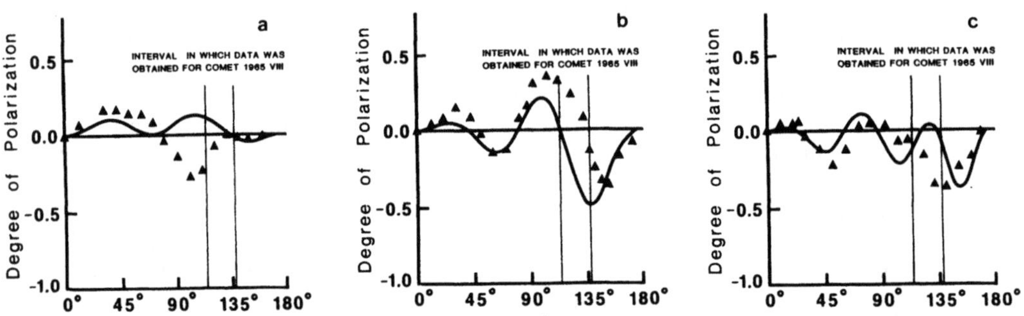

Figure 2. Scattering signatures averaged over uniform distribution of orientations of circular cylinders (triangles, measured), and prolate spheroids (solid curve, computed). In (a) both particles have surface areas corresponding to a 0.34 μm radius sphere, in (b) to 0.43 μm and in (c) to 0.51 μm.

technique (Schuerman, 1980). The dimensions of the target and the wavelength are scaled by the same amount. The index of refraction ($m = 1.61 - i\ 0.004$) of the target material (acrylic) at the laboratory frequency 9.417 GHz, resembles that of silicates at optical frequencies. Scattering by the spheroids was computed using a code developed by Schaefer (1980).

In Figure 2, the degree of polarization of scattering from three sets of sizes 4:1 aspect ratio silicate cylinders and prolate spheroids are compared. The strong dependence of the angular distribution of the degree of polarization on shape and size makes the correlation between observations (Figure 1) and scattering by 0.4 μm diameter cylinders remarkable (Figure 2b). A size increase towards high scattering angles, similar to the one postulated by Swamy would further improve this fit. With similar surface area spheroids the neutral point occurs at much lower scattering angles. The 20% (linear dimensions) smaller particles (Figure 2a) produce negligible negative polarization. With 20% larger particles (Figure 2c) there is little positive polarization in the backscattering hemisphere. Similar sized chains of spheres (Figure 3) produce a much weaker angular dependence of the degree of polarization in the observed interval where it remains negative.

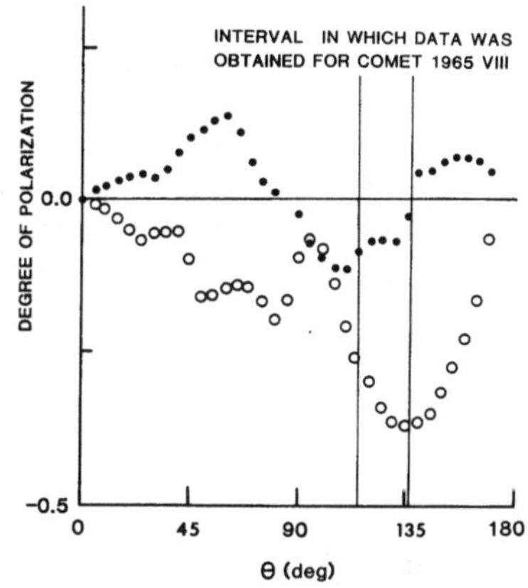

Figure 3. Polarization data averaged over uniform distribution of orientation for chains of five contacting spheres. The radii of the surface equivalent spheres are 0.41 μm (dots) and 0.58 μm (circles).

4. CONCLUSIONS

Circular cylindrical particles of materials, sizes and elongations corresponding to four times the estimated core size of interstellar dust particles (Greenberg, 1978) could have been an optically important dust-component in the tail of comet Ikeya-Seki (1965 VIII) at the time it was observed by Weinberg and Beeson (post-perihelion). Although the overall shapes of the model particles were similar, the scattering signatures efficiently discriminated amongst their detailed shapes. The observations fit the scattering by dust with sharp edges (which are expected from a breakup event). Comparisons to the scattering by these and several other particles indicate that the polarization in the tail of comet Ikeya-Seki may be relatively efficient at discriminating amongst particle shapes and material. Nevertheless, silicate cylinders are not claimed to be a unique alternative to Krishna Swamy's interpretation. The combination of an unusually strong dependence of the observed degree of polarization on scattering angle and the sensitivity of this parameter to particle shape and size (over a wide range), appear to imply a narrow size and shape distribution of the dust or a well defined variation along the tail. The observations do not fit the polarization produced by similarly sized chain-like aggregates.

5. ACKNOWLEDGEMENT

The experimental data were obtained as part of a continuing effort initiated by the late Dr. D.W. Schuerman to systematically investigate the optical properties of non-spherical particles. Dr. R.T. Wang has generously extended much help and enlightening discussions regarding lightscattering experiments and theory. The author is also indebted for helpful discussions with Drs. J.L. Weinberg and J.M. Greenberg.

6. REFERENCES

Greenberg, J.M., 1978, 'Interstellar dust', Cosmic Dust, (ed. J.A.M. McDonell), John Wiley and Sons, New York, 187.

Krishna Swamy, K.S., 1978, 'On the observed polarization of comet Ikeya-Seki (1965 VIII)', Astrophys. Space Sci., **57**, 491.

Schaefer, R.W., 1980, 'Calculations of the light scattered by randomly oriented ensembles of spheroids of size comparable to the wavelength', Ph.D. Thesis, State U. New York at Albany.

Schuerman, D.W., 1980, 'The microwave analog facility at SUNYA: capabilities and current programs', Light Scattering by Irregularly Shaped Particles, (ed. D.W. Schuerman), Plenum Press, New York and London, 227.

Schuerman, D.W., Wang, R.T., Gustafson, B.Å.S., and Schaefer, R.W., 1981, 'A systematic study of light scattering: I. Particle shape', Applied Optics, **20**, 4039.

Sekanina, Z., and Farrell, T.A., 1980, 'Evidence for fragmentation of strongly nonspherical dust particles in the tail of Comet West 1976 VI', in Proc. IAU Symposium No. 90, Solid Particles in the Solar System, (eds. I. Halliday and B.A. McIntosh), Ottawa, 267.

Wang, R.T., and Gustafson, B.Å.S., 1983, 'Angular scattering and polarization by randomly oriented dumbbells and chains of spheres, in Proc. 1983 CSL Scientific Conference on Obscuration and Aerosol Research', U.S. Army, Aberdeen, MD.

Weinberg, J.L., and Beeson, D.E., 1976a, 'Polarization reversal in the tail of comet Ikeya-Seki (1965 VIII)', Astron. Astrophys. **48**, 151.

Weinberg, J.L., and Beeson, D.E., 1976b, 'Photoelectric polarimetry of the tail of comet Ikeya-Seki (1965 VIII)', in Proc. IAU Symposium No. 25, The Study of Comets, (ed. B. Donn et al.), NASA SP-393.

Zerull, R.H., 1985, 'Laboratory investigation and optical properties of grains', in Proc. IAU Colloquium No. 85, Properties and Interactions of Interplanetary Dust, this volume.

PRELIMINARY RESULTS OF A DUST SCATTERING EXPERIMENT

P. BLIEK
P.L. LAMY
G. COURTES
Université de Provence, Marseille
Laboratoire d'Astronomie Spatiale, Marseille

ABSTRACT. An experimental device, of the nephelometer type, to study the scattering properties of dust particles is presented. A fluidized bed generator produces a continuously flowing aerosol which is illuminated by either a He-Ne laser or a Xenon are lamp with interference filters. The size of the dust particles ranges between 1 and 30 μm. The scattering properties of dust particles are measured in the two directions of polarization. The first results are presented.

1. EXPERIMENTAL PRINCIPLE AND SET-UP

1.1. Description

The principle retained for our experimental investigation consists in generating a continuously flowing aerosol, a section of which is illuminated in order to study its scattering properties over the 0-180° interval of scattering angles. It can therefore be considered as a nephelometer whose key part is a fluidized - bed generator which receives a mixture of the dust to be studied (in the form of powder) and of "large" glass spheres having diameters of 100 to 200 μm. An air flow forces the mixture to "boil", the collisions between the glass spheres desagglomerate the powder and the liberated dust particles are transported by the air flow. The "boiling" chamber is fed by an endless screw which continuously enrichs the mixture so as to compensate for the loss of transported particles. It can be shown that the operation of the generator is governed by a very simple equation which related the ponderal concentrations of the mixture and the flow rates to the concentration of the aerosol. After a transitory phase of approximately 10 mn, the stationary regime given by the equation is reached and was experimentally shown to be remarkably stable. A nozzle concentrates the aerosol in a column having a diameter of 5 mm; it is then allowed to flow freely over a length of 20 mm before being taken over by a suction device. The free part is of course used for the optical measurements. It is illuminated by a collimated beam from either an He-Ne laser or a Xenon arc-lamp. The optical part of the detection system is based an classic principles of photometry

R. H. Giese and P. Lamy (eds.), Properties and Interactions of Interplanetary Dust, 231–234.

with two doublets and a fiels diaphragm (Fig. 1). A rotating polaroid polarizer is mounted at the entrance pupil. The scattered light is measured by a photomultiplier tube working in the analog mode. Two identical optical detection systems have been set up, one fixed to provide a reference so as to cancel any variation in the aerosol or in the illuminating souce and one mounted on a rotation table moved by a stepping motor. The whole experiment is under computer control. For a given scattering angle, we take the average of typically 3 measurements, each one having been previously divided by the corresponding reference signal.

1.2. Characteristics

The size range of dust particles allowed by the fluidized-bed generator depends upon the density of the dust but is typically 1 to 40 μm. The scattering volume has a diameter and an height of 5 mm and contains approsimately 100 to 1000 particles. Two illuminating sources are used:
i) a 7 mW Helium-Neon laser (6328 Å) with a beam expander yielding a useful diameter of 5 mm;
ii) a 150 W Xenon arc lamp with an optical system giving a collimated beam having an internal angular dispersion of ± 2° (given by the size of the arc) and interference filters centered at 4470, 5500, 6450, 7060, 7440 and 8240 Å (typical bandpass ∿ 450 Å). The two detection systems have an angular aperture of 2° and are equipped with Hamamatsu R 378 PMT whose sensitivity extends over the interval 1250-8500 Å. However, a cut-off at 3250 Å is introduced by the lenses. Going further down in the ultra-violet would simply requires changing the lenses and using a Xenon lamp with a quartz envelope. Finally, the mechanical set-up has been optimized to allow a range of scattering angles as large as possible, 5 to 174 deg.

1.3. Main advantages

The presence of a large number of grains in the scattering volume directly average the measurements over the various parameters characterizing them (e.g., roughness). The flow being turbulent, the average further extends over the randow orientations of the grains. The scattering light is also sufficiently intense to allow spectral measurements in a wide spectral domain with a source of reasonable power. The system has the capability to directly measure the volume scattering function for a dust population having a selected size distribution. However the main drawback is the absolute calibration which requires the exact determination of the concentration in the aerosol. Although the required equipment is commercially avaible, we did not have yet the possibility to acquire it.

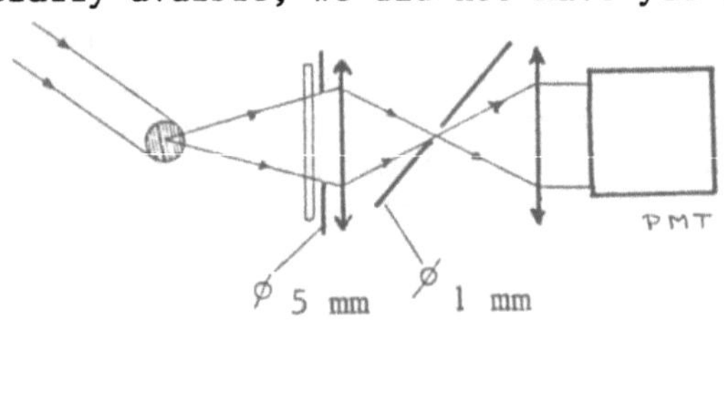

Fig. 1.: Optical layout of the detection system

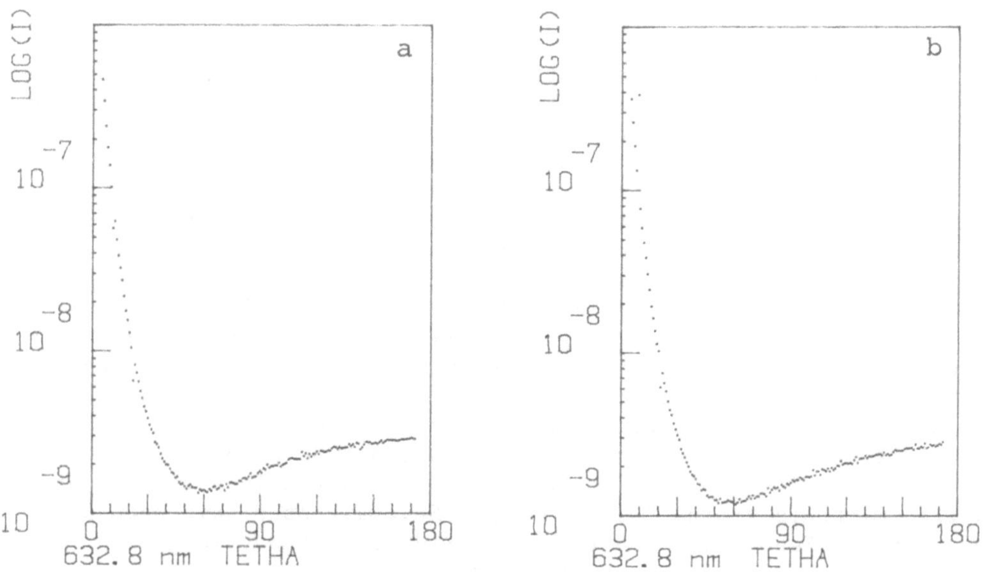

Fig. 2 : Total intensity scattered by iron grains under He Ne laser illumination as a function of scattering angle. This illustrates the stability of the measurements as the two results a and b were obtained at a one day interval.

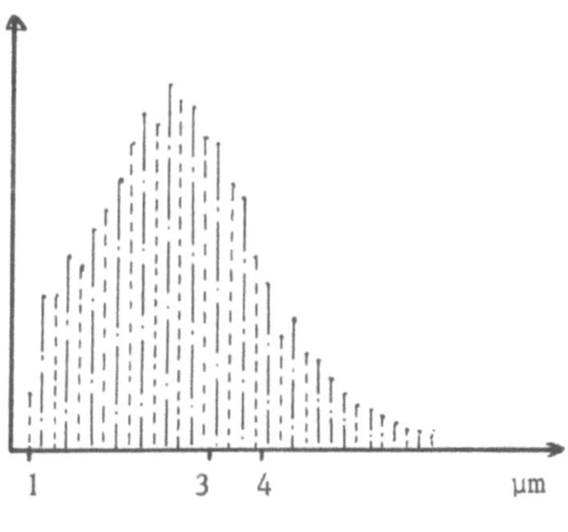

Fig. 3: The size distribution function of silicon dioxyde grains

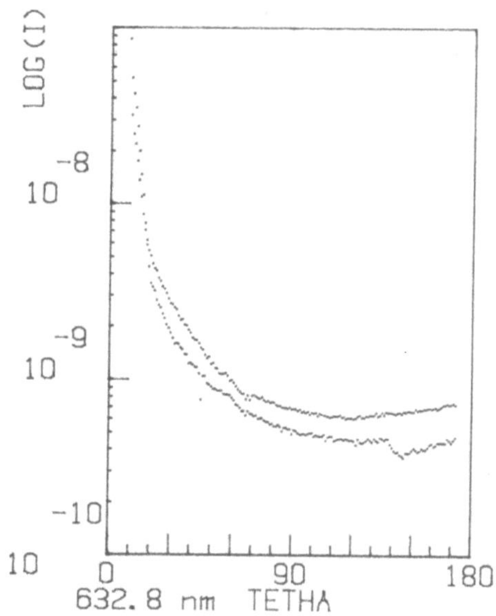

Fig. 4 : The polarized intensities scattered by silicon dioxyde grains under He Ne laser illumination

2. PRELIMINARY RESULTS

The results presented here were obtained with commercially available powders and with the He-Ne laser. Fig. 2 is intended to demonstrate the reproductibility of the measurements: the same iron powder was used for two different runs at a one-day interval. The difference is well within the error bars as given by the dispersion of the data points. This proves that the method of normalization using the reference signal is working very well. A silicon dioxyde powder was also studied. Its size distribution was measured independently with a Coulter counter and peaks at a diameter of 2.33 μm (Fig. 3). The silicon dioxyde aerosol was illuminated by He-Ne laser light polarized in the parallel and perpendicular directions w.r.t. the scattering plane. The results appear in Fig. 4: the upper curve corresponds to the parallel case and the lower one, to the perpendicular case. However, they have been obtained with two different runs and therefore, should not be compared on an absolute basis. The qualitative behaviour is nevertheless typical of dielectric micronic grains (Giese, 1971). Several tests have been performed with the Xenon lamp and results for different wavelengths will hopefully be presented in the near future.

REFERENCE

Giese, R.H.: 1971, Tabellen Von Mie - Streu Funktionen I. Gemische dielektrischer Teilchen, BMBW-FB W 71-23 Ruhr-Universität Bochum

INFLUENCE OF SURFACE ROUGHNESS ON SCATTERING PATTERNS

S. Mukai and T. Mukai
Kanazawa Institute of Technology
Nonoichi, Ishikawa 921
Japan

ABSTRACT. The influence of surface roughness on scattering properties of light is examined in terms of our previous mathematical method(Mukai et al. 1982). We found that albedo for single reflected light increases with an increase of surface roughness. In addition, a higher order of reflection light becomes effective for stronger rough particles.

1. INTRODUCTION

Scattering pattern of light by a rough particle has been investigated by several authors with different theoretical methods, e.g. Wolff (1975), Mukai et al.(1982), and Perrin and Lamy(1983). In Mukai et al.(1982), we presented the mathematical treatment based on the stochastic process of multiple reflection of light by surface roughness, and succeeded to explain the observed evidences of scattering patterns by rough particles, i.e. a backward enhancement of intensity, and reduction somewhat of polarization below a scattering angle of 90°compared with the expected results with a smooth surface particle.

Here we shall examine the influence of degree of roughness on the scattering properties of light, which might relate to an albedo of interplanetary dust grain.

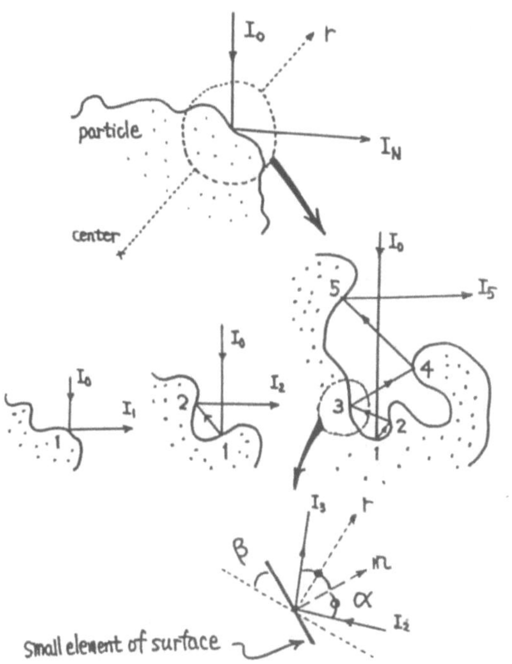

FIG.1. Definition of multiple reflection by surface roughness

R. H. Giese and P. Lamy (eds.), Properties and Interactions of Interplanetary Dust, 235–238.

2. ALBEDO FOR SINGLE REFLECTION

The complete mathematical analysis has already been presented in Mukai et al.(1982). Here we therefore simply summarize the model. In our treatment we assume that a particle is larger than a wavelength of light and rough surface of a particle consists of many small elements with different slopes.

In addition, each surface element is also larger than a wavelength of light and a particle has a large absorption coefficient. Then, the reflection of light by each of the surface elements is treated by Fresnel reflection and the refracted light inside the particle is absorbed there.

Incident light on a particle is successively reflected by rough surface. In each reflection, the incident angle α and the angle of slope β play an important role, where β is measured from the average surface in the reference system(see figure 1). A reflection pattern of small elements on the surface is defined by surface slope distribution function $P(x)$, where $x=\tan\beta$, and Fresnel reflection pattern $\rho(\alpha)$. For surface slope distribution, we apply a gaussian distribution as

$$P(x)=\exp\{-(x/\sigma)^2\}/(\pi^{\frac{1}{2}}\sigma), \qquad (1)$$

where σ is a parameter to define a degree of roughness. For example, as shown in figure 2, surface roughness increases with σ.

A single reflection pattern is obtained in terms of both surface slope distribution and Fresnel reflection. We assume this reflection pattern as a single scattering pattern for a successive order of scattering. Namely, we treat the successive reflection of light by small surface elements as the multiple scattering of light by particles with a single scattering pattern derived above. Using the method of successive orders of scattering to solve the equation of radiative transfer, we sum up each order of reflection.

Since the reflected intensity depends on an incident angle through a Fresnel reflection pattern, an albedo for single reflection ω by each surface element depends on the slope of the surface element, i.e. the degree of surface roughness. This is derived from the fact that Fresnel reflection of a larger incident angle produces higher

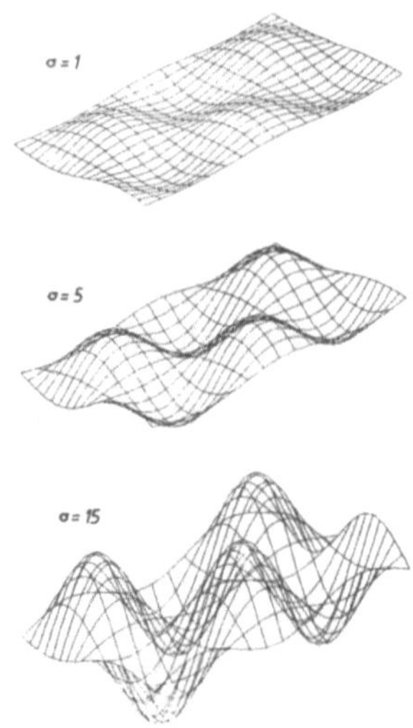

FIG.2. Sample of surface roughness, where σ is a parameter to define the roughness shown in eq.(1).

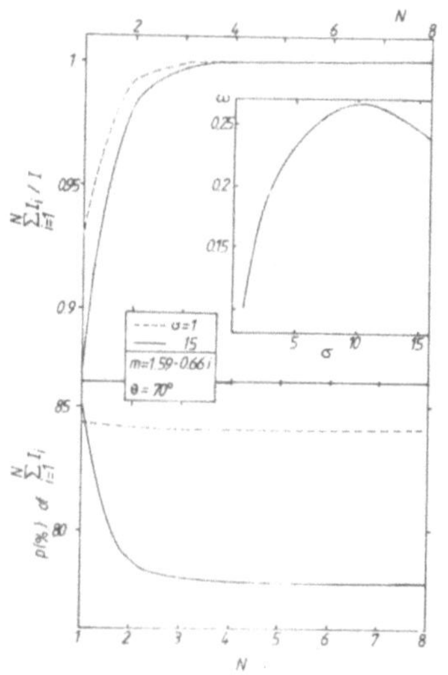

FIG.3. Intensity I and polarization p of multiple reflected light. Small figure shows an albedo ω of single reflection with roughness σ.

reflectance. In general, surface roughness causes an increase of reflection of a larger incident angle, therefore an albedo increases with σ as shown in figure 3.

In order to examine a contribution of higher order reflected light on total reflected intensity, we show in figure 3 the intensity and polarization as a function of order of reflection N.

The higher order of reflection becomes significant as roughness increases because an albedo increases with roughness. For example, the fourth-order of reflected light by surface roughness cannot be neglected for σ=15. Subsequently, polarization for σ=15 is clearly reduced by a contamination of highly ordered, reflected light because the degree of polarization becomes smaller after reflection.

3. SCATTERING BY A LARGE ROUGH PARTICLE

In order to compare the above results with experimental data, we should integrate the reflected intensity by each local place on the particle

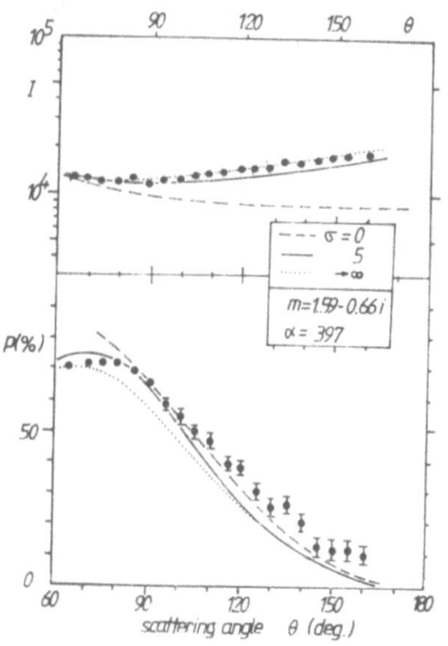

FIG.4. A comparison of expected scattered light by rough particle with experimental data by Weiss-Wrana(1983)(filled circles).

surface all over the surface.

Figure 4 shows the intensity and polarization of scattered light as a function of scattering angle θ. Below $\theta=60^0$, diffraction of light becomes dominant. We have omitted that region here to simplify discussion.

It is found in figure 4 that as roughness increases, a higher order of scattered light increases a backward enhancement of intensity and reduces the degree of polarization. Therefore, the degree of roughness should be considered as one of important factors for scattering by a particle with rough surface.

REFERENCES

Mukai,S., Mukai,T., Giese,R.H., Weiss,K., and Zerull,R.H.:1982, Moon and Planets 26, p. 197.

Perrin,J.M.and Lamy,P.L.:1983, Optica Acta 30, p. 1223.

Weiss-Wrana,K.:1983, Astron. Astrophys. 126, p. 240.

Wolff,M.:1975, Appl. Opt. 14, p. 1395.

DISCUSSION

ZERULL: Is your theoretical approach also able to reproduce the branch of negative polarization? If not, could it be modified appropriately?

MUKAI: Negative polarization occurs when the multiple reflected light becomes important. For absorbing particle, roughness on the surface of particle can produce negative polarization, but smooth surface never make this. For dielectric particle, the refracted light penetrated inside the particle produces negative polarization due to internal multiple reflection. In addition, roughness on the surface enhances this feature. Our model treatment can provide the reduction of polarization for absorbing particles and will be able to reproduce the negative polarization when it is extended to dielectric particles.

ZODIACAL LIGHT AND THE SPATIAL DENSITY OF INTERPLANETARY GRAINS

P.L. Lamy
J.M. Perrin
Laboratoire d'Astronomie Spatiale
Marseille, France

ABSTRACT. Using our model of light scattering by large rough grains, the compatibility of the observed properties of the zodiacal light with models of the spatial density of interplanetary grains is investigated. The agreement is not yet satisfactory and probably calls for further revision of the density distribution function. Our previous conclusion that submicronic grains gives a non-negligible contribution (Lamy and Perrin, 1980) is confirmed.

I. INTRODUCTION. The possibility to derive the differential spatial density ("DSD") of interplanetary grains from lunar microcraters and direct space measurements has prompted several investigations of its compatibility with the observed properties of the zodiacal light. The analysis performed by Giese and Grün (1976) and Giese et al. (1978) relied on the microcrater distribution function of Fechtig et al. (1974). Lamy and Perrin (1980) tested the DSD obtained by Le Sergeant and Lamy (1980). Likewise the above studies, ours was limited by the lack of an appropriate theory for the scattering of light by non-spherical grains and the subsequent use of approximations. Since 1980, we have devoted considerable effort to this aspect of the problem and proposed a new model which applies to the scattering by large rough spherical grains (Perrin and Lamy 1983). The present analysis closely follows our previous one (Lamy and Perrin, 1980) but incorporates our new scattering model and also tests the new DSD of Grün et al. (1984 and this volume).

II. THE DIFFERENTIAL SPATIAL DENSITY AT 1 AU

The DSD used here are therefore those of Le Sergeant and Lamy (1980) and Grün et al. (1984) illustrated in Figure 1. In comparison with the former solution, the latter one differs on the following points:

i) a larger density of micronic grains resulting from different absolute calibrations (solar flares lunar clock and Pegasus data);

ii) a lower density of submicronic grains;

R. H. Giese and P. Lamy (eds.), Properties and Interactions of Interplanetary Dust, 239–243.

iii) a steeper gradient for large grains (s $\gtrsim$ 100 μm) resulting from a different calibration law for the ratio of crater to grain diameter.

However, it can be noted that the solution of Le Sergeant and Lamy (1980) is compatible with the large errror bars of the Heos and Pegasus data as given by Grün et al. (1984).

III. THE OBSERVED AND CALCULED VOLUME SCATTERING FUNCTION (VSF) AND POLARIZATION

The "observed "VSF" and polarization are deduced from the observed brightness of the Zodiacal light assuming a power law $r^{-\nu}$ for the dependence of the spatial density upon the heliocentric distance (see, for example, Dumont and Sanchez, 1975, and Leinert et al. 1976).

We performed a separate analysis of a set of carefully selected data (Lamy and Perrin, in preparation) and obtained an acceptable VSF - as well as polarization - only for $\nu = 1$. These results at λ = 6328 Å are plotted in figures 2 to 3 in order to be compared with the calculated models.

Several assumptions must be made for the calculations of the VSF. We split the size distribution in two parts by introducing a boundary at a radius s_0 = 3 μm. The purpose is twofold: first, this corresponds to the two populations 1 and 2 introduced by Le Sergeant and Lamy (1980) and second, to the ranges of validity of the scattering models.

i) large grains (s > 3 μm) are rough being aggregates of submicron grains as observed by Brownlee (1978). We felt that the scattering properties of a grain extracted from the Allende matrix experimentally studied by Weiss-Wrana (1983) and successfully modelled by Perrin and Lamy (1983) should be appropriate for this population. A refractive index of 1.803 - 2.5 x 10^{-3} i at 6328 Å was finally retained (see Egan and Hilgeman, 1979) and the integration over the size distribution function extended to s = 100 μm.

ii) small grains (s < 3 μm): we consider two cases.

First, following Le Sergeant and Lamy (1980), this population is considered independent and connected to the homogeneous, nearly spherical FSN grains collected by Brownlee (1978). The closest analog is probably pyrrhotite, a non-stoichiometric iron sulfur often found in meteorite, whose refractive index

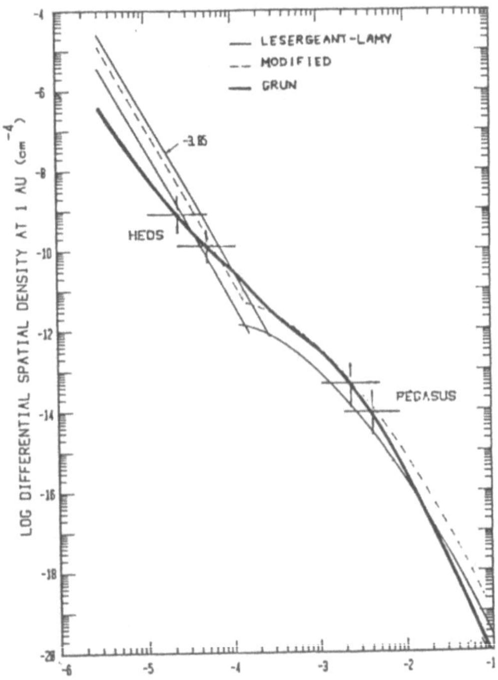

Figure 1. The differential spatial density.

was measured by Egan and Hilgeman (1977) and is 1.77 - 1.94 i at 6328 Å. As a second alternative, we follow Grün et al. (1984) and consider that this population is dominated by collisional fragments. Its composition is therefore the same as the population of large grains, namely the above grain of the Allende matrix.

In both cases, these small particles are very likely to be smooth (at least, with respect to the visible wavelength of interest here) either resembling the FSN collected grains or being collisional fragments. We felt therefore that the Mie theory should apply to them. Finally, the integration over the size distribution extends from 3 μm down to 0.025 μm.

IV. RESULTS AND DISCUSSION

Figure 2a shows the result for the VSF using the Grün et al. (1984) model. An important point is that both populations contribute significanly to the VSF: the contribution from small grains even dominates at scattering angles below ≃ 75° while the situation is reversed above. At intermediate angles, the total calculated VSF falls below the observed one by a factor 2. Let us suppose that we allow for this correction on the DSD; then, the forward and, mainly, the backward enhancements appear too strong, the second effect being entirely caused by the small silicate grains. As a matter of curiosity, we performed the calculation assuming that the submicronic grains ($s < 3$ μm) are composed of pyrrhotite (figure 2b). The contribution from the large grains dominates at scattering angles larger than 50° but is equal to that of small grains below. The discrepancy with the observation is still a factor of 2 but the overall behaviour appears more satisfactory than the first solution, except for too strong forward scattering. Figure 3a shows the polarization for both solutions: none appears really satisfactory and one has the feeling that some "intermediate" solution may improve the fit.

Turning now to the Le Sergeant - Lamy model, we essentially confirmed our preliminary results (Lamy ant Perrin, 1980). In order to analyze the implication of the observed VSF, we allowed ourselves to modify their model by

i) increasing their population 1 by a factor 4 so as to agree with the Pegasus data, and consequently, with the Grün et al. (1984) model, except for the largest grains;

ii) decreasing their population 2 by a factor 2 (Figure 1). This correction results in a very good agreement with the observed VSF as illustrated in Figure 2c. We further note that the scattering at angles below 120° is dominated by the population of small pyrrhotite grains, which becomes equal to that of large grains beyond 120°.

However, the results for polarization (Figure 3b) is not satisfactory. Although the theoretical tools are now available, an in-depth analysis is beyond the scope of the present article. We may however present several directions for further investigations:

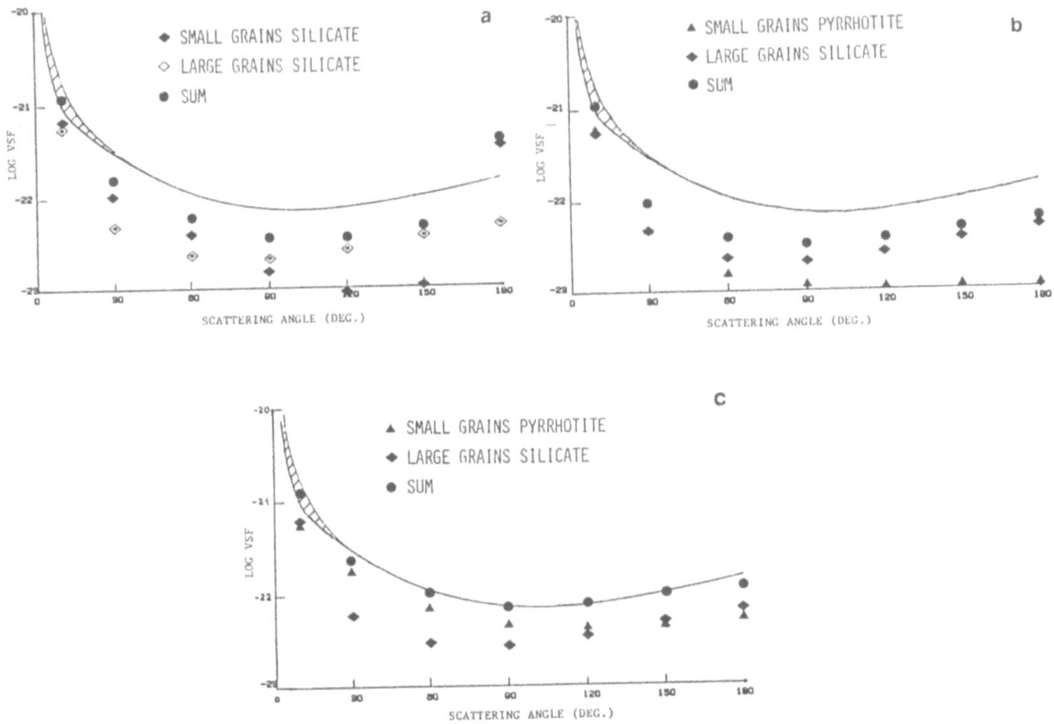

Figure 2. Comparison of the observed VSF (solid line and hatched band) in unit of $cm^2\ st^{-1}\ cm^{-3}$ with different models of DSD
a) and b) Grün et al. (1984)
c) Le Sergeant and Lamy (1980)

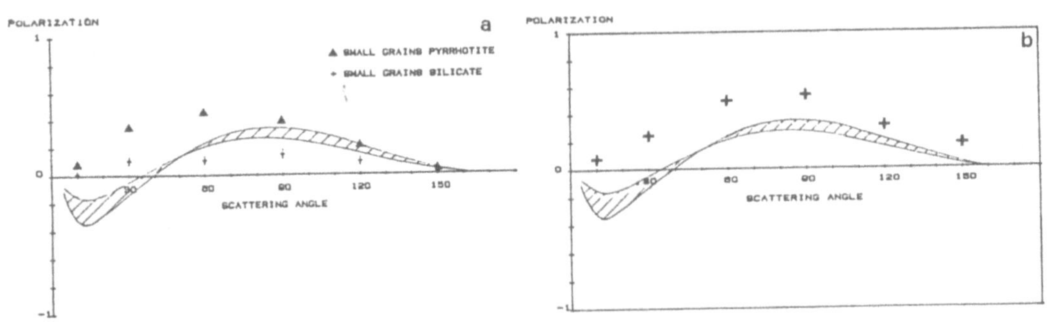

Figure 3. Comparison of the range of observed polarization (hatched band) with different models of DSD
a) Grün et al. (1984)
b) Le Sergeant and Lamy (1980)

i) the importance of the contribution of small particles to the VSF as already established by Lamy and Perrin (1980) is fully confirmed, even for the low flux of Grün et al. (1984). The exact relative importance of the contributions from the two populations is, in part, a matter of assumptions in particular on the albedo of the grains. Our scattering model gives a geometric albedo of the order of 5 % for the large rough grains, a value consistent with the dark appearance of collected aggregates (Brownlee et al. 1976) and the measurements on Cl and C2 carbonaceous chondrites of Johnson and Fanale (1973).

ii) the upward correction proposed by Grün et al. (1984) on the basis of the Pegasus data considerably eases the compatibility with the brightness of the Zodiacal Light; a further increase may even be necessary as seen above.

iii) the pyrrhotite solution for the small grains gives a better fit to the observed VSF than the silicate solution but the polarization still poses problem. We feel that a combination of the two solutions may improve the situation. This is supported by various evidences as silicate collisional fragments are certainly present while FSN grains are collected and an "absorbing" component is implied by the infra-red observations of comets (Liu &Kimura,this vol.).This alternative would indicate that population 2 is not entirely composed of collisional fragments. The model of Grün et al. (1984) would then be too low and the true flux would lie somewhere between their curve and the modified Le Sergeant - Lamy model.

REFERENCES

Brownlee, D.E.: 1978, Cosmic Dust (Wiley) p. 295
Dumont, R., Sanchez, F.: 1975, Astron. Astrophys. 38 p. 405
Egan, W.G., Higelman, T.: 1979, Optical Properties of Inhomogeneous Materials (Academic Press)
Fechtig, H., Gentner, W., Hartung, J.B., Nagel, K., Neukum, G., Shneider, E., and Storzer, D.: 1974, Proc. Soviet American Conf. Cosmochemistry Moon and Planets, Moscow (NASA SP-370, 1976), pp. 585-604
Giese, R.M., Grün, F.: 1976, Lecture Notes in Physics 48 p. 135
Giese, R.M., Weiss, K., Zerull, R.H., Ono, T.: 1978, Astron. Astrophys. 65 p. 265
Grün, E., Zook, H., Fechtig, H., Giese, R.H.: 1984 submitted to Icarus
Johnson, T.V., Fanale, F.P.: J. Geophys. Res. 78 p. 8507
Lamy, P.L., Perrin, J.M.: 1980, Solid Particles in the Solar System (Eds, J. Halliday and B.A. Mc Intosh) p. 75
Leinert, C., Link, H., Pitz, E., Giese, R.H.: 1976, Astron. Astrophys. 47 p. 221
Le Sergeant d'Hendecourt, L.B., Lamy, P.L.: 1980, Icarus 43 p. 350
Perrin, J.M., Lamy, P.L.: 1983, Optica Acta 30 p. 1223
Weiss-Wrana, K.: 1983, Astron. Astrophys. 126 p. 240

OPTICAL PROPERTIES OF ROUGH GRAINS : A THEORETICAL STUDY

J.M. PERRIN and Ph. L. LAMY
Laboratoire d'Astronomie Spatiale
Traverse du Siphon
13012. Marseille
France

ABSTRACT. Recent work on the light scattering by large rough particles has led us to propose a model based on the high-energy approximation, the laws of geometrical optics and a mathematical description of the properties of the particle roughness. The influence of the various parameters of the model including those characterizing the roughness on the total intensity and the polarization is presented.

Collections of micrometeoroïds in the stratosphere show that a part of interplanetary dust is composed of large rough particles w.r.t. optical wavelenghts (Brownlee, 1978 and this volume). Light scattering by these particles can be interpreted on the basis of several independent scattering processes which may be treated separately : forward diffraction, Fresnel reflection, non-polarized reflection and transmission (Giese et al, 1977). Perrin and Lamy, (1983) proposed a model which describes these different processes. The forward diffraction is treated in the context of the high-energy approximation, based on a high-energy formulation of relativistic quantum scattering, where the expression of the scattering potential involves the complex index of refraction of the grain and a Fermi type distribution of the amplitude of the asperities; away from the forward direction, this approximation is supplemented by the Fresnel reflection and the transmission of light through the particle which are treated by the method of Wolff (1975 and 1980) adapted to our problem, using a geometrical description of the rough curved surface. The reflected intensity has two partially polarized components: one from the single reflection and the other one from the double reflection. The description of the roughness of the surface involves the following parameters (Fig.1) :

- a : mean radius of the grain.
- d_p : mean amplitude of asperities.
- ω_p : mean angular aperture of roughness.
- x_F, x_B, $x_L = x_R$: effective angular parameter for respectively forward, backward, lateral left and right double reflection.

R. H. Giese and P. Lamy (eds.), Properties and Interactions of Interplanetary Dust, 245–248.

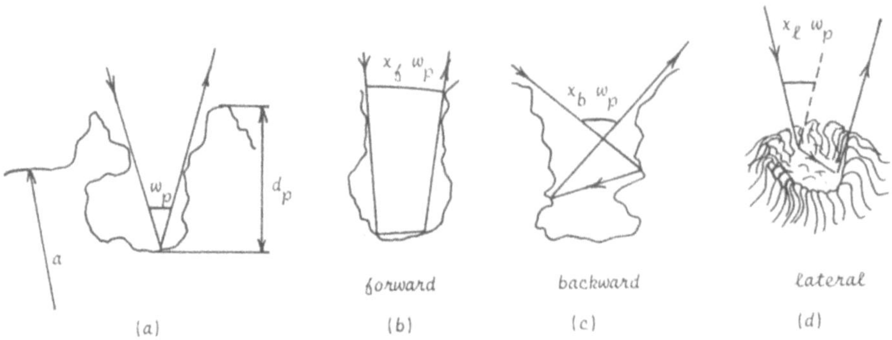

Fig.1 Parameters of the roughness

The parameters d_p and w_p characterize the large scale roughness, for instance pits on the surface of the particle. For a pit at a given phase angle, it is then possible to check if a single reflection may take place and to obtain the analytic expression for the reflected ray. The parameters x characterize the small scale roughness inside the pits and control the double reflection. For example, in the case of the forward double reflection (Fig. 1b), the roughness inside the pit causes a double reflection such that, the angle between the incident and reflected beams outside the pit is equal to $w_p/2$, so $x_F = 0.5$. Obviously, when $x_F = 1$, the double reflection becomes a single reflection. For a real particle, these parameters can be estimated using, for example, a SEM photograph. The good agreement between experimental results (Weiss, 1981; Weiss-Wrana, 1983) and the calculations (Perrin and Lamy, 1983) illustrated in Fig. 2, shows that this description of the particle roughness is sufficient to account for the properties of the scattered light intensity and polarization provided the complex part of the imaginary index is less than 0.05.

We already started to study the influence of the above parameters on the efficiency factors of rough particles (Lamy and Perrin, 1983). In this present contribution, we start to investigate their influence on the intensity as well as the polarization of the scattered light. Each parameters (including further the index of refraction) is allowed to vary in its range of validity, the others being kept constant. We have specifically investigated the following important features :

- the relative importance of the diffraction lobe,
- the maximum value of polarization and the corresponding scattering angle,
- the negative branch of polarization,

for the nomical case of a particle characterized by :

a = 20 µm d_p = 0.5 µm w_p = 90°

x_F = 0.5 x_B = 10 $x_L = x_R$ = 4

$n = 1.803 - 2.5 \times 10^{-3}\,i$ for λ = 6328 Å

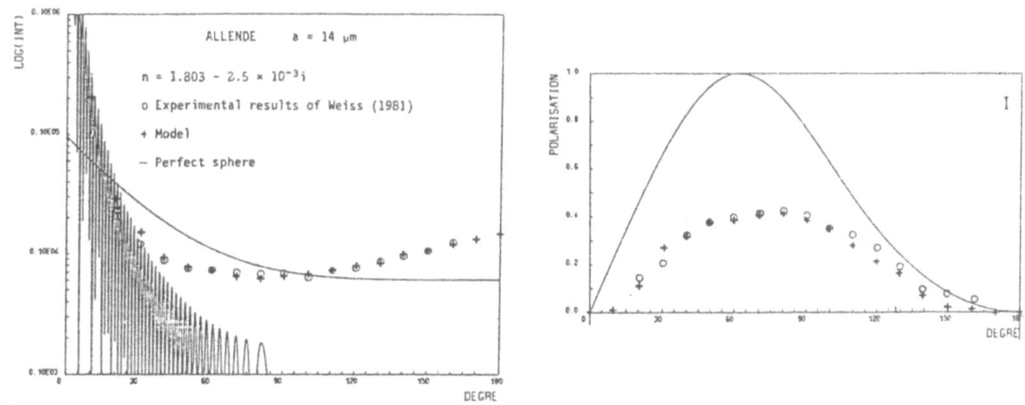

Fig.2 The square of amplitude function and the polarization for a rough particle of Allende meteorite of mean radius equal to 14µm. The solid line gives the result of Fraunhofer diffraction and of Fresnel reflection for a perfect sphere having the same radius and the same index of refraction. The error bar for the experimental polarization values appears in the upper right-hand corner.

We characterize the diffraction lobe by the ratio of the scattered intensities at 10° and 30°. The parameter which has the greatest influence is the mean amplitude of asperities d_p as illustrated in Fig.3; the ratio I(10°)/I(30°) increases dramatically with increasing d_p.

Several parameters control the location θmax and the value of maximum polarization Pmax such as the real part of the refractive index and the radius of particle. However, the most critical parameter is the imaginary part n_i of the refractive index. For our nominal case, Fig.4 shows the rapid decrease of the angle θmax from 100° to 80° when n_i increases from 1×10^{-4} to 5×10^{-3} correlated with a substantial increase of Pmax, from 0.05 to 0.7 approximately. There seems to be some sort of "saturation effect" outside the 1×10^{-4} - 5×10^{-3} interval as both θmax and Pmax vary only slightly.

The negative branch being the consequence of the double reflection, it is controlled by the parameters x. In fact, it is caused by the lateral double reflection. When $x_L = 1$, this component does not exist and the polarization remains positive. When $x_L > 1$, the negative branch appears as illustrated in Fig.5 obtained for the nominal case ($x_F = 0.5$ and $x_B = 10$). However, the magnitude of the negative branch, as characterized by the minimum (negative) polarization P_{min} and the angle of null polarization θ_o, depends also upon the relative importance of the forward and backward double reflections. Fig.6 obtained for $x_L = 4$ shows how the negative branch evolves rapidly as a function of x_F. When the forward double reflection decreases, the relative importance of the lateral component increases and P_{min} may reach quite large negative value ($\simeq$ -5%) while θ_o decreases drastically.

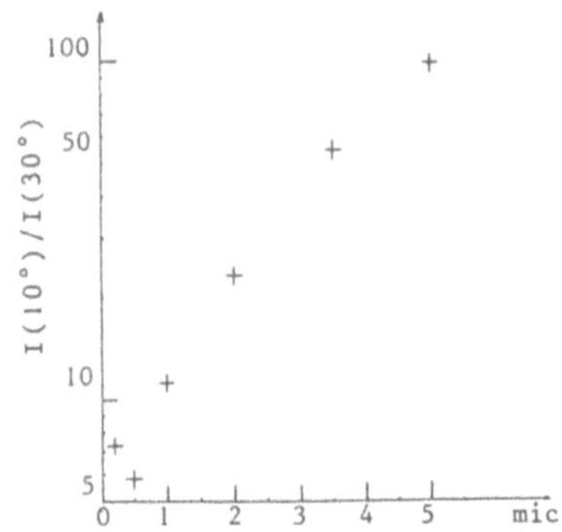

Fig. 3 Diffraction vs roughness

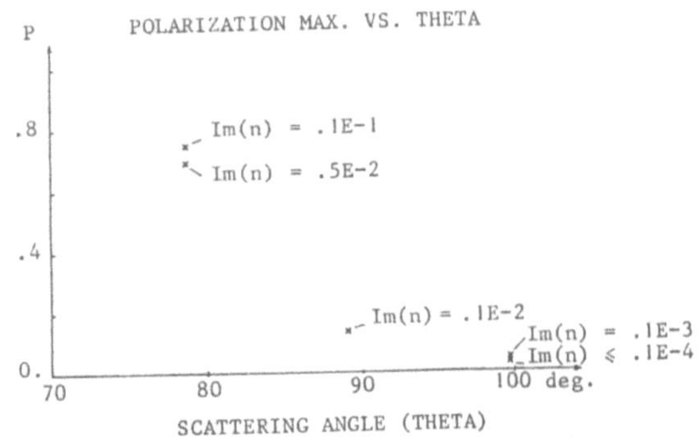

Fig. 4 Maximum polarization vs theta

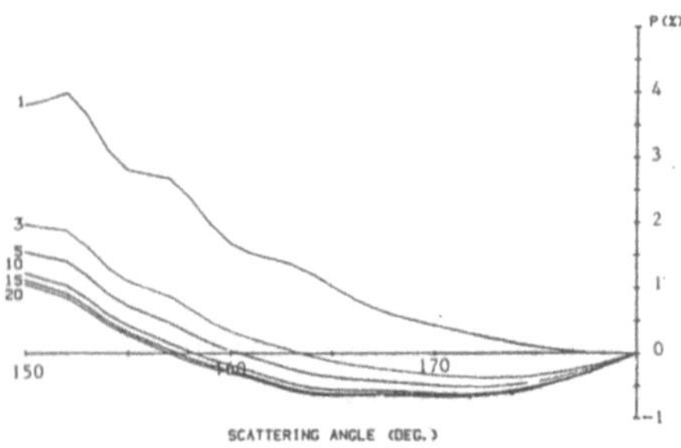

Fig. 5 The negative branch of polarization for various x_L

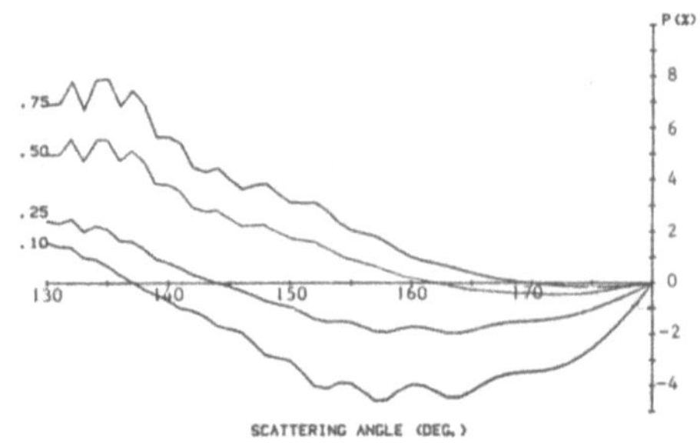

Fig. 6 The negative branch of polarization for various x_F

Although this study is far from being complete, it appears quite possible, in an inverse way, to deduce the values of the parameters from the observed scattered intensity and polarization and thus, to infer the physical characteristics of the particle.

REFERENCES

Brownlee,D.E.,1978: Cosmic dust,ed. J.A.M. McDonnell, Wiley, Chichester, pp. 295-336.

Giese,R.H., Weiss K., Zerull, R.H.,Ono,T., 1977 : Astron. Astrophys.,65 p.265.

Lamy,P.L.,Perrin,J.M.,1983 : Asteroïds Comets Meteors (Eds C.I. Lagerkvist and H. Rickman), p.273.

Perrin,J.M.,Lamy,P.L.,1983 : Optica Acta,30 p.1223.

Weiss,K.,1981 : Ph.D. Thesis Ruhr-Universität Bochum and Report BMFT-FB-W 81-047.

Weiss-Wrana,K.,1983 : Astron. Astrophys., 126 p.240.

Wolff,M.,1975 : Appl. Optics, 14 p. 1395.

Wolff,M.,1980 ; Icarus, 44 p.780.

WAVELENGTH DEPENDENCE OF THE ZODIACAL LIGHT

R. Schiffer and K.O. Thielheim
Institut für Reine und Angewandte Kernphysik der
Universität Kiel, Olshausenstr. 40-60, D-2300 Kiel,
Federal Republic of Germany

ABSTRACT. We calculated the scattering cross section of an ensemble of large, convex, randomly oriented particles with a slight surface roughness. If the roughness structure is described by an exponential correlation function, the degree and angular dependence of the zodiacal light reddening are well reproduced by our model.

1. INTRODUCTION

The observations in the last 20 years and especially the experiments on Helios 1 and 2 appreciably improved our knowledge of the interplanetary dust, and there is a consensus about the essential properties of those dust particles that produce most of the visual zodiacal light: they are much larger than wavelength and show up a very irregular structure. These irregularities lead to shadowing and multiple scattering at the surface of the particles, effects which cause the "Gegenschein" and the low polarization values.

Several particle models have been suggested, and most of them are based on geometrical optics (Mukai et al., 1982; Schiffer and Thielheim, 1982; Perrin and Lamy, 1983); that means that the radii of curvature of the surface irregularities are assumed to be much larger than wavelength. In spite of this restriction these model particles can very well reproduce the essential features of the zodiacal light, and so they possibly represent a rather realistic description of a great part of the interplanetary dust.

But scattering models based on geometrical optics lead to wavelength independent scattering - if we neglect the wavelength dependence of the refractive index for the moment. The measurements carried out on the Helios space probes, on the other hand, show a slight reddening of the scattered intensity with respect to the solar spectrum, and this reddening decreases with increasing elongation; moreover the degree of polarization becomes lower with increasing wavelength (Leinert et al., 1981). A similar type of reddening has been found for other atmosphereless bodies of our planetary system. It should, however, be emphasized that not all zodiacal light observations support the evidence of this

R. H. Giese and P. Lamy (eds.), Properties and Interactions of Interplanetary Dust, 249–253.

colour effect.

As a reddening in the visual range of the spectrum cannot well be explained by a wavelength dependent refractive index (Röser and Staude, 1978), we suspected that it might be produced by the surface structure of the particles, and we shall show in the following that a slight surface roughness, a roughness structure of the order of wavelength and smaller, can in fact explain the observed colour effects; this slight roughness might be superposed on the large irregularities mentioned above.

2. THE PARTICLE MODEL

In order to simplify the calculations we constructed a model particle which shows up only the slight surface roughness. This model does not represent a realistic interplanetary dust particle, because the large irregularities are neglected, but it is sufficient for a discussion of colour effects. Our model consists of an ensemble of large, convex, randomly oriented particles with a slightly rough surface. The material is slightly absorbing so that light refracted into the particles will not escape - so only the reflected component has to be considered. The particles are assumed to be so large that results for light reflection from planar rough areas may locally be applied to the surface of the particles.

The problem of light reflection from slightly rough planar areas can be solved in form of a perturbation expansion, the so-called Rayleigh-Rice-approach (Lord Rayleigh, 1907; Rice, 1951). Given the rough plane and the incident plane wave, the refracted and the reflected light are expanded in plane waves, and the coefficients of these terms are determined by the perturbed boundary conditions. Explicit results are then obtained as an expansion in hk (k: wavenumber, h: mean height deviation of the roughness structure), as shown here for the reflected electric field:

$$\vec{E}_r = \vec{E}_r^{(0)} + \vec{E}_r^{(1)} + \vec{E}_r^{(2)} + \ldots ,$$

and if h is much smaller than wavelength, we may neglect terms of higher than second order in hk.

Averaging the intensity over the statistical properties of the surface then gives, up to the second order in hk,

$$\langle |\vec{E}_r|^2 \rangle = |\vec{E}_r^{(0)}|^2 + \langle |\vec{E}_r^{(1)}|^2 \rangle + 2\,\mathrm{Re}\, \vec{E}_r^{(0)} \cdot \langle \vec{E}_r^{(2)} \rangle^* .$$

The first term represents the light reflected from a smooth surface, the second is the incoherent contribution produced by the roughness, and the third term is the coherent component - it is for the most part negative and represents the intensity diminution of the zero order term. Explicit expressions have been given e.g. by Barrick and Peake (1967) and Valenzuela (1967).

The decisive quantity, which has most influence on the scattering behaviour, is the correlation function $\rho(r)$ of the roughness structure; for a rough surface $z=f(x,y)$ ($\langle f \rangle=0$) we have

$$\langle f(x,y)\ f\ (x+\Delta x,\ y+\Delta y) \rangle = h^2\ \rho(\sqrt{(\Delta x)^2+(\Delta y)^2}) \quad \text{with} \quad \rho(0) = 1,$$

where the bracket denotes the average over (x,y). In the following we choose an exponential function for $\rho(r)$:

$$\rho(r) = \exp(-r/l)$$

l is called the correlation length of the roughness structure.

These results can now be applied to the surface of our model particles. Here we make use of a well known theorem which states that the scattering pattern of large, convex, randomly oriented particles is identical with the scattering pattern of a large sphere with the same surface conditions. So for calculating the reflected light we may assume our model particles to be spherical.

3. RESULTS AND DISCUSSION

Calculations yield, for unpolarized incident light and a dielectric constant $\varepsilon = 5$, the curves in Figs. 1 and 2 for intensity and polarization of the reflected light as a function of the scattering angle Θ; the infinite value of the parameter kl corresponds to a perfectly smooth surface (Schiffer and Thielheim, 1984).

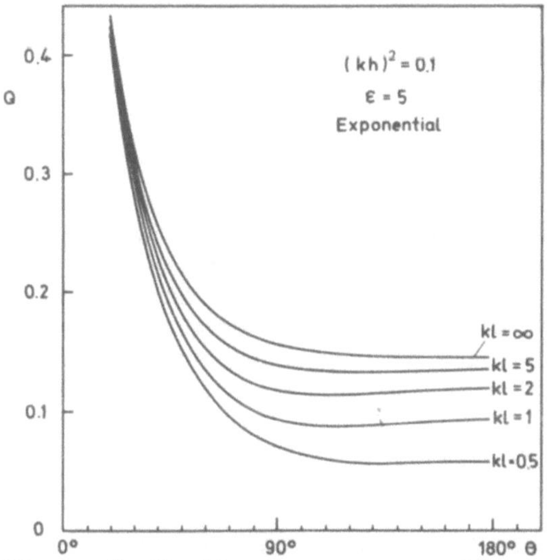

Figure 1. Scattering cross section of a slightly rough sphere.

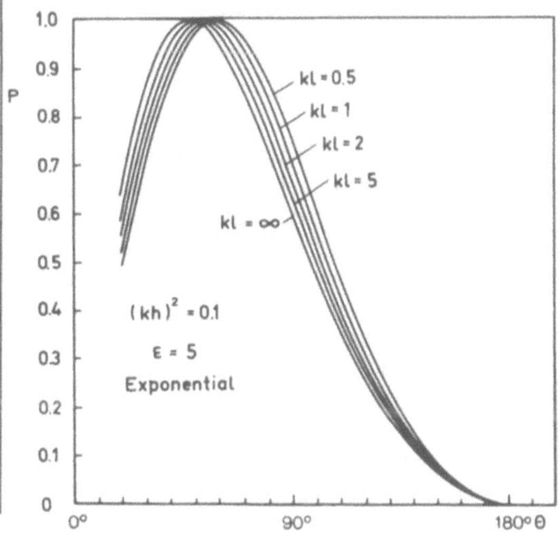

Figure 2. Polarization curve of a slightly rough sphere.

These plots show that stronger roughness, that means smaller l, reduces the reflectivity of the surface and shifts the polarization maximum to larger scattering angles.

Fig. 3 shows the wavelength dependence of the scattered intensity between 4000 and 8000 Å for several scattering angles, a dielectric constant $\varepsilon = 2.25$, and a correlation length l = 5000 Å; the curves have been normalized to 1 for $\lambda = 4000$ Å. Obviously the scattered light is reddened for all scattering angles.

For a comparison with measurements we calculated the ratio of intensities at visual and blue wavelengths as a function of the elongation, as measured by Helios. For achieving this we have to integrate our results

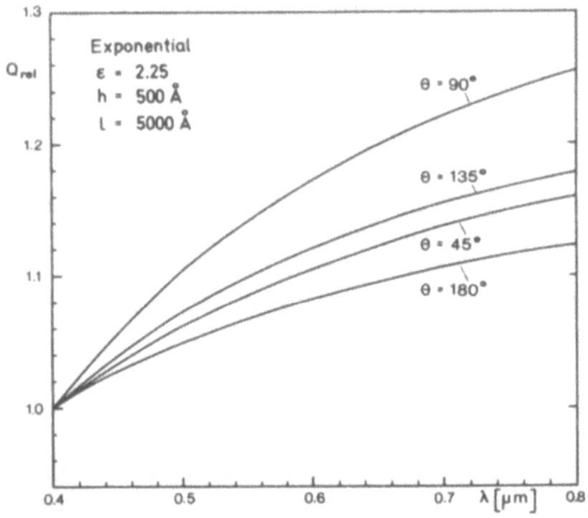

Figure 3. Normalized scattering cross section as a function of wavelength for several values of the scattering angle θ.

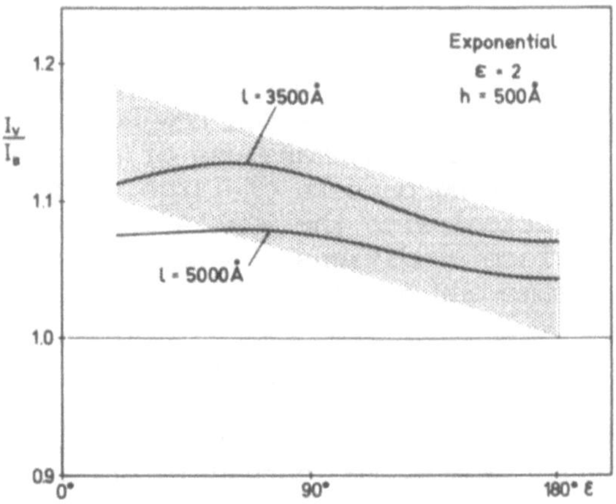

Figure 4. Ratio of intensities at wavelengths λ_V = 5290 Å and λ_B = 4250 Å as a function of elongation; shaded area: range of measured data.

over the line of sight, and assuming a particle density proportional to $R^{-1.3}$ (R: distance from the sun) we obtain curves as shown in Fig. 4; the shaded area indicates the range of the measured data. One may conclude that the observed reddening of the zodiacal light can well be produced by slight surface irregularities of the type considered here.

However, we so far neglected the strong roughness, which is certainly the dominant type of irregularity for the interplanetary dust particles. But closer inspection shows that this strong roughness, since it does not lead to wavelength dependent effects, does not change these curves appreciably. So particles with a composite surface combining both roughness types possibly represent a realistic model of the zodiacal light particles.

The superposition of both roughness types also provides an explanation of the observed flattening of the polarization curve with increasing wavelength: the polarization values are in this model mainly determined by the light scattered multiply at the surface of the particle, which is an effect of the strong type roughness; enhanced multiple scattering leads to smaller polarization values. When the reflectivity of the surface increases for larger wavelengths, this especially enhances the multiply scattered intensity and so leads to a further flattening of the polarization curve.

Maybe the observed reddening of the scattered light from other atmosphereless objects of the planetary system can at least in part be explained by a slight surface roughness of the type considered above - it seems plausible that bodies exposed to the environment of the solar system exhibit a similar microstructure of their surfaces.

<u>Question (Zerull)</u>: Does your theoretical approach also reduce the

polarization reversal, or could it be modified to do this?

Answer (Schiffer): No, our model does not lead to negative polarization values near backscattering, at least not for dielectric materials. We believe the polarization reversal to be an effect of the strong type roughness.

REFERENCES

Barrick,D.E. and Peake,W.H., 1967, Battelle Memorial Institute, Report No. BAT-197-A-10-3, AD 662 751.
Leinert,C., Richter,I., Pitz,E., Planck,B., 1981, Astron. Astrophys. 103, 177.
Lord Rayleigh, 1907, Philos. Mag. 14, 70.
Mukai,S., Mukai,T., Giese,R.H., Weiß,K., Zerull,R.H., 1982, Moon Planets 26, 197.
Perrin,J.-M. and Lamy,P.L., 1983, Opt. Acta 30, 9, 1223.
Rice,S.O., 1951, Commun. Pure Appl. Math. 4, 351.
Röser,S. and Staude,H.J., 1978, Astron. Astrophys. 67, 381.
Schiffer,R. and Thielheim,K.O., 1982, Astron. Astrophys. 116, 1.
Schiffer,R. and Thielheim,K.O., 1984, J. Appl. Phys., in press.
Valenzuela,G.R., 1967, IEEE Trans. Antennas Propag. AP-15, 4, 552.

OPTICAL MODELS OF THE THREE DIMENSIONAL DISTRIBUTION OF INTERPLANETARY DUST

R.H. Giese, G. Kinateder, B. Kneissel, U. Rittich
Ruhr-Universität Bochum
Bereich Extraterrestrische Physik
P.O.Box 10 21 48
4630 Bochum 1, F.R.G.

ABSTRACT. Several models of three dimensional distributions of the interplanetary dust number densities are compared with observational values. It is shown that a comparatively satisfying fit can be achieved by distributions similar to the "Sombrero-Model" of Dumont and that multilobe models as recently proposed by Buitrago et al. can be rejected due to discrepancies with rocket photometry of the inner zodiacal light.

1. INTRODUCTION

In general, there is agreement that the zodiacal cloud is concentrated towards a plane (or surface) of symmetry, which is close to but not identical with the ecliptic plane. Furthermore one assumes rotational symmetry with respect to an axis (z) originating in the sun and perpendicular to the symmetry plane. But it is still in question how the interplanetary dust is distributed off the plane of symmetry referred to above. In the following the models in discussion are compared and one model will be definitely ruled out. For this discussion we neglect possible variations of the dust population (size distribution, physical characteristics cf. Schuerman, 1980; Dumont, this volume) and the difference between the ecliptic plane and the symmetry plane. The local dust density $n(\vec{r})$ is described in helioecliptical coordinates $(r, \beta_\odot)$ of the position vector $\vec{r}$, which has its origin in the sun. In this coordinates the ecliptical plane (here: plane of symmetry), is $z = r \sin\beta_\odot = 0$ since $\beta_\odot = 0$ (see Fig.1).

Because of the rotational symmetry with respect to z (pole axis) the introduction of a 3-rd coordinate can be avoided (cf. Fechtig et al., 1981).

2. MODELS

From interpretation of earth based observations (Dumont, 1976), measurements by Helios 1/2 (Leinert, 1978) and Pioneer 10/11 (Hanner

R. H. Giese and P. Lamy (eds.), Properties and Interactions of Interplanetary Dust, 255–259.

et al., 1976) a power law $n(r) \sim r^{-\nu}$ ($1{,}2 \leq \nu \leq 1{,}5$) was derived as a first one-parameter-approach for the density distribution in the ecliptic plane. For the three dimensional distribution $n(\vec{r})$, most authors adopt an analytical form $n(r,\beta_\odot)=n_o \cdot (\frac{r}{A})^{-\nu} \cdot f(\beta_\odot)$, where n_o is the number density near the earth's orbit ($r=A=1$ A.U. and $f(\beta_\odot=o)=1$). Layer models ($n=n\, r^{-\nu} f(z)$) are ruled out by rocket photometry of the inner zodiacal light (Leinert et al., 1976). Models presently most under consideration are based on different analytical expressions for $f=f(\sin\beta_\odot)$. Fig. 1 presents some examples. In the following discussion $\nu=1.3$ (cf. Leinert 1981) is adopted. The numerical value of the parameters are selected to achieve some optimum fit to the observations of the zodiacal light.

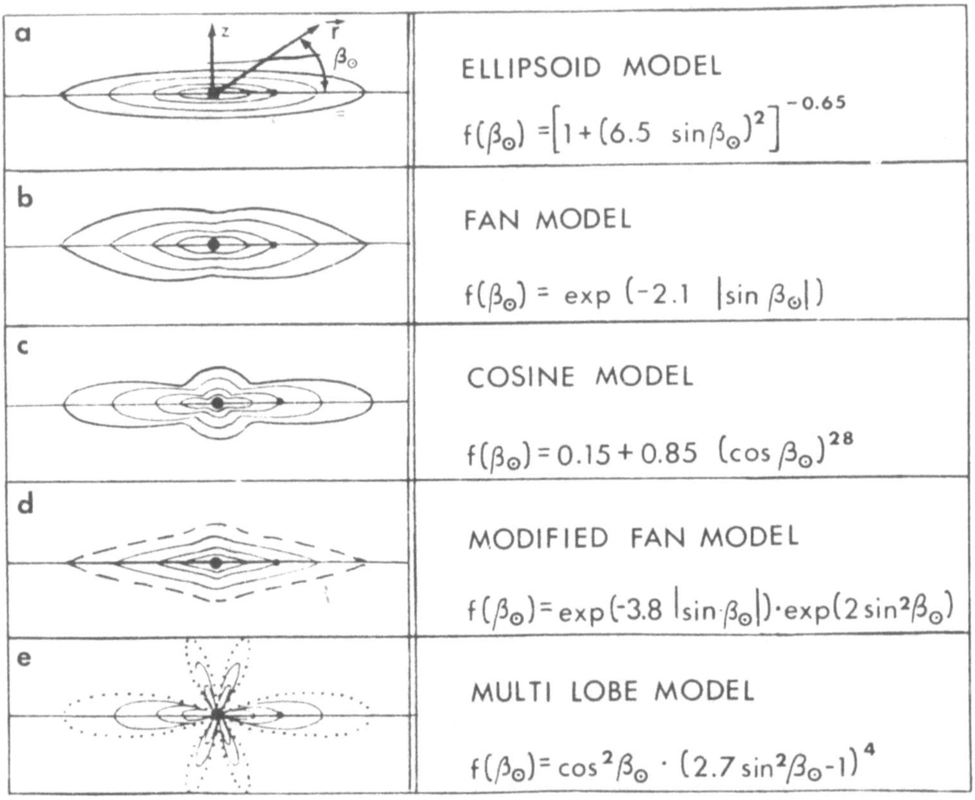

Figure 1. Isdodensity curves of particle number densities for different models.

In the "Ellipsoid Models" (Giese and Dziembowski 1969, Dumont 1976) the surfaces of equal number density are ellipsoids. They are shown in Fig. 1a (same in b, c, d, e) for $n=2n_o$, $n=1n_o$, $n=0.5n_o$, and $n=0.3n_o$.

"Fan-Models", like Fig. 1b, were proposed by Leinert (1978) as a best fit to observations of the inner zodiacal light and for interpretation of the Helios results. Furthermore they have been adopted for theoretical studies (e.g. Fahr at al. 1981, Leinert et al. 1983).

To improve the fit to zodiacal light observations Dumont first proposed at the IAU General Assembly 1976 and later modified (Dumont, 1984) his "Sombrero-Model", which is characterized by a bulge at the z-axis caused by an additional, radially symmetric component. Even though the "Cosine-Model" (Fig. 1c) proposed in the present paper is based on a more simple analytical form it can be considered to be a Sombrero-Model.

A "modified Fan-Model" (Fig. 1d) was very recently proposed by Lumme and Bowell (Icarus, in print). Contrary to the original Fan-Models it shares with the Sombrero-Models the enhancement of $n(\vec{r})$ close to the z-axis.

All models referred to above have in common a monotonic decrease of $n(\vec{r})$ with increasing z. In contrast to this Buitrago et al. (1982) derived from zodiacal light data in the helioecliptical meridian for elongations $\varepsilon > 90°$ a bimodal function $f(\sin\beta_{\odot})$. They solved the inversion of the zodiacal light brightness integral by transforming it to a Volterra integral of the second kind and arrived at a model similar to Fig. 1e (Multilobe model), which we defined here by means of an analytical expression and using $\nu=1.3$ instead of $\nu=1$ and the numerical table as Buitrago et al. provide. If models of this type could be confirmed, the orbital distribution of the zodiacal grains should have in addition to the usual population of micrometeoroids concentrated towards the ecliptic plane a second component with inclinations clustered about some angle i with $45°<i<90°$, a very challenging situation.

3. COMPATIBILITY WITH OBSERVATIONS

3.1 Viewing directions along selected great circles

To compare the different types of models integration along the line of sight was performed using Leinerts empirical volume scattering function (Leinert, 1978). The surface brightness $I(\ell,\beta)$ of the zodiacal light was calculated as a function of ecliptic latitude β and longitude ℓ (with respect to the sun as seen from the observers position). It was then compared with observational values (Levasseur-Regourd and Dumont, 1980) along the helioecliptic meridian plane ($\ell = 0$) and for a circle $\ell = 90°$ from the ecliptic to the ecliptical pole. In the latter directions the ellipsoid model provides an exellent fit. Also the deviations of the "Cosine-Model" and of the modified Fan-Model stay within 10% of the observed values. On the other hand the multilobe Model yields maximum deviations of about 20% at ecliptic latitudes about $\beta=60°$ and the Fan-Model predicts too much brightness from $\beta \approx 30°$ increasing to a deviation of about 40% near the pole.

In the helioecliptic circle the brightness of the Cosine-Model stays everywhere within about 10% of the observed values. The Ellipsoid-Model yields too low intensity for elongations $\varepsilon = \beta < 90°$ (40% near $\varepsilon=45°$) and the Fan-Model too high intensities between $\varepsilon=60°$ and $\varepsilon=130°$. For $\varepsilon < 60°$ the fit of the Fan-Model is exellent (5%).

The modified Fan-Model stays within deviations of 10% up to $\varepsilon = 120°$ but then deviates by more than 20%. The multilobe Model generally stays within 20% but produces much too low intensities (about 40%) near $\varepsilon = 120°$. Generally the Sombrero-Models (cf. Cosine-Model) yield the best fit. Fan-Models are satisfying for the inner zodiacal light, but not elsewhere. The fit by the multilobe Model is not worse than the behaviour of the celebrated Fan- or Ellipsoid-Models in unfavourable viewing directions.

3.2 Inner zodiacal light

The agreement of different models with a rocket photometry of the inner zodiacal light along circles at $\varepsilon = 15°$, 21°, and 30° about the sun was discussed by Leinert et al., 1976 (cf. their Fig. 2). It was found, that only the Fan-Model yields a satisfying fit in this region.

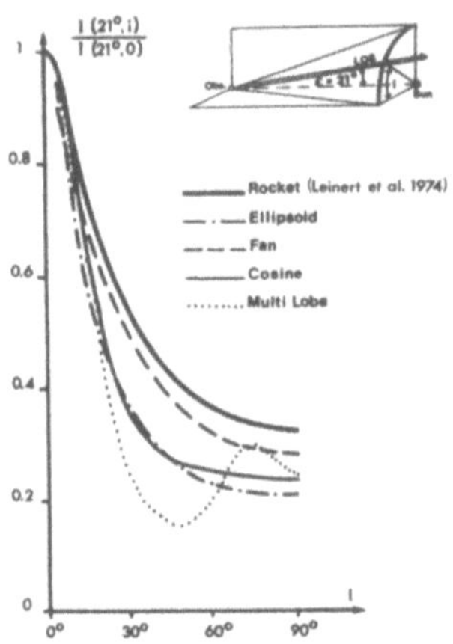

Figure 2. Measured Intensity $I(\varepsilon, i)$ of the inner zodiacal light compared to predictions from models.

It is tempting to investigate now if multilobe models show diagnostic signatures in this domaine. In viewing direction on the great circles as discussed in section 3.1 the line of sight passes through both lobes of dense and well illuminated material. Therefore the drastic differences between multilobe and ordinary models are smoothed out because of the integration along the line of sight. On the other hand at low elongations and moderate inclinations (i) of the scattering plane with respect to the ecliptic one expects that the line of sight misses to penetrate the second lobe of an additional concentration of material outside the ecliptic plane. For viewing directions towards these gaps a decrease would be expected in the zodiacal light. As a test we calculated the intensity along a circle with $\varepsilon = 21°$ about the sun (corresponding to Leinert et al., 1976, Fig. 2) expected for the multilobe Model (Fig.2). Contrary to the monotonic decrease of brightness from i=0 to 90° as found for all other models, only the multilobe Model shows a strong minimum close to i=45° and a secondary maximum near i=75°. This definitely is in disagreement with the rocket observations of Leinert et al., 1976.

4. CONCLUSIONS

None of the existing models shows a perfect agreement with the brightness distribution of the zodiacal light. However, for a first approach sombrero type models provide a best compromise for a simple approximation of $n(\vec{r})$ by few parameters. Multilobe models as proposed by Buitrago et al. are in strong contradiction to observations of the inner zodiacal light and should be rejected. The preliminary results referred to above are presently complemented by investigations including other viewing directions and possible observations from space probes.

5. REFERENCES

Buitrago, J., Gomez, R., and Sanchez, F., 1982, in: Sun and Planetary System (Fricke, W. and Teleki, G. Eds.), Reidel Dordrecht, p. 345-347.

Dumont, R., 1976, in: Interplanetay Dust and Zodiacal Light (Elsässer H. and Fechtig, H. Eds.), Lecture Notes in Physics, **48**, Springer-Verlag, Heidelberg, p. 85-100.

Dumont, R., 1984, private communication.

Fahr, H.J., Ripken, H.W., and Lay, G., 1981, Astron. Astrophys., **102**, p. 359-370.

Giese, R.H., and Dziembowski, C., 1969, in: The Zodiacal Light and the Interplanetary Medium (Weinberg, J.L., Ed.), NASA SP-150, p. 271-276.

Fechtig, H., Leinert, C., and Grün, E., 1981, in: Landolt-Börnstein, Volume 2, Astronomy and Astrophysics, Subvolume a (Schaifers, K. and Voigt, H.H., Eds.), Springer-Verlag, Berlin, Heidelberg and New York, P. 229-243.

Hanner, M.S., Sparrow, J.G., Weinberg, J.L., Beeson, D.E., 1976, in Interplanetary Dust and Zodiacal Light (Elsässer, H. and Fechtig, H., Eds.), Lecture Notes in Physics, **48**, Springer-Verlag, Heidelberg, p. 29-35.

Leinert, C., Link, H. Pitz, E., and Giese, R.H., 1976, Astron. Astrophys., **47**, p. 221-230.

Leinert, C., 1978, Die räumliche Verteilung des interplanetaren Staubes nach Zodiakallichtbeobachtungen von Helios 1 und 2, Habilschrift, Heidelberg.

Leinert, C., Richter, I., Pitz, E., and Planck, B., 1981, Astron. Astrophys., **103**, p. 177-188.

Leinert, C., Röser, S., and Buitrago, J., 1983, Astron. Astrophys., **118**, p. 345-357.

Levasseur-Regourd, A.C., and Dumont, R., 1980, Astron. Astrophys., **84**, p. 277-279.

EQUILIBRIUM TEMPERATURES OF H_2O, CO_2 and NH_3 ICE GRAINS IN THE SOLAR SYSTEM

J.F. CRIFO
C. N. R. S.
B. P. n° 10
91371 Verrières le Buisson Cedex
France

ABSTRACT. The equilibrium temperatures of spherical grains of pure H_2O, CO_2 and NH_3 ice in the solar system are computed at heliocentric distances ranging from 1 AU out to (> 1500 AU) where solar illumination becomes negligible compared to galactic and extra-galactic light ; the grain size range is from submicron to tens of meter in diameter. These results are based on recently published optical constants and on the use of the "complex angular momentum" (CAM) theory for computing the interaction of the grains with light.

1. INTRODUCTION

Ices of simple molecules are expected to be present in the heliosphere under the form of "grains" with sizes ranging from fractions of micron in the environment of comets, to kilometers in the remote "Oort cloud" of comets, with intermediate sizes there, as well as in planetary rings and, possibly, in the interplanetary medium. Since the pioneering works of Watson, Murray and Brown (1963), many theoretical studies have appeared, focussed on H_2O ice at moderate sizes and heliocentric distances. We present here partial results of a work undertaken to cover the full range of solar system conditions ; a detailed account of it, with comprehensive bibliography and detailed discussion of the results will be published in another place.

2. BASIC PRINCIPLES

The temperature, T, of a spherical grain with given radius at a given heliocentric distance is obtained by solving for T the grain energy budget equation : see for instance Mukai and Schwehm (1981). Following these authors, the contribution to this budget from galactic and solar charged particles can be ignored at our locations. The energy input is thus absorption of solar, galactic and extragalactic light, and the energy loss is the sum of the sublimation heat loss and of thermal radiation from the grain. We neglect all thermal time constants, i.e. the grains are assumed isothermal.

R. H. Giese and P. Lamy (eds.), Properties and Interactions of Interplanetary Dust, 261–266.

3. INPUT PARAMETERS

The grain thermodynamic constants needed in computing the vaporization loss are taken from standard Physics Handbooks. For the solar flux we use available reference solar spectra (extreme U.V. fluxes being taken at solar minimum values). Galactic light is represented by a sum of diluted blackbodies and extragalactic light by a uniform 3K blackbody.

Our sources for the ice optical constants are the following : for H_2O, we use the remarkable work of Warren (1984) ; for NH_3, the compilation of Martonchik (1984) is adopted ; for CO_2 , we use our own compilation, which interpolates between the following data : Egan and Spagnolo (1969) from near U.V. to near I.R., Tsujimoto (1983) at thermal I.R. wavelengths, and Ron and Schnepp (1967) for the two quadrupole-induced bands in the far-I.R..

4. GRAIN OPTICAL MODEL

Integration of the absorbed and radiated emission over wavelength is made for H_2O from 0.04 to 1000 μm, for NH_3 from 0.15 to 200 μm, and for CO_2 from 0.3 to 1000 μm. A specially developped integration routine based on the trapezium method is used : it varies the wavelength step and minimizes the interval of integration in order to insure a preset numerical accuracy and a minimum computing time.

For small values of the diameter to wavelength ratio x, Rayleigh and Mie theory are used according to standard practice ; beyond x = 22, we use the CAM theory (Nussenzveig and Wiscombe, 1980) which gives the Mie values to a high level of accuracy, has no limit in x, and is computationally very efficient.

5. RESULTS

Figures 1 to 3 present the equilibrium temperatures of pure H_2O, NH_3 and CO_2 ice grains, respectively. Each curve corresponds to a given heliocentric distance. The curve labeled "∞" is the temperature at negligible solar illumination (distances beyond several 1000 AU). The computations were made with preset accuracy, selected in accordance with the quality of the input optical constants. This limited accuracy combined with the finite radius increment selected for the plotting produces the slope discontinuities present in some curves. The upper limit of the error in temperatures is 2.5 % for ice grains ; for NH_3, it is 2.5 % at grain radius a larger than about 10 micron, and 5 % below ; for CO_2 it is 2.5 % at a greater than about 0.5 mm, 10 % for a between 60 micron and 0.5 mm, and gradually increases as a decreases, up to about 25 % for submicron grains. This behaviour is due to the fact that small grains radiate and absorb in narrow lines, the shape of which is not accurately known.

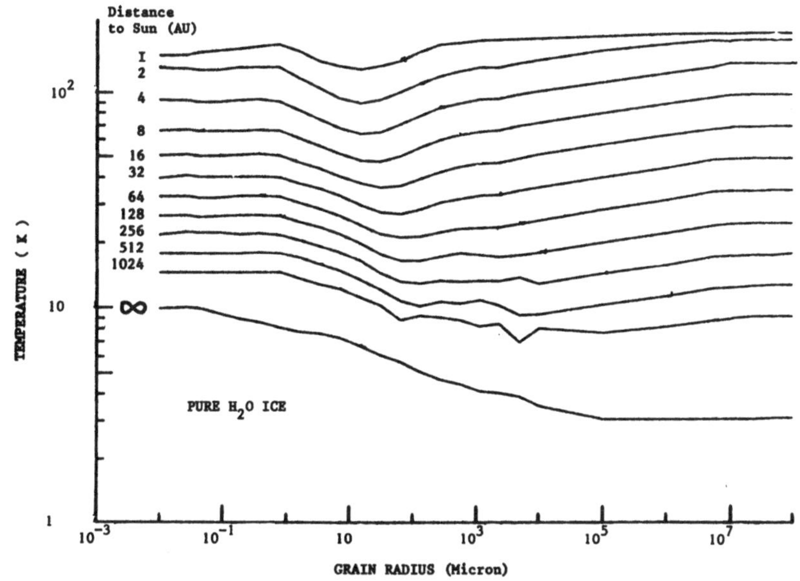

Figure 1 : Equilibrium temperatures of pure H_2O ice grains

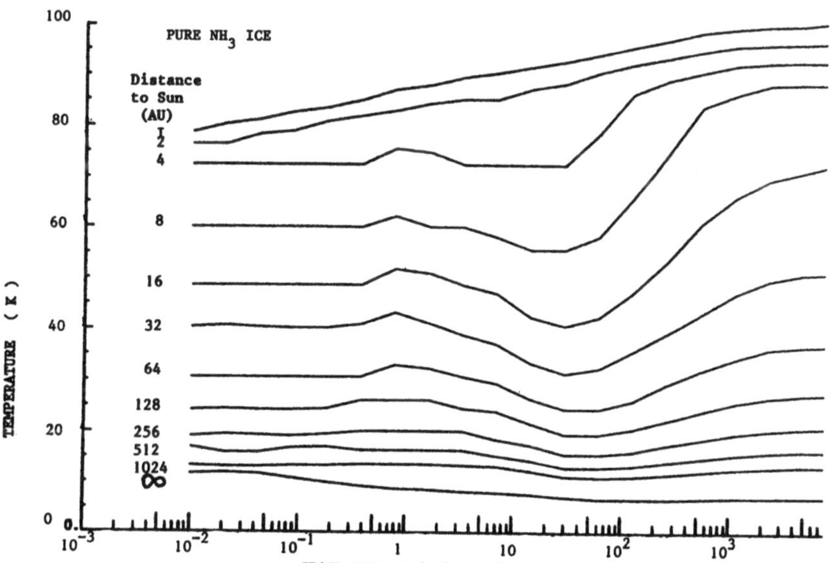

Figure 2 : Equilibrium temperatures of pure NH_3 ice grains

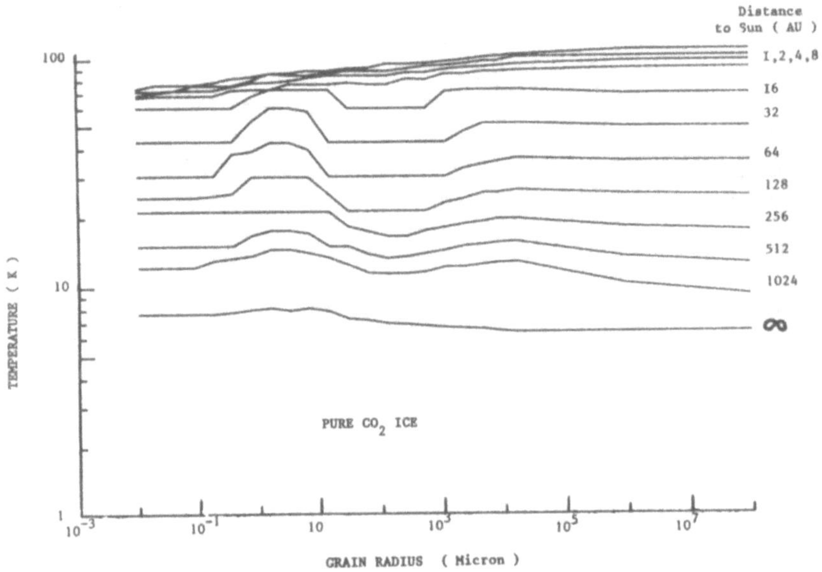

Figure 3 : Equilibrium temperatures of pure CO_2 ice grains

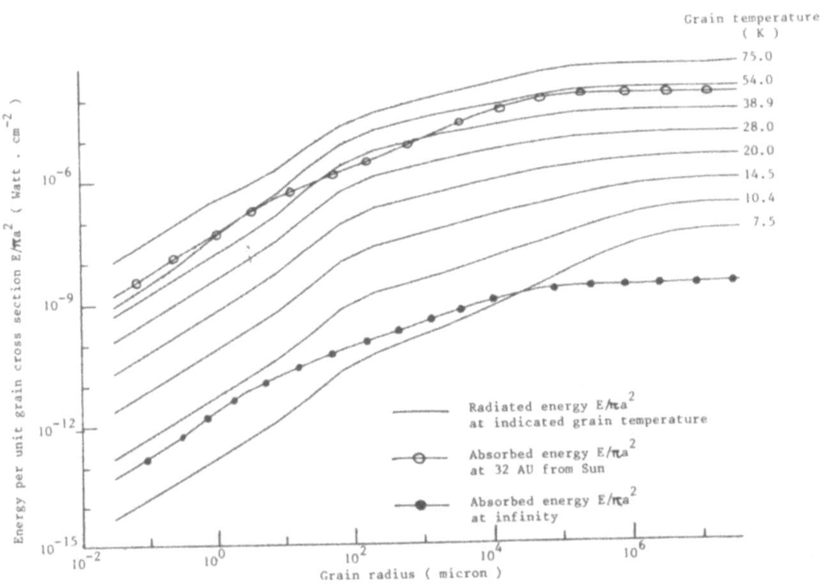

FIGURE 4 : Size variation of the energy absorbed and radiated by CO_2 ice grains

6. DISCUSSION

There has been a substantial number of works on the temperature of water ice grains, among which Greenberg (1971) for grains in the local interstellar medium, and Patashnick and Rupprecht (1975), Lamy and Jousselme (1976), Mukai and Schwehm (1981) for grains in interplanetary space. The present results agree reasonably with these authors when comparison is possible, but strictly identical results are not to be expected since the physical input parameters used here are based on new experimental data. Detailed comparison will be made in another place.

Figure 1 extends the previous knowledge of the size variation of water ice grain temperatures towards large sizes. The fact that this variation ceases only for "grains" larger than a meter in radius is a consequence of the very high transparency of ice as revealed by recent experimental data.

Examination of Figures 1-3 shows that CO_2 and NH_3 temperatures have an even stronger variation with size than H_2O ones do. It may be interesting to find out precisely where these variations come from.

First, consider small heliocentric distances (e.g. 2 AU for CO_2 or NH_3). There, the vaporisation energy loss Evap(a,T) is found to dominate completely the radiation loss Erad(a,T). Consequently, T is obtained by balancing the energy absorbed Eabs(a) with Evap(a,T). It thus varies very little with size and/or heliocentric distance since Evap is a very strongly increasing function of temperature : T(a) curves are flat and closely packed. Similar flat curves do not appear on Figure 1 because pure vaporisation equilibrium occurs at heliocentric distances smaller than 1 AU for water ice.

On the other hand, at large heliocentric distances where Erad(a,T) > Evap(a,T), T is given by equating Eabs(a) and Erad(a,T). Figure 4 shows as an example the set of curves Erad(a,T) appropriate for CO_2 ice and two curves Eabs (a) corresponding to positions at 32 AU from Sun and at infinity. All these curves have an a^3 dependance at small sizes where the particle is in the Rayleigh regime for all wavelengths, and become independant of a for particles large enough to be optically thick to all wavelengths. In between, there are inflexions which reveal the fact that the absorption bands come into play one after the other as a increases. Multiple crossings between a pair (Eabs, Erad) are possible and explain the shapes of the curves in Figure 3. This phenomenon appears to be quite general for transparent materials and, for instance, has been shown to exist for olivine by Röser and Staude (1978).

To conclude, let us note that the meaning of the present results is tied mainly with the following initial assumptions : (a) that the grains are isothermal ; (b) that their surface is smooth to all scales up to ~ 1 mm ; and (c) that they are chemically pure. All these assumptions can be dropped, but at the cost of introducing many additional parameters, a task which may prove justified when experimental results will require it.

ACKNOWLEDGEMENTS

Drs. G. GREHAN and G. GOUESBET, and Dr. W.J. WISCOMBE, are thanked for having supplied their SUPERMIDI and QABSASY Codes, on which part of the routines used here are based.

REFERENCES

Egan, W.G., Spagnolo, F.A., 1969, Applied Optics, **8**, 2359.

Greenberg, J.M., 1971, Astron. Astrophys., **12**, 240

Lamy, Ph., Jousselme, M.F., 1976, Proceedings of IAU Colloquium 31, edited by H. Elsässer and H. Fechtig, Springer Verlag, p. 443-447

Martonchik, J.V., Orton, G.S., Appleby, J.F., 1984, Appl. Opt., **23**, 4, 541.

Mukai, T., Schwehm, G., 1981, Astron. Astrophys., **95**, 373.

Nussenzveig, H.M., Wiscombe, W.J., 1980, Phys. Rev. Lett., **45**, 18, 1490.

Patashnick, H., Rupprecht, G., 1975, Astrophys. J., **197**, L79

Röser, S., Staude, S.J., 1978, Astron. Astrophys., **67**, 381

Ron, A., Schnepp, O., 1967, J. Chem. Phys., **46**, 10, 3991.

Tsujimoto, S., Konishi, A, Terada, N., Kunimoto, T., 1983, Cryogenics, **23**, 5, 251.

Warren, S.G., 1984, Appl. Opt;, **23**, 8, 1206.

Watson, K., Murray, B.C., Brown, H., 1963, Icarus, **1**, 317.

THE OPTICAL PROPERTIES OF CIRCUMSTELLAR AND INTERSTELLAR DUST IN THE MID-IR

B. PEGOURIE and R. PAPOULAR
CEN-Saclay, Service d'Astrophysique
91191 Gif-sur-Yvette, Cédex, France

ABSTRACT. We deduce the optical efficiency $Q(\lambda)/a$ of the circumstellar dust material over the range 8-30 µm from twelve M and G giants and supergiants and that of the dust present in the general ISM and molecular clouds from thirteen protostars. We then discuss the differences between these two profiles in the light of the condensation theory.

1. INTRODUCTION

The IR spectrum of oxygen-rich stars is characterized by two important bands at 10 and 20 µm ascribed to silicates (Woolf and Ney, 1969). These features are visible in absorption in HII regions, molecular clouds and protostars spectra, and in the general interstellar extinction curve. The absorption profile is then narrow and peaks at 9.5 µm. One finds them in emission in young stars, M giants and supergiants, novae and supernovae. In these cases, the emission features are generally wider (FWHH from 2.5 to 3.6 µm) and their wavelength of maximum intensity is around 10 µm.

Up to now, neither meteoritic nor laboratory materials permit to precisely reproduce the emission observations. That is why we adopted the inverse attitude: the empirical determination of the optical properties of dust between 8 and 30 µm, from objects with a simple structure to permit without ambiguity the Q/a profiles extraction.

In Sect.2, we present quickly the choice criteria for the sources and the method we used. We then present in Sect.3 two profiles deduced from emission and absorption observations and show in Sect.4 how condensation and thermodynamic computations permit to understand the observed differences.

2. DESCRIPTION OF THE METHOD

2.1. Each object consists of a central star surrounded by a spherical dust shell. The envelope contribution to the total flux between 8 and 30 µm (i.e. the IR excess) is obtained by subtracting from the total flux the stellar continuum extrapolated from the photometric HKLM points.

R. H. Giese and P. Lamy (eds.), Properties and Interactions of Interplanetary Dust, 267–271.

The large variety of emission profiles (fig.2) can be understood if large grain radii are allowed (Papoular and Pégourié, 1983). Figure 1 shows the silicates ' optical efficiency between 8 and 13 µm for different grain sizes as predicted by theoretical computations which also predict that, for a grain radius $\lesssim 0.3$ µm,the optical properties in that range of wavelength become independent of a. The variation of the shape of the band with a is clearly similar to that of the IR excess of the three stars shown in fig.2.

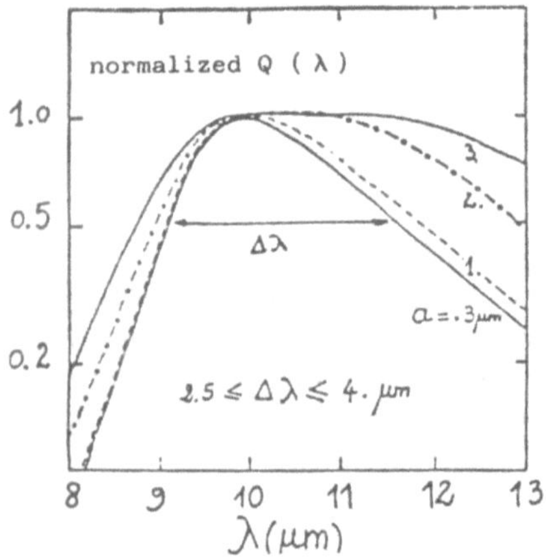

Fig.1. : Mie computations of silicates' optical efficiency for a grain radius varying from 0.3 to 3 µm.

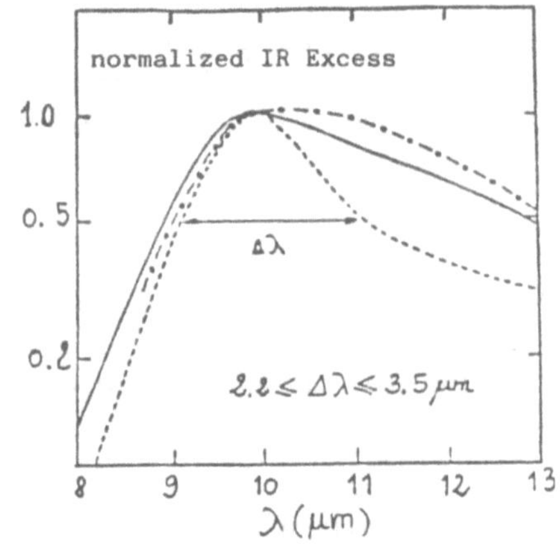

Fig.2: IR excess between 8 and 13 µm for ---TZ Cas, —— R Leo, and -.-. RCnc.

Many M supergiants exhibit a strong and narrow excess (identical to that of TZ Cas) indicating small grains and small optical depth, i.e. the best conditions to have a strong similarity between the IR excess and the dust's optical properties profile.

2.2. Similarly, favorable conditions for the analysis of absorption features will be found in the case of young objects like protostars embedded in dense molecular clouds which are expected to reflect the properties of dust in the local ISM, at least in composition.

2.3. We used a spherical shell model with power law distributions for grain density and temperatures:

$$n(r)=n_o\,(r/R_o)^{-2} \qquad \text{and} \qquad T(r) = T_o\,(r/R_o)^{-.4}$$

as suggested by the rigorous models (Rowan-Robinson and Harris, 1982). For the present purposes, the more important parameters are :

- The colour temperature of the central star: T_*,
- The inner temperature of the shell : T_o,
- The optical thickness at 10 µm: τ_{10},
- The optical efficiency profile of dust: $Q(\lambda)$.

Equilibrium condensation computations of a gas of cosmic composition (Grossman, 1972) show that the first materials to condense will be SiO_4Mg_2 , then SiO_3Mg. We used their optical properties as measured by Day (1979) as zero-order approximation to those we want to deduce from the observations.

For each object, we first determine the parameters T_*, T_0 and τ_{10}, using the optical properties of SiO_3Mg or SiO_4Mg_2 and minimizing the quantity:

$$S^2 = \Sigma_\lambda \left\{ \left[F(\text{data}) - F(\text{model}) \right] / F(\text{data}) \right\}^2$$

We then modify the Q (λ)/ a values by an iterative process until we obtain a common profile for all the emission features and a common profile for all the absorption ones, which permit to best fit the individual spectra.

3. COMPARISON OF THE TWO PROFILES

Figures 3 and 4 show the comparison between data and model for the twelve objects in emission, and the thirteen in absorption. In every case, the residual discrepancy S is smaller than 15%. The resultant optical efficiency profiles are shown fig.5.

The a and b profiles are clearly distinguished by the wavelength of their maximum emissivity peak and by the depth of the 14 μm trough. Their principal properties are summarized in the table 1 below.
The average properties deduced from absorption spectra (b) are identical to those of amorphous enstatite: SiO_3Mg (residual discrepancy < 9%), whereas those deduced from emission features (a) are close to those of amorphous forsterite: SiO_4Mg_2 (c), except between 12 and 16 μm where they are considerably stronger. Note that these differences must have a physical meaning since they are larger than the dispersion of individual profiles about their average (the length of one vertical bar is 2σ of the mean).

	CS shells (emission)	ISM (absorption)
principal component	SiO_4Mg_2	SiO_3Mg
λmax (μm)	10	9.5
HHFW (μm)	2.5	2
Q(10 μm)/Q(14 μm)	6	20

Table 1: Characteristics of observational Q (λ)/ Ao profiles.

4. DISCUSSION

Grossman (1972) shows that, in a cooling atmosphere, solid SiO_4Mg_2 condenses before solid SiO_3Mg then reacts with the gas to transform into the latter if enough time and solid-gas interface is available. The difference between the two profiles we deduced above is understood in the light of this result:

CIRCUMSTELLAR SHELLS:
Rather short residence time (10^3 years) ....
Large radii (i.e. small area/volume for soldi-gas interaction) } SiO_4Mg_2
Relative movement of dust w.r.t. gas ....

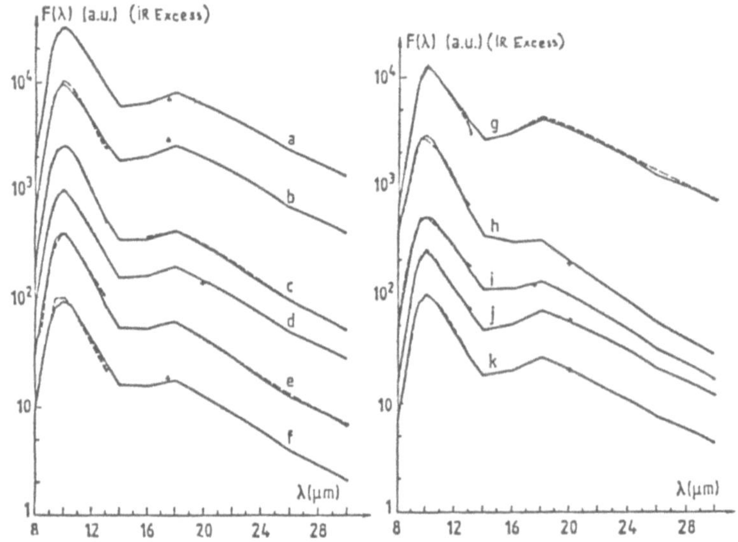

Fig.3 : Model (full line) and data (dashed line) for: a: V 915 Sco,
b: HR 5171 A,
c: μ Cep (1,2),
d: RW Cyg (1),
e: α Ori (1,2),
f: K W Sgr,
g: PZCas (2),
h: VXSgr (1),
i: UYSct,
j: RSCnc (1),
k: SLep.
Data are from: (1) Merrill et al (1976) and (2) Forrest et al. (1979).

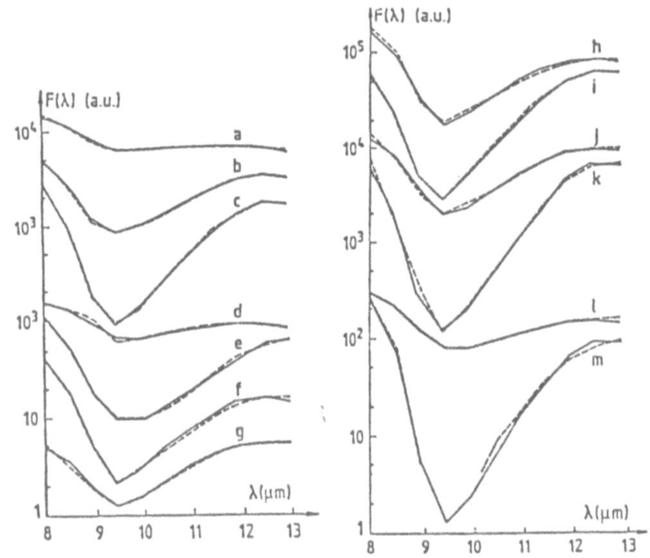

Fig.4: Same as fig.3 for absorption features:
a: AFGL989,
b: AFGL 2591,
c: NGC 7538/IRS 1,
d: AFGL 2059,
e: S255/IRS1,
f: AFGL 2136,
g: AFGL 2884,
h: NGC 7538/IRS 9,
i: G 45.07 + 0.13,
j: NGC 2170/IRS 3,
k: W3/IRS5,
l: BN,
m: W 33.
Data are from Willner et al (1982). Curves are arbitrarily displaced in the vertical direction for convenient display.

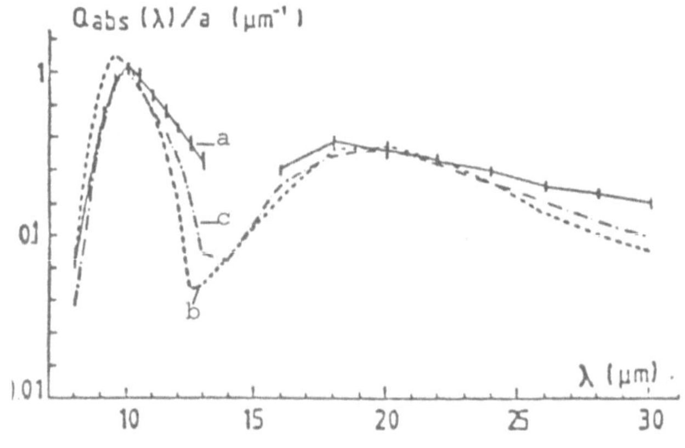

Fig.5 : Resultant Q (λ)/a profiles:
(a) : emission features,
(b) : absorption features Identical to pure SiO_3Mg,
(c) : pure SiO_4Mg_2.

GENERAL ISM:

Long residence time ---------------------------------- } SiO_3Mg
Small grains (i.e. large area/volume for solid-gas interaction)---- }

INTERMEDIATE CASE:

It is interesting to note that several cold oxygen-rich stars, with thick CS shells, display a narrow and shallow dip at ≈ 9.5 um super-imposed upon a wide emission feature of the SiO_4Mg_2 type. This could be interpreted as an equilibration of gas and dust at the envelope periphery resulting in the transformation of SiO_4Mg_2 into SiO_3Mg.

Herndon and Suess (1977) show that, in the condensate of a gas of solar composition, the larger the proportion of SiO_4Mg_2 is w.r.t. SiO_3Mg, the larger is the proportion of iron in the silicate phase. Our findings agree with this result because:

- The optical efficiency profile from emission spectra is consistent with a mixture of SiO_4Mg_2 and SiO_4Fe_2 taking into account the respective abundances of iron and magnesium.

While the optical efficiency profile from absorption spectra excludes other materials exhibiting strong spectral features between 8 and 30 μm, it permits the presence of metallic iron or FeS or any other material having a monotonous optical efficiency profile in this range of wavelength.

REFERENCES

Day, K.L., 1979, Astrophys. J., 234, 158.
Forrest, W.J., Mc Carthy, J.F., Houck, J.R., 1979, Astrophys. J., 233, 611.
Forrest, W.J., Gillett, F.C., Houck, J.R., Mc Carthy, J.F., Merrill, K.M., Pipher, J.L., Puetter, R.C., Russell, R.W., Soifer, B.T., Willner, S.P., 1978, Astrophys. J. , 219, 114.
Grossman, L., 1972, Geochim. Cosmochim. Acta, 36, 597.
Herndon, J.M., Suess, H.E., 1977, Geochim. Cosmochim. Acta, 41, 233.
Mc Carthy, J.F., Forrest, W.J., Briotta, D.A., Houck, J.R., 1980, Astrophys. J., 242, 965.
Merrill, K.M., Stein, W., 1976, Publ. Astron. Soc. Pacific, 88, 285.
Papoular, R., Pegourie, B., 1983, Astron. Astrophys., 128, 335.
Rowan-Robinson, M., Harris, S., 1982, Monthly Notices Roy. Astron.Soc., 200, 197.
Seki, J., Hasegawa, H., 1981, Progr. Theo. Phys., 66, 903.
Willner, S.P., Gillett, F.C., Herter, T.L., Jones, B., Krassner, J., Merrill, K.M., Pipher, J.L., Puetter, R.C., Rudy,R.J., Russell, R.W., Soifer, B.T., 1982, Astrophys. J., 253, 174.
Woolf, N.J., Ney, E.P., 1969, Astrophys. J., 155, L181.
Woolf, N.J., 1973, in "Interstellar Dust and Related Topics", IAU Symposium n° 52, ed. J.M. Greenberg and H.C. Van de Hulst, (Reidel), 485.

LABORATORY INFRARED TRANSMISSION MEASUREMENTS OF INTERPLANETARY DUST AND IMPLICATIONS FOR REMOTE OBSERVATIONS OF COSMIC PARTICLES

S.A. Sandford and R.M. Walker
McDonnell Center for the Space Sciences
Washington University
St. Louis, Missouri 63130
U.S.A.

ABSTRACT

Laboratory infrared absorption spectra of interplanetary dust particles show that most fall into one of three spectral groups (1) designated as "pyroxenes", "olivines", and "hydrated silicates".

Comparisons with 10 μm emission data from Comet Kohoutek (2) show that no single spectral type matches the cometary data. However, a reasonable match is obtained by coadding equal amounts of pyroxene and hydrated silicate type spectra.

The possibility that the 10 μm feature seen in some comets is due to a combination of these two components can be tested in two ways. First, the pyroxene type spectra show detailed features of variable position and strength within their 10 μm bands. Based on the laboratory data we estimate that comet spectra taken at a resolution of $\lambda/\Delta\lambda \sim 200$ would be sufficient to find these features, if present. The magnitude of these features is expected to be ≤ 0.1 of the maximum strength of the 10 μm band.

Second, the hydrated silicate particles usually have bands at 3.0, 6.2, and 6.9 μm. The first two bands are probably linked to the presence of water while the 6.9 μm band is probably due to carbonates. The 3.0 μm feature is broad and might be hard to separate from thermal background. However, spectra taken at resolutions of $\lambda/\Delta\lambda \sim 50$ could see the 6.2 and 6.9 μm bands. The strengths of these bands (which are also seen in the spectra of CM meteorites) vary (one particle has a carbonate band that is actually stronger than its corresponding 10 μm silicate band), but are typically ~0.1 the depth of the silicate features. A second, narrow carbonate band at 11.4 μm can also sometimes be seen in particle spectra but its strength is less than that of the 6.9 μm band. This the search for the 6.2 and 6.9 μm bands in astronomical objects, while not requiring high resolutions, would require good S/N.

REFERENCES

(1) P. Fraundorf, R.I. Patel, S.A. Sandford, and R.M. Walker, 1983, LPS, **XIV**, p. 207.
(2) M.K. Merrill, 1974, Icarus, **23**, p.566.

R. H. Giese and P. Lamy (eds.), Properties and Interactions of Interplanetary Dust, 273.

PLANS FOR RELEASE OF SIMULATED INTERPLANETARY MATERIALS INTO LOW EARTH ORBIT

I. B. Strong, R. R. Brownlee, E. H. Farnum,
W. F. Huebner, T. D. Kunkle, J. R. Stephens,
Los Alamos National Laboratory
Mail Stop D436
Los Alamos, New Mexico 87545
USA

M. F. Bode
Department of Astronomy
University of Manchester
Oxford Road
Manchester M13 9PL
England

ABSTRACT. A number of experimental releases into low earth orbit of small, but detectable, quantities of very fine particulates are planned for the next several years. The principal materials are chemically and physically similar to those generally assumed to be the major constituents of interplanetary and circumstellar dusts (silicates, silicon carbide, graphitic materials). Later releases will include multi-phase releases of non-volatile particulates and volatile ices, simulating cometary material. Both ground-based and Space-Shuttle-based observations will be made in the visible and infra-red to study particle-plasma interactions, cometary physics and cloud dynamics in orbit.

Since opportunities for observing the behavior of such well-characterized materials in space are likely to be rare, and the effective duration of significant concentrations will be very short, we wish to have early input from the scientific community so as to maximize the scientific return. We will be receptive to all suggestions including collaborations in making the observations.

R. H. Giese and P. Lamy (eds.), Properties and Interactions of Interplanetary Dust, 275–278.

1. INTRODUCTION

The program described here was presented orally at I.A.U. Colloquium No. 85 while it was in its formative stages. The purpose in presenting it at such an early time was twofold.

First, the objection has been raised that perhaps not much may be learned from such simple observations that cannot be deduced from laboratory measurements. There is some truth to this in the sense that it would be very surprising to find significant differences between measured scattering of the solar illumination and that measured in laboratory apparatus such as the scattering chamber described below. On the other hand, the behavior of the grain cloud under the influence of the ionizing component of the solar radiation, in a very tenuous atmosphere containing magnetic and electric fields is not easily predictable. Desirable diagnostics include determination of alignment, coagulation and plasma interaction. Such simpler studies, during which we will develop methods for material preparation, storage, release and diagnosis are necessary preliminaries to projected later releases of grains combined with ices and plasma-forming materials. Spectroscopic studies of the behavior of such simulated cometary materials cannot easily be reproduced in the laboratory. On the whole, the response to the proposed program has been encouraging.

The second reason for an early presentation was to elicit input from others, whether positive or negative. We have had a certain amount of both, which has been valuable to us. There have been suggestions for ensuring the dispersion of the grains, details of earlier failures and objections based on possible contamination of other experiments in LEO and the upper stratosphere by material tailored to mimic interplanetary grains. Our proposed tagging of the grains by modifying isotopic ratios of some of the composing elements was considered difficult to detect by many workers. The suggestion, by Professor Lamy, of incorporating small, but easily detectable amounts of unlikely elements has been adopted. Simple equipment already in wide-spread use will easily distinguish our grains from all other natural and artificial materials. One must also keep in mind that the quantities to be released (~ few kgm) should be compared to the naural infall of meteroids ($\gtrsim 10^5$ kgm/day).

2. GRAIN COMPOSITION AND PREPARATION

In part, the impetus for beginning this work was due to the existence at Los Alamos, in the Materials Science and Technology Division, of equipment for producing uniform, superfine powders of refractory substances in useful quantities--hundreds of grams per day, using plasma chemical synthesis techniques. Volatile precursors (e.g., $Fe(CO)_5$, SiH_4, CH_4, etc.) are injected into a radio-frequency plasma heating system, where they react before reaching an expansion chamber in which grain nucleation takes place. These grains, 5 nm to 50 nm in size, coagulate into tangled chains of overall dimensions ~10 μm. Similar grain structures are seen in many condensation processes.

Specific impurities are inserted with the feed gases by injecting volatile precursors containing the elements of interest.

3. GRAIN CHARACTERIZATION

The grains will be characterized for physical and optical properties in the laboratory. Physical characterization, e.g., the size distribution, elemental composition and crystalline forms will be measured using transmission electron microscope and x-ray diffraction techniques, and these properties correlated with the optical properties.

Optical properties will be measured using a light-scattering apparatus whose construction is now almost complete. The grains will be suspended in air in a scattering chamber, and illuminated by a solar simulator light source. The detector, a monochrometer-optical multichannel detector assembly, is on a rotating arm pivoted about the scattering chamber, allowing wavelength and angle resolved detection of the scattered light. The aim of the optical characterization is to quantify the scattering properties of irregular-shaped grains, which are not amenable to theoretical calculation.

Instrument operation and the taking of data is completely computer-controlled. Improvements now being investigated include polarization sensitive detection, monitoring aerosol mass using a quartz element mass monitor, and extending the wavelength coverage into the UV and IR.

4. PACKAGING AND RELEASE MECHANISMS

The most difficult part of the program is to devise a method of packaging the grains for later release in space. There are a number of constraints and other requirements. A cloud of grains must be formed rapidly which will scatter the solar illumination to the observing station(s) on the ground. Each grain must scatter independently with minimal multiple-scattering. The cloud must be observable from its initial appearance in the west until it sets in the east several minutes later, without growing too large.

There is every reason to believe that a compact mass of grains simply released into orbit will remain compact, the individual grains aggregating in clumps. To avoid this we are investigating packaging grains into a matrix consisting of a sublimating solid or low-boiling point liquid, which will inhibit coagulation and aid in the slow dispersal of the grains in space. Binder dispersal may be enhanced by heating the binder-aerosol combination to control the aerosol release rate.

The first release attempts will be made from small sub-satellites, with a total mass of around 50 kg, ejected from a Get-Away-Special (G.A.S) canister on a Space Shuttle. This will take place near the end of a mission, the sub-satellite being triggered from the ground to eject the grains after the Shuttle has de-orbited, obviating Shuttle environment contamination.

5. GROUND AND SPACE BASED OBSERVATIONS

Only ground-based observations of the optical and dynamical behavior of the aerosol clouds will be made during the initial particle releases. The releases will be timed so that the cloud is illuminated by the sun, but will be passing the observing station just before dawn or just after dusk so that the background sky is dark. The optical observations will be made from the DARPA Maui Optical Station (AMOS) and the Maui Optical Tracking and Identification Facility (MOTIF) on the summit of Mt. Haleakela, H.I. The telescopes there are designed to track rapidly moving objects which cannot, in general, be followed by more slowly moving astronomical telescopes. The principal instruments are a 1.6 meter Cassegrain telescope (AMOS) and twin 1.2 meter telescopes on a common mount (MOTIF). Associated instrumentation will probably include a multi-spectral photopolarimeter and an infra-red polarimeter operating in the 10 μm to 30 μm range. A laser beam director is available to probe the clouds using LIDAR techniques.

In the late phases of the program it is hoped that observations can be made directly from the Shuttle. The quantities of material needed would be much smaller, and spectral observations could be extended well into the ultra-violet and infra-red.

6. CURRENT STATUS

The basic scattering apparatus for optical characterization of the grains is nearly completed. A large environmental test chamber has been acquired for laboratory studies of grain dispersion and aggregation. Authorization has been received to reserve spaces in GAS canisters, and sub-satellite design has begun.

We hope to perform the first release which will primarily an engineering, dispersal and observationals systems test, during 1986.

ON THE BIMODAL NATURE OF THE PARTICLE-SIZE-DISTRIBUTION FUNCTION OF COMETARY DUST

C.P. Liu§ and H. Kimura§
Max-Planck-Institut für Aeronomie
D-3411 Katlenburg-Lindau
Federal Republic of Germany

ABSTRACT. The bimodal nature of distribution function $f(\gamma)$ is confirmed by analysing the optical isophotal maps of comets Kohoutek 1973 XII and Bennett 1970 II. Extending the dynamic analysis to the infra-red data of two comets, Kohoutek 1973 XII and IRAS-Araki-Alcock 1983 d, we demonstrate that the secondary component $f_2(\gamma)$, which dominates in the range of large particles, plays a much more important role in emitting the thermal radiation (IR) than in the case of light scattering (optical). The relative weight of two components, measured by their height ratio, g_2/g_1, should be increased by a factor of about 20, in order to obtain a good fitting to the infra-red data.

1. INTRODUCTION

Based on the mechanical theory, we have analysed the brightness distribution of comet Arend-Roland (1957 III) and obtained a distribution function $f(\gamma)$ of bimodal nature (Kimura and Liu, 1975; Liu and Kimura, 1982). Our result is different from the results by other authors (e.g., Sekanina, 1980; Campins and Hanner, 1982), in its bimodal nature, or in the unique manifestation of the secondary mode $f_2(\gamma)$. The bimodal size distribution might correspond to the two-component model suggested by the infrared observation of comet West 1976 VI (Oishi, et al. 1978). From this point of view, IRAS data of comet IRAS-Araki-Alcock, published quite recently (Walker, et al. 1984), and an earlier infrared (λ = 3.5 μm) data of comet Kohoutek (Ney 1974) are interesting. In the present work, in order for clarifying the nature of bimodal size distribution, the infrared data mentioned above have been studied by the method of dynamical analysis.

§ On leave from Purple Mt. Observatory, Nanjing, China

R. H. Giese and P. Lamy (eds.), Properties and Interactions of Interplanetary Dust, 279–282.

2. THE DYNAMICAL ANALYSIS AND RESULTS

The method of analysis and the model of dust emission, described and used in earlier works (Kimura and Liu, 1977; Liu and Kimura, 1982), are generalized to include the anisotropy effect of particle ejection from the inner coma region. Anisotropy is represented by adding a lobe type flux enhancement to the spherical (isotropic) flux of ejected particles.

At the date of Ney's observation, IR radiation of comet Kohoutek in the wavelength range $\lambda = 3.5$ μm was essentially due to thermal emission as indicated by the multiband photometry. If the bimodal distribution $f(\gamma)$ represents the existence of two components of cometary dust, we might reconcile the Ney's IR profile and the previous result of $f(\gamma)$ from the optical studies, simply by changing the relative weight of two components of $f(\gamma)$, namely, the ratio of two peak heights g_1 and g_2. In Fig. 1 the calculated intensity profiles in the main and sunward tails are shown for different values of the ratio g_2/g_1. Comparing with the observational data, it appears that $g_2/g_1 \cong 10$ gives the best result. The bimodal nature of $f(\gamma)$ is supported by our recent analyses of optical data for two comets, Kohoutek 1973 XII, and Bennett 1970 II. The details will be discussed in a separate article (Liu and Kimura in preparation), but the results on $f(\gamma)$ are summarized in Table 1.

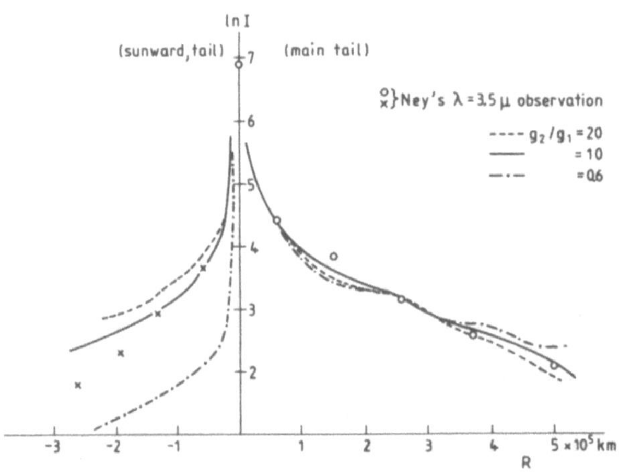

Figure 1. The IR($\lambda = 3.5$ μm) intensity profile along the tail axis of comet Kohoutek, on Jan 1.7 (U.T.), 1974. The data from Ney's IR($\lambda = 3.5$ μm) observation are shown by open circles (main tail side) and by crosses (sunward tail part). Three profiles calculated by assuming different peak height ratio, i.e., $g_2/g_1 = 0.6$, 10, and 20, are depicted by dash-dot, full, and broken lines, respectively.

TABLE 1 The parameters in the distribution function $f(\gamma)$

$$f(\gamma) = f_1(\gamma) + f_2(\gamma)$$

$$f_i(\gamma) = g_i(\gamma/\gamma_i)^{\beta_i} \exp\{(\beta_i/\alpha_i)\,|\,1 - (\gamma/\gamma_i)^{\alpha_i}\,|\}$$

Comet	Spectral Band	γ_1	γ_2	(g_2/g_1)
Bennett	orange	0.3	0.01	0.6
Kohoutek	orange	0.10	0.01	0.6
Kohoutek	IR(λ = 3.5 μm)	0.10	0.01	10
IRAS-A-A	IR(λ = 25 μm)	0.10	0.01	10
Arend-Roland	orange	0.10	0.01	0.6

Note 1. $\alpha_1 = 1.0$, $\alpha_2 = 0.5$, and $\beta_1 = \beta_2 = 4.0$ are common to all calculations treated in the present work.
2. γ_i: the γ-value at the i-th peak of the distribution.
3. g_i: the height of the i-th peak.
4. γ: repulsive force in unit of solar gravity.

Among the published IR data of comet IRAS-Araki-Alcock (Walker, et al., 1984), an isophot measured at λ = 25 μm was fitted by using a similar distribution function $f(\gamma)$ as that used in calculating the Ney's IR profile of comet Kohoutek, but with cutting off the contribution from particles in the range of γ larger than 0.3. This cutoff is intended to be a simplified treatment of the size effect for thermal emission from small particles. After incorporating the anisotropy effect in particle ejection, a fairly good agreement has been achieved, as is shown in Fig. 2, between the calculated isophotal contours and the observed ones.

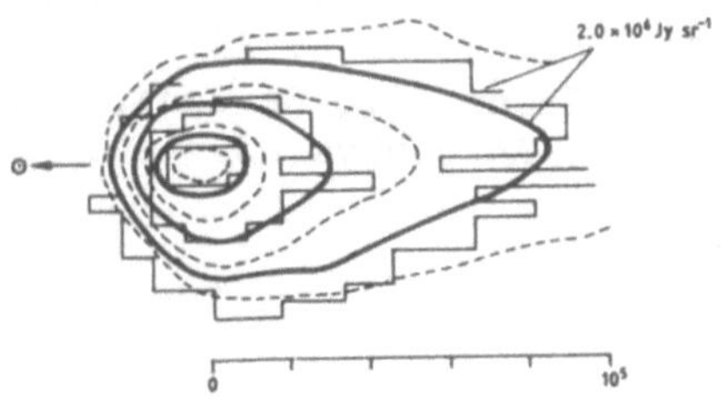

Figure 2. Brightness distribution of comet IRAS-Araki-Alcock, 1983 d. The isophotal contours from the λ = 25 μm observation are shown by thin broken and thick lines with flux level interval of a factor 2. The calculated contours are shown by thin lines with the flux level interval of a factor 4, which are to be compared with the thick line contours from observation. Scale is in km.

3. DISCUSSION AND CONCLUSION

The bimodal nature of distribution function $f(\gamma)$, first obtained from the study of comet Arend-Roland, is confirmed by analysing the optical data of comets Kohoutek and Bennett.

Extending the dynamic analysis to the infrared data (of comets Kohoutek, and IRAS-Araki-Alcock), we have found that the contribution from the particles belonging to the second component, $f_2(\gamma)$, is very important in emitting thermal radiation. Numerical results show that the ratio of the thermal emission efficiency to the light scattering efficiency of the second component particles is higher than that of the first component, and $(g_2/g_1)_{IR} \sim 20\ (g_2/g_1)_{op}$. This suggests that the $f_2(\gamma)$ particles (large particles) are much more effective thermal emitters, i.e. presumably absorbing metallic particles.

Acknowledgements. We wish to thank K. Jockers and W.-H. Ip for discussions and S. Deutsch for preparing the manuscript. We acknowledge Visiting Fellowships from the Max-Planck-Society during this work.

REFERENCES

Campins, H., and Hanner, M.S. (1982), in "Comets", ed. L.L. Wilkening, p. 341

Hanner, M.S., Aitken, D.K., Knacke, R., McCorkle, S., Roche, P.F., Tokunaga, A.T. (1984) JPL-preprint, Comet. Sci. Team PP.S. No. 49

Ishida, K., and Kosai, H. (1971) Tokyo Astr. Bull., 2nd ser., No. 204

Kimura, H., and Liu, C.-P. (1977) Chinese Astr., 1, 235

Liu, C.P., and Kimura, H. (1982), Chinese Astron. Astrophys. 7, 11

Ney, E.P. (1974) Icarus 23, 551

Oishi, M., Okuda, H., and Wickramasinghe, N.C. (1978) Publ. Astr. Soc. Japan 30, 161

Sekanina, Z. (1980) in "Solid Particles in the Solar System, IAU Symp. No. 90," eds. I. Halliday and B.A. McIntosh, p. 251

Walker, R.G., Aumann, H.H., Davies, J., Green, S., de Jong, T., Houck, J.R., and Soifer, B.T. (1984) Astrophys. J. Lett. 278, L 11

ON THE DYNAMIC BEHAVIOUR OF DUST STRIAE OBSERVED IN COMETS WEST 1976VI, AND MRKOS 1957V

H. Kimura §, C.-P. Liu §, and K. Jockers
MPI für Aeronomie
D-3411 Katlenburg-Lindau
Federal Republic of Germany

ABSTRACT. The dynamic properties of the dust striae, observed in comets West 1976VI, and Mrkos 1957V, are reexamined in terms of the high speed emission model (HS model) proposed by Notni (1966). It is demonstrated that most striae can be well fitted to modified synchrones but this kind of fitting does not lead to a conclusive model; in most cases, the dynamic solution is not unique, and, furthermore, the derived high values of the ejection speed appear to be unrealistic, particularly for distant (or old) striae. The results suggest that the required initial velocity could be better interpreted as indicating the presence of nonradial acceleration.

1. INTRODUCTION

The striae observed in the dust tail of several brighter comets are an interesting topic to be reexamined on the basis of mechanical theory. This is because, on one hand, the physical interpretation of striae formation is still an unsettled problem (Sekanina 1980; Fernândez and Jockers 1983). On the other hand, the mechanical theory which involves a large number of parameters to describe the particle dynamics, has not been worked out well.

Though the possible electromagnetic nature of the dust striae has been suggested by several authors (Vsekhsvyatskii 1959, Lamy and Koutchmy 1979), quantitative dynamical analysis has so far been made only within the framework of mechanical theory. Notni (1964) suggested that the striae might be interpreted as the result of intermittent high speed ejection of particles caused by a strong coupling between the dust and high temperature plasma. The dynamical fitting appeared successful, but the difficulty to produce a hot dense plasma in a cometary environment has been pointed out (Sekanina 1980). On the other hand, Sekanina and Farrell (1980, 1982) proposed a particle fragmentation model (PF model) and carried out a detailed analysis of the dust striae observed in two comets. PF model fitting agrees with

§ On leave from Purple Mt. Observatory, Nanjing, China

R. H. Giese and P. Lamy (eds.), Properties and Interactions of Interplanetary Dust, 283–286.

the observations as well, but no fragmentation mechanism is known to be consistent with the required trend of model parameters, for example, the large cometocentric distance of the fragmentation sites of the order of 10^6 km.

2. REEXAMINATION OF THE HIGH-SPEED EJECTION MODEL (HS MODEL).

An attempt was made to fit the data published by Sekanina and Farrell (1980, 1982) with an HS model calculation. Some results are shown in Figs. 1 and 2 for comets West and Mrkos, respectively. In the framework of the HS model the fitting parameters are ejection speed, v_e (or $w = v_e/V_c$, V_c denoting the orbital speed of the comet), age of the synchrone, τ (represented here as the true anomaly α_τ of the comet at the time of ejection) and azimuthal ejection angle Ψ_A. This angle is measured in the comet orbit plane from the prolonged radius vector into the direction opposite to the comet's orbital motion. For simplicity it was assumed that the vector of the ejection velocity lies in the comet orbit plane. To demonstrate the quality of

TABLE 1 Model Parameters of boxes shown in Figs. 1 and 2

Comet	Grid	True Anomaly α_τ	Ejection Velocity w	Ψ_A	γ (*)
West	W	+40°, +45°	0.30	20°, 40°	1.0, 2,5
	X	+45°, +50°	0.20	10°, 30°	"
	Y	+50°, +60°	0.10	0°, 30°	"
Mrkos	Q	+35°	0.15,0.20	0°, 20°	"
	R	+40°	0.05,0.10	0°, 20°	"
	S	+10°, 35°	0.05	0°, 20°	0.3, 0.8

(*) γ : the repulsive force in unit of solar gravity

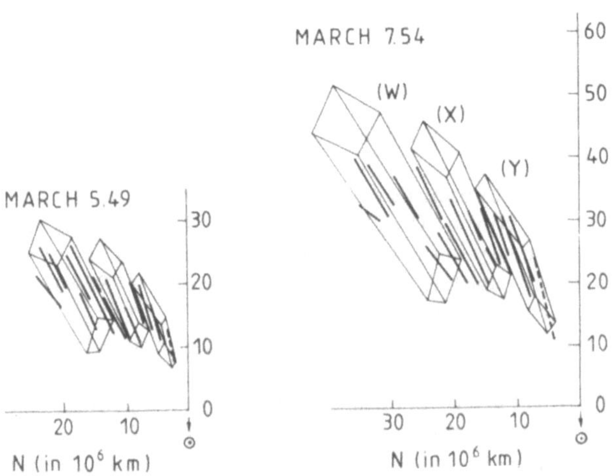

Figure 1. Time development of the striated tail of comet West 1976VI, is shown in the sky-plane projection. Observational data are taken from Sekanina and Farrell (1980). Calculated parameter boxes of displaced synchrones are superposed (for an explanation see text).

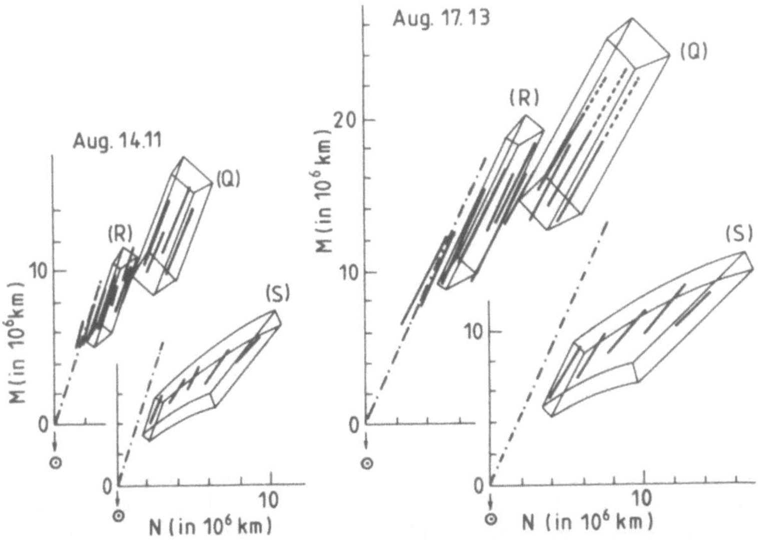

Figure 2. The same as Fig. 1 but for comet Mrkos 1957V. The obs. data are taken from Sekanina and Farrell (1982). The bottom graphs refer to the second family of striae observed in this comet.

the fit, in Figs. 1 and 2 in addition to the observed striae (heavy lines) "parameter boxes" are shown, of which the parameters are listed in Table 1.

While the overall fit seems quite satisfactory it is worth noting that in case of comet West the required ejection speed and angle gradually increase with increasing age of the stria (Y → X → W). This may indicate, that, contrary to our assumption, the ejection direction is not restricted to the comet orbit plane. During the relevant period of observations, the angle between line of sight and comet orbit plane was 35° for comet West in contrast to 65° for comet Mrkos.

In Fig. 3 the speed of particle ejection required to fit the striae in comets West and Mrkos is plotted vs heliocentric distance of the comet. For comparison, the orbital velocity of the comets are also shown.

3. DISCUSSION AND CONCLUSION

In comparison with the PF model of Sekanina and Farrell (1980, 1982) the following results emerge from this study: (A) Both the PF and HS model require high values of the repulsive force parameter γ (Table 1), i.e., absorbing submicron particles (0.1 - 0.3 μm in radius) as the main component of striae. (B) In both models the derived initial conditions are physically unrealistic, the too large ejection speed in the HS model as well as the too large cometocentric distance of the fragmentation site in the PF model. (C) The derived parameters for the striae show a remarkably large systematic variation with increasing age (see Figs. 3 and 4).

According to the current idea on the electrostatic charging of

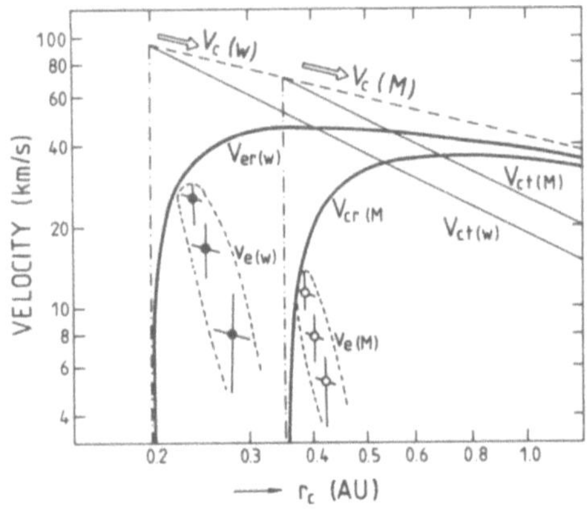

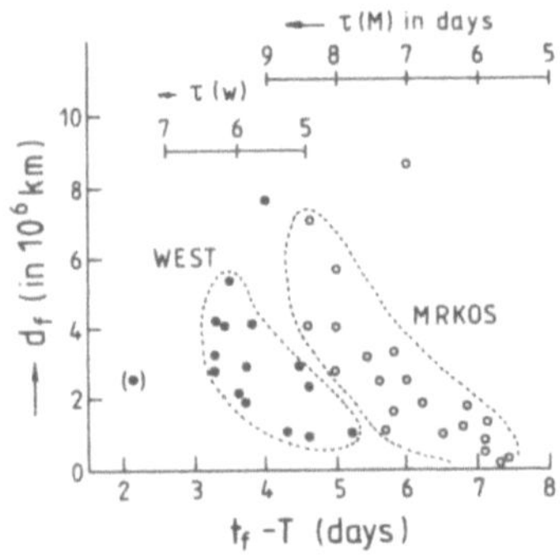

Figure 3. (left) The ejection speed v_e of the particles required for striae formation is plotted vs heliocentric distance of the comet. The magnitude V_c and the radial and transverse components V_{cr} and V_{ct} of the orbital velocity of the two comets are also shown. Subscripts M and W refer to comets Mrkos and West,respectively.

Figure 4. (right) The cometocentric distance of fragmentation, as required in the PF model is plotted as a function of time relative to perihelion passage (adapted from Sekanina and Farrell 1980, 1982).

dust grains, the submicron size particles in interplanetary space are very likely to be affected by Lorentz force acceleration (Wallis and Hassan 1983). From this viewpoint, statements (A) and (B) indicate the invalidity of models based solely on the central forces of the inverse square law (mechanical theory). Furthermore, statements (A) and (C) suggest the presence of nonradial acceleration; the magnitude of this acceleration may be roughly estimated as the ratio of v_e/τ obtained in the HS model. This value amounts to about 0.3 g_o in the case of comet West, and to about 0.5 g_o for comet Mrkos, where g_o denotes the solar gravitational attraction. These values are in the reasonable range of Lorentz-force acceleration acting on charged dust particles (Ip, Kimura and Liu in this volume).

REFERENCES

Fernández, J.A., and Jockers, K. (1983) Rep. Progr. Phys. 46, 665
Ip, W.-H., Kimura, H., and Liu, C.-P. (1985) this volume.
Lamy, P., and Koutchmy, S. (1979) Astron. Astrophys. 72, 50
Notni, P. (1964) Veröff. Sternw. Babelsberg, 15, No. 1, p. 1
Sekanina, Z. (1980) in "Solid Particles in the Solar System, IAU Symp. No. 90", ed. by I. Halliday and B.A. McIntosch, p. 237
Sekanina, Z., and Farrell, J.A. (1980) Astron. J. 85, 1538
Sekanina, Z., and Farrell, J.A. (1982) Astron. J. 87, 1836
Vsekhsvyatskii, S.K. (1959) Sov. Astron. J. 3, 490
Wallis, M.K., and Hassan, M.H.A. (1983) Astron. Astrophys., 121, 10

SELECTION EFFECTS AGAINST SMALL COMETS

H.A. Zook[1], J.A. Fernández[2], and E. Grün[3]

[1] NASA Johnson Space Center, Houston, Texas 77058 USA

[2] Observatorio do Valongo, U.F.R.J. Ladiera Pedro Antonio, 43, 20.080-Rio de Janeiro, Brazil

[3] Max-Planck-Institut für Kernphysik, Postfach 10 39 80, 6900 Heidelberg, F.R.G.

Abstract

There are three factors that inhibit detecting small comets relative to large comets: 1. They must pass closer to the Earth than large comets to be detected. 2. The resulting higher angular velocity for nearby comets means that, at comparable apparent magnitudes, there is less time available to discover a small comet relative to a large comet. 3. Small comets physically decay and vanish faster than do large comets. With these three factors, it can be shown that if the small comet input distribution g_i (R_n) varies as $R_n^{-\beta_i}$, where R_n is the nuclear radius of the comet, then the observed distribution will vary as $g_i(R_n)\ R_n^{5.5} \sim R_n^{5.5-\beta_i}$.

1. Introduction

The fainter a comet is, the more difficult it is to discover. This obvious fact gives rise to a catalogued bias against faint comets that needs to be corrected before the true magnitude number distribution can be obtained. Everhart (1967) showed that the probability of detecting a comet is jointly proportional to how far (in magnitudes) the apparent magnitude of a comet remains above some threshold magnitude for detection and to how long it remains above that threshold magnitude. The optical threshold magnitude depends both upon observing conditions and upon telescopic technique but, under good conditions, is around $H_t = 12$ for most amateur comet searchers.

We adopt Everhart's formulation in order to determine how the probability of discovering a comet depends on its nuclear radius. We concentrate our analysis upon comets that are so small (not more than two or three hundred meters in diameter) that they would normally only be optically detectable near the Earth. Heliocentric variations in cometary brightness can then be neglected. We will find that there is such a strong observational bias against

R. H. Giese and P. Lamy (eds.), Properties and Interactions of Interplanetary Dust, 287–291.

detecting small active comets that it is not surprising that small comets are not observed -- even though their true spatial density may be high.

We assume here that "small" (as defined above) active comets are detected only by reflected sunlight from their dust comas, and not from the gaseous part of the coma, nor by direct detection of the nucleus. The reason the gaseous part of the cometary coma should practically be invisible for small nuclear radii is due to the following considerations: The C_2 Swan band emissions, that are primarily responsible for the visual brightness of the gaseous coma, increase in integrated luminosity with increasing distance from the cometary nucleus up to about 3×10^4 km at one AU. Inside this "creation" distance for the C_2 bands, the coma brighness is nearly constant and depends on the square of the nuclear radius. When the nuclear radius is too small, no part of the the gaseous coma brightness will be visible against the celestial background light. The dust coma brightness, however, will still be above the visibility threshold near the nucleus of a small active comet.

2. The Dust Coma

The spatial mass density, $p(R')$, due to outflowing dust, at distance R' from the center of the comet is given by

$$p(R') = \frac{\dot{m}}{4\pi\,(R')^2\,\nu_t} = \frac{f\,\dot{R}_n}{\nu_t}\left(\frac{R_n}{R'}\right)^2 , \qquad (1)$$

where ν_t is the terminal radial velocity of dust grains due to drag by outflowing gas, $\dot{m}$ and $\dot{R}_n$ are the time rate of change of cometary mass and radius, respectively, and f is the fraction of the cometary surface that is actively releasing gas and dust. ν_t is related to cometary radius by $\nu_0 R_n^{\delta}$ where, from Probstein (1968) and Delsemme and Miller (1971), one can derive that $\delta \approx 0.5$ for small comets, and ν_0 is a constant.

The brightness of the coma at an angle ϕ from the nucleus is proportional to the optical depth of dust grains along the line of sight (see fig. 1). Integrating over the coma surface brightness out to the angle at which the brightness vanishes into the background light, one obtains the integrated luminosity, L, of a comet: L is then related to comet nuclear radius, R_n, and comet-Earth distance, Δ, as

$$L \sim \frac{R_n^{\,3}}{\Delta^2} . \qquad (2)$$

This relation assumes $\delta = 0.5$. To a given limiting apparent magnitude (or integrated luminosity), the distance to the comet varies with comet nuclear radius as $\Delta \sim R_n^{3/2}$.

3. Discovery Probability Versus Cometary Radius

From Figure 7 in Everhart (1967), the probability of

discovering a comet is given by

$$P = S_0/80 \tag{3}$$

where S_0 is defined by

$$S_0 = \int_{t_1}^{t_2} (H_t - H)\, dt \,. \tag{4}$$

H_t is the maximum apparent magnitude, or threshold magnitude, above which the comet is unlikely to be discovered by visual comet searches, and H is the apparent magnitude of the comet at time t. t_1 and t_2 are the times, in weeks, when the comet first becomes brighter or, respectively, dimmer, than H_t. The detection geometry for a hypothetical comet is depicted in Figure 2. Depicted there are the threshold distance Δ_t at which the comet first becomes detectable, the distance Δ_{min} of closest approach to Earth, and an intermediate distance Δ.

Space limitations preclude a rigorous mathematical treatment here, but one can qualitatively deduce much of the logic from inspection of Figure 2. The probability that our hypothetical comet will pass within the threshold distance, Δ_t, for detection is proportional to Δ_t^2. In addition, it is possible to show, using equations (3) and (4) that once it passes within Δ_t of the Earth, the probability that the comet will be detected is proportional to Δ_t. One can see this result in qualitative terms, by noting that for a given chord across the circle in Figure 2, that the time of passage along that chord is proportional to Δ_t(assuming constant comet-Earth relative velocity). This is the same time that appears in the sum in equation (4).

Thus, for a comet whose discovery sphere has radius, Δ_t, its probability of discovery is proportional to Δ_t^3. However, from equation (2) we note that $\Delta_t \sim (L_t)^{1/2} R_n^{3/2}$, where L_t, corresponding to H_t, is the threshold integrated luminosity of a comet that can be detected at distance, Δ_t. Therefore the probability, $P(R_n)$, of detecting small, active comets whose nuclear radius, R_n lies in the range dR_n, is given by the proportionality

$$P(R_n) \sim R_n^{4.5} g(R_n)\, d R_n \,, \tag{5}$$

where $g(R_n)$ is the true (as opposed to observed) differential distribution of comets of radius R_n.

4. The Effect of Cometary Physical Decay and Conclusions

Another factor concerning small comets is that they physically decay faster than large comets. This factor also affects their probability of being detected. It is not difficult to show, and we state here without proof, that if comets are continuously input into the inner solar system with a differential input distribution,

g_i (R_n), of the type

$$g_i(R_n) \sim R_n^{-\beta i} \quad , \tag{6}$$

then a temporal equilibrium distribution $g(R_n)$ under steady cometary decay is given by

$$g\,(R_n) \sim R_n^{-\beta i+1} \quad . \tag{7}$$

It is assumed in this analysis, that the radii of all comets are constantly being reduced at the rate $\dot{R}_n$.

Comets that that have been input into the inner solar system with an initial size distribution, $R_n^{-\beta i}$, will be observed with a size distribution

$$g_i(R_n)\,R_n^{5.5} \sim R_n^{5.5\,-\,\beta i} \tag{8}$$

We see that, unless the input distribution is very steep ($\beta_i \gtrsim 5.5$), small comets would be very difficult to detect with the classic search technique. Small comets could, indeed, contribute quite significantly to the meteoritic complex without our being aware of it.

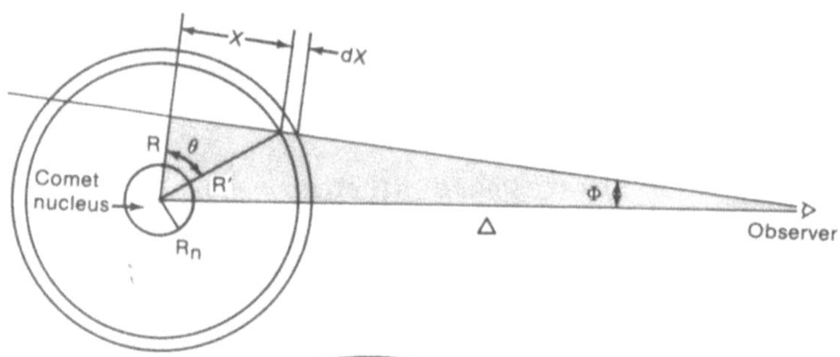

Figure 1. Geometry for observing the cometary coma brightness from distance, Δ.

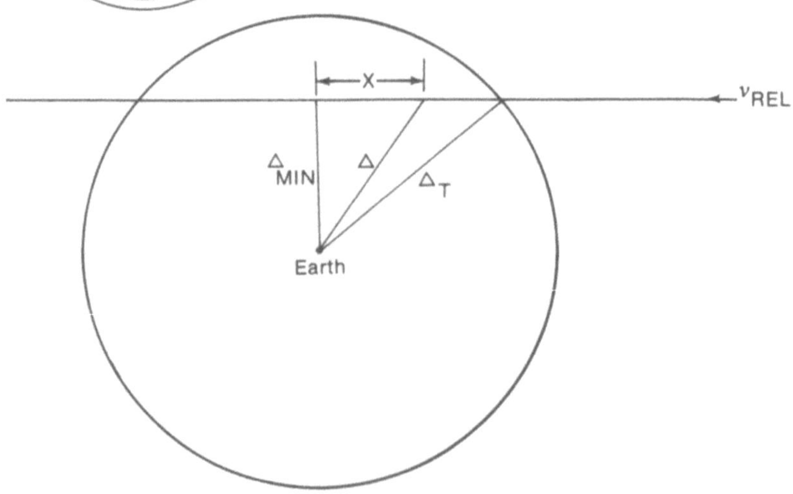

Figure 2. Geometry for a comet passing within the detection threshold distance, Δ_t, from Earth.

6. References

Delsemme A.H. and Miller D.C. (1971) Physico-chemical phonomena in comets--III The continuum of Comet Burnham. Planet. Space Sci. 19, 1229-1257.

Everhart E. (1967) Comet discoveries and observational selection. Astron. J. 72, No. 6, 716-726.

Probstein R.F. (1968) Dusty gasdynamics of comet heads. In: Problems of Dynamics and Continuum Mechanics, SIAM, Philadelphia, 568-583.

OPTICAL DETECTION OF LARGE METEOROIDS IN SPACE

H. A. Zook and A. E. Potter
NASA Johnson Space Center, Houston, Texas 77058, USA

Abstract

CCD sensors placed across the focal plane of large Schmidt telescopes have great potential for detecting and measuring the very low flux in space of meteoroids with diameters larger than 1 meter. With the Palomar "Big Schmidt", a detection rate of 1.4 per hour is obtained for meteoroids between 0.6 and 200 meters in diameter. For the Baker-Nunn "Satellite Tracking Camera", the corresponding rate is about 0.8 per hour. The key to obtaining such high detection rates derives from approximately setting the sensor integration time equal to the time it takes a meteoroid to cross a pixel field of view. This minimizes signal to noise problems and is accomplished, in practice, by multiple summing of short integration time data records to obtain data records of longer effective integration times.

1. Introduction

The Prairie Network -- a network of cameras that recorded meteor fireball trajectories for several years over about 10^6 square kilometers of the midwestern U.S. -- recorded the flux of meteoroids with masses up to about 10^6 g (McCrosky, 1968). Although estimates of the meteoroid masses causing these fireballs have since been revised significantly downward (e.g., Wetherill and Revelle, 1981), the data obtained from this network and from the similar Canadian Meteorite Observation and Recovery Project (Halliday, et al., 1984) have given us a fair estimate of the flux of moderately large (1 to 10^6 g) meteoroids. Little is known, however, about the near-Earth flux of meteoroids with masses much above 10^6 g until one reaches sizes corresponding to the telescopically observed Apollo asteroids (e.g., Shoemaker, et al., 1979). Masses of the latter objects are mostly upward of 10^{15} g.

The recent advent of CCD (charge coupled device) technology, however, now makes it very possible to telescopically detect an extremely low flux of very large meteoroids (or small asteroids and comets) with ground-based optical telescopes. Gehrels (1981) and McMillan and Stoll (1982) have already used CCD detectors to search for faint comets and for asteroids. In this paper, we describe a

R. H. Giese and P. Lamy (eds.), Properties and Interactions of Interplanetary Dust, 293–298.

promising technique of using CCD's as meteoroid detectors, and derive an expected telescopic detection rate for meteoroids between 0.6 and 200 meters in diameter. The method can easily be extended to objects outside this size range by adjustment of optical and electronic parameters.

2. CCD Detector Performance

Our analysis will be based on the published performance of an 800x800 pixel array CCD image sensor designed by investigators at the Jet Propulsion Laboratory and at Texas Instruments, Inc. (TI) and built by TI (Blouke, et al., 1983). This sensor is back-illuminated and has a thinned sensor active surface area of 1.22 cm x 1.22 cm with elementary pixel diameters of 15 μm. The exterior dimensions are about 1.7 cm x 1.7 cm x 0.2 cm. Four of these chips are to be used in the wide field/planetary camera on the space telescope. We chose this sensor because of its exceptionally low read-out noise and to illustrate the kind of performance that can be obtained from an existing device. Blouke, et al. (1983) have described the performance of this sensor in excellent detail, and we only repeat some pertinent facts.

The dark current noise is neglible when the sensor is sufficiently cooled. The sensor can be employed as four independent 200 x800 arrays of pixels. We will use the middle two arrays as the imaging section for our analysis and, after an exposure, will rapidly shift the data outward to the two outlying 200x800 arrays which will be shielded from incoming light. While a new image is being built up on the center two arrays, data in the two outlying arrays are to be respectively read to each of two outlying serial shift registers and subsequently read out through each of two read-out amplifiers. It is in the read-out amplifiers that read-out noise is generated. This noise is about 15 electrons at a 50 kpixel/s data rate. We will conservatively assume a read-out noise of 20 electrons for our analysis and will require that our data read-out rate not exceed 50 kpixel/s (so as not to pay an increased read-out noise penalty).

3. Data Acquisition and Analysis Procedure

Consider a CCD sensor placed at the focus of an optical telescope to detect sunlight reflected from objects passing through the sensor field of view. We state as a theorem that the maximum signal to noise ratio for a moving object is obtained when the sensor integration time, t, is such that the image of the object just moves across one pixel (picture element) in time t. This theorem is rather similar to the "dwell-in-the-cell" concept of R. Weber (1979) and so is not a new idea. We now examine the special application of the theorem to the use of CCD's in looking for large meteoroids passing rapidly through a telescope's field of view.

In actual fact, because different meteoroids will have different angular velocities through the telescopic field of view, it is not possible to integrate for just the right length of time such that each meteoroid image will cross just one pixel during an integration time. Thus we cannot strictly satisfy the conditions of our theorem.

However, we can approximately satisfy the theorem by the following procedure:

1. Integrate for some short time t, read out, integrate again, and so on, producing a record of many frames, each frame of 200x800 pixels and each representing an integration time t.

2. Sum the frames together into a second record consisting of adjacent pairs of frames. In this second record, each pixel in each frame will now have an effective integration time 2 t and the entire record will consist of half as many frames as the first record.

3. Form a third record similarly from the second record for an effective integration time of 4 t. Continue this process to obtain records up to any desired integration time.

4. Examine each record separately for meteoroids crossing the field of view of the sensor. When many sensors are placed across a diameter at the focal surface, this procedure must be separately carried out for each sensor.

In the procedure given above, the records differ from each other by factors of 2 in integration time. Thus, no matter what the angular velocity of the meteoroid, the meteoroid image will move across the sensor at such a velocity that, in some one of the above records, it will not be far from satisfying our theorem; namely, the meteoroid image will not move much more or much less than one pixel in one integration interval. This will be the record in which our theorem will be best satisfied and through which we will best be able to discover that meteoroid which is moving at the correspondingly appropriate angular velocity.

Although the records could be digitally analyzed to discover meteoroids crossing the sensor field of view, we will here consider the video technique successfully used by investigators at the M.I.T. Lincoln Laboratory, (Weber, 1979; Taff et al., 1984) to detect earth orbiting objects with the optical telescopes at the ETS (Experimental Test Site) in New Mexico. They simply play the record back through a video monitor at 30 frames/s and, while viewing it, look for objects streaking across the field of view. They state that objects with a signal to noise ratio of 1.2 are easily detected.

4. Analysis

Space limitations prevent a full analysis being presented here. We therefore present only our fundamental starting equations and our assumptions.

The number of electrons generated and stored in a pixel during the base integration time, t (in seconds), due to background light is given by b where

$$b = C_1 B \delta \eta \alpha^2 D^2 t. \qquad (1)$$

B is the assumed background light brightness in 21st magnitude stars per square arc sec. We assume B = 1.6. This accounts for a considerable infrared airglow component while looking at 45° with respect to the zenith. δ is the sensor efficiency for transforming photons to conduction electrons in the CCD semi-conductor. We assume that $\delta = 0.5$ over the wavelength range 0.3μm to 1.0μm. η is the fraction of light that makes it through the telescope optics to reach the sensor. We chose η = 0.6 for these Schmidt telescopes. α is the field of view (in arcsec) for each "effective pixel" and depends on the telescope used. Each "effective pixel" chosen as most efficient for the Palomar Big Schmidt consists of a 4x4 array of elementary pixels and is 60 μm across. Because on-chip summing of pixels can be done free of noise before amplifying in the read-out amplifiers, read-out noise relative to signal can thus be reduced. D is the telescope diameter in meters (given in Table 1). With these definitions and units, C_1 has the value C_1 = 194.8.

The number of signal electrons, S, stored in a pixel in time t, (resulting from a meteoroid image in that pixel), is given by

$$S = C_2 T \delta \eta p \left(\frac{R_0}{R}\right)^2 \left(\frac{d}{\Delta}\right)^2 D^2 E t . \qquad (2)$$

T is the atmospheric transmission. T=0.8 is assumed. p is the geometric albedo. We assume p = 0.05 (a fairly dark object) and that we can always observe approximately in the anti-sun direction. R_0 is the distance of the Earth from the sun and R is the meteoroid distance from the sun. d is the particle diameter and Δ is the distance of the particle from an observer on Earth. δ, η, and D are as previously defined. With these units, $C_2 = 6.02 \times 10^{20}$. E corrects for the fact that the meteoroid signal integration time in a pixel will not normally exactly match the background integration time. E can vary from about 0.75 to 1. E = 1 is assumed in this analysis. The number of noise electrons, N, generated during an integration interval is assumed to be given by

$$N = \sqrt{b + R_n^2} , \qquad (3)$$

where R_N is the read-out noise (R_N = 20 assumed).

In this analysis, we will detect meteoroids by processing each record at 30 frames/s through a video terminal and will look for streaks. We conservatively assume that we will detect all objects with a signal to noise ratio, S/N, greater than 2. From a number of computer analyses we have found that if we set the frame integration time to t = 0.25 sec, we get good results for meteoroids larger than about one meter in diameter with the optics we have choosen.

The distance Δ_m out to which objects will be best observed (i.e., best satisfy our theorem) during integration time t, is given by

$$\Delta_m = \frac{C_3 \nu_t t}{\alpha} \qquad (4)$$

where v_t is the meteoroid velocity transverse to the viewing direction, α is the pixel field of view in arcsec, and $C_3 = 2.0626 \times 10^3$ arc sec/rad. We assume that $v_t = 10^4$ m/s, is a reasonable weighted average.

We assume that the meteoroid flux, F(d), is given by

$$F(d) = \begin{cases} 10^{-19.4}\ d^{-1.86} & d<1 \text{ meter} \\ 10^{-19.4}\ d^{-3} & d>1 \text{ meter} \end{cases}$$

where F is in number per square meter per second striking one side of a flat surface, and d is the meteoroid diameter in meters. F(d) represents a good fit to both the meteor fireball data of Halliday, et al. (1984) and the observed flux of Apollo asteroids (Shoemaker, et al., 1979).

The smallest meteoroids d_1, that can be detected at distance Δ_{m1} is given by solving equations (1), (2), (3) and (4), and using our assumptions on the parameter values. This meteoroid will also be detected at all lesser distances. We can similarly solve for the smallest meteoroid, d_2 that can be detected out to distance Δ_{m2} (obtained from eq. (4) when t is replaced by t = 2 t). We then obtain a lower limit for the number of detections per unit time for meteoroids between diameters d_1 and d_2 by multiplying the flux of meteoroids in this interval, $F(d_1) - F(d_2)$, times twice the area over which all meteoroids in this size interval will be detected. This area is given by $0.5\Omega\ \Delta_{m1}^2$, where Ω is the sum of all the angles subtended by each sensor at the focal surface. A "twice" factor comes in because meteoroids can approach from either of two directions. However, a gravitational decrease factor of 2 on F(d) makes the net factor equal to one. The "lower limit" applies because meteoroids larger than d_1, but not as large as d_2, can actually be detected at a greater distance than Δm_1. Similar analyses are now done for the next meteoroid diameter interval, and so on.

5. Results

The results of these analyses are shown in Table 1 for the Palomar Big Schmidt. The total detection rate for detecting meteoroids between 0.6 and 206 meters in diameter using 20 sensors at the telescope focus is about 1.4/hr. Distances at which they are detected and their apparent magnitudes are also shown. The corresponding rate for detecting meteoroids between 0.34 and 71 meters in diameter using 14 sensors at the focus of the Baker-Nunn Satellite Tracking Camera turns out to be about 0.8/hr. The latter telescope is much smaller than the Big Schmidt, but gives rise to only a modest loss in count rate.

The count rates shown in Table 1 will more than double if we can detect streaks with a signal to noise ratio as low as 1. They will also increase by about a factor of 2 due to certain of our approximations. There will, however, be a minor reduction in count rate when E is realistically set to less than 1. The reduction is about 30% at E = 0.8.

The tremendous capability of this technique is demonstrated when we consider that the entire Prairie network with its many cameras covering approximately 1 million square kilometers would only observe about one object larger than 1 meter in diameter every 4 or 5 calendar years.

Table 1. Counts/hr for Palomar "Big Schmidt" with 20 sensors and SN = 2

Diameter Range (m)	Distance (m)	Apparent Magnitude	Counts/hr
0.61--1.02	$1.28x10^8$	19.6	0.15
1.02--1.72	$2.56x10^8$	20.0	0.27
1.72--2.91	$5.11x10^8$	20.4	0.23
2.91--4.92	$1.02x10^9$	20.8	0.19
4.92--8.39	$2.05x10^9$	21.1	0.16
8.39--14.5	$4.09x10^9$	21.5	0.13
14.5--25.6	$8.19x10^9$	21.9	0.10
25.6--47.4	$1.64x10^{10}$	22.3	0.075
47.4--94.0	$3.28x10^{10}$	22.6	0.049
94.0--206	6.55x1010	23.0	0.026

Telescope and sensor parameters: Effective objective diameter = 1.22 meters. Effective focal length = 3.07 meters. Diameter of focal surface = 35 cm. Pixel diameter = 60 μm. Pixel field of view = 4 sec/pixel.

References

Blouke M.M., Janesick J.R., Hall J.E., Cowens M.W., and May J.P. (1983) 800x800 charge-coupled device image sensor. Optical Engineering, 22 (5), 607-614.

Gehrels T. (1981) Faint comet searching. Icarus 47, 518-522.

Halliday I., Blackwell A.T., and Griffin A.A. (1984) The frequency of meteorite falls on the Earth. Science 223, 1405-1407.

McCrosky R.E. (1968) The distribution of magnitudes, masses, and energies of large meteoritic bodies. SAO Special Report 280, 13 pages.

McMillan R.S. and Stoll C.P. (1982) Software simulations of the detection of rapidly moving asteroids and charge coupled devices. SPIE 331, Instrumentation in Astronomy IV, 104-112.

Shoemaker E.M., Williams J.G., Helin E.F., and Wolfe R.F. (1979) Earth-crossing asteroids: Orbital classes, collision rates with Earth and origin. In Asteroids (Ed. T. Gehrels) Univ. Arizona Press, p. 253-282.

Taff L.G., Beatty D.G. Yakutis A.J., and Randall P.M.S. (1984) Low altitude, one centimeter, space debris search at Lincoln Laboratory's Experimental Test System. Space Research, in press.

Weber R. (1979) Passive ground-based electro-optical detection of artificial Earth satellites. Optical Engineering, 18, 82-91.

Wetherill G.W. and Revelle D.O. (1981) Which fireballs are meteorites? A study of the Prairie Network photographic meteor data. Icarus 48, 308-328.

METEOROID ORBITS DETERMINED BY SOUTHERN HEMISPHERE RADAR

D. I. Steel and W. J. Baggaley
Department of Physics
University of Canterbury
Christchurch
New Zealand.

ABSTRACT. A radar meteor orbit facility of novel design has been developed at the University of Canterbury. Synchronous digital noise averaging, impulse suppression and meteor echo recognition using a microprocessor system, along with high gain antennas, allow a limiting radar magnitude of about +13 to be achieved, making this the most sensitive meteor orbit radar to date. An exhaustive survey of faint sporadic meteors will be carried out, along with an investigation of the little-known Southern showers. Since the observatory is ideally located with respect to the Eta Aquarid radiant, a programme of observations during the Comet Halley perihelion epoch is planned.

4. INTRODUCTION

Direct spacecraft measurements of meteoroidal impacts have the advantage of being unaffected by the vagaries of the Earth's atmosphere, and can also investigate the interplanetary environment away from 1 A.U. However, the detectors only give crude measures of meteoroidal velocity and orbit, and since the flux of the dust is low the small detector area in a spacecraft experiment leads to a tiny sample of the inter-planetary dust complex.

Meteor observations, which use the atmosphere as a collecting area, do not suffer from these limitations. A variety of optical methods have been used to measure orbits down to +7.5 magnitude (Hawkes et al. 1984) and fluxes down to +12 (Cook et al. 1980). With the radar technique, in which the samples may number thousands of meteors per hour, a few orbits down to +12 magnitude and many to +10 have been measured by the Harvard survey (Cook et al. 1972), and fluxes to a limiting magnitude of +15 (Eshleman and Gallagher, 1962).

The presently described radar apparatus will determine the orbital parameters of meteors mostly in the magnitude range +8 to +13. The limiting meteoroid diameter is about 50 microns: this will therefore be the first orbital survey to have a significant overlap with those particles causing the zodiacal light and gegenschein, and will hence allow the dynamics of the zodiacal cloud to be directly investigated.

R. H. Giese and P. Lamy (eds.), Properties and Interactions of Interplanetary Dust, 299–303.

2. PREVIOUS RADAR TECHNIQUES

A variety of radar techniques have been used to determine meteor orbits since the first demonstration by McKinley and Millman (1949). The group at Jodrell Bank, England, used three spaced receivers to determine the meteor trajectory with the velocity being calculated from the instantaneous oscillation frequency of the Fresnel diffraction echo at any one of them (Davies and Gill, 1960). This method is limited by the small fraction of meteors which render usable diffraction echoes.

Most subsequent experiments have been broadly based upon the spaced receiver method. The Harvard group in the United States used eight separate sites (Cook et al. 1972), whilst a rotatable antenna was used at Kazan in the Soviet Union to reach magnitude +8 (Andrianov et al. 1970). Previous meteor orbital surveys in the Southern hemisphere have been those at Adelaide, South Australia (Nilsson, 1964; Gartrell and Elford, 1975), reaching magnitudes +6 and +8 respectively. There is little known of the meteoric complex fainter than magnitude +8 in the Southern hemisphere.

3. THE CANTERBURY METHOD

The spaced-receiver technique employed to date suffers from the severe drawback that few meteors give measurable diffraction oscillations. This is especially important in that it is a biased selection effect, with certain velocities being partially excluded. The resultant orbital distribution of sporadic meteors would not, therefore, be a true measure of the interplanetary population.

In the Canterbury method three stations, each separated by about 3km, are again used. However, the two components of the velocity (North-South and East-West) are determined from the relative timings of meteor echoes at the three sites, rather than by the Fresnal oscillation method. Indeed, the oscillations are lost from the recorded echoes due to temporal averaging of the echoes, employed to reduce the effect of random cosmic noise.

The specular reflection causing a radar echo from the meteor trail (see e.g. McKinley, 1961) means that the time-lag between two stations is exactly half of the meteor travel time for that direction and separation. Typical time-lags would be of the order of 0.05s; since the radar operates at a pulse repetition frequency of 450 Hz, this corresponds to about 20 pulses. By using digital cross-correlation analysis, relative delays precise to a quarter of a pulse period are possible, and the velocity can be found in the vast majority of cases to better than 5%. The radiant coordinates are limited by the antenna beam width and elevation data (as discussed below) to an accuracy of about $\pm 2^{\circ}$.

The velocity components in two orthogonal directions are now known. To determine the complete trajectory, the specular reflection constraint is again applied. The phase coherent radar transmitter operates at 26.36 MHz, delivering 25 kW peak pulse power. The antenna gains are high: about 26 dB for the transmitter and 18.6 dB for the receiver, compared to an isotropic radiator. These correspond to linear antenna dimensions of about 300 metres. The beam widths are narrow in azimuth ($\pm1^{o}.6$ for the transmitter, $\pm2^{o}.4$ for the receiver arrays, centered on due South) but fairly wide in elevation (or zenith) angle (elevation maximum at $30^{o}\pm15^{o}$). The relative beam widths in azimuth for the transmitter and receiver mean that secondary lobes are suppressed.

The echo plane (in which the meteor trail must form if it is to give an echo) is hence constrained in azimuth but not in elevation; by measuring the elevation with a phase pair of antennas (separation 4.75 wavelengths) the meteor velocity and trajectory are found. Any possible ambiguity in the elevation angle due to the multiple lobes of the phase pair is removed by inspection of the echo decay time which depends upon the ambient diffusion coefficient and hence the height of the meteor above sea-level. The antenna pattern of the transmitter should result in very few meteors being detected outside of the main phase-pair lobe centered at 30° elevation.

3.1 Synchronous data handling

Data is transmitted back from the two remote receiver sites via frequency-modulated HF links to the main site, where the transmitter and phase pair antennas are located. Daisy chains of microprocessors then synchronously analyse the digital data from five separate channels: three amplitudes (the two remote sites and one of the two phase antennas) and two trigonometric functions (the sine and cosine of the phase pair). The firing of the transmitter is also administered by the main microprocessor control system so that the range, in 5 kilometre bins, is known from the echo timing.

Separate microprocessor algorithms suppress impulses due to extraneous sources, smooth the signal in each range bin from sweep to sweep, average adjacent range bins for additional noise damping, and then discriminate meteor echoes from the background noise. Recognition of an echo in any one of the three amplitude channels results in the data from all five channels being stored on magnetic tape. Since meteors occur about once every five seconds at peak influx, and their duration is up to 0.5s for the purposes of this experiment, the tape is storing data for about 10% of the time. At 450 p.r.f., and with 5 kilometre range bins going out to 333 km, the processing rate is 30 kHz.

The digital tapes are removed for later analysis on a main-frame computer although real-time analysis is technologically feasible: it is hoped to implement such a system in the future.

3.2 Data analysis

The meteoric velocity and trajectory differ from the original heliocentric values due to: (i) Deceleration in the atmosphere; (ii) Acceleration by the Earth's gravitational field; (iii) The rotation of the Earth; (iv) The Earth's orbital motion.

In this experiment no direct measurement of atmospheric deceleration is made, unlike the Harvard survey (Verniani,1973). A uniform correction of about 2% is made to the measured velocities, based upon previous data for the deceleration, the ablation coefficient, and meteoritic densities. This gives the pre-atmospheric geocentric velocity which is then corrected for the other effects (Earth attraction, rotation, motion) so that the meteoroid's heliocentric velocity and true radiant are determined. From these the orbital elements can be found, given the time of observation.

Particularly important in finding the distribution of the interplanetary dust are the semi-major axis, eccentricity and inclination. Although individual meteor orbits are not of high precision, the large number of meteors measured during a limited observation period (say, 24 hours per month) permit measurements of the gross orbital parameters, and how these vary with the size of the meteoroid.

It is well known that meteor radars are selective in certain respects. Firstly the 'echo-ceiling' operative at short radio wavelengths limits the probability of detection for meteors ablating at higher altitudes, these being fast meteors (McKinley, 1961). The amount of ionization produced is also a strong function of the meteor velocity, so that the probability of actually detecting a meteor depends upon a number of factors, including the antenna patterns at the ablation height of each meteor.

Secondly, the effective collector area, taking into account gravitational focussing, will also involve the antenna patterns. In general the meteor's velocity is not perpendicular to the collector area so that the radiant is required to implement this calculation.

Thirdly, the probability of a particular meteoroid colliding with the Earth depends upon its orbital parameters, with low inclination orbits being favoured.

These three factors together lead to a 'cosmic weighting' for each detected meteor, and thus an estimate of the number of Earth-crossing meteoroids in particular types of orbits.

4. RESEARCH PROGRAMME

The programme of observations is directed towards four major topics: the orbital characteristics of sporadic meteors and how these vary with mass, so that the processes affecting the evolution of cometary debris may be more completely understood; the identification of the many Southern Hemisphere showers indicated by the high fluxes in January and February, and also in June and July; the relationship between orbit type and ablation characteristics (Baggaley, 1981); and a survey over the rest of this decade of the Eta Aquarid shower to see how it changes as the parent comet, Halley, passes perihelion. This shower (May 4-6) has already been used in preliminary checks of the system and sensitivity calibrations. The full orbital programme is expected to begin in late 1984.

REFERENCES

Andrianov, N.S., Pupysev, V.A., and Siderov, V.V., 1970, *Mon. Not. Roy. astr. Soc.,* 148, 227-237.

Baggaley, W.J., 1981, *The Observatory,* 101, 9-12.

Cook, A.F., Flannery, M.R., Levy, H., McCrosky, R.E., Sekanina, Z., Shao, C.Y., Southworth, R.B., and Williams, J.T., 1972, *NASA Contractor Report,* NASA CR-2109.

Cook, A.F., Weekes, T.C., Williams, J.T., and O'Mongain, E., 1980, *Mon. Not. Roy. astr. Soc.,* 193, 645-666.

Davies, J.G., and Gill, J.C., 1960, *Mon. Not. Roy. astr. Soc.,* 121, 437-462.

Eshleman, V.R., and Gallagher, P.B., 1962, *Astron. J.,* 67, 245-248.

Gartrell, G., and Elford, W.G., 1975, *Australian Journal of Physics,* 28, 591-620.

Hawkes, R.L., Jones, J., and Ceplecha, Z., 1984, *Bull. Astron. Inst. Czech.,* 35, 46-64.

McKinley, D.W.R., 1961, *Meteor Science and Engineering,* McGraw-Hill, New York.

McKinley, D.W.R.,and Millman, P.M., 1949, *Canadian Journal of Research,* A27, 53-67.

Nilsson, C.S., 1964, *Australian Journal of Physics,* 17, 205-256.

Verniani, F., 1973, *J. Geophys. Res.,* 78, 8429-8462.

DUST-PLASMA-GAS INTERACTIONS IN THE HELIOSPHERE

H.J. Fahr and H.W. Ripken
Institute f. Astrophysics and Extraterrestrial Research,
University Bonn, Auf dem Huegel 71, 5300 Bonn 1, F.R. Germany

ABSTRACT. Several different forms of how neutral gases and dusty media can interact with a plasma have been discussed in the literature. Common to all of these interactions is that they in principle lead to changes in both the composition and dynamical state of all interacting media. The most important scenarios in the heliosphere involving these kinds of interactions are described in this paper. Special emphasis is placed on the solar wind interaction with the zodiacal dust. In regard to the dust, this leads to severe complications with respect to its dynamics due to the plasma Poynting- Robertson effect and due to Lorentz scatterings. As for the plasma, this causes the appearance of dust-generated neutrals and secondary ions. An influence on the plasma dynamics, at least in the present solar system, can be neglected. The surface chemistry of the plasma-dust reactions is described in some detail here, and the densities and fluxes of exotic, dust-generated ions not originally belonging to the coronal solar wind are calculated. Observabilities of these neutrals and ions are investigated and it is discussed how physical, chemical, and dynamical parameters of the zodiacal dust cloud could be derived from these observations.

1. GENERAL INTERACTION SCENARIOS IN THE HELIOSPHERE

It is well known that neutral gases and magnetized plasmas cannot freely penetrate each other. The reason why a relatively strong plasma-gas coupling occurs is a rapid pick-up of newly created secondary ions by means of the magnetic fields which are frozen into the plasma and thus reflect the plasma dynamical state. Secondary ions are generated from neutrals either via photoionization, charge exchange, electron impact ionization, or critical velocity ionization (see, e.g., Holzer, 1977; Petelski et al., 1980; Formisano, Galeev and Sagdeev, 1982). Hence they belong, immediately after their creation, to the dynamical state of the neutral gas. Their incorporation into the plasma via electromagnetic re-acceleration to the plasma bulk speed results in a corresponding change of the momentum density of the plasma, leading to a deceleration and a nonthermal heating of the plasma (discussed by Holzer and Axford, 1971;

R. H. Giese and P. Lamy (eds.), Properties and Interactions of Interplanetary Dust, 305–323.

Fahr, 1971; Axford, 1972; Fahr, 1973; Holzer and Leer, 1973). For the plasma, this can be described as a local effect inducing a systematic change of the plasma properties. In contrast, the local neutral gas properties are not affected by secondary neutrals because of their large mean free paths. This will in general lead to the existence of two independent populations of neutrals in velocity space.

A large series of interaction phenomena in the heliosphere are characterized by processes based on momentum and energy exchanges between gases and plasmas. For instance, in the context of planetary exospheres, plasma-gas interactions take place in two different forms:
(a) The interaction of neutral atmospheres and ionospheres of planets causes an enhanced escape of neutral gases from the planetary gravitational field. This topic has just recently been reviewed by Fahr and Shizgal (1983). While its importance for the terrestrial case has again been stressed recently in a paper by Nass and Fahr (1984), the extraordinary role of this interaction for the non-terrestrial exospheres has been emphasized in papers by Wallis (1978), Hodges and Tinsley (1981), Cloutier et al. (1978), Eviatar et al. (1979), Ip and Axford (1980), Goertz (1980), Brown and Schneider (1981), Goertz and Ip (1982), and Ip (1982).
(b) The interaction of the neutral atmosphere of a comet or an unmagnetized planet, as for instance the planet Venus, with the approaching solar wind plasma leads to a solar wind mass loading, a plasma flow deceleration and possibly a shock-free transition to a subsonic solar wind flow (Biermann et al., 1967; Wallis, 1974). This, on the other hand, leads to an additional escape of atmospheric constituents (Bertaux et al., 1978; Richter et al., 1979).

More diffuse forms of a plasma-gas interaction, spread out in space over several 100 AU, are represented by the interaction of the local neutral interstellar medium with (a) the perturbed plasma regions ahead of the heliosphere and (b) with the supersonic solar wind in the heliosphere. Such interaction phenomena have been quantitatively treated for case (a) by Wallis (1974), Ripken and Fahr (1983), Wallis (1984), Fahr and Ripken (1984); and for case (b) by Axford (1972), Holzer (1977), Fahr et al. (1981a), Petelski et al. (1980), and Fahr et al. (1981b). It has been shown by the latter authors that a rigorous treatment of these interaction problems requires the simultaneous solution of a set of coupled differential equations describing the continuities of mass, momentum, and energy fluxes of both neutral and ionized media. The plasma and gas dynamics are formulated in a hybrid form of magnetohydrodynamic equations containing terms derived in a kinetic form that takes account of the exchanges of mass, momentum and energy between the two media. The terms that couple the two sets of differential equations describe the corresponding exchange rates per unit time and volume connected with the newly created ions and neutrals. A plot of typical solar wind and neutral gas properties resulting from this mutual interaction is given in Figure 1. Due to the large mean free paths of neutrals with respect to collisions with their own species, primary and secondary neutrals behave as independent populations. Thus, in this case, no momentum change of the primaries occurs. Thus for the neutrals we have given in Fig. 1 values only for the corresponding densities.

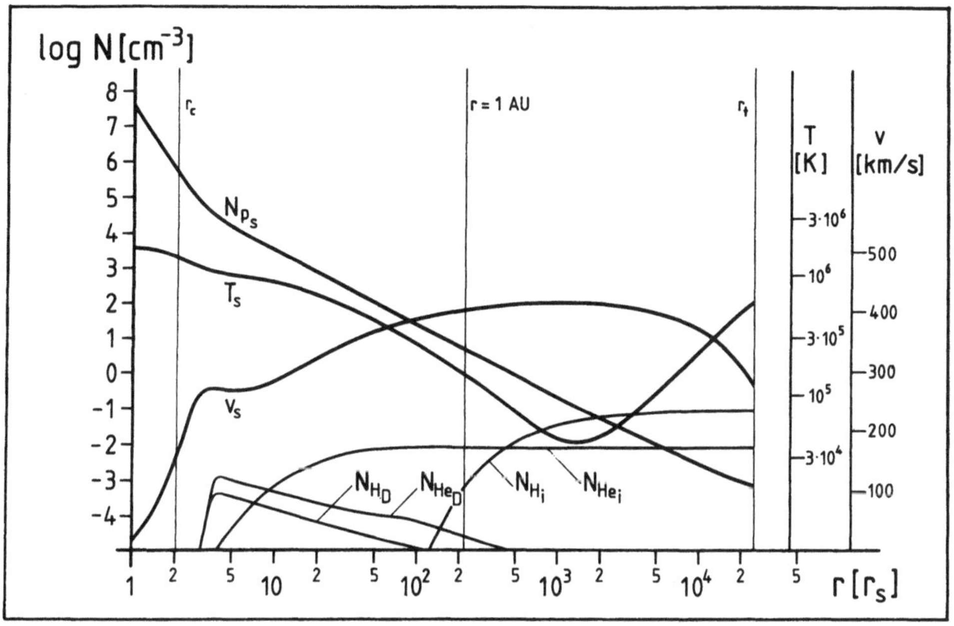

Figure 1. Solution for a solar wind-interplanetary neutral gas interaction model. Solar wind parameters shown are the velocity v_S, temperature T_S, and proton density N_{PS}. Four density curves of the neutral gas are given: interstellar hydrogen N_{H_i} and helium N_{He_i}, dust-generated hydrogen N_{H_D} and helium N_{He_D} (Fahr et al., 1981b).

We now address the role of dust particles in the gas-plasma interaction scenario. It has become evident by measurements of the zodiacal light (Giese, 1977; Leinert et al., 1977; Schwehm and Rohde, 1977) and the F-corona of the sun (Koutchmy and Magnant, 1973; Beard, 1979), and by infrared measurements of dust emissions and by direct particle detection of dust particles, that the sun, even in its closest environment, is surrounded by dust particles of different physical, chemical, morphological, and dynamical properties. These dust particles are known to have amorphous, fluffy structures rather than a highly organized cristalline consistency, and mostly they are of a silicate or carbonate type. Within the solar wind regime, these particles give rise to a plasma-solid body interaction. The most evident form of such an interaction is an ion and electron interception by the dust surfaces due to the impinging solar plasma. Seen from this aspect, dust particles should simply serve as sinks in the plasma flow, leading to an extraction of some plasma species from the flow. The ion and electron bombardment of the dust surfaces, in competition with the photoelectron emission and thermionic emission (see e.g. Millet et al., 1980; Lafon et al., 1981),will soon lead to an equilibrium electric charge of the dust particle and to a balanced rate by which, in response to the impinging ions, some neutral products will desorb from the surface. The details of this process are considered in section 3 of this paper. It can be already mentioned here that whenever saturation of the outer cristalline lattice of the dust particles with trapped neutralized solar wind ions is reached, these surface layers act as a prompt converter of ionic species into neutral species. Dust particles therefore act as sources of neutrals in the solar wind regime.

If the dust particles would act only as a sink of ions, no influence on the solar wind bulk velocity could be expected since the kinetic energy per mass would not be changed by this process. However, due to the fact that dust particles act as sources of neutrals, most probably as sources of hydrogen and helium atoms (as is discussed later), a momentum exchange of these neutrals with the solar wind plasma flow occurs that will influence the solar wind dynamics in just the same way as the interstellar neutrals do.

However, one can easily estimate that under the present conditions of the tenuous zodiacal dust cloud no significant change of the unperturbed solar wind momentum flow is possible. Assuming an average dust cross sectional area per unit volume of G = 1.0E-18 cm-1 (see, for instance, Banks, 1971; Holzer, 1977) at the Earth's orbit and adopting a radial dust density dependence according to $n = n_E(r_E/r)^{1.3}$, one arrives at the following probability of converting an original solar wind ion on its way from the corona through the zodiacal dust cloud into one with vanishing radial velocity

$$W = 1 - \exp\left|-\int_{2r_\odot}^{\infty} (r_{E/r})^{1.3} dr\right| = 2\cdot 10^{-2} \qquad (1)$$

Thus only about two percent of the outflowing coronal solar wind ions will be converted due to reactions at zodiacal dust surfaces.

For the T-Tauri phase of the sun conditions may have been very different because in the phase immediately following the collapse of the protosolar cloud, primordial dust which preferentially settled down into the ecliptic plane (see, e.g., Morfill, 1983) may have caused much higher values of G there. Thus it may be very likely that the T-Tauri-phase solar wind was choked off totally near the ecliptic plane and only could expand to larger distances only at higher ecliptic latitudes. In addition, as was shown by Fahr (1980) and Ripken and Fahr (1980), the solar wind may not have had developed at all during the early phases of the solar system when the protosolar core still was surrounded by high density gases and dust. Later on, however, after the development of a supersonic solar wind phenomenon, the passage of the heliosphere even through a very dense interstellar dust or gas cloud cannot lead to a choke-off.

2. REACTION KINETICS AND THE MINOR IONS PROBLEM

Neutral gas in the solar system is being generated by plasma interaction with saturated solid body or dust grain surfaces, as was discussed in the previous section. Solar wind ions with energies in the keV range impinge onto the surfaces and penetrate about 10 to 100 nm into the mineral (Lord, 1968; Ducati et al., 1973). At the end of the particle trajectory, neutralization via electron capture occurs. The newly created neutrals are temporarily retained in the dust grain, undergoing diffusion processes and chemical reactions. Depending on the type of mineral, hydrogen atoms react to form OH, water, and hydrogen molecules (Zeller et al., 1966; Zeller and Ronca, 1968; Cadenhead and Buergel, 1973; Fastie et al., 1973), as well as various carbon and silicon hydrids. Helium, generated from deionized solar wind alpha particles, will stay in its atomic form. The grains cannot accumulate intercepted particles indefinitely; saturation occurs on very short time scales compared to the typical Poynting-Robertson residence times in the inner solar system (Fahr et al., 1981b). After saturation is reached, on the average one molecule or atom will be released from the dust grain surface for each impinging solar wind ion.

For particles released from the surface by an evaporative process, the initial orbital dynamics essentially are the same as those of the parent grains and can be calculated in an analogous manner (Banks, 1971; Burns et al., 1979; Hughes, 1980). These particles are immediately subject to strong interaction processes with the solar wind and with the solar radiation field. Photoionization, charge exchange reactions, electron impact ionization, and molecular dissociation take place. Ionized particles will be "picked up" by the magnetized solar wind plasma and form singly-charged secondary solar wind constituents (protons, helium ions, and ionized molecular hydrogen). In addition, singly-charged hydroxyl, carbon and silicon hydrids, and other molecular ions are formed in this process, depending on the type of the parent grain. Only very few dust-generated neutrals will reach the orbit of the Earth as low-energy neutral particles.

Besides these surface-evaporated particles, the processes of erosion, disruption, sputtering, vaporization, and sublimation of dust grains generate atoms, molecules, and clusters of molecules not containing solar wind material as prime constituents. With observed average erosion rates of about 0.001 nm/yr (at the orbit of the Earth) and an average sputtering yield of 0.03 atoms per incident ion, the dust surfaces are saturated and fully reduced (see, e.g., McDonnell and Ashworth, 1972; McDonnell et al., 1976; Flavill et al., 1980). The neutralized solar wind particles will predominantly desorb as molecular hydrogen from the grains. The non-solar wind material, in its singly-charged ionic forms, is subsequently incorporated into the solar wind and can be traced as very minor constituents. The most prevalent ones are ions of oxygen, silicon, iron, and carbon compounds.

The generation of solar wind minor ions by means of plasma-dust interaction processes takes place predominantly in regions very close to the sun (heliocentric distances of 4 to 20 solar radii) where erosion, sputtering, and desorption processes are strongest and where the dust particle density is highest. Figure 2 shows the mean densities of solar wind ions (protons and alpha particles), of interstellar neutral gas (hydrogen, helium, and oxygen), and of dust-generated singly-charged solar wind minor ions. Singly-charged helium is also being generated by ionization of interstellar neutral helium; this causes the double-hump signature of the N_{He^+} density curve. Analysis of ionized dust grain molecules (not shown in Fig. 2) can give substantial insight into the physical and chemical composition of the zodiacal dust cloud, whereas an analysis of the elemental abundances of converted solar wind material yields information on sputtering and erosion rates.

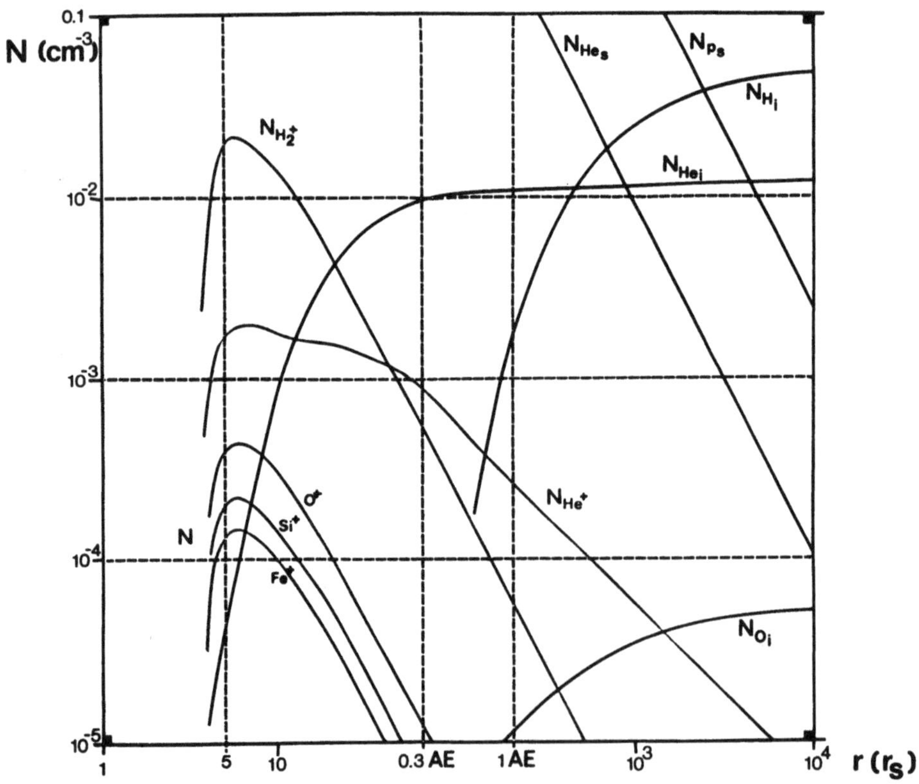

Figure 2. Shown versus solar distance in units of r_s (solar radii) are densities of various interplanetary neutral and ionic species. N_{O_i}, N_{H_i}, N_{He_i} are interstellar oxygen, hydrogen, and helium densities. N_{p_s} and N_{He_s} are solar wind proton and alpha particle densities. The remaining curves present densities of secondary ions in the solar wind regime that are due to dust-plasma-gas interactions as described in section 2: H_2^+, He^+, O^+, Si^+, Fe^+ ions.

In the past, several observations of exotic minor ions in the solar wind have been accomplished. Recently Bame (1983) reviewed the more important measurements. Singly-charged helium is an invaluable tracer of the solar wind-dust grain interaction process, and has been observed for example by M/Q mass spectrometers onboard IMP 7, VELA 5A, and HELIOS I. While normal or hot spectra show adequate fits of count rates by calculated ionization state envelopes, cold plasma can exhibit prominent peaks of surplus ions, i.e. singly-charged helium, and double- and triple-charged oxygen. Threshold limits of the present instruments prohibit the detection of the minor ion background of dust-generated origin; the background is expected to reach about 10% to 15% of the observed anomalous surplus component discussed above.

3. PLASMA-INDUCED DYNAMICS OF DUST PARTICLES

As was already noticed many years ago (e.g. Haug, 1958; Briggs, 1962; Divari, 1968), the interplanetary dust density distribution is closely connected with the interplanetary dust dynamics and the relevant dust particle sources inside and outside of the solar system. This connection results from the fact that the dust density is expressed as a 3-dimensional velocity space integral of the velocity distribution function $f_{\mathfrak{D}}(\underline{r},\underline{v})$ of interplanetary dust particles which in turn is determined by Boltzmann kinetic approaches employing the dynamical trajectories of dust particles. Except for the protosolar collapse phase, a kinetic approach to the problem is always justified in the later stages of the solar system evolution. This is correct even in view of stochastic scattering processes of electrically charged dust particles with the varying solar wind magnetic fields (first studied by Parker, 1964) and in view of systematic drag effects of the solar wind plasma sweeping over the dust grains (treated, e.g., by Lamy, 1974a).

Therefore the Boltzmann integro-differential equation will always yield a proper formulation of the function $f_{\mathfrak{D}}(\underline{r},\underline{v})$ dependent on the dynamical variables $\underline{r}$ and $\underline{v}$, i.e. the position and velocity vectors of the dust particles. Together with the boundary conditions for the dust, a solution of this function can then be generated provided that the dust dynamics is known. Many attempts to determine the dust dynamics under a variety of complicating conditions connected with the radiation and plasma environment of the dust have been undertaken in the past (the most important contributions were given by Briggs, 1962; Parker, 1964; Lamy, 1974a,b; Burns et al., 1979; Mukai, 1981; Mukai and Yamamoto, 1982; Morfill and Grün, 1979a,b; Barge et al., 1982b; Mukai and Giese, 1984). In a paragraph later in this section we will briefly review these attempts.

If the dynamical trajectories of individual dust particles can be given in a mathematically satisfactory form for each set of dynamical variables, it is advised to transform the Boltzmann equation into its characteristics form (see, e.g., Hays and Liu, 1965). Here the invariance of $f_{\mathfrak{D}}$ along the dynamical trajectory is formulated, assuming no dust particle sources and sinks. This then allows a solution for the function f_D if at some boundary surface the value for $f_{D,\infty}$ is known.

Rigorous formulations of this problem in terms of the Boltzmann kinetic approach in its original form have been given by Morfill et al. (1978), Völk et al. (1978), and Fahr et al. (1981b). Fokker-Planck type treatments of the problem have been used by Morfill and Grün (1979a,b), Consolmagno (1980), Barge et al. (1982a, b), and Mukai and Giese (1984) in order to include the additional difficulty of Lorentz-force-induced stochastic scattering of dust particles by the solar wind magnetic fields.

For a realistic solution of these types of equations both the appropriate boundary conditions of the dust problem and the dust dynamics have to be known. For the interstellar dust particles one would need information about the function $f_{D,\infty}$ at the border of the solar system, e.g. on a closed boundary surface surrounding the sun at a sufficiently large distance. For the zodiacal dust problem some complication arises from the fact that the origins of this dusty material are not securely known. As for the component connected with the primordial dust of the collapsing solar cloud still entering the inner solar system from beyond the asteroid belt, one needs to know the accretion rate and distribution of inclination angles of the orbital planes. The contribution to the zodiacal dust cloud from asteroid belt material requires information about density, mass, and size distributions of this material, while the initial orbital elements of its motion can be assumed to be known. In regard to the cometary dust release, one needs to know the source functions of dust particles (a) as functions of time and position in interplanetary space and (b) as functions of size and orbital elements of generated particles. In addition to these somewhat open boundary conditions, the solution of the dust problem requires a determination of the dust particle dynamics. This problem has a variety of aspects that can only briefly be touched here.

Dust particles are generally assumed to move at each instant of time along specific Keplerian orbits around the sun, either hyperbolic, elliptic (ballistic), or parabolic. These particles, however, are subject to non-conservative forces which cause secular temporal changes of the orbital elements such as total energy, angular momentum, eccentricity, and semimajor axis. Among these forces, the classical Poynting-Robertson effect (see Robertson, 1937) and its plasma analogon, the plasma Poynting-Robertson effect, have to be considered as the most important ones. The P-R effect for orbiting dust particles, simply expressed, results from the fact that solar material (photons and/or ions) with a specific velocity vector is adsorbed at the dust grain surface per unit time; this is equivalent to a temporal change of the dust particle momentum. This obvious principle of mass charging has not always been presented clearly in the literature. This has its roots in the somewhat confusing and in its quantitative form incorrect presentation of this effect by Robertson (1937). As pointed out by Burns et al. (1979), the radiative P-R effect is caused only by the momentum change due to an absorption of solar photons by the dust grains. It does not have a second contribution from a later photon reemission from the dust particle when the grain is in radiative equilibrium. It is evident that an isotropic emission from a moving particle cannot cause its deceleration and will not lead to a net drag force in the solar rest frame.

Both the radiative and the plasma-induced P-R effect lead to non-conservative forces with a radial and an orbital force component connected with radiation and plasma pressure and with an angular momentum loss. Quantitative calculations on the basis of Mie scattering theory lead Mukai and Mukai (1973), Lamy (1974b, 1976), Schwehm (1976), and Burns et al. (1979) to the conclusion that the P-R pressure (except for iron and magnetite grains of sizes between 0.2 and 0.06 microns) does not compensate gravity. Iron particles smaller than 0.06 microns will again be subject to an attractive solar force field. Furthermore, it is found by Lamy (1974b), Burns et al. (1979), Mukai and Schwehm (1981), and Mukai and Yamamoto (1982) that the plasma-induced P-R effect is only a small addition to the radiative effect, the radial component being totally negligible and the orbital component attaining up to 20% of its radiative counterpart. These results remain unmodified whether or not the plasma ions impinging on the dust surfaces stick there, or are deflected, or cause in addition some sputtering or evaporation of dusty material.

For the temporal change of the orbital elements of particles moving in quasicircular orbits, one arrives at the following results:

$$\frac{de}{dt} = \frac{H}{gmM} \left| F_r \cdot \sin \phi + F_o(\cos \phi + \cos \Psi) \right| \tag{2}$$

$$\frac{da}{dt} = 2 \left(\frac{a^2}{mH}\right) \left| F_r \cdot e \cdot \sin \phi + F_o(1+e \cos \phi) \right| \tag{3}$$

where m = dust particle mass, M = solar mass, a = semimajor axis, e = eccentricity, H = angular momentum per mass, g = gravitational constant, ϕ = true anomaly, and Ψ = eccentric anomaly. F_r and F_o are the radial and orbital components of the total P-R force. Since no normal (latitudinal) component of this force exists, the inclination angle i of the dust particle orbit is not subject to a first order temporal change. A change of this quantity, however, results from stochastic Lorentz force scatterings that electrically charged particles at small and moderate ecliptic latitudes suffer while moving on their main trajectories in interplanetary magnetic fields of alternate polarities. This kind of plasma-induced diffusion of charged dust particles moving near the ecliptic plane has been carefully treated by Morfill and Grün (1979a,b), Consolmagno (1980), Barge et al. (1982a,b), and Mukai and Giese (1984). For an as realistic as possible determination of this diffusion effect, Morfill and Grün (1979b) calculate values of the needed Fokker-Planck diffusion coefficients in velocity space. However, in this effort they have used the incorrect assumption that the electrically charged particles, while moving in the interplanetary magnetic fields, do not suffer any kinetic energy changes. This is true only in a reference frame co-moving with the solar wind plasma in which the magnetic fields are frozen-in. In contrast, in the solar reference system also electrical forces will be noticed by the dust particles with charge Q and a velocity $\underline{v}_{rel}$ relative to the plasma flow. Thus the kinetic energy L will change in time by

$$\frac{dL}{dt} = \frac{Q}{c}\left[\underline{v}\cdot(\underline{v}_{rel} \times \underline{B})\right] \tag{4}$$

where $\underline{v}$ is the orbital velocity of the dust particle. Thus the changes of the parallel and perpendicular velocity components are not related anymore by

$$d\,v_{\parallel} = -\,d\,v_{\perp}\,\frac{v_{\perp}}{v_{\parallel}} \tag{5}$$

as used in Morfill and Grün (1979b), unless particles with orbits in the ecliptic plane are considered for which the scalar triple vector product vanishes. This fact had already been noticed in the work of Parker (1964), in which the Fokker-Planck type diffusive treatment was first applied to the problem of charged zodiacal dust grain motions influenced by Lorentzian forces due to interplanetary magnetic fields. As could be shown in this early approach, an appreciable increase of the inclination angles i of dust particle orbits can be expected for dust particles of diameters less than 1 micron that move in the ecliptic plane.

A much more thorough and comprehensive treatment of this problem, on the basis of the Fokker-Planck equation in the diffusive approximation derived by Chandrasekhar (1943), was published recently by Barge et al. (1982a, b). In their first paper, the authors give a very fundamental derivation of the Fokker-Planck equation describing diffusive motions in 3-dimensional parameter space for particles orbiting in first-order Keplerian orbits with eccentricity e, semimajor axis a, and inclination i. The diffusion formalism is developed on a very general physical basis and can in this form even be applied to very different stochastic diffusion processes, as for instance the cosmic ray propagation in the fluctuating interplanetary magnetic fields. The diffusion terms are obtained as functions of the time-integrated components of the Lorentzian perturbation forces. However, this theoretical approach is only viable if the orbital revolution periods of the dust grains in their first-order orbits can be considered to be long in comparison to the characteristic fluctuation periods of the perturbation forces. Thus, this approach does not seem to be applicable for close solar orbits with $a < 40\ r_s$ if magnetic sector passages are to be taken as indicative for the fluctuation periods.

In their second paper (Barge et al., 1982b), the authors make use of their very general formalism, applying it to the concrete problem of the stochastic Lorentz force scatterings of dust grains with diameters between 1 micron and 0.2 micron. As is pointed out here for the first time, the Lorentz scattering process is composed of three different contributions due to independent fluctuations of the three quantities Q, $\underline{v}_{rel}$, and $\underline{B}$ appearing in Eq. (4). As will be mentioned later in the text, the electric charge of the dust grain, Q, fluctuates in accordance with solar wind density and velocity changes with a small time delay of about one day (i.e. the charging time scale; Hill and Mendis, 1979). Thus, Q and $\underline{v}_{rel}$ fluctuate in the rhythm of successive high speed and low speed solar wind signatures. The stochastic diffusion thus is driven by magnetic sector structures corotating with the sun and by the temporal changes of the solar wind plasma properties.

In their calculations of the time-averaged changes of the first order orbital elements a, e, and i an integral scattering time Δt of 10 years has been used by the authors. This time interval Δt may sometimes become a critical quantity, since the Fokker-Planck treatment requires it to be both very long in comparison to the fluctuation periods and very short in comparison to the long period orbit evolution, i.e. the P-R residence times.

An interesting result of the study of Barge et al. (1982b) is the detection of the resonant character of the diffusion. the diffusion in parameter space is greatest for orbits with revolution periods that are multiples of the relevant fluctuation periods. Thus, it must be noted that diffusion ceases to operate inside the corotation orbit ($a < 37\ r_s$). Therefore the general result of the paper, that the main diffusion effect concerns the inclination angle i and occurs at solar distances between 20 to 60 r_s, has to be taken with some caution. It remains open whether or not the observed zodiacal dust distribution at high inclination angles (Leinert et al., 1976) can be ascribed to the operation of stochastic Lorentz scatterings on primary dust grains in the ecliptic plane slowly spiralling inwards. If stochastic Lorentz scatterings are indeed negligible, one is left with the alternate hypothesis that the original dust population already contains an appreciable fraction of high inclination orbits. At least for dust particles with radii larger than 0.5 micron this seems to be an unavoidable conclusion because they are nearly unaffected by the solar wind and interplanetary field environments.

The charge to mass ratio of the dust particles is strongly varying with the dust material and the environmental conditions of the dust particles. As a general fact it can be stated (Consolmagno, 1980) that the electric charges are very much size-dependent and may vary between electric voltages of +40 V and −10 V.

More careful studies of the specifics of the dust grain electric charging under different radiative and plasma environmental conditions have been carried out by Millet et al. (1980) and Lafon et al. (1981). Both papers use the concept of free orbital motions of electrons in the electric field of the charged dust grains. The field is assumed to be a radially symmetric Coulomb field; the space charge induced field perturbations in the Debye surroundings of the dust grains are not taken into account. The Coulomb potential barrier with respect to the plasma potential is then consistently determined, adopting no net electric current flowing onto the grain in the steady equilibrium state. This is based on the fact that charging time scales are on the order of 1 day (Hill and Mendis, 1979). Free electron trajectories in the Debye sphere between the dust grain surface and the unperturbed plasma region are divided into elliptic, ballistic and hyperbolic categories. Knowing the velocity distribution functions for the electrons at these two boundary sides, one can determine that specific grain potential which suppresses the net electric flow onto the grain surface.

For the solar wind ions and electrons simple Maxwellian and bi-Maxwellian velocity distribution functions have been adopted in the solar wind rest frame. The distribution functions for electrons leaving the dust grain surfaces are determined by the photoelectric emission (Lafon et al., 1981) and the thermionic emission from the surface (Millet et al.,1980). As is shown by the latter authors, the thermionic emission starts to play an appreciable role in the circumsolar regions inside 10 r_S where the grain temperatures according to Mukai and Yamamoto (1979) or Lamy (1974) increase to values beyond 1000 K. If the thermionic emission of the electrons out of the solid grain surface is treated in the Sommerfeld approximation, i.e. no electron coupling with the crystalline lattice, distibution functions can be generated relatively easy. With the use of this approximation Millet et al. (1980) can show that the electric charging effect by the dense solar wind plasma in the close solar environment, yielding according to earlier publications (Lefevre, 1975) negative potentials of -40 V to -120 V, is very much reduced in this region and even inverted to positive potentials inside solar distances of $r < 7\ r_S$. An electric disruption of the grains in this region thus does not appear very likely. The effect of intercepted solar wind ions is also studied by Lafon et al. (1981) who can show that, dependent on the grain temperature, a strong dependence of the grain potential on the solar wind bulk speed relative to the grain is apparent. For higher bulk velocities the intercepted ion current is strongly enhanced and the net negative potential of the grain is thus effectively decreased. These results also show that the grain potentials will vary strongly with the signature of the ambient solar wind, i.e. high speed or low speed wind, moving compression regions etc.

As was pointed out by Morfill and Grün (1979b), systematic effects resulting in a focusing or defocusing of dust particle orbits with respect to the ecliptic plane occur for particles moving at higher ecliptic latitudes under unipolar field conditions. For particles orbiting within the ecliptic plane, a stochastic scattering occurs due to repeated passages through magnetic sectors of opposite polarity. The overall effect of this kind of diffusive motion in configuration space and velocity space on a steady state zodiacal dust density distribution has recently been considered in some detail by Mukai and Giese (1984).

They find that the enforced diffusive motion perpendicular to the ecliptic plane results in a steady increase of the dispersion of the inclination angles i. Thus, the steady state radial density gradient obtained by $n = n_E(r_E/r)^{1.3}$ for the P-R effect alone steepens somewhat for particles with sizes between 30 to 50 microns. These particles contribute most to the zodiacal light intensity. This result depends on the inherent assumption that Lorentz scattering in the ecliptic plane unilaterally leads to a diffusive loss of particles out of the unperturbed orbits without compensation by diffusive gains from orbits at higher ecliptic latitudes.

Drawing a conclusion from all this, it can be stated that for those dust particles that predominantly contribute to the light scattering from zodiacal dust, i.e. particles of sizes between 30 to 50 microns (Hughes, 1980) a relatively simple motion in quasicircular Keplerian orbits can be assumed in first order for most of their lifetimes in the inner solar system. This conclusion is possible since the secular changes of the orbital elements have very long time periods compared to the orbital

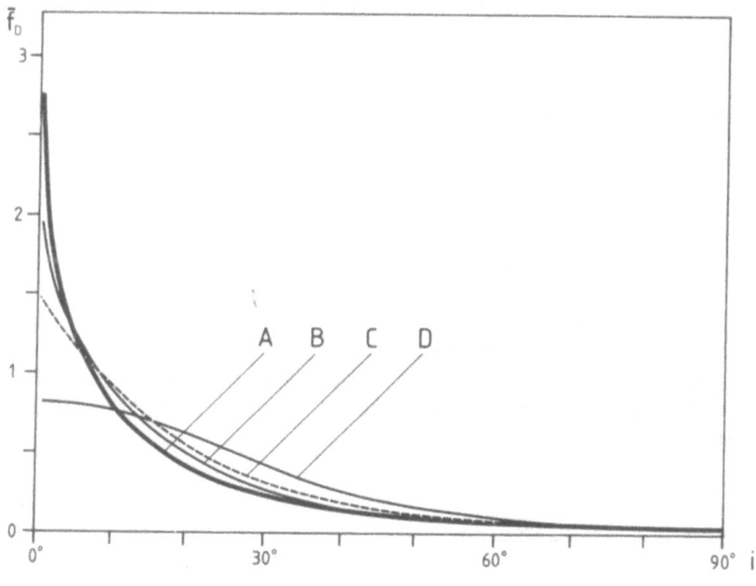

Figure 3. Distribution function $f_D(i)$ of dust particle orbital inclinations i at an arbitrary point in the ecliptic. The solid lines give three solutions for $f_D(i)$ based on different dust density distributions. Parameters are for curve A: $C = 2.1$, $H = 1$ (Leinert et al., 1978;) for curve B: $C = 2.6$, $H = 1.3$ (Leinert et al., 1976); for curve D: $C = 2.1$, $H = 2.0$. The dashed curve C shows $f_D(i) \sim \exp(-3i)$ (Bandermann, 1967).

revolution periods, for particles not too close to the sun. In the close solar environment where rapid changes of mass and size of the particles due to enhanced melting, sublimation, and sputtering rates occur, the dynamics may be much more complicated (Lamy, 1974b; Mukai, 1981).

On the basis of a zodiacal dust density distribution as derived by Leinert et al. (1978) from Helios zodiacal light measurements, one can derive a solution for the dust distribution function f_D in the form of specific integral equations. One will obtain integral equations of the Abel type or of the Volterra type (Fahr et al., 1981b; Buitrago et al., 1983) if a motion of the relevant dust particles in quasicircular orbits, with a specific separable distribution function for the orbital inclination angles i, is assumed. Under these assumptions one can show that the distribution function $f_D(\underline{r},\underline{v})$ can be separated into a product of functions in the form

$$f_D(\underline{r},\underline{v}) = f_{De}(e)\cdot f_{Da}(a)\cdot f_{Di}(i) \qquad (6)$$

as was done by Haug (1958) and Divari (1968). If in addition only circular orbits are admitted, this representation of f_D can be still more simplified (see Leinert et al., 1976; Fahr et al., 1981b):

$$f_D(\underline{r},\underline{v}) = n_{oE}\left(\frac{r}{r_E}\right)^{-1.3}\cdot \bar{f}_D(\cos\tau\cdot\cos\varphi) \qquad (7)$$

where φ is the ecliptic latitude angle and τ is the azimuthal angle of $\underline{v}$, measured in the local tangential plane of the circumsolar sphere with radius r.

Assuming a loss-free motion on these circular orbits as first order approximation, one is led to an Abel type integral equation that can be solved numerically and yields an explicit solution of the function $f_D(i)$ (Fahr et al., 1981b):

$$f_{Di}(i) = \frac{e^{-c}}{\pi}\left(\frac{1-J\cdot\cos^2 i}{1-\cos^2 i}\right) \quad \text{with} \qquad (8a)$$

$$J = e^c \int_0^{\cos i} \frac{\left[1-h\cdot c(\alpha^2+\sin^2 i)^{\frac{h}{2}}\right]\exp\left(-c(\alpha^2+\sin^2 i)^{\frac{h}{2}}\right)}{\cos i}\, d\alpha \qquad (8b)$$

In Figure 3 we have shown different alternative representations for this function f_D obtained under various assumptions.

One new interesting access to the dust distribution function $f_D(\underline{r},\underline{v})$ is offered by the observation of resonance radiation emitted from neutral gases, like hydrogen and helium, that desorb from the dust surfaces (see section 2). Assuming that these atoms desorb with velocities small in comparison to the orbital velocities of the dust grains, one can calculate the spectral profiles of the resonance luminence of dust-desorbed H and He atoms. One can then set the velocity distribution function of the dust particles to be equal to that of the resonantly scattering desorbed gases. If one thus uses the distribution function for f_D given in Equ. (8) as the velocity distribution function for dust-generated helium and hydrogen atoms, one can calculate exact spectral profiles of the He I 58.4 nm emission and the H I 121.6 nm emission. In Figures 4 and 5 such theoretical spectra are shown. The observation of these spectra can serve as a tracer to the dust dynamics in the close solar environment and is the aim of the EUV rocket experiment INTER-ZODIAK scheduled for launch in March 1985.

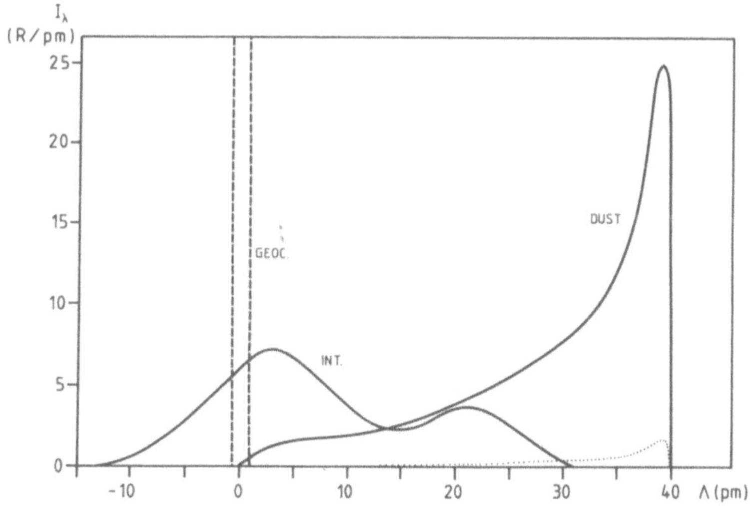

FIGURE 4: Spectral intensity distribution I_λ of resonantly scattered Ly-α radiation in Rayleigh per pm, versus distance from line center, Λ, in pm. The vertical dahsed lines mark the position of the geocoronal Ly-α intensity peak. The dust curves are given for ε = 0.9 (solid line) and for ε = 0.05 (dotted line) (see Fahr et al., 1981b)

The experimental demands, to resolve with satisfactory resolution power the spectra shown in Figs. 5 and 6 and to suppress the geocoronal radiation contaminations encountered during a rocket flight, can only be fulfilled by application of resonance absorption spectrophotometric techniques. These techniques consist of the use of resonant atomic absorbers distributed along a specific part of the radiation ray path to the detecting photometer. For the spectral analysis of the Lyman alpha resonant emission profile one uses atomic hydrogen in the ground state, for the analysis of the helium emissions at 53.7 nm and 58.4 nm one uses helium atoms as resonant absorbers. The absorber gas is contained in a hermetically sealed cell. The cell is traversed by the relevant photons that enter and leave the cell through two oppositely mounted broad band filter windows, in a hydrogen cell made out of MgF_2 material, in a helium cell made out of a tin foil. These resonance cell techniques meanwhile are sufficiently well known and were described recently by Bertaux and Lallement (1984) for the hydrogen case and by Delaboudiniere and Carabetian (1975), Freeman et al. (1977), Fahr et al. (1977), Lay et al. (1978), and Crifo et al. (1979) for the helium case.

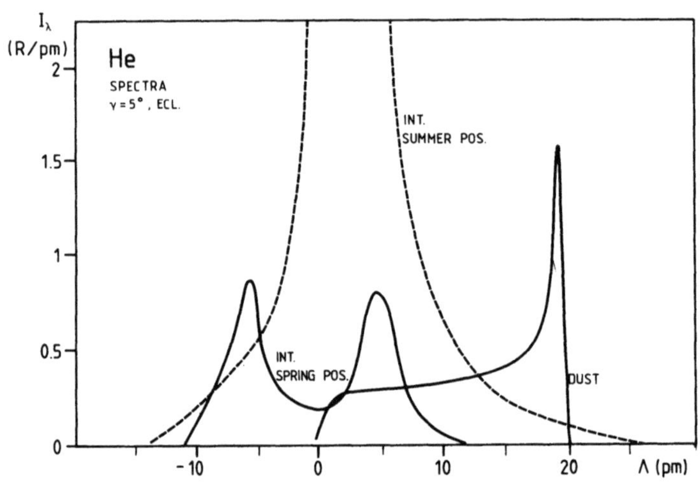

FIGURE 5: Spectral intensity distribution I_λ of resonantly scattered He I 58.4 nm radiation, in Rayleigh per pm, versus distance from line center, Λ, in pm. The interstellar radiation intensities are calculated for a geocentric observer on March 21 ("Spring Pos.") and on June 21 ("Summer Pos."). The dust-generated signal is independent of observation time.

The emissions from dust-generated hydrogen and helium atoms are strongest in the close solar environment. Thus an experiment aiming at the identification of these emissions has to observe under as close as possible solar offset angles, only large enough to keep the solar coronal emissions at a contamination level that is tolerable. An experiment adhering to these restricting conditions has been conceived and optimized by Fahr et al. (1980) and is technically described in detail by Neumann et al. (1983). The experiment INTER-ZODIAK is aimed at the spectral analysis and total intensity measurement of resonance emissions from dust-generated hydrogen and helium atoms in the "ecliptic" and "out-of-ecliptic" 6-degree-solar-offset region around the sun, employing the resonance spectral absorption techniques.

Since the physico-chemical reactions between dust grains and the solar wind plasma relevant to the rate of helium atom production is known with the accuracy of the solar wind parameters, one will first evaluate the He I 58.4 nm spectra in order to derive basic characteristics of the circumsolar zodiacal dust cloud within 30 r_s. The absolute intensities of the resonance emission from dust-generated helium will yield a very good measure of the mean dust cross sectional area per unit volume and thus will help to find the absolute value for zodiacal dust density model fits, as for instance for that of Leinert et al. (1978). The intensity asymmetry apparent on a circular scan around the sun with a 6° solar offset will immediately reveal the particle size average of the inclination angle distribution $f_D(i)$. Furthermore, the spectral location of the intensity peak of the He I 58.4 nm emissions of dust-generated helium atoms will give a clear indication of the relevant dynamics of the dust in a 30 r_s circumsolar sphere. The ratio of the integrated intensities of H I 121.6 nm and He I 58.4 nm emissions will finally contain direct information about the chemical and physico-chemical conditions at the gas-desorbing dust grain surfaces.

REFERENCES

Axford, W.I.: 1972, in: Solar Wind, NASA SP-308, p.609

Bame, S.J.: 1983, in: Solar Wind 5, NASA CP-2280, p. 573

Bandermann, L.W.: 1967, Thesis, Univ. of Maryland

Banks, P.M.: 1971, J. Geophys. Res. 76, 4341

Barge, P., Pellat, R., Millet, J.: 1982a, Astron. Astrophys. 109, 228

Barge, P., Pellat, R., Millet, J.: 1982b, Astron. Astrophys. 115, 8

Beard, D.B.: 1979, Astrophys. J. 234, 669

Bertaux, J.L., Blamont, J.E., Marcelin, M., Kurt, V.G., Romanova, N.N.: 1978, Planet. Space Sci. 26, 817

Bertaux, J.L., Lallement, R.: 1984, Astron. Astrophys. (in print)

Biermann, L., Brosowski, B., Schmidt, H.U.: 1967, Solar Phys. 1, 244

Briggs, R.E.: 1962, Astron. J. 67, 710

Brown, R.A., Schneider, N.M.,: 1981, Icarus 48, 519

Buitrago, J., Gomez, R., Sanchez, F.: 1983, Planet. Space Sci. 31, 373

Burns, J.A., Lamy, P.L., Soter, S.: 1979, Icarus 40, 1
Cadenhead, D.A., Buergel, W.G.: 1973, Science 180, 1166
Chandrasekhar, S.: 1943, Rev. Mod. Phys. 15, 1
Cloutier, P.A., Daniell, J.R., Dessler, A.J., Hill, T.W.: 1978, Astrophys. Space Sci. 55, 93
Consolmagno, G.J.: 1980, Icarus 43, 203
Crifo, J.F., Fahr, H.J., Seidl, P., Wulf-Mathies, C.: 1979, Rev. Sci. Instrum. 50, 1047
Delaboudiniere, J.P., Carabetian, C.: 1975, Space Sci. Instrum. 1, 91
Divari, N.B.: 1968, Soviet Phys. Astron. 11, 1048
Ducati, H., Kalbitzer, S., Kiko, J., Kirsten, T., Müller, H.W.: Moon 8, 210
Eviatar, A., Siscoe, G.L., Mekler, Y.: 1979, Icarus 39, 450
Formisano, V. Galeev, A.A., Sagdeev, R.Z.: 1982, Planet. Space Sci. 30, 491
Fahr, H.J.: 1971, Astron. Astrophys. 14, 263
Fahr, H.J.: 1973, Solar Phys. 30, 193
Fahr, H.J.: 1980, Mitt. Astron. Ges. 47, 233
Fahr, H.J., Shizgal, B.: 1983, Rev. Geophys. Space Sci. 21, 75
Fahr, H.J., Crifo, J.F., Wulf-Mathies, C., Seidl, P.: 1977, Astron. Astrophys. 58, L21
Fahr, H.J., Ripken, H.W., Lay, G.: 1980, ESA SP-152, p.449
Fahr, H.J., Petelski, E.F., Ripken, H.W.: 1981a, in: Solar Wind 4, Report MPAE-W-100-81-31, p. 543
Fahr, H.J., Ripken, H.W., Lay, G.: 1981b, Astron. Astrophys. 102, 359
Fahr, H.J., Ripken, H.W.: 1984, Astron. Astrophys. 139, 551
Fastie, W.G., Feldman, P.D., Henry, R.C., Moos, H.W., Barth, C.A., Freeman, J., Paresce, F., Bourges, S., Lampton, M.: 1977, Astrophys. J. 208, 747
Thomas, G.E., Donahue, T.M.: 1973, Science 182, 710
Giese, R.H.: 1977, J. Geophys. 42, 705
Goertz, C.K.: 1980, J. Geophys. Res. 85, 2949
Goertz, C.K., Ip, W.H.: 1982, Planet. Space Sci. 30, 855
Hang, U.: 1958, Z. Astrophys. 44, 71
Hays, P.B., Liu, V.C.: 1965, Planet. Space Sci. 13, 1185
Hill, J.R., Mendis, D.A.: 1979, Moon Planets 21, 3
Hodges, R.R., Tinsley, B.A.: 1981, J. Geophys. Res. 86, 7649
Holzer, T.E., Axford, W.I.: 1971, J. Geophys. Res. 76, 6965
Holzer, T.E., Leer, E.: 1973, Astrophys. Space Sci. 24, 335
Holzer, T.E.: 1977, Rev. Geophys. Space Phys. 15, 467
Hughes, D.W.: 1980, Nature 283, 331
Ip, W.H., Axford, W.I.: 1980, Nature 283, 180
Koutchmy, S., Magnant, F.: 1973, Astrophys. J. 186, 671
Lafon, J.P.J., Lamy, P.L., Millet, J.: 1981, Astron. Astrophys. 95, 295
Lamy, P.L.: 1974a, Astron. Astrophys. 33, 191
Lamy, P.L.: 1974b, Astron. Astrophys. 35, 197
Lamy, P.L.: 1976, in: Interplanetary Dust and Zodiacal Light, Springer-Verlag, Berlin, p. 437
Lay, G., Fahr, H.J., Wulf-Mathies, C.: 1978, ESA SP-135, p. 277
Lefevre, J.: 1975, Astron. Astrophys. 41, 437

Leinert, C., Link, H., Pitz, E., Giese, R.H.: 1976, Astron. Astrophys. 47, 221
Leinert, C., Pitz, E., Hanner, M., Link, H.: 1977, J. Geophys. 42, 699
Leinert, C., Hanner, M., Pitz, E.: 1978, Astron. Astrophys. 63, 183
Lord, H.C.: 1968, J. Geophys. Res. 73, 5271
McDonnell, J.A.M., Ashworth, D.G.: 1972, Space Res. 12, 333
McDonnell, J.A.M., Ashworth, D.G., Flavill, R.P.: 1976, Space Res. 16, 951
Millet, J., Lafon, J.P.J., Lamy, P.L.: 1980, Astron. Astrophys. 92, 6
Morfill, G.E., Röser, S., Tscharnutter, W., Völk, H.J.: 1978, Moon Planets 19, 211
Morfill, G.E., Grün, E.: 1979a, Planet. Space Sci. 27, 1269
Morfill, G.E., Grün, E.: 1979b, Planet. Space Sci. 27, 1283
Morfill, G.E.: 1983, MPE Preprint 21, (ISSN 0340-8922)
Mukai, T., Mukai, S.: 1973, Publ. Astron. Soc. Japan 25, 481
Mukai, T.: 1981, Astron. Astrophys. 99, 1
Mukai, T., Schwehm, G.: 1981, Astron. Astrophs. 95, 373
Mukai, T., Yamamoto, T.: 1982, Astron. Astrophys. 107, 97
Mukai, T., Giese, R.H.: 1984, Astron. Astrophys. 131, 355
Nass, H.U., Fahr, H.J.: 1984, J. Geophys. (in print)
Neumann, G., Boison, M., Henkel, R.: 1983, ESA SP-183, p.421
Parker, E.N.: 1964, Astrophys. J. 139, 951
Petelski, E.F., Fahr, H. J., Ripken, H.W., Brenning, N., Axnäs, I.: 1980, Astron. Astrophys. 87, 20
Richter, E., Fahr, H.J., Nass. H.U.: 1979, Planet. Space Sci. 27, 1163
Ripken, H.W., Fahr, H.J.: 1980, Mitt. Astron. Ges. 50, 46
Ripken, H.W., Fahr, H.J.: 1983, Astron. Astrophs. 122, 181
Robertson, H.P.: 1937, Mon. Not. Roy. Astron. Soc. 97, 423
Schwehm, G.: 1976, in: Interplanetary Dust and Zodiacal Light, Springer-Verlag, Berlin, p. 459
Schwehm, G., Rohde, M.: 1977, J. Geophys. 42, 727
Völk, H.J., Jones, F., Morfill, G.E., Röser, S.: 1978, Moon Planets 19, 221
Wallis, M.K.: 1974, Mon. Not. Roy. Astron. Soc. 167, 103
Wallis, M.K.: 1978, Planet. Space Sci. 26, 949
Wallis, M.K.: 1984, Astron. Astrophys. 130, 200
Zeller, E.J., Ronca, L.B., Levy, P.W.: 1966, J. Geophs. Res. 71, 4855
Zeller, E.J., Ronca, L.B.: 1968, The Surface Geochemistry of Solid Bodies in Space, in: Origin and Distribution of the Elements, Pergamon Press, Oxford

INTERACTION OF THE COMETARY DUST WITH THE SOLAR WIND AND COMETARY PLASMA

W.-H. Ip, H. Kimura§ and C.-P. Liu§
Max-Planck- Institut für Aeronomie
D-3411 Katlenburg-Lindau
Federal Republic of Germany

§ On leave from the Purple Mountain Observatory,
Nanking, China

ABSTRACT. Along the lines of previous studies on the cometary dust-plasma interaction processes, we evaluate the electromagnetic effect of submicron dust in the coma and in the dust tail. The possible occurence of electrostatic disruption of charged dust in the ion tail is also discussed from theoretical and observational points of view. The relevance of these processes to the formation of striae in the dust tails of several comets is addressed and the conclusion is made that trajectory study of the submicron dust in the striae must take into account the action of the interplanetary electric field.

It has long been recognized that electrostatic charging effect of interplanetary dust of micron or submicron size could lead to strong perturbation of their orbital motions (Parker, 1964; Levy and Jokipii, 1976; Consolmagno, 1979; Morfill and Grün, 1979). As estimated by Mukai (1981), an equilibrium of photoelectron current and the currents of the solar wind electrons and ions to the surface of the dust particles produces surface electrostatic potential (Φ) on the order of +0.5 to +2.9 volts for graphite particles and +4.2 to +6.5 volts for silicate particles during average solar wind condition at 1 AU. Note that these are values pertinent to particles of radii > 1 µm, and for particles with smaller radii up to 0.03 µm the corresponding values of Φ will be larger due to their increasing absorption cross sections of solar radiation.

Since the electron charge on a dust grain with surface potential Φ is

$$Q = 2.4\mathrm{x}10^{-4}\ \Phi.r.e \qquad (1)$$

with Φ in V and the particle radius r in cm, the Lorentz force experienced by the charged dust particle can be written as

$$\begin{aligned} \underline{f}_L &= Q.(\underline{E} + \underline{V}_d \mathrm{x} \underline{B}_i) \\ &= 2.4\mathrm{x}10^{-4}\ \Phi.r.e\ (\underline{E} + \underline{V}_d \mathrm{x} \underline{B}_i) \qquad (2) \end{aligned}$$

where $\underline{V}_d$ is the dust velocity relative to the comet, $\underline{B}_i$ is the inter-

R. H. Giese and P. Lamy (eds.), Properties and Interactions of Interplanetary Dust, 325–328.

planetary magnetic field, and $\underline{E} = -\underline{V}_{sw} x \underline{B}_i$ is the interplanetary electric field acted on the charged dust particles. Because $V_d \leq 10$ km/s and $V_{sw} \approx 400$ km/s we have

$$\underline{f}_L = 2.4x10^{-4} \Phi.r.e \underline{E} \quad (3)$$

and the corresponding acceleration is simply

$$\underline{a}_L = 5.7x10^{-5} (e\Phi/r^2 d) \cdot (\underline{V}_{sw} x \underline{B}_i), \quad (4)$$

where m_d is the mass of the dust particle and d its average density. For typical values of Φ ($\approx$ 5 V), V_{sw} ($\approx$ 400 km/s), B_i ($\approx 10^{-4}$ gauss), d (≈ 1 g/cm^{-3}) we find $a_L \approx 0.06$ cm/s^2 for r $\approx$ 1 μm and $a_L \approx 5.5$ cm/s^2 for r $\approx$ 0.1 μm. In comparison, the solar gravitational acceleration at 1 AU is $a_g = 0.6$ cm/s^2. It thus can be seen that the motion of submicron dust particles with the ratio $a_L/a_g > 1$ are strongly influenced by their interaction with the interplanetary electric field.

The above consideration implies that the motions of submicron cometary dust, after their release from the cometary nuclei, could be affected significantly by the electrodynamic force of the solar wind plasma (Wallis and Ip, 1982; Wallis and Hassan, 1983). In the vicinity of the cometary comas, the motions of the dust particles relative to the cometary nuclei can be approximated as:

$$\dot{\underline{V}}_d = \underline{a}_p + \underline{a}_L, \quad (5)$$

where a_p is the radial acceleration due to the solar radiation pressure. Now assuming $\underline{a}_p \perp \hat{a}_L$, it can be seen that - as was first pointed out by Wallis and Hassan (1983) - the trajectories of the charged dust particles will be deflected from the Sun-comet line by an angle θ given by $\tan\theta = a_L/a_p$.

Another effect of interest concerns the plasma drag effect of the cometary ions on the charged dust particles in the vicinity of the cometary ionosphere or ion tail. From a simple photochemical model, it has been found that the plasma density in the central axis of the ion tail cannot be larger than 10^3 cm^{-3} for a medium-bright comet like Comet Kohoutek 1973 XII at 1 AU solar distance. This small plasma density then limits the plasma drag effect to be quite negligible as first stated by Notni (1964, 1966). Even so, the ion tail could still have very strong coupling with the dust coma in case auroral-type activities could take place there as was postulated before (Ip and Mendis, 1976). Impact by an intense flux of energetic electrons could lead to electrostatic fragmentation of the cometary dust of fluffy structure (Hill and Mendis, 1980; see also Fechtig et al. 1979; Rhee, 1976; Öpik, 1956).

Note that Larson and Sekanina (1984) have found one event of narrow dust jet formation in the anti-sunward direction in the coma of Comet Halley - during its last perihelion passage. The association with electrostatic disruption of the dust particles is by no means clear; but it should be considered as one possible interpretation. The least could be said, in the present context, is that the interesting study by Larson and Sekanina has demonstrated the possibility of correlating the fine structures of cometary comas with energetic plasma processes in the ion tails.

If the dust jet detected in Comet Halley's tail contained a large amount of submicron particles (or hypothetical dust jets created by electrostatic fragmentation in the ion tail) their subsequent motions would be determined by electromagnetic force as well as the solar radiation pressure. In order to present the simplest possible picture, we have considered the case in which the interplanetary electric field at 1 AU is the same throughout the whole ion tail and the solar wind; also the dust jet is supposed to have a linear structure with uniform distribution of particles of radii between 1 µm and 0.1 µm. The 0.1 µm particles are assumed to experience a solar radiation pressure acceleration of 1 cm/s^2 and a Lorentz force term of 0.6 cm/s^2 in the perpendicular direction; the 1 µm particles are assumed to have a_p of 0.5 cm-s^2 and a_L of 0.1 cm/s^2. The differences in the values of a_p and a_L would lead to the gradual separation of the particles of different sizes as shown in Fig. 1. Under steady state condition, the action of the interplanetary electric field is to spread a linear dust jet into a thin sheet of highly elongated configuration.

Without electrostatic charging effect, the dust particles would move purely according to their corresponding solar radiation pressure forces, and mechanical theory alone needs to be employed to interpret the brightness distribution (i.e., synchrons and syndyns) in the dust tails (Finson and Probstein, 1968a; Kimura and Liu, 1977). However, when the Lorentz force term is significant as appropriate to submicron particles, great care must be taken in using the mechanical theory to trace the trajectories of the dust particles. The implication on the study of the evolution of the striae in the dust tails of Comets West 1976 VI, Mrkos 1957 V and others as investigated by Sekanina and Farrell (1980, 1982) is clear as these banded structrues are composed mostly of submicron dust particles.

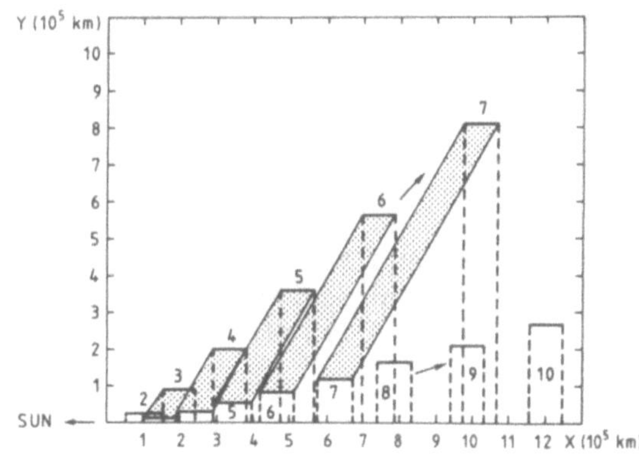

Fig. 1 An illustration of how a narrow jet of dust particles in the tail might evolve into a thin sheet with the particle sizes being arranged according to the values of a_L/a_p. The numbers denote the time intervals in days after the initial injection of the dust particles; and the bottom line represents the particle beam of 1 µm radius and the upper line represents the particle beam of 0.1 µm radius. All the dust particles share the same initial radial speed of 0.3 km/s. If f_L = 0, the dust jet will be elongated in the radial direction only

In summary, in consideration of the dust-plasma interaction in the vicinity of cometary comas and ion tails, we have assessed the effects of:

a. Lorentz force of the interplanetary electric field (important);
b. plasma drag by the cometary ionospheric plasma (insignificant);
c. electrostatic disruption of cometary dust (possible);
d. electromagnetic perturbation of submicron particles in the dust tails (important).

Recent results of investigation of the fine-scaled structures of cometary comas using image enhancement method (Larson and Sekanina, 1984) have demonstrated the feasibility of monitoring the process of electrostatic fragmentation of dust particles in the comas and ion tails of comets. A simple calculation of the trajectories of submicron dust particles initially confined in a narrow jet in the anti-sunward direction allows us to estimate the long-term effect of the Lorentz force. On the basis of this study, it could be stated with some certainty that tracing of the motions of the striae in the dust tails of several comets must incorporate the pertinent electromagnetic effect due to interaction with the solar wind plasma.

REFERENCES

Consolmagno, G.L. (1979) Icarus, 38, 398.
Fechtig, H., Grün, E. and Morfill, G. (1979) Planet. Space Sci., 27, 511.
Finson, M.L. and Probstein, R.F. (1968) Astrophys. J., 154, 327.
Hill, J.R. and Mendis, D.A. (1980) Astrophys. J., 242, 395.
Ip, W.-H. and Mendis, D.A. (1976) Icarus, 29, 147.
Kimura, H. and Liu, C.-P. (1977) Chinese Astron., 1, 235.
Larson, S.M. and Sekanina, Z. (1984) Astron. J., 89, 571.
Levy, E.H. and Jokipii, J.R. (1976) Nature, 264, 423.
Morfill, G. and Grün, E. (1979) Planet. Space Sci., 27, 1269.
Mukai, T. (1981) Astron. Astrophys., 99, 1.
Notni, P. (1964) Veröff. Sternwarte Babelsberg, 15(1).
Notni, P. (1966) Mem. Soc. Roy. Sci. Liege, 5°, ser. 12, 379.
Öpik, E.J. (1956) Irish Astron. J., 4, 84.
Parker, E.N. (1964) Astrophys. J., 139, 951.
Rhee, J.W. (1976) in Interplanetary Dust and Zodiacal Light, eds., H. Elsässer and H. Fechtig, Springer-Verlag, Berlin, p. 238.
Sekanina, Z. and Farrell, J.A. (1980) Astron. J., 85, 1538.
Sekanina, Z. and Farrell, J.A. (1982) Astron. J., 87, 1836.
Wallis, M.K. and Hassan, M.H. (1983) Astron. Astrophys., 121, 10.
Wallis, M.K. and Ip, W.-H. (1982) Nature, 298, 229.

OPTICAL AND PHOTOELECTRIC PROPERTIES OF MATERIALS OF ASTROPHYSICAL INTEREST

A. Quemerais, A. Seignac, M. Priol, J. Lefèvre*
Laboratoire de Spectroscopie, Université de Rennes,
35031 Rennes, France.
* Observatoire de Nice, BP 139, 06007 Nice, France.

ABSTRACT. The charge of cosmic grains can be calculated if the optical constants, the photoelectric yields and the photoelectron energy distributions for a large range of energy of incident photons are known. Our measurements are made on several materials which are possible constituents of interstellar and interplanetary dusts. We present here our results for fused quartz, iron and carbon.

1. INTRODUCTION

The electric charge of interstellar or interplanetary grains depends on their optical and photoelectric properties. We present here our results for several materials which are considered as the more common components of grains. The reflectivity has been measured for silica, iron and carbon in the energy range 3-27 eV and the photoelectric measurements were extended to 40.8 eV. The photoelectron energy distribution curves show the importance of secondary electrons.

2. EXPERIMENTAL METHODS.

All experiments were performed under ultra high vacuum conditions ($p < 10^{-9}$ torr). The optical constants were obtained by reflectivity measurements under various incidences. The real and imaginary parts of the refractive index are calculated by Kramers-Krönig analysis and multiangle method. The quantum yield was measured by using a calibrated photodiode and the energy distributions of photoelectrons were obtained by means of a retarding potential, the photocurrent being amplified by microchannel plates (Priol et al., 1980). The wavelengths given by the specially realized monochromator cover the photon energy range from 3 to 50 eV. The acquisition, the treatment of the data and the control of the apparatus were realized by a computer.

3. EXPERIMENTAL RESULTS.

For our sample of fused silica, the reflectivity (Fig.1 was found

R. H. Giese and P. Lamy (eds.), Properties and Interactions of Interplanetary Dust, 329–333.

larger than the values given by Philipp)1971) and Lamy (1977). The resulting complex index of refraction is given in Fig. 2.

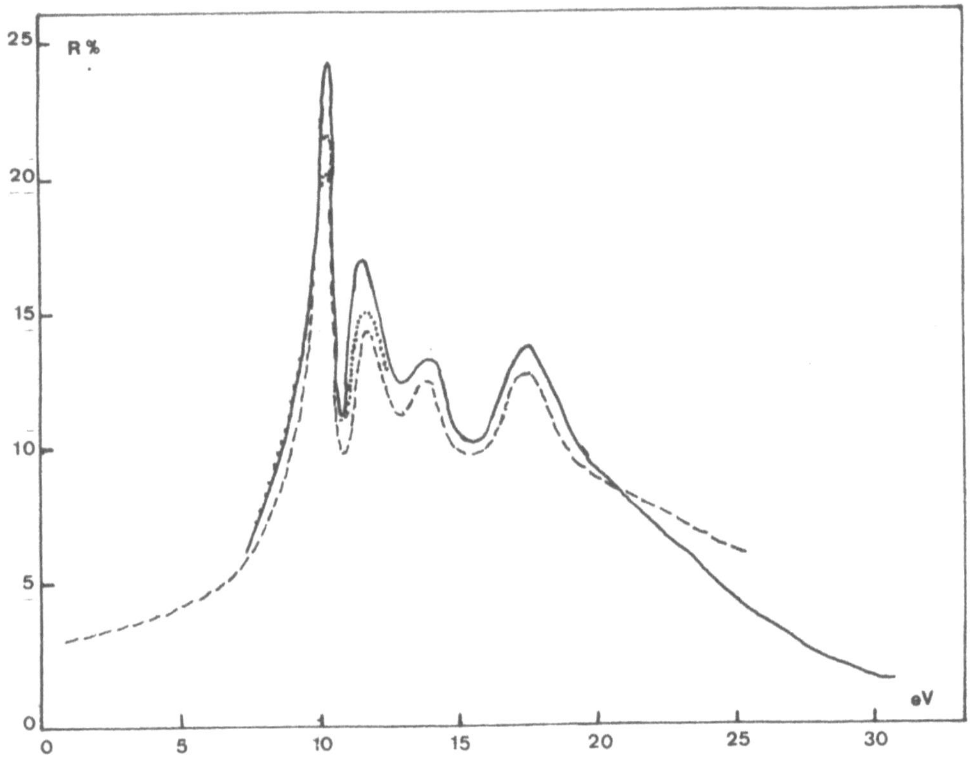

Fig. 1. Reflectivity of fused silica. - : our results ; --- : Philipp (1971) ; .... : Lamy (1977).

The photoelectric yield per absorbed photon (Fig. 3) was obtained with a thin film of SiO_2 on a silicon (111) substrate. A small thickness was necessary (220 A°) to keep a fixed value of the electric potential of the target.

The photoelectron energy distribution curves have been plotted (Fig.4) for three different photon energies (16.8, 21.2, 26.8 eV). Results are also available for other energies. The maximum at low energies is due to secondary electrons.

Finally we present the reflectivity (Fig. 5) and the quantum yield (Fig. 6) of iron thin films and of polished glassy carbon. A pronounced minimum at 7 eV for glassy carbon is due to interband transitions. For iron, the shoulder at about 8 eV is related to the surface plasmon.

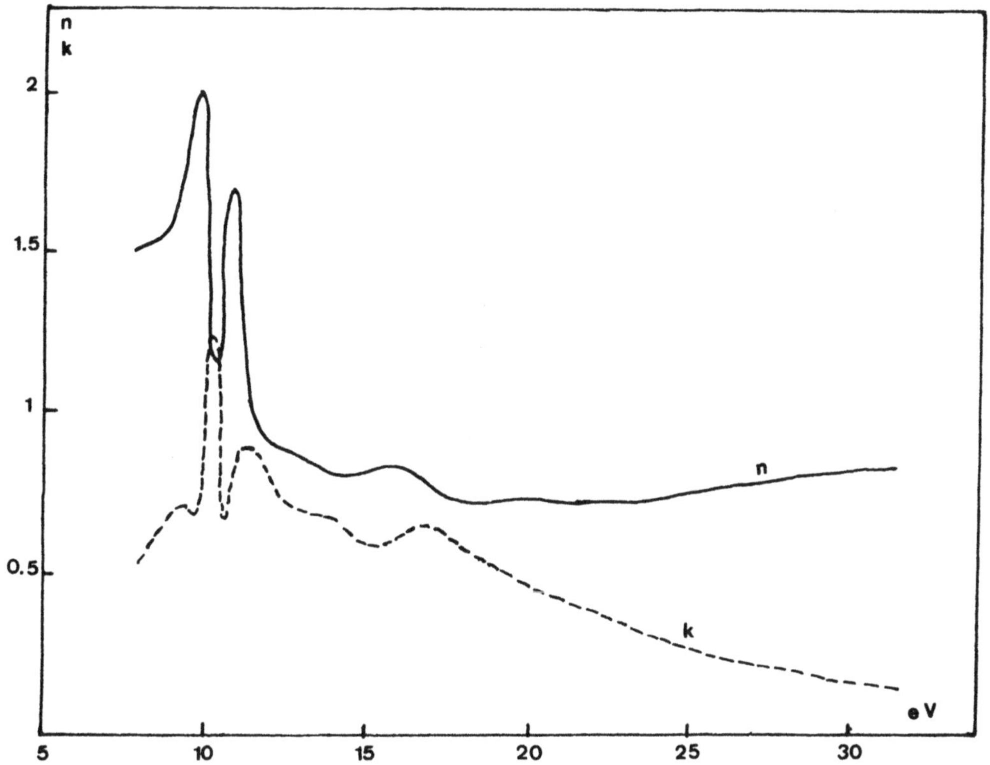

Fig. 2. Refractive index of fused silica. - : real part ;
---- : imaginary part.

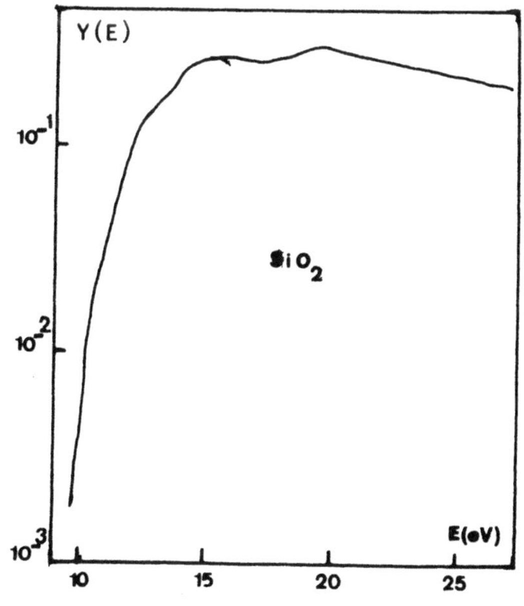

Fig. 3. Photoelectric yield per absorbed photon for silica.

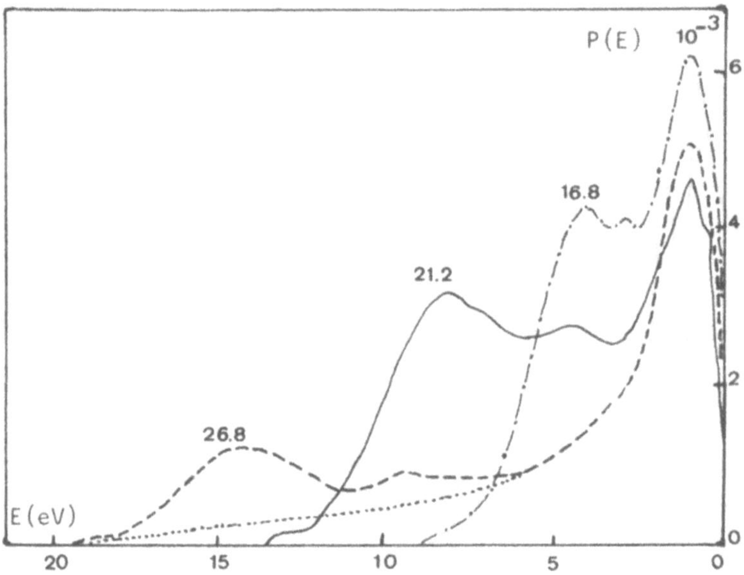

Fig. 4. Photoelectron energy distribution curves for silica at different photon energies. The ordinate is the number of emitted electrons per incident photons and the abscissa is the energy of photoelectrons. Each curveis labelled with the energy of incident photons. On the 26.8eV curve the relative importance of primary and secondary electrons is shown.

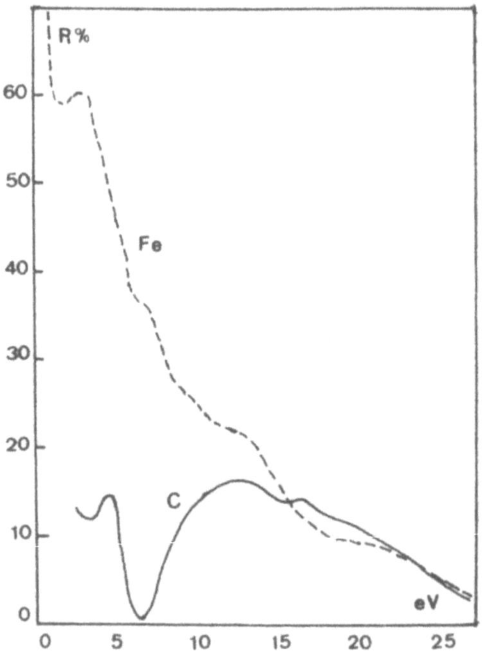

Fig. 5. Reflectivity for iron thin film (—) and polished glassy carbon (---)

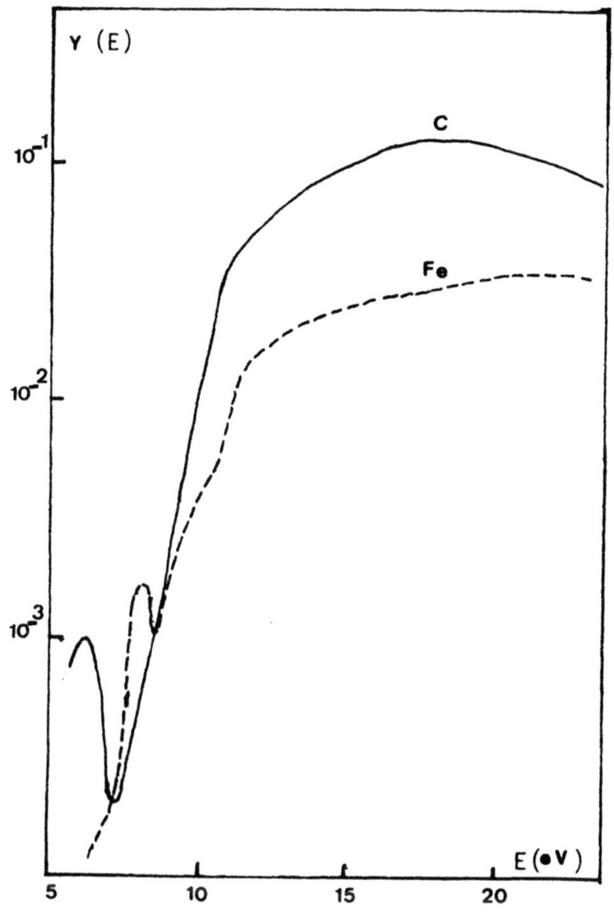

Fig. 6. Photoelectric yield per absorbed photon for iron and carbon.

REFERENCES.

Lamy P., 1977, Appl. Optics, 16, 2212.
Philipp H.R., 1971, J. Phys. Chem. Solids, 32, 1935.
Priol M., Alexandre T., Thomas J., Robin S., 1980, Le Vide les Couches Minces, 203, 89.

ELECTROSTATIC CHARGE OF INTERPLANETARY DUST GRAINS : NEW RESULTS

P.L. LAMY
Laboratoire d'Astronomie Spatiale,
Marseille
J. LEFEVRE
Université de Nantes
J. MILLET
Université de Provence et Laboratoire d'Astronomie Spatiale
Marseille
J.P. LAFON
DESPA, Observatoire de Meudon

ABSTRACT. A continuing program aimed at improving the determination of the charge of interplanetary as well as cosmic grains is presented. Recent data have been combined to generate two high-resolution solar spectra corresponding to the maximum & minimum of activity. The energy distribution of photoelectrons emitted by quartz grains under solar irradiation is calculated using the new laboratory measurements of Quemerais et al. (this volume). Finally, the method of Lafon, Millet and Lamy (Astron. Astrophys., 95,295, 1981) is used to find the electrostatic potential of these grains.

1. PHOTOELECTRIC EMISSION FROM FUSED-QUARTZ GRAINS

The differential flux of photoelectrons emitted by fused-quartz grains under solar radiation H (w) as a function of energy w in eV is expressed following Grard (1973) as

$$H(w) = Y(w)\ Q_{abs}(w)\ S(w)$$

where Y (w) is the photoelectric yield per <u>absorbed</u> photon, Q_{abs} (w) is the efficiency factor for absorption calculated with the Mie theory using the complex index of refraction of fused-quartz (Lamy, 1977) and S (w) is the differential solar photon flux.

Considerable progress has been achieved in measuring S (w) since the work of Grard (1973). The full-disk solar irradiance in the spectral range 1150 - 3200 Å is systemically measured by Mount and Rottman (1983) using rockets. These most recent results have benefited from improved calibrations and we retain those obtained on May 17, 1982, half-way in time between solar maximum and minimum but already representative of the <u>minimum</u> of the solar cycle. Solar <u>maximum</u> values were obtained on the July 15, 1980 flight.

R. H. Giese and P. Lamy (eds.), Properties and Interactions of Interplanetary Dust, 335–339.

They are characterized by a large increase - by a factor 2 approximately - at wavelenghts below 1900 Å while they are equal to solar minimum values above. In the interval 300 - 1150 Å, we use the AFGL data which have a satisfactory overlap with the minimum values of Mount and Rottman. Figure 11 of Schmidtke (1981) allows to determine an average increase by a constant factor of 2.3 to generate solar maximum values. This approximation is not critical since this interval does not contribute significantly to the photoelectric effect. Figure 1 gives the combined data average in 20 Å intervals as a function of energy in the corresponding energy range 4 - 40 eV.

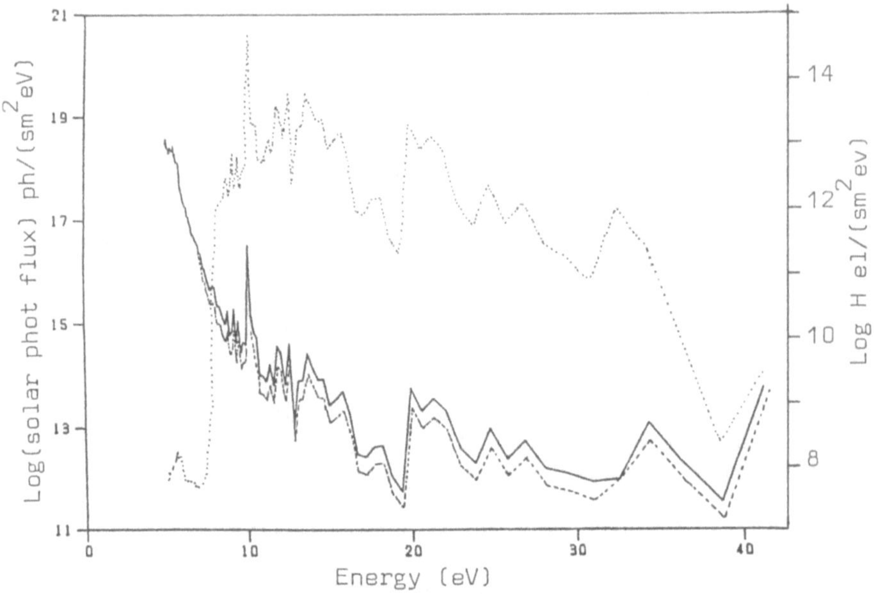

Figure 1. The differential solar photon flux for the maximum (solid line) and minimum (broken line) of activity (left scale) and the differential flux of photoelectrons H (w) (dotted line, right scale)

Using the laboratory measurements of Y (w) by Quemerais et al. (this volume), we obtain H (w) as illustrated in figure 1 for the case of solar minimum. These authors also measured the energy distribution functions of emitted photoelectrons f_w (ψ) for several energies w of incident monochromatic light. A best-fit analysis of their data has allowed us to obtain a parametric representation of the surface f (w,ψ).

After normalizing according to

$$\int_0^\infty f\ (w,\psi)\ d\psi = 1$$

we obtain the photoelectron energy distribution of fused quartz grains under solar irradiation

$$p\ (\psi) = \frac{1}{Is} \int_0^{\infty} f\ (w,\psi)\ \ H\ (w)\ dw$$

where Is, the saturation current density is expressed by

$$Is = \int_0^{\infty} H\ (w)\ dw$$

2. THE ELECTROSTATIC POTENTIAL OF INTERPLANETARY FUSED-QUARTZ

The problem of building a completely self-consistent model for a spherical body, for any surface effects and plasma compositions, with no approximation concerning the sheath was solved by Lafon, Lamy and Millet (1981). The effects due to the finite distance of the body from the enlightening stars were included. It was shown that the determination of the collected currents in the "Orbital Motion Limited" regime (Lafon, 1975) is exact for a grain at rest in the surrounding plasma and quite accurate in presence of drift velocities or of a disymmetry in the emission (or reemision) of particles by the body. The floating potential was shown to be unique in any case and its value was found whichever the preferential directions which can be different for particles of various species. Finally, the limits of validity of the model were discussed.

In the present article, we limit ourselves to the case of fused-quartz grains at 1 AU whose charge is the result of two competiting effects, the photoelectric effect and the collection of solar wind particles. The numerical program of Lafon et al. (1981) requires the following information:

2.1. Photoelectric effect

i) the electron volume density at the surface of the grain

$$No = \frac{8m}{e}\ Is \int_0^{\infty} \frac{p(\psi)}{\sqrt{\psi}}\ d\psi$$

where e and m are the charge and the mass of the electron.

Figure 2 shows how No depends upon the grains'radius (via Q_{abs}) in the interval 0.1 - 10 μm and upon solar activity as there is a factor 2 difference between the cases of solar minimum and maximum.

ii) the perpendicular velocity distribution function of the photoelectrons

$$F\ (v) = \frac{1}{2\pi} \left(\frac{m}{e} \right)^2 \frac{Is}{No} \frac{p(\psi)}{\psi}$$

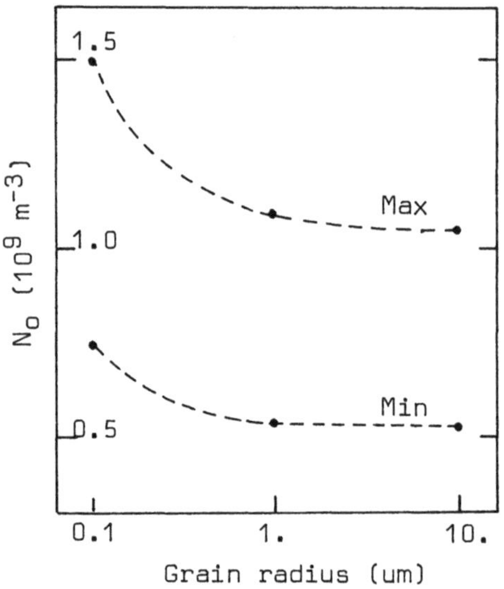

Figure 2. The electron volume density at the surface of fused-quartz grains

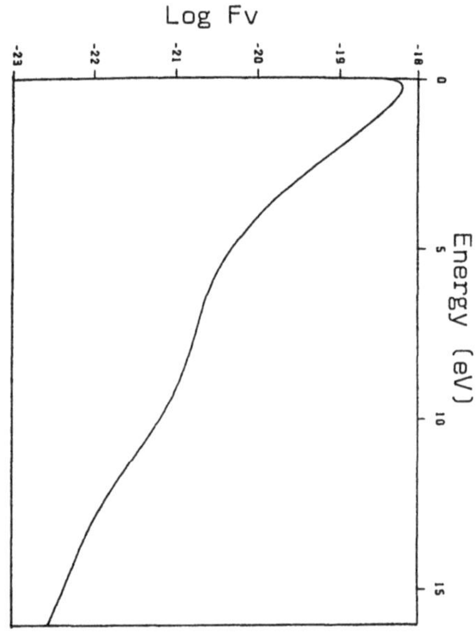

Figure 3. The perpendicular velocity distribution function of photoelectrons.

where the velocity is expressed as $v = (2e\psi/m)^{1/2}$. It is convenient to assume a maxwellian distribution and calculate the corresponding temperature

$$T = \frac{2e}{3k} \frac{\int_0^\infty \sqrt{\psi}\, p(\psi)\, d\psi}{\int_0^\infty \frac{p(\psi)}{\sqrt{\psi}}\, d\psi}$$

We found $T = 1.2 \times 10^4$ K, constant within a few percents, i.e, independent of grains'size and of solar activity. The graph of the function F (v) (Figure 3) reveals that the assumption of a maxwellian distribution is strictly not correct. This will be improved in a forthcoming analysis, using for instance, two maxwellians.

2.2. Solar wind

The spatial density, the bulk velocity and the velocity distribution function for the solar wind particle species (electrons, protons and α - particles) are required. We retain the synthetic model of Millet et al. (1980) who considered the two "classical" situation of quiet solar wind (Q S W) and high-speed streams (H S S).

The final results for the potential of fused-quartz grains at 1 AU appear in table 1. One sees the influence of the size of the grain, of the solar activity and of the solar wind. Whatever these conditions, the potential lies in the range 2.5 - 5 Volts. It appears fairly constant in the size interval 1 - 10 μm but increases noticeably at 0.1 μm. There is typically a difference of a factor 1.7 between the minimum and the maximum conditions for both the sun and the solar wind.

Grain radius (μm)	0.1		1		10	
Solar Activity	MAX	MIN	MAX	MIN	MAX	MIN
Solar Wind HSS	4.93	4.08	4.72	3.83	4.68	3.79
Solar Wind Quiet	3.91	3.0	3.67	2.72	3.63	2.68

Table 1: electrostatic potential of Fused-quartz grains at 1 AU.

REFERENCES

Grard, R.J.: 1973, J. Geophys. Res. 78, p. 2885
Lafon, J.P.: 1975, J. Plasma Phys. 17, p. 741
Lafon, J.P., Lamy, P.L., Millet, J.M.: 1981, Astron. Astrophys. 95, p. 295
Lamy, P.L.: 1977, Appl. Opt. 16, p. 2212
Millet, J.M., Lafon, J.P., Lamy, P.L.: 1980, Astron. Astrophys. 92, p. 6
Mount, Rottman: 1983, J. Geophys. Res 88, p. 5403
Schmidtke, G.: 1981, Solar Phys. 74 p. 251

ON THE ELECTRICAL EFFECTS OF THE PRESENCE OF FLUFF ON THE SURFACE OF COSMIC DUST GRAINS

J. P. Lafon° and J. M. Millet°°

° Observatoire de Paris-Meudon
92195 Meudon Principal Cedex (France)

°°Laboratoire d'Astronomie Spatiale
Traverse du Siphon
Les Trois Lucs
13012 Marseille (France)

ABSTRACT. We give a n analytical model describing the effects of "fluff" on the potential and the electric field on and close to a charged spherical body embedded in an astrophysical plasma. The consequences are investigated for dust grains biased at positive or negative potentials but large enough for electron or ion field emission to be active.

1. INTRODUCTION

It is usual to determine the electric charge carried by cosmic dust grains embedded in astrophysical plasmas using the assumption that the grain is spherical. However, departure from spherical symmetry can produce some perturbations of the equipotential surfaces that have been underestimated. Thus, we consider a quasi-spherical rough grain with protrusions and recesses; it is assumed that the typical length scales of these irregularities are smaller than the average radius of the quasi-spherical grain and we discuss the topology of the equipotential surfaces and the value of the electric field in the environment of the grain. Since the scale length characterizing the roughness is smaller than the average grain radius, we analyse the effect of each protrusion (or recess) separately, considering that the total effect is the result of the superposition of all individual effects.

2. TWO POINT CHARGES MODEL FOR A SPHERE WITH A PROTRUSION

It has been shown elsewhere (Lafon and Millet, 1984a,b) that a simple and fairly good model for a spherical metallic grain with one protrusion is that delimited by one of the

R. H. Giese and P. Lamy (eds.), Properties and Interactions of Interplanetary Dust, 341–346.

equipotential surfaces of a system of two point charges, one small αQ at some point D and the other larger $(1-\alpha)Q$ ($0 < \alpha << 1$) at some point O separated by the distance $R(1+d)$ ($d << 1$). The surface S of the grain is that at that at the potential $V = 4\pi\varepsilon_o RQ$ of the sphere of radius R carrying the charge Q; it is practically identical to the sphere centered at the point O (greater charge) with radius R, except close to the small charge, where it is strongly perturbed. The perturbation can be characterized by the maximum difference between the distance from O of any point of S and the radius R of the"equivalent"spherical grain, in units of R: x; x is the scale length characterizing the protrusion. $(1+x)R$ is the distance from O of the end the end of the protrusion, which is on the axis of symmetry OD. The degree of attachment of the protrusion can be characterized by a parameter $g = x^2/\alpha$, that can be of the order of a few units.

Of course, if R and Q are used as units of length and charge, α and d are sufficient to fully characterize the model; conversely, for given x and g

$$\alpha = x^2/g \qquad d = x(g-1)/(g+x)$$

Thus, a couple x,g is also sufficient to fully characterize the model.

However, g must be smaller than some g_m , which is a function of x practically constant and close to 5.8 if $x << 1$). Otherwise, the equipotential surface is no longer a deformed sphere, but it is made of two separated lobes, which has no longer any physical meaning for the problem of the protrusion.

Fig. 1 illustrates the meridian sections of the equipotential surfaces of a system of two point charges, in the neigh bourhood of the small charge (the straight line OD defined by the two point charges is an axis of symmetry for the system). In this case $x = 0.05$ and all lengths are measured in units of R in the case where $g = 7$. The lower index of the curves is the potential V_s of the equipotential surface, in units of $V = Q/4\pi\varepsilon_o R$. The upper index is the corresponding g. There is a scaling law that says that the equipotential surfaces indexed by values of $g \neq 7$ and corresponding $V_s \neq 1$ can be changed into equipotential surfaces at potential 1 by scaling lengths by a factor $1/V_s$ (Lafon and Millet, 1984a). Small g correspond to weakly protruding bumps, whereas high g correspond to "loosely attached" protrusions.

Fig. 2 shows the geometry of the "two point charges" model . λ denotes the distance of any point M from the small point charge αQ at D.

At any point on the surface of the grain, the electric field can be expressed as a function of λ (note that though the derivations are correct the final expression of E^2 is not correct in Lafon and Millet, 1984a)

$$E^2 = \frac{1}{(1-\alpha)^2\lambda^4}\{(\lambda-\alpha)^4 + \alpha^2(1-\alpha)^2 - \frac{\alpha(1+d)^2(\lambda-\alpha)^3}{\lambda^2} + \alpha(\lambda-\alpha)((\lambda-\alpha)^2 + (1-\alpha)^2)\}$$

where E is measured in units of $Q/4\pi\varepsilon_0 R^2$.

At the end of the protrusion, P, on the axis of symmetry, this yields

$$E_P = \frac{1-\alpha}{(1+x)^2}\left(1 + \frac{(x+\alpha)^2}{\alpha(1-\alpha)}\right)$$

$$E_P \simeq (1-2x)(1+g) \qquad \text{for } x \ll 1 \text{ and } g = O(1)$$

so that g characterizes also the enhancement of the electric field on the protrusion.

Finally, note that, as explained in Lafon and Millet 1984a, this model, which is based on the theory of electric images, is strictly valid for metals but it is also practically valid with high accuracy for materials with high enough relative permittivity ε, "high enough" meaning $\varepsilon>3$, i.e. for most of the insulators usually expectable in astrophysical grains.

3. DISCUSSION OF THE MODEL

Assuming that the small charge is a point charge we have described the fundamental features due to a protrusion with only two parameters , x for the radial departure from the spherical symmetry and g for the shape of the bump. Now we investigate the effects of this simplification using more complicated models of the same type.

First assume (Fig. 3) that the small charge αQ is replaced by a small charge $\alpha_N Q$ evenly distributed along a needle with length 2L aligned on the axis of symmetry OD, and centered at the same distance OD = R(1+d) than the point charge. Assume also that the resulting projection is salient in a similar way, i.e. the distance (1+x)R between O and the of the protrusion on OD is not changed. Then a tedious but straigthforward analytical discussion of the equations leads to the following results (the method is sketched in Lafon and Millet, 1984b)

$\alpha_N < \alpha$

The maximum admissible g, g_m is slightly reduced

For small needles $(L \ll x-d)$, $\alpha_N = \alpha - d\alpha$ with $d\alpha = O(L^2/(x-d)^2)$. Moreover, the electric field at the end of the protrusion is increased by a small amount of the second order in L

$$dE = \frac{2\alpha}{3(x-d)^2}\frac{L^2}{(x-d)^2}$$

In a similar way, assume (Fig.4) that the small

charge αQ is replaced by a small charge $\alpha_D Q$ evenly distributed over a disc with radius L, perpendicular to the axis of symmetry OD , centered at the same distance OD = (1+d)R as the point charge. Assume also that the resulting protrusion is still salient in a similar way : OP = (1+d)R unchanged. A similar analysis leads to the following results :

$\alpha_D > \alpha$

The maximum admissible g, g_m is slightly reduced

For small discs ($L \ll x-d$) $\alpha_D^m = \alpha - d\alpha$ with $d\alpha = O(L^2/(x-d)^2)$. Moreover, the electric field at the end P of the protrusion is reduced by a small amount of the second order in L :

$$dE = -\frac{1}{2}\frac{\alpha}{(x-d)^2}\frac{2}{(x-d)^2}$$

To summarize, more sophisticated charge distributions will describe protrusions with different shapes but the description will be qualitatively unchanged and quantitatively weakly perturbed. The parameters x and g are sufficient to account for the fundamental physical phenomena.

4. THE ELECTRICAL EFFECTS OF FLUFF

The main results of the previous sections is that even small protrusions ($x \ll 1$) can produce large enhancements of the electric field, (multiplication by a factor of the order of 1+g , where g is independent of x and can be of the order of a few units). On the other hand, an asymptotic analysis (Lafon and Millet, 1984a) shows that the equipotential surfaces surrounding a grain with a protrusion characterized by x,g becomes spherically symmetric at distances from O of the order of R(1 x(1 g^-)) i.e. close to the grain surface, independently of x.

Now, note that a recess can be described with a similar model but with $\alpha < 0$. However, the electric field is reduced close to a recess, which produces less noticeable effects.

Finally, the important parameters for a fluffy grain are the greatest x and g characterizing the sizes and the degrees of attachment of its protrusions.

The protrusions will be important for charging/discharging processes sensitive to the local electric field, like the field emission(for which a small enhancement of the local field increases the emitted current by orders of magnitude), but not for those sensitive to the potential structure(like the collection of plasma particles). In particular, the presence of fluff can enlarge significantly the range of particle sizes for which the field emission fixes an upper bound for the magnitude of the potential. The field emission limited potential of a grain with protrusions characterized by x,g will be lower by a factor (1-2x)(1+g) ≃ 1+g than for a smooth grain. Fig. 5 (from Lafon and Millet. 1984 a)illustrates this effect under magnetospheric conditions.

Of course, the violent discharge of fluffy charged grains by field emission (Lafon and Millet, 1983) can be triggered for larger grains(with sizes larger by a factor ≃ 1+g).

For positively charged grains, the emission of ions produces mass loss from the emitting site. Strong field emission reduces the lifetime of a grain. In other words, the the grain can survive so far as ion field emission is not active. Fluffy grains will tend to loose their fluff and have smooth surfaces; they can even be destroyed if their potential becomes excessive for their size. Smaller grains will be sensitive to such effects when fluffy.

REFERENCES

Lafon J.-P. J.. Millet J. M., Astron. Astrophys., 1983, 123 73-79

Lafon J.-P. J., Millet J. M., Astron. Astrophys., 1984a, 134 296-300

Lafon J.-P. J., Millet J. M., Advances in Space Research, 1984b, Meeting of Cospar ISCB, in press.

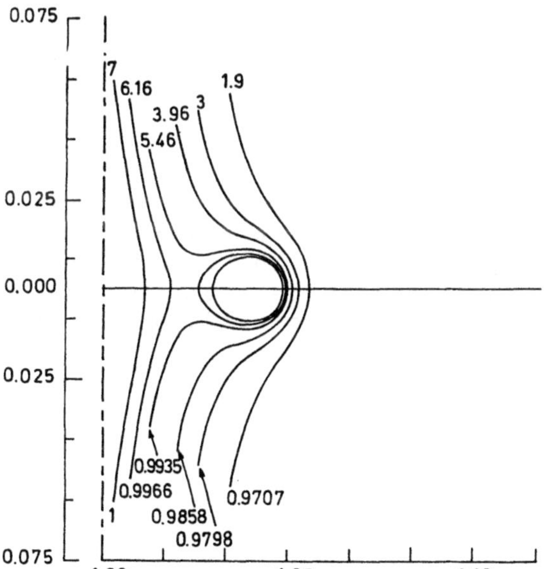

Fig. 1 Meridian sections of the equipotential surfaces around a sphere with a protrusion for fixed x, g (or x, α) and various d indexed by the value of the potential V_s (lower part of the figure). They are also the meridian section of the surface of spheres with protrusions for various g and same α after scaling by the factor $1/V_s$. The indexes of the upper part of the figure are the corresponding values of g. Here $x=0.05$, $g=7$

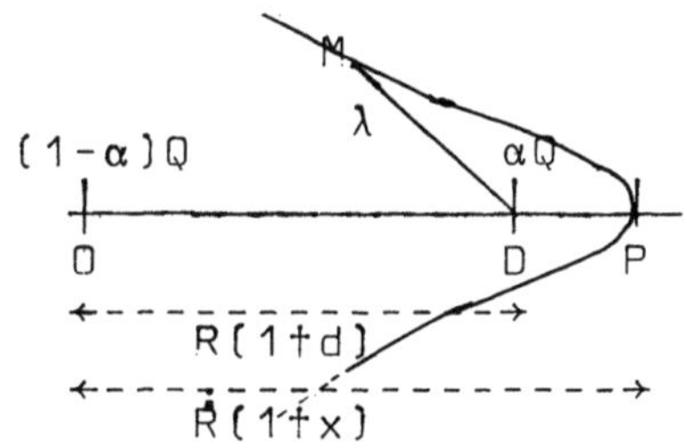

Fig. 2

Fig. 3

Fig. 4

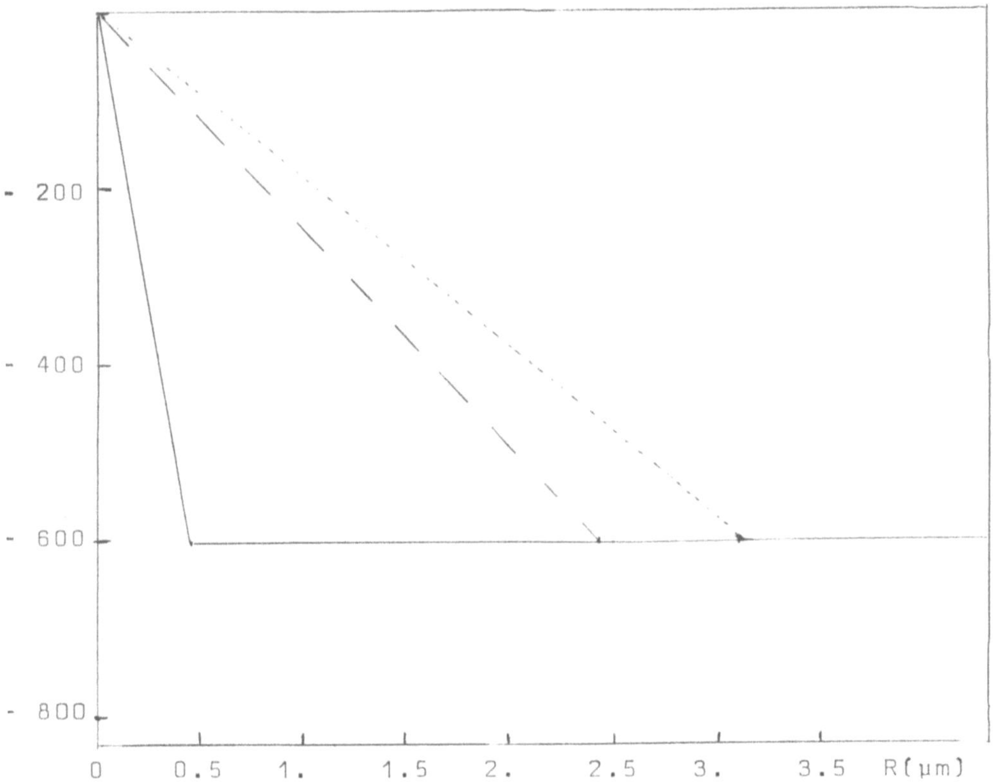

Fig. 5 Potential of fluffy (dashed or dotted curves) and smooth (solid curve) spherical dust grains as functions of their radius R under the conditions of the terrestrial magnetosphere. g is taken equal to 4 (dashed curve) and 5.5 (dotted curve). x = 0.05 From Lafon and Millet, 1984 a,b

SECONDARY EMISSION FROM INTERPLANETARY DUST GRAINS

J.M. MILLET
Université de Provence
Laboratoire d'Astronomie Spatiale, Marseille

J.P. LAFON
DESPA, Observatoire de Meudon

ABSTRACT. Secondary emission can play an important role in the charging of dust grains. This process is very sensitive to their size and to the physical properties of the solid. It is particularly important in the case of insulators in which an intense electric field is present at the surface and can accelerate the excited electrons.

As a consequence, the secondary emission yield δ does not depend on the energy of primaries and multiple solutions of the potential due to the usual non-monotonic form of δ versus the energy of primaries disappear.

1. INTRODUCTION.

When an electron impinges upon the surface of a metallic or dielectric material, it will suffer elastic reflection or be scattered by electrons in the material; it may also induce an excitation of the inner electrons which are ejected out. This last mechanism is known as the "true secondary emission".

In this paper, we present a specific new aspect of the true secondary emission taking place in porous dielectric materials such as those present in the interplanetary medium.

2. MAIN FEATURES OF THE ORDINARY TRUE SECONDARY EMISSION

In this mechanism, electrons excited to energies greater than the ionization energy, can escape through the potential barrier of the material.

The secondary yield δ - the number of secondary electrons induced by one primary electron - has a maximum value ranging from 1 (for metals and semi conductors) to 20 and more for insulators.

This ratio depends upon three factors: the energy of primaries, their angle of incidence and the nature of the material (Whipple, 1981).

The shape of the secondary yield curve as a function of primary energy is an universal curve when normalized to the maximum yield δ_m and the primary energy at this maximum E_m (Dekker, 1958).

After averaging w.r.t. the angle of incidence, the resulting analytic formula

R. H. Giese and P. Lamy (eds.), Properties and Interactions of Interplanetary Dust, 347–350.

$$\delta(E) = 7{,}5\ \delta_m\ E/E_m\ \exp\left[-2(E/E_m)^{1/2}\right]$$

is in good agreement with experimental data. (Sternglass, 1954). Some typical values are given in the following table.

	δ_m	E_m (eV)
Fe	1,3	350
C	1	250
SiO_2	2,1 - 4	400
Al_2O_3	2 - 9	350 - 1300
MGo	3 - 25	400 - 1500

Secondary electrons are emitted with a Maxwellian velocity distribution having thermal energy $E_s = KT_s \simeq 3$eV. (Chung-Everhart, 1974).

3. THE CASE OF POROUS MATERIALS : FIELD DEPENDENT SECONDARY EMISSION (FDSE)

In a first paper, Jacobs (1951) noticed that high secondary emission (ie, 1000 to 1) were due to positive ions formed at the surface of the material by electron bombardment.

Experiments on insulators, (Jacobs, Friely, Brand, 1952) showed that this mechanism is sensitive to the structure of the material and the most porous, the largest the secondary yield. Due to the high resistivity of the material, positive charges stay at the surface for a long time and so, high fields exist and electrons can be liberated by field emission. This electric field affects the secondary emission in such a way that the yield is independent of bombarding energies and is exponentially dependent upon the field across the dielectric during bombardment.

This led Jacobs et al. (1952) to the hypothesis that the FDSE process is similar to the pre-sparking mechanism in gas-discharge.

Porosity allows bombarding electrons to penetrate inside the material. Secondary electrons released within the material are now accelerated toward the surface under the influence of the high field; they travel freely through these pores and gain sufficient energy to create further electrons and so on.

From the Townsend equation, we have; $\delta = e^{\alpha x}$ where α is the number of electrons generated along a unit path by electron and x, the depth within the material at which the initial secondary electrons are excited. α is related to the mean free path L_e, the strenght E of the electric field and the work function V_i fo the bulk material by

$$\alpha = \frac{2.4}{\pi} \cdot 10^3 \ p \ L_e \ E \exp(- 16V_i/\pi^2 L_e E) \ (1 + 16V_i/\pi^2 L_e E)$$

where p is the equivalent pressure of colliding centers in the material leading to the same mean free path in gases. For example, to $L_e = 10^{-5}$cm corresponds $p = 4 \cdot 10^5$mm Hg.

This mechanism appears if excited electrons gain between each collision an energy greater than the ionization energy; that is $E \ L_e > V_i$. In common cases, $V_i \sim 5$V and $E > 10^6$ V cm^{-1}, and the application of this mechanism is limited to grain with radius $R > L_e > 0,5$ µm.

The statistical properties of such electrons are the same than that of gas-discharge: the energy distribution is Maxwellian, with a temperature T_s related to the mean free path L_e on which electrons can be accelerated without collisions (0,1µ) by $T_s = e \ L_e \ E/1,64 \ R$. Typical values are $\delta \sim 300 - 1000$ and $T_s \sim 10 - 20$ eV.

4. THE POTENTIAL OF THE GRAIN

The determination of the grain potential when immersed in a plasma (H_{II} regions, stellar wind) is completely solved by Lafon et al. (1982) in the case of species of ionized particles originally independent.

An approximation which balances the total flux of particles impinging and leaving the surface of the grain has been used previously by Feuerbacher et al. (1973).

In the case of secondary emission the flux of particles leaving the surface is dependent upon the energy of the primaries. Prokopenko et al. (1980) and Meyer - Vernet (1982) used the following expressions for the secondary current.

$$\Phi < o \qquad I_{sec} = \int_{-e\Phi}^{\infty} (1+e\Phi/E) \ 2\pi E/m_e^2 \ n_e \ (m_e/2\pi KT_e)^{3/2} \exp(-E/KT_e) \ \delta \ (E+e\Phi)dE$$

$$\Phi < o \qquad I_{sec} = I_s \exp(e \ /KT_s) \ (1 + e \ /KT_s)$$

where n_e, m_e, T_e are the density, mass, mean temperature of the primaries, Φ the grain potential, e the absolute value of the electron charge and

$$I_s = \int_0^{\infty} (1+e\Phi/E) \ 2\pi E/m_e^2 \ n_e \ (m_e/2\pi KT_e)^{3/2} \exp(-E/KT_e) \ \delta \ (E+e\Phi) \ dE$$

Using the above approximation, Meyer - Vernet (1982) obtained the floating potential of interplanetary grains by solving for the zeros of the function

$$I_t(\Phi) = I_i + I_{sec} - I_e$$

I_e and I_i are the electronic and ionic currents driven from the plasma to the grain. She showed (her figure 2) that the potential is not unique and that the charge of grain can oscillate (flip-flop) under appropriate conditions.

In the case of real grains with porous surface, the yield of secondary emission does not depend upon the energy of primaries and so, multiple roots do not appear. On other hand, the secondary emission being enhanced, the total current $I_T(\phi)$ cannot zeroed wathever .

The approximation which balances the fluxes is not applicable and we have to numerically calculate the sheath (Lafon et al, 1981) which limits the currents from the grain or to introduce other process able to limit the high voltage such as disruption.

REFERENCES

Whipple, E.C.: 1981, Rep. Prog. Phys, 44, p. 1197
Dekker, A.J.: 1958, Solid State Physics, vol 6, Academic New-York
Sternglass, E.J.: 1954, Phys Rev, 95, p. 345
Meyer - Vernet, N.: 1982, Astron. Astrophys, 105, p. 98
Chung, M.S. and Everhart, T.E.: 1974, J. of applied Physics, 45, p. 707
Jacobs, H.: 1951, Phys. Rev, 84, p. 877
Jacobs, H., Freely, J. and Brand, F.A.: 1952, Phys. Rev, 88, p. 492
Lafon, J.P.J., Lamy, PH.L. and Millet, J.: 1981, Astron. Astrophys, 95, p. 295
Feuerbacher, B. Willis, R.F. and Fitton, B.: 1973, Astrophys. J, 181, p. 101

COMPUTER CALCULATION OF COLLISION CASCADES BY ENERGETIC PARTICLES PENETRATING DUST GRAINS

K. Rössler and G. Eich
Institut für Chemie 1 (Nuklearchemie) der Kernforschungsanlage Jülich GmbH, Postfach 1913
D-5170 Jülich
Federal Republic of Germany

ABSTRACT. The penetration of energetic particles from stellar winds, cosmic rays and fast moving gas clouds in interstellar dust grains induces a variety of physical and chemical changes. Detailed information on penetration ranges, number of secondary projectiles produced by knock-on and their energy distribution is obtained from computer simulation of collision cascades with the program MARLOWE. Model targets are polycrystalline H_2O-ice, NH_4Cl, and SiO_2, projectiles are 10^2-10^4 eV H, He, and C. Depending on the energy and the projectile/target mass ratio up to 10^2 secondary projectiles are created per primary impact with energies ranging from a few to some 10^3 eV. In composite grains, condensed gases and ice mixtures the energetic secondaries can undergo hot chemical reactions leading to a series of precursors for larger organic and biomolecules.

1. INTRODUCTION

Interplanetary and interstellar dust grains are bombarded by energetic particles such as electrons, protons, alphas and heavier ions from local radiation belts, solar (stellar) wind, cosmic rays and fast moving gas clouds. These impacts induce a variety of physical and chemical changes. Among them is the build-up of organic material, especially in composite grains with ice layers or frozen mixtures of gases, cf. e.g. (1,2). A new type of interactions has recently be introduced to cosmic chemistry, high energetic (hot) reactions of biogenic implants such as carbon or nitrogen (3). Hot atom chemistry (4,5) is characterized by the fact that endothermic reactions and those with high energies of activation are possible, among them atom molecule interactions. Primary implants and secondary particles knocked on inside the grain material may undergo these reactions. In homogeneous ice mixtures, e.g. H_2O-NH_3-CH_4, the products of the secondaries may be more important than those of the primaries. The hot reactions are accompagnied by radiolytic and photolytic processes which may change the initial products, cf. (6).

The physical consequences of particle impact into the grains are

R. H. Giese and P. Lamy (eds.), Properties and Interactions of Interplanetary Dust, 351–356.

not well known. Penetration depths and stopping powers are listed in table works for particles with energies $\geq$ 1oo keV/amu and for mono-elemental solids only. The lower energy regime of some 1o to some 10^4 eV, typical for accelerated particles in space, can be treated by computer simulation of collision cascades with simple codes such as TRIM (9). Detailed information on ranges, displacements and replacements, defect profiles, recombination of defects, and backscattering or sputtering from surfaces in well defined polyelemental solids can be obtained from the code MARLOWE in the binary collision approximation (1o). This code had been successfully used for studies of collision dynamics in many insulators (11-17). It is now applied to simulate cascades of light and heavier projectiles in substances representative for cosmic dust particles. Some results are presented here: mean penetration of primary and number of secondary projectiles and their energy distribution. A more detailed report will be given elsewhere (18).

2. COMPUTER CODE AND INPUT DATA

MARLOWE was used in a slightly modified version (14), however, with the standard procedure provided by the default option. Projectiles studied were ^1H, ^4He and ^{12}C with the primary energies 10^2, 10^3, and 10^4 eV. Solid H_2O, SiO_2 and NH_4Cl were chosen as model substances for ice, silicate materials and compounds composed out of light, medium and heavier elements. For H_2O and SiO_2 simplified model lattices were selected (cf. TAB. I), since their real structures under cosmic conditions are unknown. The lattice constants preserve the real densities of $\rho(H_2O,77K)$ = o.87 and $\rho(SiO_2)$ = 2.635 gcm^{-3}. For NH_4Cl the real bcc lattice with ρ = 1.527 gcm^{-3} was taken.

TABLE I. Parameters of model substances for MARLOWE calculations

	lattice type	a_o,nm	energies, eV: displacement threshold	energies, eV: binding loss	energies, eV: projectile cutoff
H_2O	antifluorite	51.6	4(H), 5(O)	2(H), 3(O)	2(H,O)
NH_4Cl	bcc	38.7	7(N),4(H),5(Cl)	3(N),2(H,Cl)	3(N,H,Cl)
SiO_2	fluorite	53.3	8(Si),6(O)	4(Si),3(O)	3(Si,O)

The displacement threshold energies for the individual atoms in TAB. I were estimated on the base of sublimation and chemical bond energies. They do not represent the classical Wigner energies for creation of stable defects, e.g. Frenkel pairs. The displacement thresholds in TAB. I represent the energy necessary for an individual displacement event, irrespective of spontaneous recombination processes of close by Frenkel pairs. Binding loss energies are substracted from the kinetic energy, a secondary atom obtains by the knock-on. The projectile cutoff energy represents the value at which a projectile is stopped by the calculation. The sum of the latter two energies should not exceed the displacement threshold energy. They are chosen empirically to about one half of the threshold value.

The projectiles were started inside the crystals. The crystal orientation was slightly changed from collision to collision by a random number generator program in order to simulate a very polycrystalline solid. No thermal vibrations were included. Only the primaries were followed to the cutoff energy. The secondaries and their energies were registered, however, their collisions were not followed for reasons of limited computation time. For 10^2 and 10^3 eV projectiles for each system 10^3 cascades were calculated, for 10^4 eV particles about 10^2 cascades.

3. RESULTS

The mean penetration of H, He and C projectiles is plotted in Fig. 1-3 versus the primary energy. The ranges decrease with increasing mass number of the projectile and with increasing density of the target. From the rather linear behaviour of the curves it seems to be possible to extrapolate to projectile energies exceeding 10^4 eV, at least for the heavier projectiles which perform an appreciable amount of elastic (nuclear) collisions.

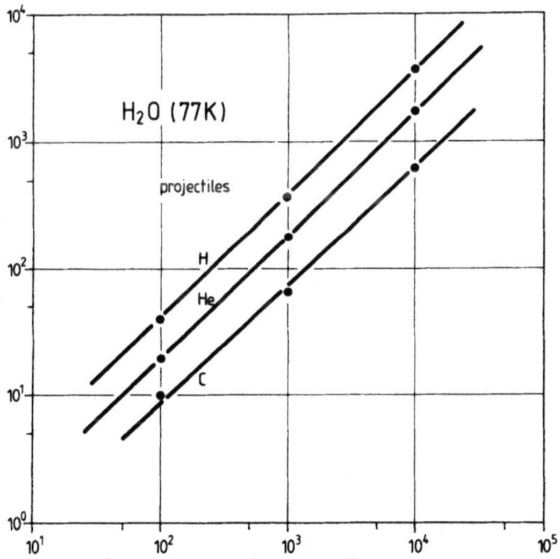

Figure 1. Mean penetration of some projectiles in H_2O-ice (77K)

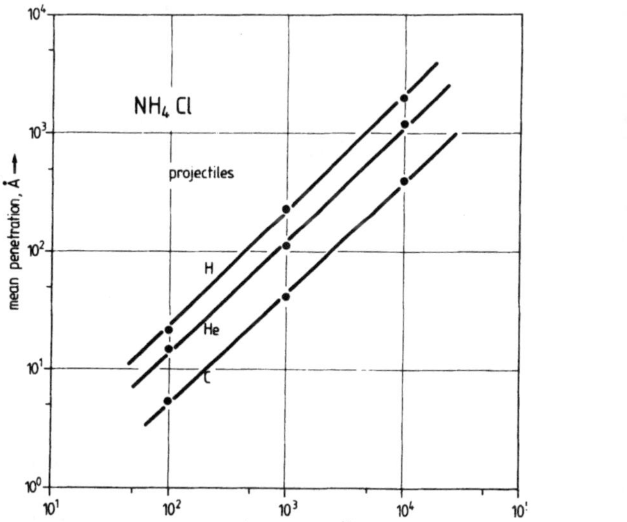

Fig. 2. Mean penetration of some projectiles in NH_4Cl

The numbers of secondary projectiles produced by knock-on processes within the solids are reported in TAB. II. The number increases with the primary energy, more steadily in the case of the

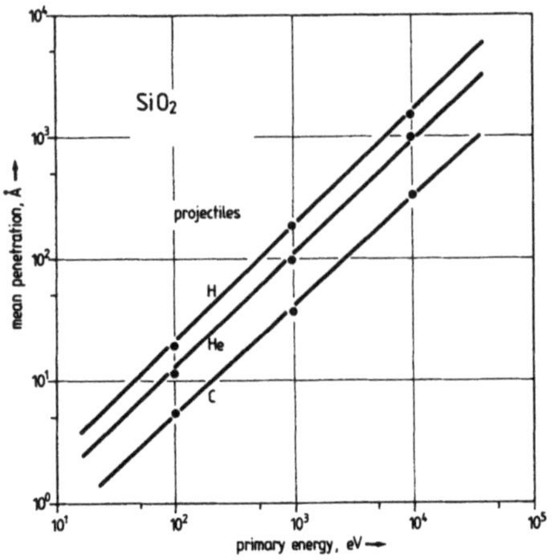

heavier projectiles than for hydrogen. It depends also on the projectile-target mass ratios. At 10^4 eV up to 100 secondaries are ejected. More information can be obtained from Figs. 4-7 which describe in more detail kind and energy distribution of the secondaries. The numbers are plotted for the individual kind of secondary ordered according to the primary energies. The distribution is given for groups ranging from 1 to 10, 10 to 10^2, 10^2 to 10^3 and 10^3 to 10^4.

Figure 3. Mean penetration of some projectiles in SiO_2(ρ=2.635 gcm^{-3}).

TABLE II. Number of secondary projectiles per cascade

prim. projectiles and energy, eV		targets H_2O	NH_4Cl	SiO_2
H	10^2	4 ± 1	3 ± 1	1 ± 1
	10^3	13 ± 3	14 ± 4	8 ± 2
	10^4	16 ± 4	27 ± 5	20 ± 5
He	10^2	5 ± 1	5 ± 2	3 ± 1
	10^3	21 ± 5	24 ± 5	15 ± 3
	10^4	76 ± 10	69 ± 10	49 ± 8
C	10^2	5 ± 2	6 ± 2	3 ± 1
	10^3	28 ± 10	24 ± 6	16 ± 5
	10^4	92 ± 21	101 ± 19	45 ± 12

4. DISCUSSION.

The calculations prove that energetic primary particles can produce an appreciable amount of secondaries, many of them with energies > 10 and even > 100 eV. Slight changes of the energetic parameters in TAB. I might somewhat affect the number of secondaries. However, the number of particles with energies > 10 eV, which are important for hot atom chemistry, will not be diminished. The higher energetic secondaries may induce displacement and, thus, create tertiary projectiles and so on. The amount of hot chemical reactions by biogenic projectiles is magnified by the creation of secondary, tertiary etc. atoms by particles which themselves do not directly form organic material such as protons, helium ions, heavier ions up to uranium, and electrons. Another possibility for the formation of

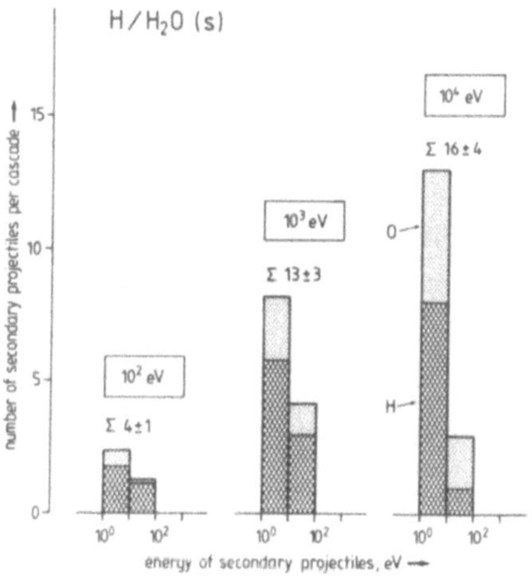

Figure 4. Secondary projetiles by H impact in poly-cryst. H_2O ice (77 K)

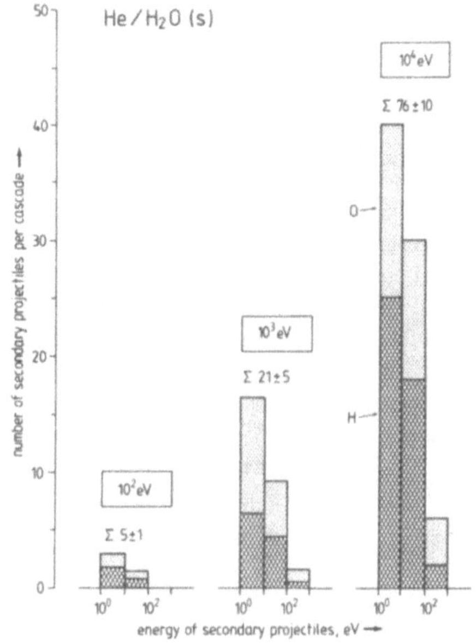

Figure 5. Secondary projectiles by He impact in poly-cryst. H_2O ice (77 K)

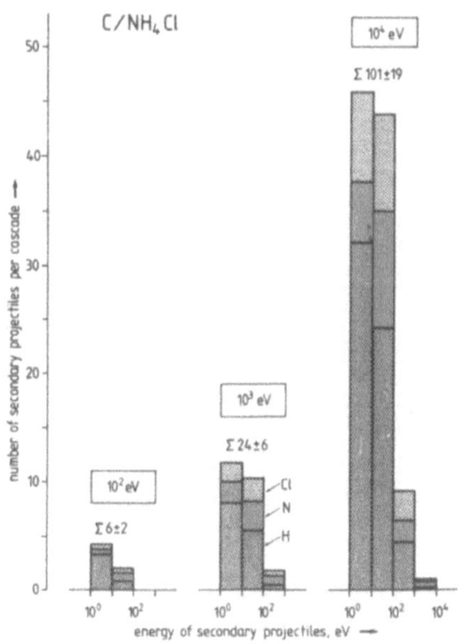

Figure 6. Secondary projectiles by C impact in poly-cryst. NH_4Cl

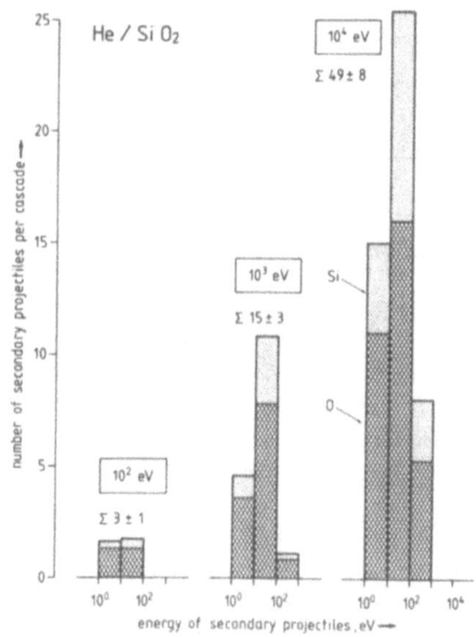

Figure 7. Secondary projectiles by He impact in poly-cryst. SiO_2

energetic particles in solids may be coulomb repulsion upon the passage of fast, ionizing rays, e.g. from cosmic radiation. The MARLOWE calculations furnish physical data on collison dynamics in dust grains which improve the probability of chemical reactions such as shown schematically in Fig. 8. Together with photolytic or radiolytic processes, hot atom chemistry may play an important role in the formation of organic material on interplanetary and interstellar dust grains, comets and surfaces of some of the moons of outer planets.

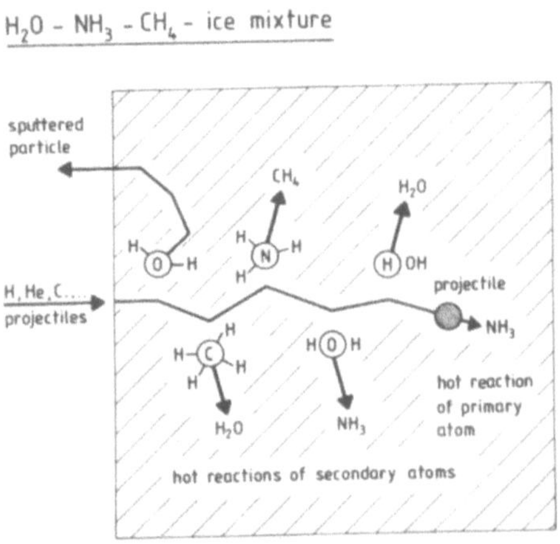

Figure 8. Schematic reactions of primary and secondary particles in an ice mixture.

5. REFERENCES

1. C. Ponamperuma (ed.), Cosmochemistry and the Origin of Life, D. Reidel, Dordrecht, 1983
2. J.M. Greenberg, in Comets, L.L. Wilkening (ed.), The University of Arizona Press, Tucson, Arizona, 1982
3. K. Rössler, H.-J. Jung, B. Nebeling, Adv. Space Res., 1985, in press
4. G. Stöcklin, Chemie heißer Atome, Verlag Chemie, Weinheim 1969 (in German); Chimie des atomes chauds, Masson et Cie, Paris 1971 (revised french version)
5. A.G. Maddock, G. Harbottle (eds.), Chemical Effects of Nuclear Transformations in Inorganic Systems, North Holland, Amsterdam 1979
6. K. Rössler, this issue
7. J.F. Ziegler, Handbook of Stopping Cross-Sections for Energetic Ions in All Elements, Pergamon Press, New York, 198o
8. U. Littmark, J.F. Ziegler, Handbook of Range Distributions for Energetic Ions in All Elements, Pergamon Press, New York, 198o
9. J.P. Biersack, L.G. Hackmark, Nucl.Instr.Meth. **174** (198o) 257/69
1o. M.T. Robinson, I.M. Torrens, Phys. Rev. B., 9 (1974) 5oo8/24
11. M.T. Robinson, K. Rössler, I.M. Torrens, J.Chem.Phys. 6o (1974) 68o/8
12. K. Rössler, M.T. Robinson, in Atomic Collisions in Solids, S. Datz, B.R. Appleton, C.D. Moak (eds.), Plenum Publ. Corp., New York 1975, 1, 237/249
13. O.S. Oen, M.T. Robinson, J.Nucl.Mat. **76/77** (1978) 37o/7
14. W. Westmeier, K. Rössler, Report Jül-1563, December 1978
15. K. Rössler, L. Pross, Rad.Eff. **48** (198o) 2o7/12
16. K. Rössler, Uranium Recoil Reactions, in Gmelin Handbook of Inorganic Chemistry, Uranium, Suppl. Vol. A6, Springer Verlag, Berlin 1983, 135/64 (139)
17. K. Rössler, A.R. Manzanares, Report Jül-1924, June 1984
18. K. Rössler, G. Eich, J.Chem.Phys., to be submitted

LABORATORY SIMULATION OF CHEMICAL INTERACTIONS OF ACCELERATED IONS WITH DUST AND ICE GRAINS

K. Rössler
Institut für Chemie 1 (Nuklearchemie) der Kernforschungs-anlage Jülich GmbH, Postfach 1913
517o Jülich
Federal Republic of Germany

ABSTRACT.
Energetic ions or atoms in space may undergo hot chemical reactions upon penetration into interplanetary or interstellar dust grains, ice layers, cometary matter, and surfaces of planetary moons. The mechanistic pathways can be different from those of classical ion molecule interactions, photolytical and radiolytical processes. The kinetic energy of the hot reactant facilitates endothermic reactions and those with high energy of activation, among them atom-molecule interactions. The conditions of hot cosmic chemistry are simulated in laboratory experiments in order to obtain insight into the nature of chemical products and the reaction mechanisms of their formation. This paper reviews the methods of ion implantation, nuclear recoil in situ, nuclear recoil implantation, secondary knock-on processes and computer simulation of collision cascades. Carbon and nitrogen impact in frozen H_2O, NH_3 and CH_4 is shown to lead to the formation and radiolytic permutation of a series of organic molecules, among them e.g. formaldehyde, methanol, methylamine, cyanamide, formamidine and guanidine which may act as precursors for biomolecules.

I. INTRODUCTION

Reactions in the bulk of interplanetary or interstellar dust grains are less frequently treated by cosmic chemistry, compared to the reactions in the gas phase. Two prominent examples are the formation of H_2 molecules from H atoms on grain surfaces (1) and the photolytic or radiolytic reactions in mixed ice layers or frosts (2). A new type of cosmic solid state reactions has been discussed recently, the interaction of accelerated ions or atoms with interstellar grains (3-7). Species with kinetic energies ranging from a few eV to some MeV are frequently encountered in space, e.g., in the radiation belts of stars and planets, in interactions with solar (stellar) winds and cosmic rays, and upon collisions of dust or gas clouds. The impact of primary energetic particles such as electrons, protons, light and heavy ions may create secondary projectiles inside a solid by knock-on processes (8) and finally may lead to sputtering (9). Chemical reactions of accelerated species are studied for more than 5o years in nuclear chemistry. The

R. H. Giese and P. Lamy (eds.), Properties and Interactions of Interplanetary Dust, 357–363.

domain of hot atom chemistry treats the reactions of particles recoiling as a consequence of nuclear processes (1o-14). Hot reactions are characterized by the fact that one of the reactants carries an appreciable amount of its kinetic energy into the reactive collision. This enables endothermic reactions and those with high energies of activation, such as atom-molecule interactions. The latter ones are only seldomly considered in cosmic chemistry higherto. Especially, the reactions of biogenic atoms such as carbon or nitrogen with interstellar dust and ice material, comets, planetary surfaces, etc. seems to be of interest with respect to the formation of organic material in space. Some first studies are concerned with the reactions of hot radioactive ^{11}C and ^{13}N with frozen H_2O, NH_3 and CH_4 at 77 K (6,15-19). These nuclear recoil techniques are compared to other solid state physical methods such as ion implantation, atomic beams, and computer simulation of collision cascades in solids. The different approaches are compared with respect to their aptitude to simulate energetic solid state processes in space.

2. CHEMICAL REACTIONS OF IMPLANTED IONS

The chemical consequences of implantation of biogenic ions into some inorganic substances have been studied by means of radioactive and stable isotopes. Besides the formation of carbides and nitrides by implantation of carbon and nitrogen ions into metals and semiconductors, many properly "chemical" systems have been reported, cf. e.g. (2o-26). Table I presents a selection of insulator systems ranging from alkali halides to targets with more cosmic relevance such as SiO_2 and water ice. In general, simple products are observed, most of them by means of optical spectroscopy. The advantage of the in situ analysis by optical spectroscopy, which applies to many systems in Table I (with exception of the radioactive implants), is however ruled out to a certain degree by the radiolytical effects of the high radiation doses delivered by the implants. Optical spectroscopy in absorption requires, even for molecules with very high (C_2^-) and high (CO_2) molar extinction coefficients in the range of $1o^4$ to $1o^5$ $cm^2 mole^{-1}$ (23), the implantation of some $1o^{15}$ to some $1o^{17}$ ions cm^{-2}. Table II. reports the radiation dose (total fluence) - in eV delivered per target molecule in the penetration range of the projectiles - for carbon implants in H_2O-ice (σ = o.87 gcm^{-3}). It can be seen, that the doses are relatively independent of the energy and that they are very high in the range of 5o to 5ooo eV per target molecule for $1o^{15}$ and $1o^{17}$ ions cm^{-2}. The second disadvantage is that also the fluxes (dose rates) are equally high. In order to obtain some $1o^{17}$ implants cm^{-2} in a reasonable time, the beam intensity cannot be much lower than o.1 to 1 µA ($6 \cdot 1o^{11}$ to $6 \cdot 1o^{12}$ ions sec^{-1}). Other methods of in situ analysis such ESR and ^{13}C-NMR require high concentration of implants, likewise. The only feasible improvement consists in optical spectroscopy via laser induced fluorescence of the chemical products. Here some $1o^{14}$ to $1o^{15}$ ions cm^{-2} would be sufficient for an effective measurement. The high energy deposition by the laser may, on the other hand, be detrimental for metastable radicals and intermediates. In order to establish a good comparison of the simulation experiments with real cosmic chemistry, temperatures of the targets should be kept in the region between 4 and 1o K by the aid of cryostats. Especially for water ice, slight changes of temperatures in the 4 to 2o K and the 77 to 115 K region result in different mobility of radiolytic H˙ or OH˙ radicals, responsible for reducing or oxidizing the primary products of the implantation, cf. (7,18,19). The major disadvantage, however, remain the too high doses and dose rates, compared to interstellar conditions.

Table I. Chemical products formed by ion implantation into some simple solids

projectile	energy keV	target	temp. K	products	analysis* method	ref.
$^{35}S^+$	6o	NaCl,KCl	298	S^{2-},S^o, SO_3^{2-},SO_4^{2-}	RC	2o
$^{32}P^+$	6o	"	298	PI,PIII,PV	"	2o
$^{12}C^+$,$^{14}N^+$	2o-2oo	KCl	298	CN^-	UV	21
$^{12}C^+$,$^{13}C^+$	25o	AX**	5	C_2^- (C_2,C_3)	VIS,UV	22,23
$^1H^+$,$^2H^+$	15	Al_2O_3	298	$Al_2O_2(OH)$	IR	24
$^1H^+$,$^2H^+$		TiO_2	298	OH,OD	IR	25,26
$^{12}C^+$	15o	MgO,Al_2O_3, SiO_2,CaF_2	5	no C_2^-	VIS,UV	22,23
$^{12}C^+$	15	minerals lunar samples	298	simple organic	MS	3,4
$^{12}C^+$,$^{13}C^+$	4o	SiO_2	298	CO_2,CO	IR	5,27,28
$^{14}C^+$	o.oo7 -o.48	H_2O	113	CO,CO_2,CH_4 CH_3OH,CH_2O HCOOH etc.	RC	29,3o
$^{12}C^+$,$^{13}C^+$	4o	H_2O	77	CO_2	IR	31
$^{12}C^+$,$^{14}N^+$	4o	H_2O	77	CNO(?)	IR	31
$^{14}N^+$	25o	H_2O	6	NO_2,HNO_2	IR	19,32
$^{11}C^+$	$1o^{-2}$-$1o^3$	C_2H_4	77	twelve ^{11}C labelled organic	RC	33

* RC = radiochem. separations, e.g. chromatography; UV, VIS, IR = optical spectroscopy in absorption; MS = mass spectroscopy of ejected ions

** A = Na, K, Rb, Cs; X = F, Cl, Br, I

3. CHEMISTRY OF SECONDARY PROJECTILES (UPON KNOCK-ON, SPUTTERING)

Energetic particles in solids create a number of secondary projectiles by knock-on processes. Provided the energy of the latter ones is large enough, tertiary projectiles are formed and so on. Computer simulation of collision cascades in SiO_2 and H_2O-ice shows that by a $1o^4$ eV hydrogen primary about 2o and by carbon between 5o and 1oo secondaries are formed with energies between some few eV and $1o^4$ eV (8). These particles can perform hot chemical reactions, which might be of special importance for frozen H_2O-NH_3-CH_4 etc. mixtures and contribute to the radiolytical reactions, cf. (3). This hot chemistry underlies the formation and chemical sputtering of molecules and molecular ions during the bombardment of ices with fast ions (9).

TABLE II. Dependance of radiation dose D^* (total fluence) delivered to H_2O-ice (ρ=o.87) by carbon implants with various energies.

energy of $^{12}C^+$, eV	mean penetration, nm cf. [8]	$n(H_2O)\,cm^{-2}$ in the impl. layer	D^*, eV per target molecule for $1o^{15}$	or $1o^{17}$ ions cm^{-2}
$1o^2$	o.7	$2\cdot 1o^{15}$	5o	5ooo
$1o^3$	6.5	$1.9\cdot 1o^{16}$	53	53oo
$1o^4$	62	$1.8\cdot 1o^{17}$	55	55oo
$1o^5$	6oo	$1.7\cdot 1o^{18}$	57	57oo

Numerous systems have been studied via mass spectrometry of the ejected ions, such as H_2O, NH_3, CO_2, SO_2, CH_4, N_2, H_2 and O_2 and mixtures thereof, cf. e.g. (9,34-42). Of interest is the observation of formaldehyde $D_2{}^{13}C^{18}O$ emerging from an ion bombarded surface of a $^{13}C^{18}O_2$-$D_2{}^{16}O$ ice film at 9 K (9). The probable mechanism of formation may be an insertion of a secondary energetic ^{13}C into the O-D-bonds of $D_2{}^{16}O$. Besides the sputtering of ices there are intersting studies on the ejection of volatiles out of carbon films during bombardment with hydrogen at energies between 1oo eV and 3o keV, cf. (43). Synergistic effects have been studied in the erosion of graphite and TiC by D^+ ions, H^o/Ar^+-molecular and ion beams and H^o/e^--molecular and electron beams (44-46) and the emission of CH_3, CH_4(CD_4) has been observed. Due to the sensitive methods of detection, irradiation doses can be kept somewhat lower than in the case of analysis of the primary's chemical fate. However, the information on the bulk chemistry is often obscured by surface effects and aggregation in the gas phase before detection.

4. NUCLEAR RECOIL IN SITU AND NUCLEAR RECOIL IMPLANTATION

Formation of energetic primaries at much lower irradiation doses ranging from $1o^{-4}$ to some $1o^2$ eV per target molecule is possible via recoil processes after nuclear reactions. Biogenic recoil atoms can be produced by a multiplicity of reactions ranging from the thermal neutron induced $^{3}He(n,p)^{3}H$, $^{14}N(n,p)^{14}C$, and $^{35}Cl(n,p)^{35}S$ to those by cyclotron accelerated particles in the multi MeV range, e.g., $^{14}N(p,\alpha)^{11}C$, $^{16}O(p,\alpha)^{13}N$ and $^{14}N(d,n)^{15}O$, cf. (7,1o). The product nuclei obtain kinetic energies from several 1oo eV to some MeV depending on the type of nuclear process. They can be projected in situ by activating a target component or implanted from the outside from another gaseous, liquid or solid source material, e.g. BN fine grains for implanting ^{11}C via the $^{14}N(p,\alpha)^{11}C$ reaction into admixed LiH grains (15,16). That the irradiation doses can be kept much lower than in the ion implantation studies, comes from the fact that the projectiles are radioactive and can be followed by radiochemical techniques such as radiogas -, high performance liquid - and thin layer chromatography. Thus $1o^{7}$ to $1o^{1o}$ implants can be analyzed for their chemical fate. The radiation dose stemms primarily from the activating particles (protons, deuterons, neutrons etc.) and not from the collision cascade of the primary as in the case of the ion implantation or sputtering studies. The disadvantages of the nuclear method are the target evaporation, melting or dissolving steps prior to the chromatographic analysis, which may change important metastable intermediates in the cold solid into different final products. Also, only reactions of single atoms can be studied. The small number of recoil atoms is not sufficient to form compounds containing two or more of them. Despite these disadvantages, the fact that (via the radioactivity) a 1oo % balance of the products of a certain implant is possible and the low dose range render the nuclear method as a powerful tool for studies of cosmic solid state chemistry. Recent examples are the systems $^{11}C/H_2O$ (77 K), $^{13}N/H_2O$ (77 K), $^{11}C/NH_3$ (77 K) and $^{13}N/CH_4$ (77 K) (7,15-19) which demonstrated the formation of products which are important for cosmic chemistry with respect to chemical evolution such as NH_3, CH_4, formaldehyde, cyanamide, methylamine, NH_2OH, CH_3OH, formamidine and guanidine. These systems showed the primary hot reactions at doses $\leq 1o^{-2}$ eV per target molecule and at increasing radiolysis, e.g. in H_2O ice, reduction by $H^{\cdot}$ radicals, and at highest doses ($>1o^{1}$ eV per target molecule) oxidation by the reactive $OH^{\cdot}$ and $O_2H^{\cdot}$ radicals. Thus, products are constantly changed by radiolysis (7). From the effects of temperature, dose and dose rate extrapolation to cosmic conditions seems to be possible.

6. CONCLUSION

The three experimental methods for studying hot reactions in solids of astrophysical interest have advantages and disadvantages with respect to information on the proper in situ reactions at lowest temperatures. Thus, they have to complete each other and proper mechanistic information should by based on intercomparison of the results obtained by at least two of the methods. Computer simulation of collision cascades provides the necessary solid state physical background information for the chemical studies.

REFERENCES

1. W.D. Watson, Rev.Mod.Phys. 48 (1976) 513/52
2. J.M. Greenberg, Sci.American 25o (1984) 96/1o6
3. J.P. Bibring, Y. Langevin, M. Maurette, R. Meunier, B. Jouffrey, C. Jouret, Earth Planet.Sci.Letters 22 (1974) 2o5/14
4. M. Maurette, Nucl.Instr.Meth. 132 (1975) 579/86
5. J.P. Bibring, F. Rocard, Radiat.Eff. 65 (1982) 159/65
6. K. Rössler, M. Vogt, 7th Int.Conf.Origins of Life, Mainz, FRG, 1o-15 July 1983, Abstract C1-26
7. K. Rössler, H.-J. Jung, B. Nebeling, Adv.Space Res. 1985, in press (COSPAR 1984, F.3.3.1 and F.5.1)
8. K. Rössler, G. Eich, this issue
9. W.L. Brown, L.J. Lanzerotti, R.E. Johnson. Science 218 (1982) 525/31
1o. G. Stöcklin, Chemie heißer Atome, Verlag Chemie, Weinheim 1969 (french version: Chimie des atomes chauds, Masson et. Ci., Paris 1971)
11. F.S. Rowland (ed.), Hot Atom Chemistry Status Report, IAEA, Vienna 1975
12. G. Harbottle, A.G. Maddock (eds.), Chemical Effects of Nuclear Transformations in Inorganic Systems, North Holland, Amsterdam 1979
13. T. Tominaga, E. Tachikawa, Modern Hot Atom Chemistry and its Applications, Springer Verlag, Berlin 1981
14. T. Matsuura (ed.), Hot Atom Chemistry, Kodansha, Tokyo 1984
15. K. Rössler, H. Lattke, C. Mathias, L.M. Al Shukri, M. Vogt, J.Lab.Comp.Radiopharm. 19 (1982) 1618/9
16. M. Vogt, Report Jül-1855 (June 1983)
17. K. Rössler, M. Vogt, Proc. 7th Int.Congr.Radiation Research, J.J. Broerse et al. (eds.), Martinus Nijhoff Publ., Amsterdam 1983, A6-05
18. B. Nebeling, Report-Jül-1973 (February 1985)
19. K. Rössler, K. Schurwanz, Report-Jül-199o (April 1985)
2o. A.G. Maddock, in (11), 33/48, and ref. cited herein
21. W.A. Metz, E.W. Thomas, Nucl.Instr.Meth. 194 (1982) 5o5/8
22. K. Rössler, A.R. Manzanares, Report Jül-1924 (June 1984)
23. K. Rössler, A.R. Manzanares, B. Stritzker, in Induced Defects in Insulators, P. Mazzoldi (ed.), Les Editions de Physique, Paris 1984, 193/9
24. D.M. Gruen, B. Siskind, R.B. Wright, J.Chem.Phys. 65 (1976) 363/78
25. M. Guermazi, P. Thevenard, P. Faisant, M.G. Blanchin, C.H.S. Dupuy Rad.Effects 37 (1978) 99
26. A. Perez, Nucl.Instr.Meth. B1 (1984) 621/7
27. F. Rocard, J.-P. Bibring, Phys.Rev.Lett. 48 (1982) 1763/6
28. J.P. Bibring, Y. Langevin, F. Rocard, J.Geophys.Res. 87 Suppl. (1982) A446/5o
29. A.G. Maddock, University of Cambridge (U.K.), private communication
3o. I.M. Evans, Ph.D.Thesis, University of Cambridge (U.K.) 197o
31. J.-P. Bibring, F. Rocard, Adv.Space Res., 1985, in press (COSPAR 1984, F.3.3.2)

32. K. Rössler, M. Batista, B. Nebeling, K. Schurwanz, Terra Cognita 5 (1985) 128
33. J. Dubrin, C. MacKay, R. Wolfgang, J.Am.Chem.Soc. 86 (1964) 645
34. W.L. Brown, W.M. Augustyniak, E. Simmons, K.J. Marcantonio, L.J. Lanzerotti, R.E. Johnson, J.W. Boring, C.T. Reimann, G. Foti, V. Pirronello, Nucl.Instr.Meth. 198 (1982) 1/8
35. V. Pirronello, G. Strazulla, G. Foti, E. Rimini, Nucl.Inst.Meth. 182/183 (1981) 315/7
36. G. Strazulla, V. Pirronello, G. Foti, Astr.Astrophys. 123 (1983) 93/7
37. J. Schou, H. Sørensen, P. Børgesen, Nucl.Instr.Meth. B5 (1984) 44/57
38. L.E. Seiberling, C.K. Meins, B.H. Cooper, J.E. Griffith, M.H. Mendehall, T.A. Tombrello, Nucl.Instr.Meth. 198 (1982) 17/25
39. R.A. Haring, A. Haring, F.S. Klein, A.C. Kummel, A.E. de Vries, Nucl.Instr.Meth. 211 (1983) 529/33
4o. R.A. Haring, A.W. Kolfschoten, A.E. de Vries, Nucl.Instr.Meth. B2 (1984) 544/9
41. A.E. de Vries, R. Pedrys, R.A. Haring, A. Haring, F.W. Saris Nature 311 (1984) 4o
42. L.J. Lanzerotti, W.L. Brown, K.J. Marcantonio, R.E. Johnson, Nature 312 (1984) 139/4o
43. J. Roth, Topics in Applied Physics 52 (1983) 91/146
44. E. Vietzke, K. Flaskamp, V. Philipps, J.Nucl.Mat. 111/112 (1982) 763/8
45. E. Vietzke, K. Flaskamp, V. Philipps, J.Nucl.Mat. 128/129 (1984) 564/9
46. E. Vietzke, K. Flaskamp, V. Philipps, J.Nucl.Mat. 128/129 (1984) 545/55o

ON TWO MECHANISMS OF X-RAY GENERATION IN COMETS

S. Ibadov
Institute of Astrophysics
734670 Dushanbe
USSR

ABSTRACT. The intensity of solar X-radiation scattered by a comet is calculated and compared to the proper X-radiation of the comet due to impacts of cometary and interplanetary dust particles. Detection of X-radiation of dusty comets at small heliocentric distances ($R \leq 1$ a.u.) is found to be an indicator of high-temperature plasma generation as result of grain collisions.

1. Introduction

The study of comets has been carried out at present in ultraviolet, optical, infrared and radio ranges (e. g. see /1-2/). Meanwhile the investigation of the cometary dust component interaction with the dusty matter of interplanetary medium reveals the possibility of arising the measureable proper X-ray emission of comets due to production of high-temperature plasma balls at heliocentric distances $R \leq 1$ a.u. /3/. X-radiation from comets may also appear in the way traditional for cold astronomical objects, namely due to scattering of solar X-rays by cometary matter.

2. Scattering of Solar X-Radiation by Comets

The intensity of cometary emission excited by incident radiation of the Sun may be presented as

$$I_s = \frac{I_o}{4R^2r^2}(4A_1 s_1^2 f_1 + A_2 r_h^2 f_2). \tag{1}$$

Here the first term in the right hand part corresponds to radiation of the optically thick zone of the cometary head

R. H. Giese and P. Lamy (eds.), Properties and Interactions of Interplanetary Dust, 365–368.

($\tau \geq 1$) and the second term - to the optically thin one ($\tau < 1$); I_0 is the intensity of solar X-radiation at the Earth's distance; R is the comet - detector distance; A_1 is the albedo of the part of cometary head having the "optical thickness" $\tau = 1$; $s_1 = \pi \sigma n_n r_n$ is the "sight distance" for solar rays passing across the cometary head with $\tau = 1$; f_1 is the phase function of the optically thick part of the cometary head; $A_2 = (4/3) a \bar{n} \sigma r_h$ and f_2 are the albedo and phase function of the optically thin part of the head; r_h is the radius of the head; $a \leq 1$ is the factor of absorption: $a = 1$ at zero absorption; σ is the cross-section of extinction for incident photons; $\bar{n}$ is the mean number density of scattering particles in the optically thin part of the head; $n_n \equiv n\,(r_n, R)$ is the number density of atomic particles near the surface of the nucleus with the radius r_n. This formula may be used for calculation both the spectral and, after corresponding integration over the spectrum, the integral intensity of radiation.

According to (1) for bright comets of Mrkos 1957V type with realistic values of $I_0(h\nu \geq 100$ eV, R= 1 a.u.)= 0,1 erg cm^{-2} s^{-1} /4/, $n(r_n$, R= 1 a.u.)= 10^{14} cm^{-3}, $r_n = 3 \cdot 10^6$ cm, $r_h = 10^{10}$ cm, $\bar{n} = 2{,}7 \cdot 10^7$ cm^{-3} /1/, $\sigma(h\nu \geq 100$ eV) $= 10^{-19}$ cm^2, $a = A_1 = 10^{-4}$ /5/, $f_1 = f_2 = 1$ we have $I_s(h\nu \geq 100$ eV, R= 1 a.u., r= $1{,}5 \cdot 10^{13}$ cm)= $7 \cdot 10^{-15}$ erg cm^{-2} s^{-1} : both terms of (1) give almost equal contribution to I_s. The value of I_s obtained is approximately one order of magnitude lower than the sensitivity threshold of modern X-ray telescopes (e.g. see /6/). Hence, comets are invisible in solar X-rays at heliocentric distances $R \gtrsim 1$ a.u.

3. New Component of Cometary Emission

At small heliocentric distances ($R \leq 1$ a.u.) the generation of proper X-radiation of comets due to interaction of cometary and interplanetary dusty matter is also possible together with the solar X-rays scattering. The intensity of this radiation I_x and the energy of quanta, corresponding to spectral distribution maximum, $h\nu_m \equiv \varepsilon_m$, we may present according to /3/ as

$$I_x(\varepsilon \geq \varepsilon_m, R, r) = I_{x1}(r_x/r)^2 R^{-(p + 3/2)}, \quad (2)$$

$$\varepsilon_m = \varepsilon_{m1} R^{-1}, \quad (3)$$

where $I_{x1} = 1{,}7 \cdot 10^{-4}$ erg cm^{-2} s^{-1} is the intensity of proper radiation of a comet at R= 1a.u. and $r = r_x$:

$$r_x = \pi n_d(r_n, R) \sigma_{dd} r_n^2 \quad (4)$$

is the radius of the X-ray active zone of the cometary head, where the interplanetary dust particles penetrating into the coma transform into plasma balls due to impacts with cometary grains; $n_d(r_n, R)$ is the number density of cometary dust particles near the cometary surface; σ_{dd} is the cross-section for collisions of cometary and interplanetary grains; p=1,5 is the parameter of spatial distribution of mass density of interplanetary dust matter; $\varepsilon_{m1} \equiv \varepsilon_m(1) =$ 65 eV is the most probable energy of quanta emitted by plasma balls at R=1 a.u.

4. Comparison of Two X-Ray Generation Mechanisms

The integral intensity of solar X-radiation may be written as

$$I_0(\varepsilon \geq \varepsilon_1) = \frac{I_0^*}{(\gamma - 1)\varepsilon_1^{\gamma-1}} . \tag{5}$$

Here $I_0^* = 10^3$ is the parameter of the solar differential energy spectrum; $2 \leq \gamma \leq 4$ is the spectral index; the numerical values correspond to energy range of solar photons $10^2 \leq \varepsilon \leq 10^3$ eV at quiet conditions according to /4/.

For registration of the proper radiation of comets the choice of passing threshold of counter as $\varepsilon_1 = \varepsilon_m$ seems most natural. In this case taking into account (3) and (5) under conditions of equal contributions of both right hand terms of (1) we get its integral presentation as

$$I_s(\varepsilon \geq \varepsilon_m , R, r) = \frac{2I_0^* A_1 f_1}{(\gamma - 1)\varepsilon_{m1}^{\gamma-1} R^{3-\gamma}} \left(\frac{s_1}{r}\right)^2 . \tag{6}$$

Here A_1 and s_1 represent the integral values of these parameters in the energy range considered.

From (6) follows that I_s weakly changes with R at $\gamma \geq 2$ while I_x by (2) quickly increases with decreasing R. On the basis of (2), (4) and (6) we may find the interval of heliocentric distances where the intensity of proper radiation of comets exceeds the intensity of scattered radiation, namely

$$R < \left\{ \frac{I_{x1}(\gamma - 1)\varepsilon_{m1}^{\gamma-1}}{2I_0^* A_1 f_1} \left[\frac{n_d(r_n,R)\sigma_{dd}}{n(r_n,R)\sigma} \right]^2 \right\}^{1/(p+\gamma-3/2)} . \tag{7}$$

Inserting into (7) realistic values of $\gamma=3$, $n_d(r_n,R)=10^4$ cm^{-3}, $\sigma_{dd}=5\cdot10^{-9}$ cm^2 and numerical values of other parameters mentioned we have $R < 3$ a.u. At the same time the initial temperature of plasma balls in the generation zone of proper X-radiation of comets according to /3/ will be $T_0(R \leq 1$ a.u.$) \geq 3\cdot10^5$ K.

Thus the registration of X-radiation of dusty comets at small heliocentric distances ($R \leq 1$ a.u.) in the definite energy range ($\mathcal{E} \geq \mathcal{E}_m$) and at quiet Sun may be an indicator of high-temperature plasma generation as a result of grain collisions, i.e. of the physical process not realized yet under terrestrial laboratory conditions.

5. Acknowledgements

The author is grateful to Prof. O.V. Dobrovolsky, Drs Yu.N. Gnedin, B.Yu. Levin, I.M. Podgorny, R.A. Sunyaev and L.M. Shul'man for stimulating discussions.

References

1. Dobrovolsky, O.V.: 1966, Comets, Nauka, Moscow.

2. Wilkening, L.L. (ed.): 1982, Comets, Proc. IAU Colloq. No. 61, Arizona Univ. Press, Tucson.

3. Ibadov, S.: 1983, in T.I. Gombosi (ed.), Cometary Exploration, Proc. Intern. Conf., Budapest, Hungary, vol. 1, p. 227.

4. Hinteregger, H.E.: 1965, Space Sci. Rev. 4, 461.

5. Blokhin, M.A.: 1957, Physics of X-rays, GITTL, Moscow.

6. Trümper, J.: 1980, The Röntgen Satellite (ROSAT), Preprint Max-Plank Institute für Physik und Astrophysik, München, FRG.

7. Grün, E.: 1980, in V.A. Dergachev and G.E. Kocharov (eds.), Cosmic Rays, Proc. V11 European Cosmic Ray Symposium, Leningrad, USSR, p. 172.

DYNAMICS AND SPATIAL DISTRIBUTION OF INTERPLANETARY DUST

Ch. Leinert
Max-Planck-Institute for Astronomy
Königstuhl
6900 Heidelberg
F.R.G.

ABSTRACT. Three attempts to explain the radial power law distribution $n(r) \sim r^{-1.3}$ of interplanetary dust are reviewed, which include the influence of Poynting-Robertson effect, collisions and interplanetary magnetic fields. Electromagnetic forces are unlikely to affect appreciably the spatial distribution. The replenishment of the cloud of interplanetary dust by the disruption of larger meteoritic particles in catastrophic collisions appears to give the most natural explanation for the observed spatial distribution.

1. INTRODUCTION

The spatial distribution of interplanetary dust is determined by its sources and the forces acting on the individual particles. Vice versa, measuring the spatial distribution should help to learn about origin and dynamics of the dust. Zodiacal light space experiments (Leinert et al. 1981, Hanner et al. 1976) yielded a power law spatial distribution somewhat steeper than the dependence $1/r$, which would result under the action of Poynting-Robertson effect alone (Wyatt and Whipple 1950). Different mechanisms have been proposed to explain this steepening: a focusing of particle orbits by inelastic collisions, leading to multiplied in-ecliptic density closer to the sun (Trulsen and Wikan 1980); addition of dust from a source region extending into the inner solar system (Leinert et al. 1983); subtraction of particles at larger heliocentric distance, where Lorentz scattering off the ecliptic would be most effective (Mukai and Giese 1984). In the following, after a comment on the observed spatial distribution, I give a critical review of these three papers.

2. OBSERVED SPATIAL DISTRIBUTION

The Helios zodiacal light experiment (Leinert et al. 1981) found the scattering cross section per unit volume to vary with heliocentric distance as power law $r^{-1.3}$ from 0.1 AU to, roughly, 1.5 AU or 2 AU.

R. H. Giese and P. Lamy (eds.), Properties and Interactions of Interplanetary Dust, 369–375.

This is compatible with Pioneer 10 observations (Schuerman 1980). For this paper and until more is known about interplanetary dust I adopt the simplest interpretation: that the radial spatial distribution is given by the above power law. Future discussions and measurements may or may not show that part of the variation is due to changing particle properties, like albedo; however, I consider the case to be strong in favour of a variation of particle number density steeper than 1/r, and this steepening has to be explained.

If one takes the strong variation of scattering cross section found by Helios and a "normal" scattering function (i.e. one with enhanced forward scattering) it is difficult to match exactly the observed brightness distribution of zodiacal light at 1 AU (Levasseur-Regourd and Dumont 1978). This is a difficulty between two different sets of measurements and not to be overcome simply by assuming another spatial distribution. But Figure 1 shows that the discrepancies are small except closer than 5° to the sun and suggests that this may be at least in part a mathematical problem resulting from finite accuracy and calibration differences of the observations.

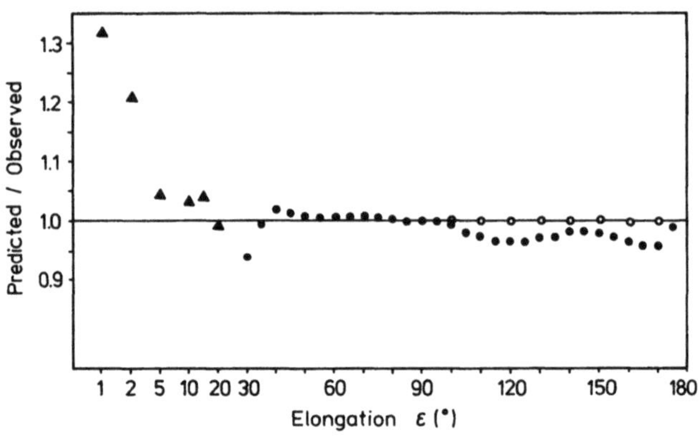

Figure 1. Zodiacal light brightness, predicted from $n(r) \sim r^{-1.3}$ and an empirical scattering function with enhanced forward scattering, in comparison to the observations summarized by Leinert (1975, o), Levasseur-Regourd and Dumont (1980, •), and Fechtig et al. (1981,▲).

3. FOCUSSING OF PARTICLES BY INELASTIC COLLISIONS

This is the basic idea of Trulsen and Wikan's (1980) paper and visualized in Figure 2. They assume a source far out. As particles are drifting inwards due to the Poynting-Robertson effect, their eccentricities are decreasing. Because collisions tend to equalize average inclination and average eccentricity, the inclinations also are decreasing, unless the initial inclinations were quite small, leading to an enhancement of density in the ecliptic by about a factor $i_o/\langle i(r)\rangle$.

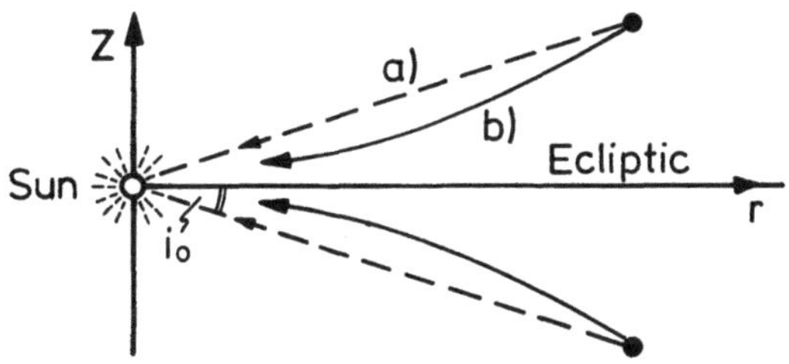

Figure 2. Drift path, under Poynting-Robertson effect, for particles with initial inclination i_o. a: unperturbed b: affected by non-destructive collisions.

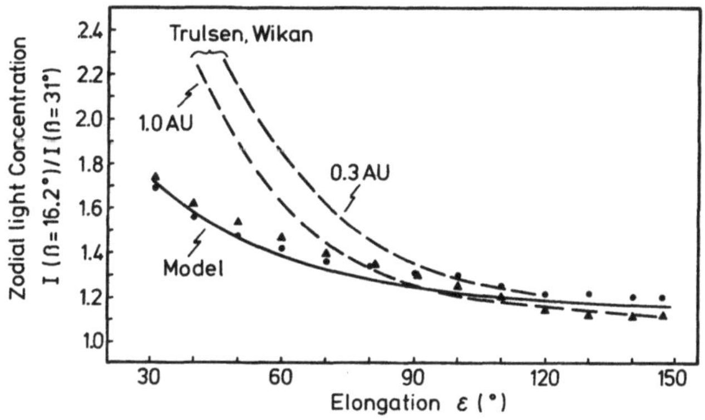

Figure 3. Zodiacal light concentration measured by Helios at 1 AU (●) and 0.3 AU (▲) compared to predictions according to Trulsen and Wikan (---, $n(z) \sim \exp[-5(z^2/r^2)r^{-0.3}]$) and model $n(z) \sim \exp(-2.1|z/r|)$.

The problem with this approach is that it assumes non-destructive collisions and therefore may not be applicable to interplanetary space where most collisions are catastrophic. Also it necessarily predicts a particular form for the z-dependence of dust distribution which does not at all fit the zodiacal light observations shown in Figure 3.

4. DENSITY INCREASE IN AN EXTENDED SOURCE REGION

Leinert et al. (1983) refer to the fact that, although a single source tends to produce a 1/r spatial distribution, a superposition of sources in an extended source region will result in a steeper distribution (Figure 4). They show that in absence of collisions and with an unbound input region an input distribution in semimajor axis of $f(a) \sim a^{-\nu}$ leads to a spatial distribution $n(r) \sim r^{-\nu}$, for $\nu \geq 1.0$. Collisional losses tend

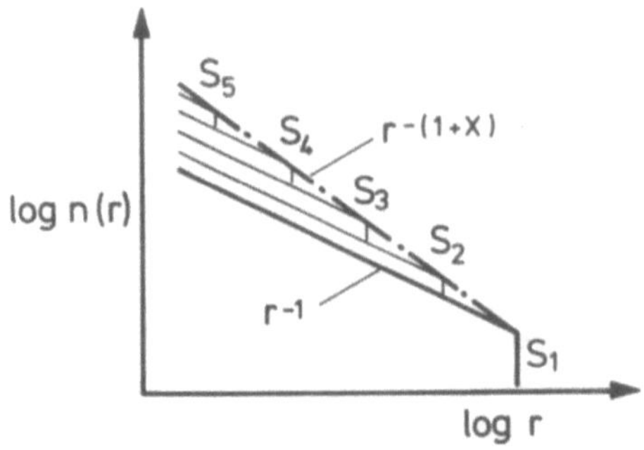

Figure 4. Steepening of the spatial distribution, resulting under Poynting-Robertson effect by a distribution of sources.

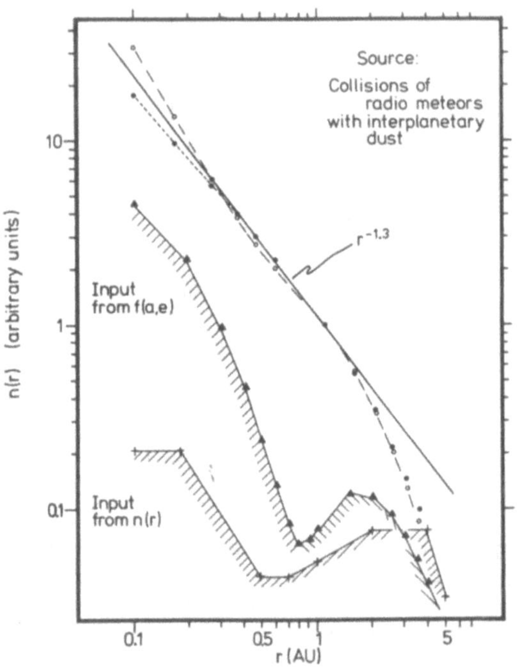

Figure 5. Relative spatial distribution expected if interplanetary dust is replenished by collisional debris from meteors (see Leinert et al., 1983).

to flatten, a limitation of the input region tends to steepen the distribution; for very strong collisional losses the simple relation holds again.

In the interplanetary complex of dust and meteors destructive collisions have the net effect that dust (i.e. particles with radii less than 100 µm) is produced at the expense of larger meteoritic particles. Taking the spatial distribution of radio meteors given by Southworth and Sekanina (1973) the calculated overall dust input is ≈ 1.2 t/sec, sufficient to maintain in equilibrium the dust density at

1 AU. There is some uncertainty regarding the spatial distribution of this dust input, because f(a,e) and n(r) for meteors given by Southworth and Sekanina are not consistent, but roughly half of the input occurs within 1 AU as needed to explain a dust distribution $n(r) \sim r^{-1.3}$ (Figure 5).

A weakness of the paper by Leinert et al. (1983) is that they do not take into account the effect of collisions on the size distribution of dust and that the calculations heavily depend on our knowledge on dust and meteor number densities.

5. RELATIVE DENSITY INCREASE BY DEFOCUSING OF PARTICLES FAR OUT

Mukai and Giese (1984) calculate that because of the slow Poynting-Robertson drift at large heliocentric distances Lorentz scattering due to the sector structure of interplanetary magnetic field may lead to higher average inclination of particle orbits than closer to the sun. The effect is a stronger reduction of in ecliptic dust densities in the outer solar system and a steepening of the dust distribution (Figure 6), e.g. to $n(r) \sim r^{-1.3}$ in the range 0.4 - 1 AU.

However, they appear to have overestimated the electromagnetic effects.

First, they refer to Consolmagno's (1979) calculations of spread in orbital inclinations. He finds for a 1 µm particle at 0.4 AU, charged to 6 V, in orbit with eccentricity e=0.5, that the root mean square change in inclination in 10 years is 4.0°, for example. Morfill and Grün (1979) predict this quantity to be 0.25°, Barge et al. (1982) even only 0.07°. It would be useful to clarify the reason for these different results, all based on the same interplanetary measurements. The true effect could be an order of magnitude smaller than Mukai and Giese assumed.

Second, for elegance of presentation and to save computer time, they estimate the reduction of in-ecliptic densities by a life-time factor $\exp(-t/\tau_L)$, where τ_L is a typical time scale for inclination

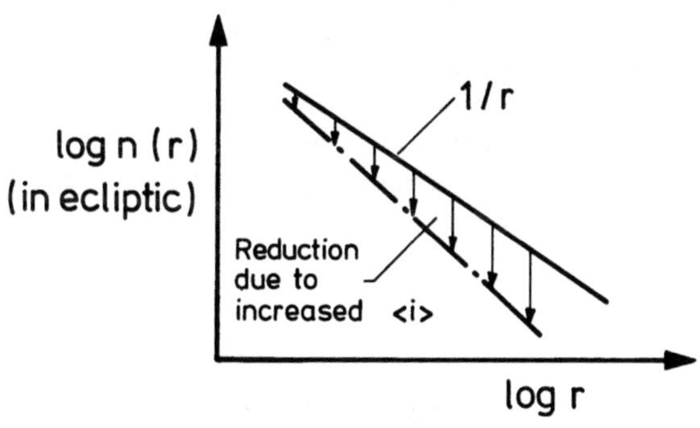

Figure 6. Steepening of spatial distribution by Lorentz scattering of interplanetary dust.

dispersion by Lorentz forces. Preferably one should calculate this reduction by considering the effect of a Gauss dispersion in inclination acting on the original inclination distribution. Figure 7 then shows that their estimate is an overestimate of the effect.

Third, their assumed average inclination of 1° is too small for interplanetary dust and overemphasizes the effect of inclination perturbations.

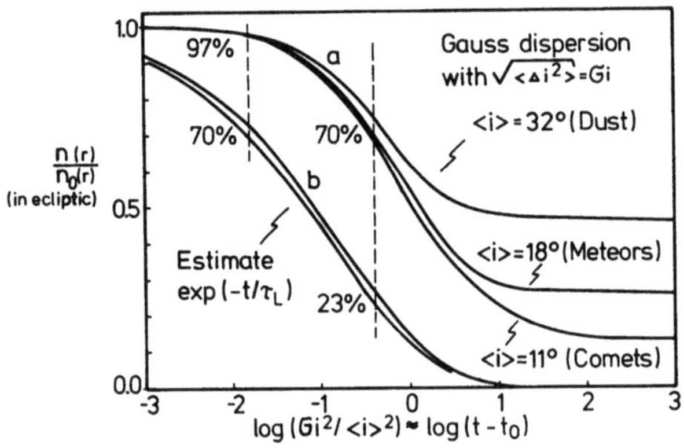

Figure 7. Reduction of spatial density in ecliptic by an added dispersion in inclination (a). The results are compared to the life time estimate of Mukai and Giese (b).

6. CONCLUSION

In view of the above, I find it unlikely that Lorentz forces appreciably affect the spatial distribution of interplanetary dust and have doubts whether the concept of non-destructive collisions may be applied to interplanetary space. On the other hand, collisions in the cloud of dust and meteors must occur all over the solar system. It appears that debris from disrupted meteors is able to balance the mass losses of the dust cloud due to Poynting-Robertson effect and collisions. The resulting extended source region is a natural explanation of the observed spatial distribution of interplanetary dust. The question how the reservoir of meteor particles is filled by comets or other sources remains open.

REFERENCES

Barge, P., Pellat, R., Millet, J.: Astron. Astrophys. **115**, 8-19

Consolmagno, G.J.: 1980, Icarus **43**, 203-214

Fechtig, H., Leinert, Ch., Grün, E.: 1981, in: Landolt-Börnstein, New Series, eds. K. Schaifers and H.H. Voigt, Springer Berlin-Heidelberg-New York, Vol. **VI 2a**, p.228-243

Hanner, M.S., Sparrow, J.G., Weinberg, J.L., Beeson, D.E.: 1976, Lecture Notes Phys. **48**, 29-35
Leinert, Ch.: 1975, Space Sci. Rev. **18** , 281-339
Leinert, Ch., Richter, I., Pitz, E., Planck, B.: 1981, Astron. Astrophys. **103**, 177-188
Leinert, Ch., Röser, S., Buitrago, J.: 1983, Astron. Astrophys. **118**, 345-357
Levasseur-Regourd, A.C., Dumont, R.: 1978, Comptes. Rend. Acad. Sci. Paris, Serie B, Vol. **286**, pp.61-64
Levasseur-Regourd, A.C., Dumont, R.: 1980, Astron. Astrophys. **84**, 277-279
Morfill, G., Grün, E.: 1979, Planet Space Sci. **27**, 1269-1282
Mukai, T., Giese, R.: 1984, Astron. Astrophys. **131**, 355-363
Schuerman, D.W.: 1980, in: Solid particles: in the solar system, IAU Symposium No. 90, eds. I. Halliday and B.A. McIntosh, Reidel Dordrecht-Boston-London, pp.71-74
Southworth, R.B., Sekanina, Z.: 1973, NASA CR-2316, Washington D.C.
Trulsen, J., Wikan, A.: 1980, Astron. Astrophys. **91**, 155-160
Wyatt, St.P.,Jr., Whipple, F.L.: 1950, Astrophys. J. **111**, 134-141

DISCUSSION

Singer: EM force referred to as Lorentz force is actually the EM force due to moving plasma (which is an order of magnitude greater).

Answer: A question of terminology. The relative velocity counts. In the frame moving with the plasma this is a Lorentz force.

Singer: With half of zodiacal cloud input within 1 AU one would expect large secular variations in the zodiacal cloud, e.g. in the dust influx to the earth. Can you make some statement on this or find further evidence?

Answer: I do not see a necessary relation between temporal variations in the cloud and spatial distribution of input. We try to model steady state. The Helios observations show the zodiacal light to be remarkably stable.

Fechtig: Don Humes' results from the Pioneer 10/11 beer can experiments show that there are particles with excentricities e=0.99, semimajor axis 10 AU, random inclinations. Those particles are most likely cometary particles. My suggestion is that one includes these results in future calculations!

Mukai: In your treatment small sized collisional debris increases with decreasing solar distance. This variation of size distribution causes a change of colour of zodiacal light. Observed results, however, show neutral and/or reddening. How do you explain this evidence?

Answer: Our model does not explicitely deal with this effect and is not self-consistent with respect to the size distribution. At present a quantitative calculation would only reflect the assumptions made on the size distribution at smaller heliocentric distances. The question is worth being pursued.

THE GRAVITATIONAL ZONES OF INFLUENCE OF THE PLANETS ACTING ON SMALL CELESTIAL BODIES

N. Y. Misconi, Space Astronomy Laboratory, University of Florida, 1810 N.W. 6th St., Gainesville, Florida 32609

E. T. Rusk, Department of Astronomy, University of Florida, Gainesville, Florida 32611

ABSTRACT. Tisserand's definition of the "sphere of action" of a planet is based on the equality of tidal vs. gravitational acceleration ratios of the sun and planet. Öpik and others based their relation on equating the differential solar and planetary forces on a particle. Neither expression was formulated to describe the zone of influence surrounding a planet when considering the small, but significant, long-term perturbative effects of the planets on a particle's orbital elements. For the purpose of determining these effects on interplanetary dust we derive a zone of influence based on equating the gravitational forces of the sun and planet.

The general expression for the radial distance to the boundary of the planetary zone of influence on small particles can be written as

$$s = k \cdot f(a_p) \cdot g(m_p, \text{ other planetary parameters}) \qquad (1)$$

where k is a constant which depends on the nature of the problem under study, f and g are functions which may also have the same dependence as k, a_p is the planet's semimajor axis, and m_p is the planet's mass. The other parameters may include the eccentricity of the planet, as well as other orbital parameters and such factors as oblateness and axial tilt. Functions f and g are generally not separable but we have found that the effect of semimajor axis can be separated from the other parameters. The equation of Tisserand (1889) and Öpik (1951) can be written as

$$s = (1/M_\odot)^{2/5} a_p m_p^{2/5}, \text{ and} \qquad (2)$$

$$s = \tfrac{1}{2}(1/2M_\odot)^{1/3} a_p m_p^{1/3}, \text{ respectively.} \qquad (3)$$

If we equate the gravitational and radiation forces of the sun and planet, a different relationship is obtained. Consider a particle of mass m at a distance s from a planet of mass m_p and semimajor axis a_p, and at a distance r from the sun. The magnitude of the force on the particle due to the planet is

R. H. Giese and P. Lamy (eds.), Properties and Interactions of Interplanetary Dust, 377–380.

$$F_p = Gm_p m/s^2 \quad , \tag{4}$$

while the magnitude of the force on the particle due to the sun is

$$F_\odot = GM_\odot m(1-\beta)/r^2 \quad , \tag{5}$$

where β is the ratio of radiation force to gravitational force. Since $a_p - s \leq r \leq a_p + s$, we can write r as $r = a_p + \delta s$ where δ is a number between -1 and +1 which depends on the planet-sun angle. Substituting the expression for r into equation (5) and substituting $M'_\odot = M_\odot(1-\beta)$ we can set the ratio of the solar force to the planetary gravitational force equal to some constant k^2, where

$$k^2 Gm_p m/s^2 = GM'_\odot m/(a_p+\delta s)^2 \text{ , which reduces to} \tag{6}$$

$$s = ka_p(m_p/M'_\odot)^{1/2}[1-k\delta(m_p/M'_\odot)^{1/2}]^{-1} \tag{7}$$

Equation (7) can be approximated by a simple power law when the mass of the planet and k are small, giving

$$s = (k/M'^{1/2}_\odot)a_p m_p^{1/2} \quad . \tag{8}$$

Compare this with the results of Tisserand and Öpik, eqs. (2) and (3).

A computer simulation technique was used to find the most applicable expression for our study. Each encounter was run using a 10,000 step Cowell's method computation over the period of one synchronous orbit. For a particle in circular orbit at 10 A.U., the errors in eccentricity and semimajor axis incurred over one orbit were less than 1×10^{-5} and 5×10^{-9} A.U., respectively. The particle was initially in a circular orbit and the planet was given a fixed orbital eccentricity. The encounter occurred at perihelion or aphelion, depending on whether the particle's orbit was inside or outside the planet's orbit. The particle's orbital plane coincided with that of the planet, so that only subsolar and antisolar encounters were studied. The distance of closest approach was compared with either a_p or m_p assuming a constant Δe or a constant percentage change in a. Figure 1 shows the effect of a_p on the zone size

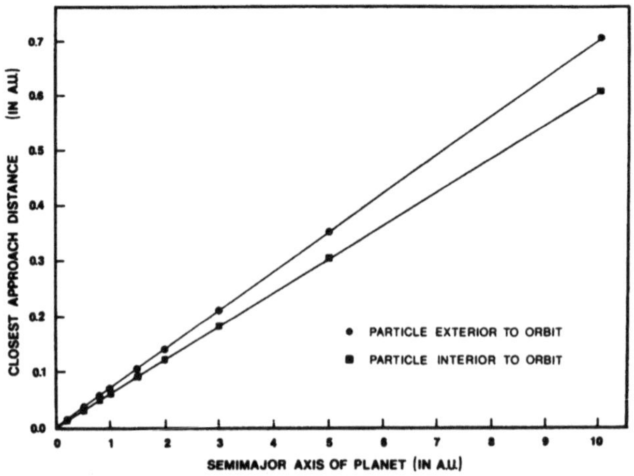

Figure 1. Relationship between a_p and s for a constant change in eccentricity (Δe=.0014). The planet's mass is 1 $M_\oplus$ and its eccentricity is 0.05.

for a constant Δe, and verifies the linear relation, common to all three derivations, between the zone boundary and the semimajor axis. This result also shows that the uncoupling of the functional forms of a_p and m_p in equation (1) is valid. The constant change $\Delta e= 0.0014$ chosen in deriving figure 1 (and later 2 and 4) is the minimum value that guarantees that the zones of influence of the inner five planets do not overlap.

Figure 2 shows the effect of the planet's mass on the zone boundary. The general trend follows a power law most closely approximated by equation (8), and diverges from the power law in the same manner as equation (7). Figure 3 is a graphic representation of equation (7). While this and the previous figure show a striking similarity, the spread due to $[1-k\delta(m_p/M_\oplus)^{1/2}]^{-1}$ is greater than the spread observed in the simulations. This is a consequence of the fact that, during an encounter, δ assumes other values than 1.

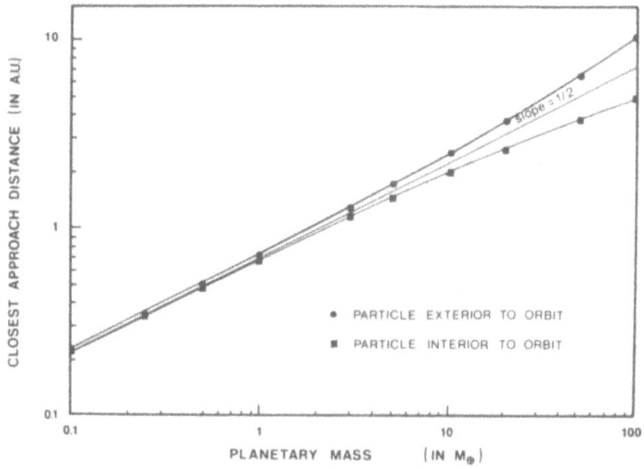

Figure 2. Relationship between m_p and s for a constant change in eccentricity (Δe=.0014). The planet's semimajor axis is 10 A.U. and its eccentricity is zero.

Figure 4 shows the effect of planetary eccentricity on the zone boundary. As implied in equation (1), this effect cannot be separated from the functional dependence on planetary mass. This is evidenced by the fact that these curves match the results for high mass planets but diverge for those of low mass. This result deserves further investigation to determine the relation between eccentricity and the size of the zone of influence.

Figure 5 shows

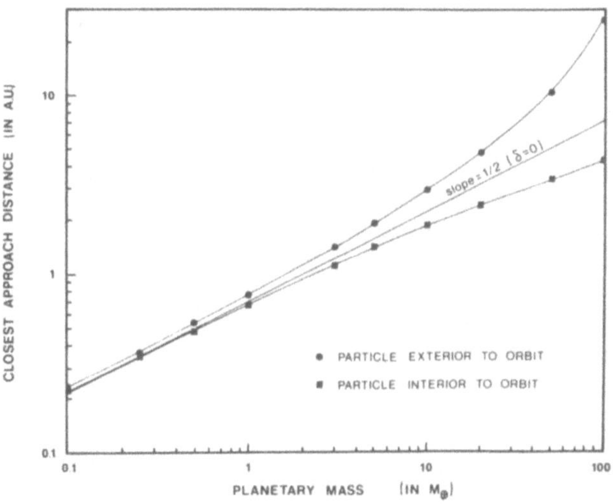

Figure 3. Theoretical curves relating m_p and s through equation (7). a_p is 10 A.U. and k is chosen to match the simulation value for $m_p = M_\oplus$. β is zero.

the relationship between planetary mass and the distance of closest approach when the desired effect is a constant percentage change in the semimajor axis of the particle, $\Delta a/a = 5\times10^{-5}$, which is the value found when $\Delta e = .0014$. Unlike the result for a constant Δe, (see fig. 3) the boundary follows the form $s \propto (m_p/M_\odot)^{3/8}$, a result which matches no theoretical formula yet derived. Note, also, the divergence from the power law for planets of high mass, similar to the form in equation (7).

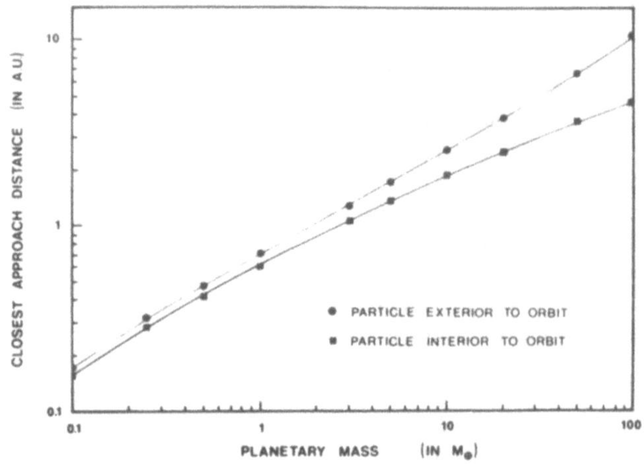

Figure 4. Relationship between m_p and s for a constant $\Delta e = .0014$. This is the same as Figure 2 except $e_p = 0.05$.

The form of the relation determining the zone of influence is highly dependent on the nature of the perturbing effect to be observed. We have shown that the expressions most applicable to interactions with orbiting particles involve powers of the planetary mass not formerly considered and that the size of the zone boundary depends upon the eccentricity. We are now studying the effect of eccentricity on the zone size for a constant $\Delta a/a$, and intend to extend this work to changes in inclination and ascending node.

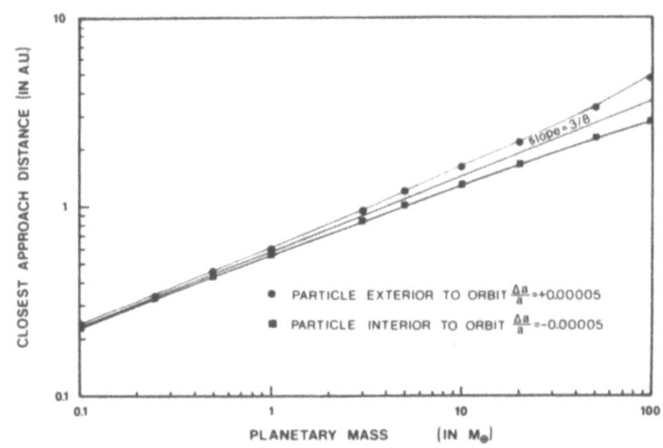

Figure 5. Relationship between m_p and s for a constant $\Delta a/a$ (= .00005). a_p is 10 A.U. and e_p is zero.

ACKNOWLEDGEMENTS:

Dr. Bo Gustafson and Dr. Carol Williams provided helpful discussions. This work was supported by NSF grant #AST-8206152.

REFERENCES:

1. Tisserand, F., *Traite de Mecanique Celeste Gauthier-Villars, Paris,* 1889.
2. Öpik, E. J., *Proc. Royal Irish Acad.* 54, 165-199, 1951.

A NEW APPROACH TO EVALUATE PLANETARY PERTURBATIONS ON A CLOUD OF DUST IN LOW ECCENTRICITY HELIOCENTRIC ORBITS

BO Å. S. GUSTAFSON
Space Astronomy Laboratory, University of Florida
1810 NW 6th Street
Gainesville, Florida 32609
U.S.A.

ABSTRACT. Dynamical perturbations on *ensembles* of particles in heliocentric orbits of low eccentricity are integrated over time. The dust is perturbed by radiation pressure, Poynting-Robertson drag, their corpuscular counterparts, and by gravitation due to any number of planets. A dust cloud is represented by a set of centroids and orbital dispersions (about the centroids). Gravitational perturbations on the centroid are derived from a single matrix, valid for any planet, in the appropriate frame of reference. After transformation of the time derivatives to a common coordinate system, the perturbation rates are summed up and integrated. The time dependence of the planets' orbital elements are evaluated inside the time integral.

1. INTRODUCTION

It is time-consuming and costly to estimate overall perturbations on a dust cloud by computing trajectories of individual dust particles. The efficiency is increased through the use of centroids to represent sets of orbits. Perturbations due to each planet are computed separately. Geometries are chosen such that the planet's perturbations on a centroid may be expressed by a *single quantity* depending only on the angle that the plane containing the centroid makes to the planet's orbital plane and the ratio of heliocentric distances. A comparatively simple interpolation-formula allows the matrix representing the perturbations to be stored online. Properly scaled, a single matrix may represent perturbations by any planet in the appropriate geometry.

2. COMPUTATIONAL PROCEDURE

2.1 The approach

Perturbations on *ensembles* of particles are evaluated as shown in the flow diagram, Figure 1. The ensemble is divided into a set of fictitious orbits, each a centroid for a set of dust trajectories. Each

R. H. Giese and P. Lamy (eds.), Properties and Interactions of Interplanetary Dust, 381–384.

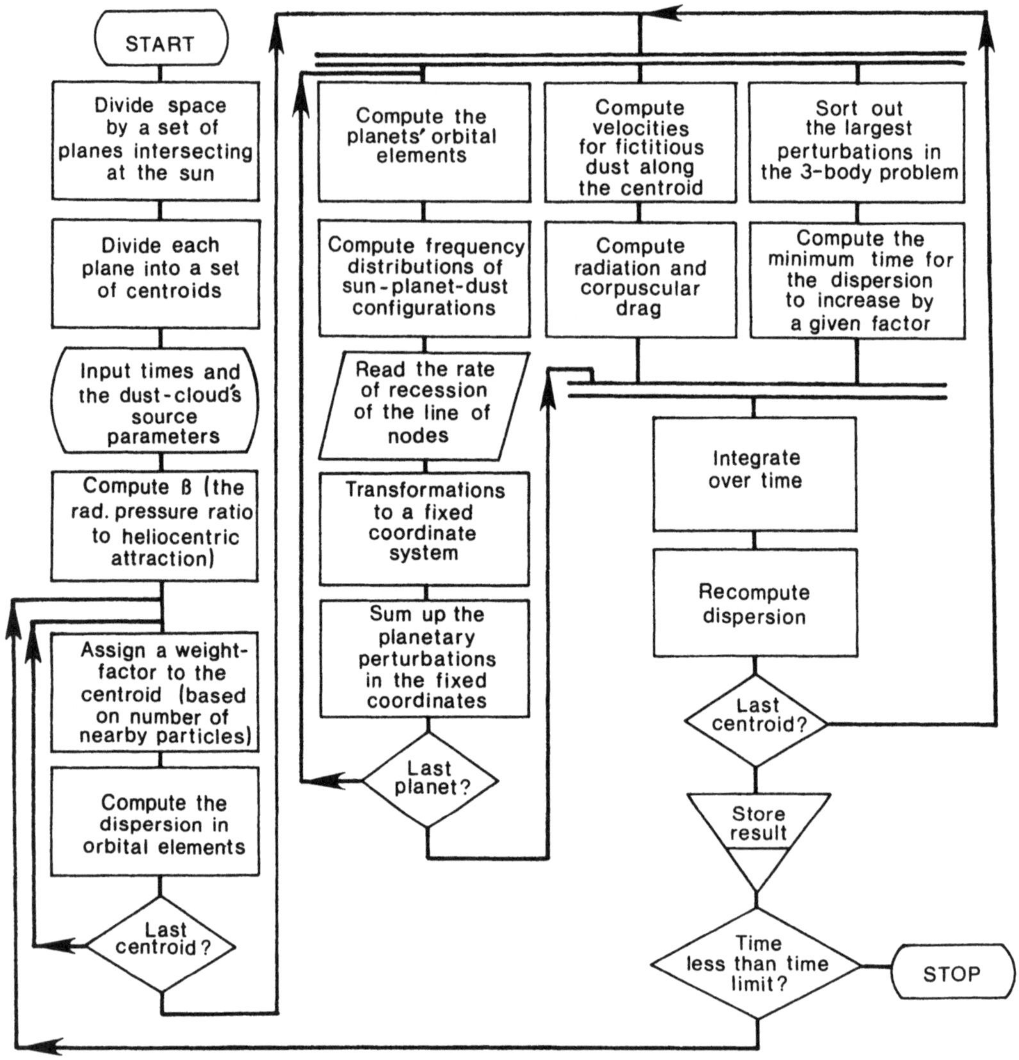

Figure 1. Flow chart diagram illustrating a methodology to evaluate the combined effects of planetary, electro-magnetic radiation and corpuscular perturbations on the shape of a cloud of interplanetary dust.

centroid is assigned a "weight" depending on the number of dust particles that it represents, and the dispersion about each centroid is computed. Mean secular perturbations are determined from the frequency distribution of sun-planet-dust configurations. This is repeated for each planet and the perturbations are summed up with other velocity-dependent disturbances, determined in a similar fashion from the distribution in orbital velocities. This procedure corresponds to a smearing out of the planets' masses along their orbits. A dispersion in orbital elements develops for actual trajectories due to differences in the sequence in which the sun-planet-dust configurations occur. This effect is evaluated for each perturbation separately, and the dispersions are combined.

2.2 The time integral

For computational convenience, the dust cloud may be divided by a set of planes intersecting at the sun. Each plane contains a set of fictitious orbits, centroids for a set of dust trajectories. The number of particles represented by a given centroid and the distribution in orbital parameter space are calculated from the starting conditions of the cloud. The rate of dispersion in orbital parameter-space is estimated as described below, and the perturbations on the centroid are integrated over a time interval in which the dispersion remains comparable to that of the original distribution. After the integration, a new set of orbits is assigned to each centroid and the process is repeated until the desired time period is covered. The planets' orbital elements may be estimated using the method of Brouwer and Van Woerkom (1950). The integration step is reduced in cases where a planet's orbit is found to change significantly over a single step.

2.3 Perturbations on the centroid

The expression for the average rate of recession $\overline{d(\Omega)/dt}$ of the line of nodes (defining the plane of the centroid) relative to the planet's orbit involves an integral that may be resolved into Jacobi's incomplete elliptic integrals of the first and second kinds. A matrix is generated in which each element is a solution to the integral at given tilt and radii of the centroid orbit. An expression of the form

$$\overline{\frac{d\Omega}{dt}} \approx \frac{K}{\sqrt{1-\beta}} \quad \frac{r(1+\cos i)}{(1+r^2-2r\ \cos^{2/3} i)^{5/4}}$$

was found to be an efficient interpolation formula, where K is a negative constant, β is the radiation force in units of the sun's gravitational attraction and i is the angle that the plane containing the centroid makes to the planet's orbital plane. The radius r of the centroid orbit is in units of the planet's heliocentric distance. The planetary mass and effects of radiation pressure intervene outside the integral. The matrix may therefore be used to evaluate the perturbation by <u>any</u> planet relative to its own orbit. It is a simple matter to transform the perturbations to a common frame of reference in which all perturbations may be summed up.

The change in orbital radii resulting from Poynting-Robertson drag is easily computed from Robertson's (1937) expressions. Similar formulae represent the corpuscular drag.

2.4 Dispersion about the centroid

The dispersion about the centroid is first computed in the restricted three-dimensional three-body problem, which is extended to incorporate radiation and corpuscular forces. This is repeated for each planet, and the dispersions are combined. An orthogonal coordinate system is introduced centered on the dust particle and is frozen into an inertial frame of reference at the moment in which the perturbations generating the dispersion are evaluated. The accelerations in this system are the sum of two three-dimensional matrices. With the approximation that the sun and all the planets revolve around the Sun-Jupiter center of mass, the first matrix represents the perturbations by the Sun-Jupiter system including radiation and corpuscular forces as a function of one of these body's coordinates. The second matrix represents perturbations by any other planet. The resulting accelerations translate into perturbations of the orbital elements, when fed into Gauss equations. Distributions about the centroid are calculated based on a frequency distribution of sequences of sun-planet-dust configurations for each planet separately.

3. SUMMARY

A method is described whereby the combined perturbations by planetary gravitation, solar-radiation and corpuscular forces on a dust cloud may be evaluated. The method takes advantage of similarities in the numerical solutions for perturbations by the planets depending on the geometry. In its present form, the method is applicable to nearly circular orbits. A generalization to orbits of arbitrary eccentricity is planned.

4. ACKNOWLEDGEMENTS

I am pleased to thank Dr. A. Ardeberg and his staff for their hospitality at Lund Observatory, where most of this work developed. I am also indebted to Mr. P.Å.J. Gustafson for discussions. Dr. J.D. Mulholland of the University of Texas, Drs. N.Y. Misconi, J.L. Weinberg, G.N. Toller, and R.B. Piercey of the Space Astronomy Laboratory reviewed the manuscript. This research is supported in part by NSF grant AST-8206152.

5. REFERENCES

Brouwer, D., and Van Woerkom, A.J.J., 1950, 'The secular variations of the orbital elements of the principal planets', Astron. Papers, XIII pt. II, 85.

Robertson, H.P., 1937, 'Dynamical effects of radiation in the solar system', Monthly Notices Roy. Astron. Soc., **97**, 423.

PLANETARY PERTURBATIONS: EFFECTS ON THE SHAPE OF A CLOUD OF DUST IN CIRCULAR HELIOCENTRIC ORBITS

BO Å. S. GUSTAFSON
Space Astronomy Laboratory, University of Florida
1810 NW 6th Street
Gainesville, Florida 32609
U.S.A.

ABSTRACT. Synthetic infrared pictures are used to illustrate changes in the shape of portions of an interplanetary dust-cloud. The dust particles, in circular heliocentric orbits, are perturbed by radiation and corpuscular forces combined with gravitational disturbances by the major planets. Dust in the inner solar system and close to the ecliptic or the orbital plane of Venus is brought to a more narrow range of ecliptic latitudes. A dust-band evolves near and inside the orbit of Venus. The cloud's shape is less affected at high ecliptic latitudes.

1. INTRODUCTION

Derivations of the shape of the zodiacal cloud based on dynamical evolution are initial-value problems. Because the source remains unknown as well as the cloud's long-term steady state or transient nature, no statements can be made regarding the absolute shape of the interplanetary dust cloud. Instead, we address the problem of *changes* in the shape of a *portion* of the cloud as the dust spirals through the inner solar system. Conventional wisdom states that dust migrates through the solar system while losing angular momentum to drag forces. It is not yet established whether transfer of momentum from the planets may affect the rate significantly. The resulting uncertainty in magnitude of the perturbations has little effect on our qualitative results. Because orbital eccentricities are neglected, momentum transfer and, by implication, dynamical life-times of the dust are not addressed in this paper.

2. OBSERVATIONS

Observations show that the symmetry plane of the inner zodiacal cloud is close to the orbital plane of Venus (Misconi, 1980, and references therein). Leinert *et al.* (1980) concluded that this symmetry extends to 1 A.U., whereas most observations in this region favor a plane close to the invariable plane of the solar system (Misconi, 1980). The brightness plane of symmetry is close to the plane of maximum particle number densities except where shifted by the dust particles' light scattering characteristics, as in the Gegenschein (Misconi, 1981).

R. H. Giese and P. Lamy (eds.), Properties and Interactions of Interplanetary Dust, 385–388.

3. MODEL CALCULATIONS

With model calculations we attempt to determine whether a change in the orientation of the plane of maximum dust density is expected from dynamical evolution alone and whether such an evolution would lead to other observable features that may help differentiate between evolutionary effects and artifacts of the initial conditions.

Dust distributed along a circular orbit about the sun is subject to permutations of nearly identical sets of perturbing forces. As long as the orbital changes are small enough to be treated as perturbations on a Keplerian orbit, the evolution may be described by a mean perturbation and a dispersion about the resulting mean orbit. The method to compute and integrate the compound perturbations by the planets Mercury through Saturn and solar radiative and corpuscular forces is described elsewhere (Gustafson, 1984). For computational efficiency, a radiation pressure efficiency 1 and ratio of radiation pressure to solar gravitation 0.5 is used throughout. The primary effect of decreasing this ratio is to increase the magnitude of integrated planetary perturbations.

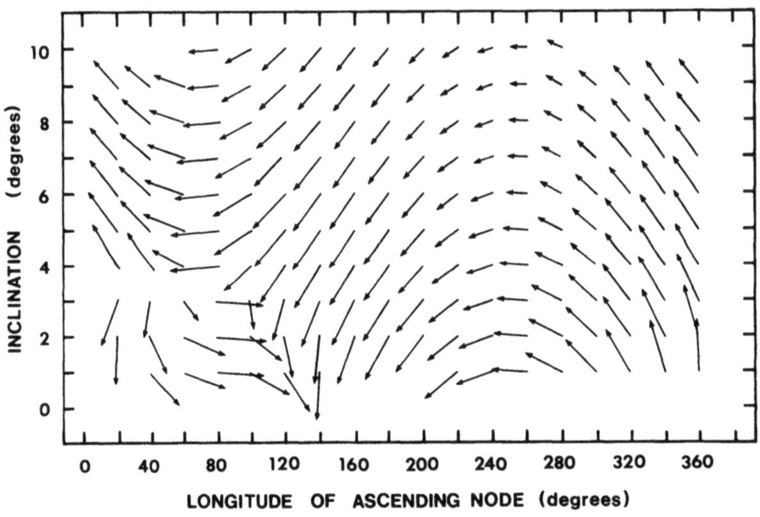

Figure 1. Arrows represent compound gravitational effects of the planets Mercury through Saturn on the orientation of 0.85 AU radius circular dust-orbits. The perturbations are averaged over a large number of orbits.

The mean change in orientation of 0.85 AU radius circular dust orbits is shown in Figure 1. The orientations are defined by the inclination i and the longitude Ω of the ascending node relative to the ecliptic of 1950. Dust in orbits whose inclination exceeds that of any of the planets', shows a recession of the longitude of the nodes when the perturbations are averaged over many orbits. For lower inclinations there is slower recession near some nodes or even an *advance* of the line of nodes. When integrated over time to simulate the evolution of a portion of a dust cloud, zones of concentration and depletion develop in plots of i versus Ω. Evolutions from orbits of four degrees inclination at

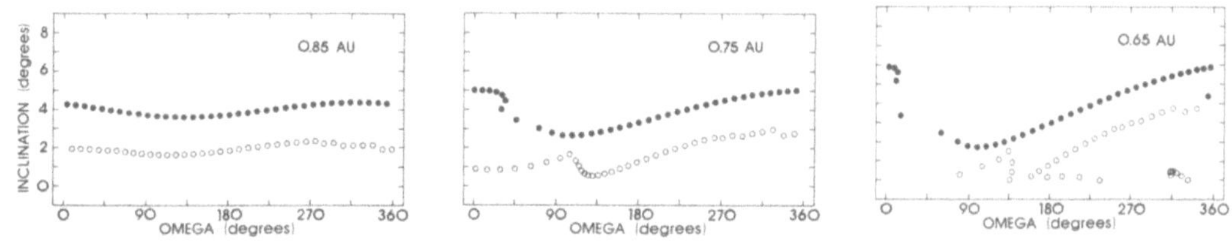

Figure 2. Evolution of the orientation of circular dust-orbits initialized at 1.25 AU with four degree inclination (dots) or two degree inclination (circles).

1.25 AU (dots in Figure 2) and two degrees inclination (circles) are representative of medium and low inclination orbits respectively. At 0.85 AU a waviness begins to develop. The cloud essentially maintains its shape while shifting orientation. Figure 1 shows rapid recession of the nodes near the orbit of Venus at medium inclinations followed by a region of inclination increase and low precession rate. A crest-like structure results in Figure 2 and appears as a dust-band in a simulation

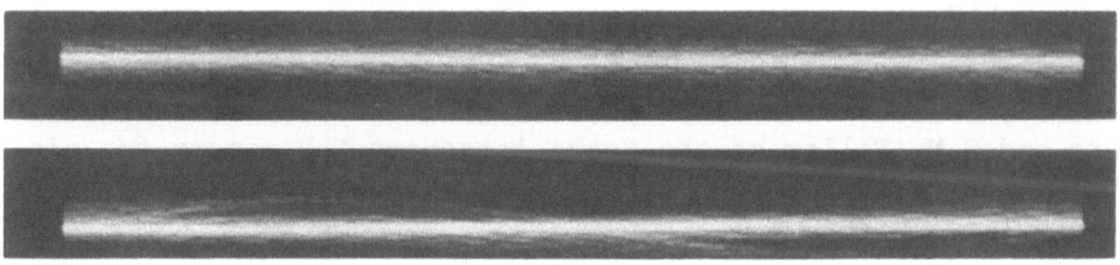

0.85 AU

0.65 AU

Figure 3. A computer simulation of a portion of a dust-cloud as seen in infrared light by an observer at the Sun. The cloud's shape at 1.25 AU is modeled on the distribution of radio-meteor's orbital elements up to six degrees inclination. The cloud is depicted at 0.85 AU and 0.65 AU from the Sun.

of a portion of a cloud depicted in Figure 3. At lower inclinations, the nodes near Venus are advancing and meet the receding ascending nodes near 100 degrees longitude. With decreasing inclinations this neutral point tends toward higher longitudes. The emerging regions of high density in orbital parameter-space also correspond to decreasing inclinations. This leads to the flattening of regions of the cloud that are below the orbit of Venus (Figure 4). Evolution from the set of four degree inclination orbits indicate a less dramatic overall narrowing of that portion of the cloud.

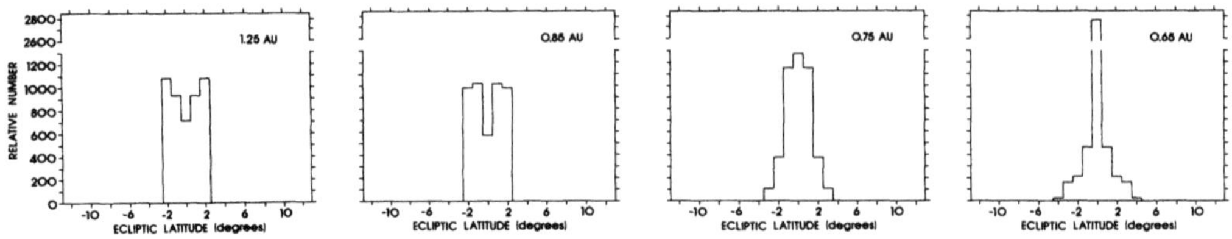

Figure 4. Evolution of the distribution of particle number densities in ecliptic latitude from two degree inclination orbits at 1.25 AU.

4. CONCLUSIONS

Dynamical effects on the shape of a cloud of dust in the inner solar system are essentially separable into effects on three regions - a flatted part close to the ecliptic or to the orbital plane of Venus, creation of a dust-band, and a small change in orientation of high latitude portions of the cloud. As a "signature" of planetary perturbations, the presence of the dust-band may help to discriminate evolutionary effects from remnants of the initial conditions.

5. ACKNOWLEDGEMENTS

Thanks are due to Dr. A. Ardeberg and his staff for their hospitality at Lund Observatory, Sweden. I am also pleased to thank Drs. N.Y. Misconi, J.L. Weinberg and G.N. Toller of the Space Astronomy Laboratory for discussions and review of the manuscript. This research is supported in part by NSF grant AST-8206152.

6. REFERENCES

Gustafson, B.Å.S., 1985, 'A new approach to evaluate planetary perturbations on a cloud of dust in low eccentricity heliocentric orbits', in Proc. IAU Colloquium No. 85, Properties and Interactions of Interplanetary Dust, this volume.

Leinert, C., Hanner, M., Richter, I., and Pitz, E., 1980, 'The plane of symmetry of interplanetary dust in the inner solar system', Astron. Astrophys., 82, 328.

Misconi, N.Y., 1980, 'The symmetry plane of the zodiacal cloud near 1 AU', in Proc. IAU Symposium No. 90, Solid Particles in the Solar System, (eds. I. Halliday and B.A. McIntosh), Ottawa, 49.

Misconi, N.Y., 1981, 'The photometric center of the Gegenschein', Icarus, 47, 265.

Dynamics of Dust Particles in the Solar System

G. Burkhardt
Astron. Rechen - Institut, Heidelberg, FRG.

This short communication is part of a thesis at the University of Heidelberg, FRG.
It deals with some dynamical aspects of small interplanetary particles, which are affected by non-gravitational forces like the Poynting - Robertson Effect.

The equations of motion of test particles are numerically integrated. The forces acting on these particles include the gravitational attraction of the Sun and the Planets as well as forces due to incident solar radiation.

The ephemerides of the Planets in the heliocentric rectangular 3-dimensional coordinate system used here are taken from the Development Ephemeris DE-102 (Newhall et al. 1983) . This restricts the time span of the integration to about 4412 years.

The strength of the non-gravitational forces in the equations of motion is given by a factor ß which can be varied for each test particle. ß is the ratio of the radiation force to the gravitational force of the sun. The Poynting - Robertson Effect and radiation pressure are introduced into the force function by

$$f = k^2 \beta \frac{M_\odot + m}{|\vec{r}|^2} \left[\frac{\vec{r}}{|\vec{r}|} - \frac{1}{c} \left(\dot{\vec{r}} + \frac{\vec{r}}{|\vec{r}|^2} \langle \vec{r} \cdot \dot{\vec{r}} \rangle \right) \right] \qquad (1)$$

(k : Gaussian constant; $M_\odot$: mass of the Sun; m : mass, $\vec{r}$: position, $\dot{\vec{r}}$: velocity of the test particle; c : speed of light).

The procedure of numerically integrating test particles in a three - dimensional coordinate system with all the Planets over a time span of 4412 years consumes very much CPU - time (on an IBM 370-168). Some first tests showed no effects neither with small values of ß (ß < 0.1), nor with

R. H. Giese and P. Lamy (eds.), Properties and Interactions of Interplanetary Dust, 389–393.

particles starting their orbits far outside the orbit of Jupiter. So the problem was restricted to

i) a $\beta = 0.0$
b Values of β between 0.1 and 0.5
ii) a not perturbed by planets
b perturbed by the Planets from Mercury through Jupiter
iii) a Orbits similar to cometary orbits
b Orbits similar to asteroidal orbits
c Orbits appropriate to Zodiacal light particles

A total of about 340 orbits have been integrated which respond to one of the 12 combinations that can be built up.

If β equals 0, we have the well-known two-body-problem (no perturbing planet), or the n - body - problem (with some perturbing planet).
If β is not zero, but there are no perturbations by planets, then the particle spirals inward to the sun. Thereby the eccentricity decreases, and the angle of perihelion moves in the orbital plane.
The lifetime of such a grain (that is the time it takes from its initial orbit to reach e.g. an orbit with semi - major axis a = 0.1 AU) can be calculated by formulae given by Wyatt & Whipple (1950) . The constants in these equations that we use result from values adopted by the IAU. The computed lifetimes are in general agreement with the time span which the integration yielded for the respective grain to spiral from its initial orbit to an orbit with a = 0.1 AU.

The most general case that we discuss here is, if β is not zero, and there are planets as perturbers. Some interesting things happen in this non - planar n - body - problem. The theoretical lifetimes are no longer in agreement with the corresponding times that the integration yield.
In the inner solar system, even with values of β as high as 0.5, most of the grains are heavily perturbed by Venus and Earth. Their orbits are drastically altered, some of the grains come so close to planets that they may be captured as satellites. This has also been stated by Briggs (1962). Others leave such close encounters with high inclinations. Using lower values of β rises the possibility that a particle comes very close to any planet.
As a result it can be said that such dust particles will not form a cloud which can be recognized as the Zodiacal dust cloud.

In the outer solar system, Jupiter dominates the motion of the test bodies. If β is zero, resonances occur when the rotation periods of two bodies are in the ratio of small integers. If ß is non zero, the resonances are shifted, depending on the value of ß :

$$a_r = \left[\frac{M_o(1-\beta)}{M_o + m_p} \left(\frac{i}{j}\right)^2 \right]^{1/3} a_p \qquad (2)$$

(i/j is the ratio of the rotation periods, m_p , a_p is the mass and the semi - major axis of the planet in resonance). E.g. the Trojans which are in the 1:1 resonance with Jupiter, would have semi - major axes of about 4.1 AU, if they were dust - sized with ß = 0.5.

Some of the particles stay in resonances for a long time (Figs. 1 through 3 show orbits for different values of ß; the initial orbits for the Figs. 1 through 3 are identical). Some of them are rejected from their way spiralling inward (Fig. 4 shows an orbit that is similar to the orbit of comet Halley, with ß=0.5; this grain here is in resonance with Uranus : 1:1 at 15 AU; 2:3 at 20 AU; 1:2 at 24 AU). Some of the grains traverse very quickly the resonance zone (Fig. 5 : ß= 0.5 , 2:1 resonance with Jupiter at 2.6 AU).

Gonczi et al.(1982) mentioned in a planar treatment that the eccentricity always grows when a body goes through a resonance. Here we see that in three dimensions this is not always true. Fig. 6 shows that the eccentricity is decreased passing the resonance (here the resonance 3:1 with Jupiter at about 2 AU for ß=0.5).

References :

Newhall, X X, Standish Jr.,E. M., Williams, J. G. :
Astron. Astrophys., Vol.125, pp 150 - 167 (1983)

Wyatt Jr., S. P., Whipple, F. L. :
Astrophys.Journal, Vol. 111, pp 134 - 141 (1950)

Briggs, R. E. : Astron. Journal, Vol.67,
Vol.67, No.10, pp 710 - 723 (1962)

Gonczi, R., Froeschlé, Ch. and Cl. : Icarus,
Icarus, Vol.51, pp 633 - 654 (1982)

Figures : Plots of the variation of the semi - major axis and the eccentricity of a grain.

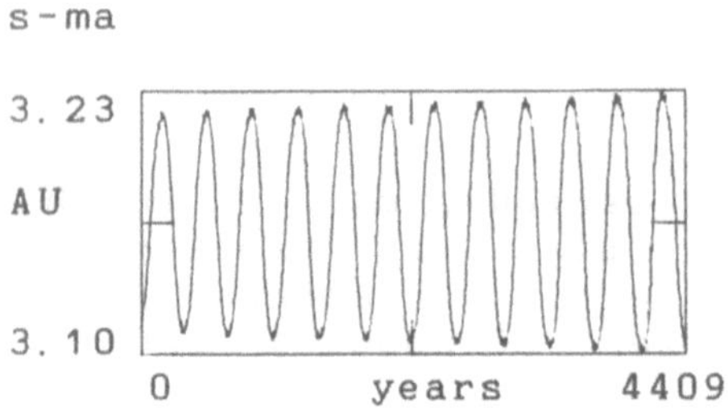

Fig. 1
β=.1

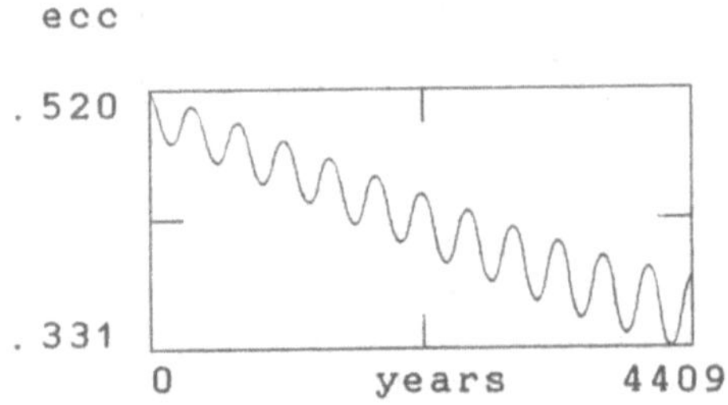

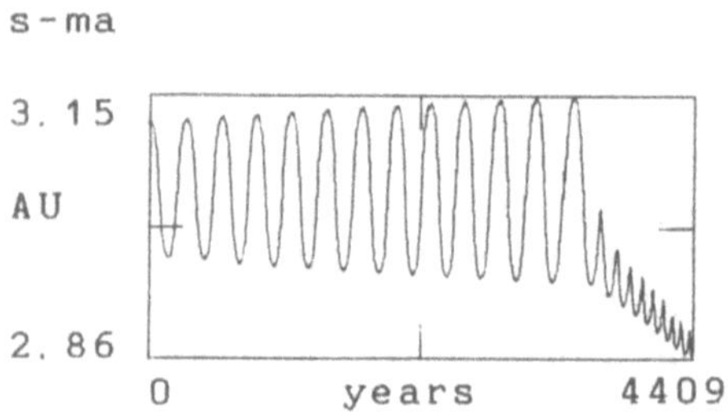

Fig. 2
β=.2

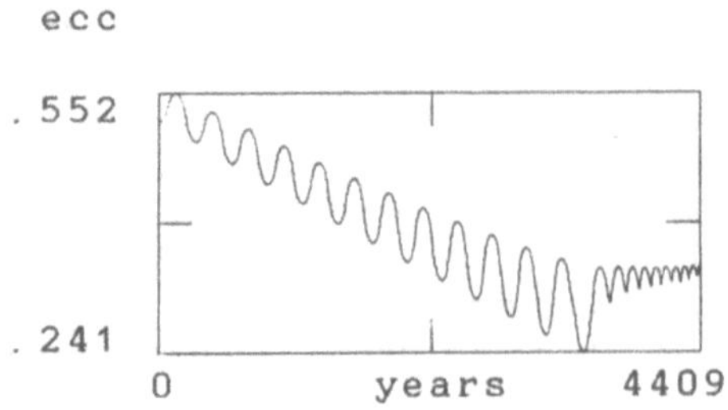

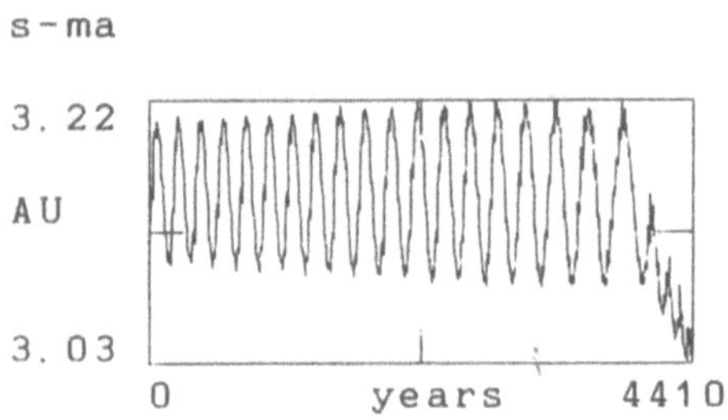

Fig. 3
β=.5

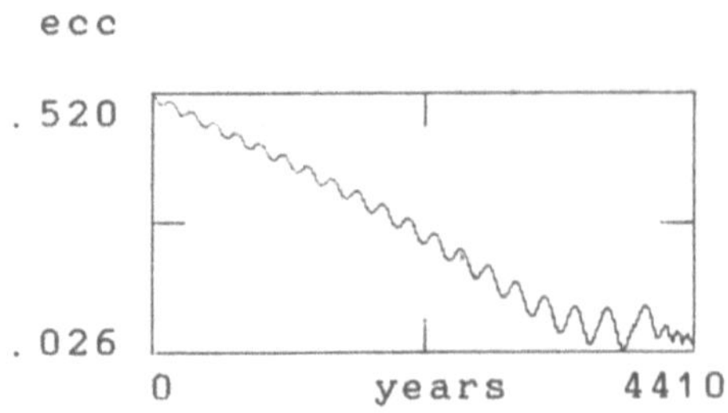

Figs. 1 - 3 : Plots of a grain which is initially in resonance with Jupiter. The orbital elements are the same for all three plots at the beginning.

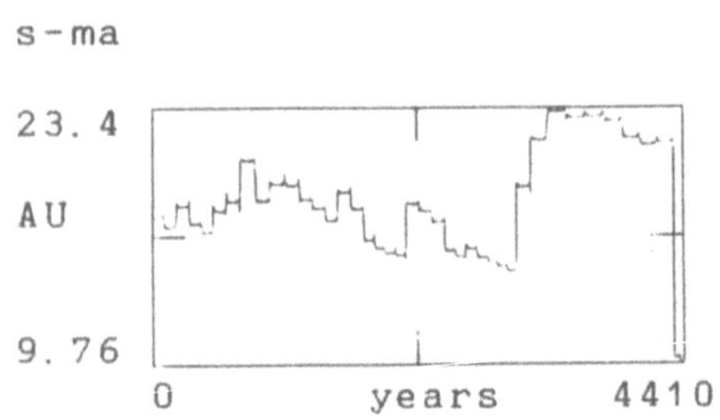

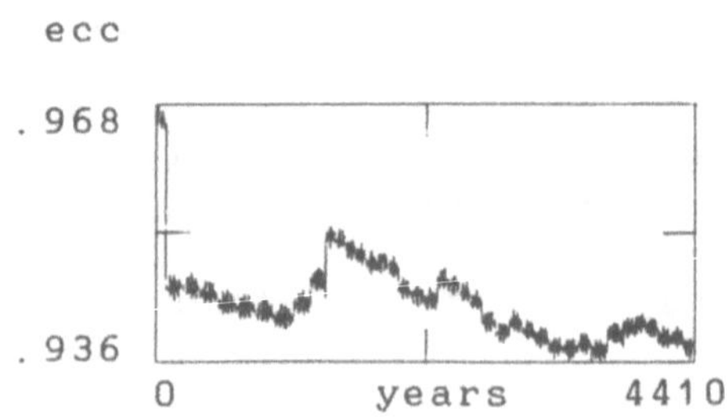

Fig. 4 : The orbit is perturbed by resonances with Uranus. β = 0.5

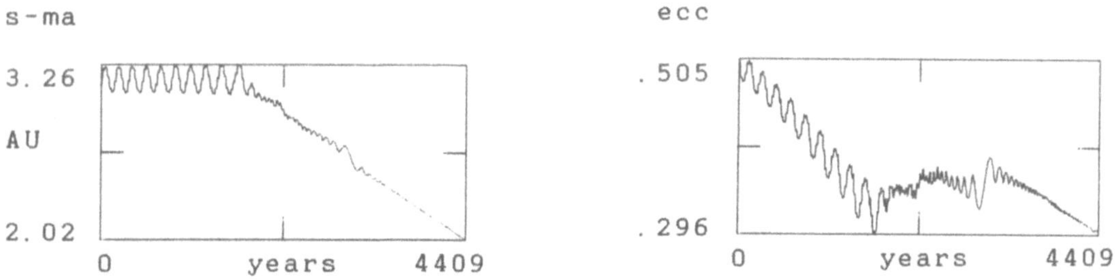

Fig. 5 : a grain traverses quickly a resonance with Jupiter (at about 2.6 AU). ß =0.5

Fig. 6 : The eccentricity decreases when the grain passes the resonance with Jupiter (at about 2 AU). ß =0.5

DISCUSSION :

Question by S. F. Singer :

How many of the orbits are stirred up by encounters?

G. Burkhardt : That depends on what you mean by 'stir up'. The inclination of an orbit is always altered when its plane has an inclination different from the gravitationally strongest planet at that time.

Question by J. D. Mulholland :

What percentage of the orbit calculations was halted due to gravitational singularities (close approaches) and what was the criterion; this relates to planetary accretion sweep-up.

G. Burkhardt : The criterion of the 'sphere of influence' is due to the stepsize of the integrator. It was chosen to be ten times the radius of the planet in question. Only fourteen (14) out of the about 340 integrations had to be halted because of this criterion.

Remark by J. D. Mulholland :

With that definition of the spheres of influence, the cross-section for accretion sweep-up by planets is several orders of magnitude smaller than the 5% rate of sphere of influence encounters. This supports the estimate that planetary sweep-up is a negligible influence on depletion of the dust cloud.

AN ANALYSIS OF IRAS' SOLAR SYSTEM DUST BANDS

S. F. Dermott, P. D. Nicholson, J. A. Burns,
and J. R. Houck
Center for Radiophysics and Space Research
Cornell University
Ithaca, NY 14853
U.S.A.

ABSTRACT. Measurements of longitudinal variations in the brightness and in the latitude of the solar system dust bands recently discovered by IRAS will determine the orbital elements of the particles involved and may discriminate between cometary and asteroidal models of the origin of these bands. The expected variations for bands of dust particles with common orbital elements and small eccentricity and inclination are calculated as functions of semimajor axis.

1. INTRODUCTION

During its all-sky survey at wavelengths of 12, 25, 60, and 100 μm, the Infrared Astronomical Satellite (IRAS)[1] discovered three narrow bands of warm (165-200K) emission circling the sky at geocentric ecliptic latitudes ~ -10°, 0°, and 10° (Low et al., 1984; Neugebauer et al., 1984). The inferred heliocentric distance of the bands is 2.7 ± 0.5 AU, suggesting an association with the asteroid belt (Low et al., 1984). The symmetry of the bands with respect to the ecliptic suggests that they are caused by at least two sets of particles, each of which has a common inclination and, more or less, uniformly distributed nodal lines. The paired bands then arise because particles in inclined orbits spend a disproportionate amount of time at the extremes of their vertical harmonic oscillations (Low et al., 1984; Dermott et al., 1984).

Following the suggestion of Low et al. (1984) that the bands could be the results of collisions between asteroids, we have shown that the bands may be associated with the principal Hirayama asteroid families, that is, the Koronis, Eos and Themis families (Dermott et al., 1984). The latitudinal distribution of the asteroids in these families nicely matches the IRAS observations, with the Eos family producing the outer

[1]The Infrared Astronomical Satellite was developed and operated by the Netherlands Agency for Aerospace Programs (NIRV), the U.S. National Aeronautics and Space Administration (NASA), and the U.K. Science and Engineering Research Council (SERC).

R. H. Giese and P. Lamy (eds.), Properties and Interactions of Interplanetary Dust, 395–409.

bands, and the Koronis and Themis families accounting for the central band. Furthermore, if we assume that the lower cutoff in the size distribution of asteroidal particles is ~1 mm (Dohnanyi, 1976), then the families could provide the optical depth ($\sim 10^{-8}$) observed by IRAS. Detailed analysis of the structure of the dust bands as revealed by the IRAS data will test the asteroid hypothesis, since measurements can be made that will determine the orbital elements of the dust particles, in particular, their semimajor axes or mean distances. Here, we outline the analysis needed to relate the perturbations of the dust particle orbits by Jupiter, Saturn, and the other planets to the longitudinal variations in brightness and ecliptic latitude that we predict will exist in the IRAS data.

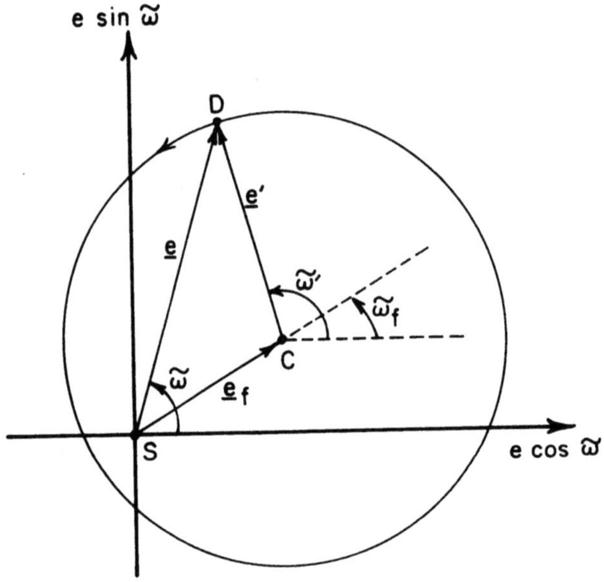

Fig. 1 Vector addition diagram relating the osculating eccentricity $\underline{e}$, the proper eccentricity $\underline{e}'$, the forced eccentricity $\underline{e}_f$ and their corresponding longitudes of pericenter: $\tilde{\omega}$, $\tilde{\omega}'$ and $\tilde{\omega}_f$.

2. SECULAR PERTURBATIONS

Perturbations by Jupiter and, to a lesser extent, Saturn and the other planets result in periodic variations in the eccentricities and the inclinations of asteroidal orbits. If the eccentricity and the inclination are small, then the eccentricity vector can be described as follows:

$$e \sin\tilde{\omega} = e'\sin(\tilde{\omega}_0' + \dot{\tilde{\omega}}'t) + e_f \sin\tilde{\omega}_f, \tag{1}$$

$$e \cos\tilde{\omega} = e'\cos(\tilde{\omega}_0' + \dot{\tilde{\omega}}'t) + e_f \cos\tilde{\omega}_f. \tag{2}$$

(see Fig. 1). Here, $\tilde{\omega}$ is the orbit's longitude of perihelion measured relative to an inertial axis, $\dot{\tilde{\omega}}'$ is the secular apsidal precession rate due to planetary perturbations, and $\tilde{\omega}_f$ is the (slowly varying) forced longitude of perihelion. The constants e' and $\tilde{\omega}_0'$ reflect initial conditions. The inclination vector can be described analogously, in terms of a proper inclination, I', and a forced inclination and nodal longitude, I_f and Ω_f. Proper orbital elements of the Koronis, Eos and Themis Hirayama families are listed in Tables 1 and 2. The forced orbital elements shown in Table 2 were calculated using the theory of Brouwer and van Woerkom (1950) which includes the effects of all the planets except Pluto, but in which terms smaller than e^2 and I^2 in the disturbing function are ignored. For the Koronis, Eos and Themis families, the errors introduced by this truncation are less than 5 percent (J. G. Williams, private communication, 1984).

TABLE 1: PROPER ORBITAL ELEMENTS OF FAMILIES

Family	a AU	e'	I'± δI' deg	$(I'/\delta I')^{\frac{1}{2}}$
Koronis	2.875	0.049	2.12 ± 0.08	5.1
Eos	3.015	0.071	10.20 ± 0.28	6.0
Themis	3.136	0.152	1.42 ± 0.22	2.5

The forced eccentricity, e_f, can be further resolved into vectorial components given by

$$e_f \begin{matrix}\sin\\ \cos\end{matrix} \tilde{\omega}_f = \sum_{j=1}^{8} \sum_{\ell=1}^{10} \frac{M_{j\ell}[0,j]}{\dot{\tilde{\omega}}' - \dot{\nu}_\ell} \begin{matrix}\sin\\ \cos\end{matrix} (\dot{\nu}_\ell t + \nu_{0\ell}) , \qquad (3)$$

where

$$\dot{\tilde{\omega}}' = \sum_{j=1}^{8} (0,j) , \qquad (4)$$

the (0,j) and the [0,j] are the Lagrange parentheses of the jth planet, the $\dot{\nu}_\ell$ are the eigenfrequencies of the solar system associated with the motions of the pericenters of the planetary orbits and the $M_{j\ell}$ are coefficients in the expansion of the planetary eccentricities (Brouwer and Clemence, 1961). Similar expressions can be written for the forced inclination, I_f.

There are two important circumstances in which a single term or group of terms can dominate the double summation shown in equation (3). One occurs when an asteroid orbits close to a massive planet. As $a \rightarrow a_j$, where a and a_j are the semimajor axes of the asteroid and the jth planet, respectively, both (0,j) and [0,j] increase as $[1 - (a/a_j)^2]^{-2}$.

TABLE 2: FORCED ORBITAL ELEMENTS

Family	I_f deg	Ω_f deg	e_f	$\tilde{\omega}_f$ deg	$a(1-e') \pm ae_f$ AU
Koronis	1.16	96.1	0.037	6.2	2.73 ± 0.11
Eos	1.19	97.1	0.037	7.6	2.80 ± 0.11
Themis	1.22	97.8	0.038	8.7	2.66 ± 0.12

If (0,j) is large and $\dot{\tilde{\omega}}' \gg \dot{\nu}_\ell$ for all ℓ, which is the case for asteroids orbiting close to Jupiter, then, as $a \to a_j$,

$$e_f \begin{matrix}\sin\\ \cos\end{matrix} \tilde{\omega}_f \to \frac{[0,j]}{(0,j)} e_j \begin{matrix}\sin\\ \cos\end{matrix} \tilde{\omega}_j , \tag{5}$$

$$= \frac{b_{3/2}^{(2)}(\alpha)}{b_{3/2}^{(1)}(\alpha)} e_j \begin{matrix}\sin\\ \cos\end{matrix} \tilde{\omega}_j , \tag{6}$$

where the $b_s^{(k)}(\alpha)$ are Laplace coefficients with $\alpha = a/a_j$ for $a < a_j$. Furthermore, as $a \to a_j$, $b_{3/2}^{(2)}(\alpha)/b_{3/2}^{(1)}(\alpha) \to 1$, so that $e_f \to e_j$ and $\tilde{\omega}_f \to \tilde{\omega}_j$. Similarly, $I_f \to I_j$ and $\Omega_f \to \Omega_j$; that is, the forced orbit locks onto and duplicates the orbit of the dominant planet. As seen in Figs. 2 and 3, this occurs for asteroidal orbits when $a \gtrsim 3$ AU.

If Jupiter were the only important influence on the asteroidal orbits, then e_f and $\tilde{\omega}_f$ (and I_f and Ω_f) would be poor indicators of distance, since they would vary little across the belt. However, resonant terms for which $\dot{\tilde{\omega}}' \approx \dot{\nu}_\ell$ (or $\dot{\Omega}' \approx \dot{\nu}_\ell$ in the analogous equations describing the motions of the nodes) can also dominate equation (3). Since three of these resonances occur in, or near, the main asteroid belt, the forced elements can be used to determine a. For small e and I, the $\dot{\nu}_{16} = \dot{\Omega}'$ and the $\dot{\nu}_6 = \dot{\tilde{\omega}}'$ resonances are located at 1.94 and 2.03 AU while the $\dot{\nu}_{10} = \dot{\tilde{\omega}}'$ resonance is located at 2.64 AU. The complete variations of e_f, $\tilde{\omega}_f$, I_f, and Ω_f across the main belt are shown in Figs. 2 and 3. For a < 2.7 AU, the forced orbital elements vary markedly with semimajor axis and thus, in the inner half of the asteroid belt, at least, they are good indicators of distance.

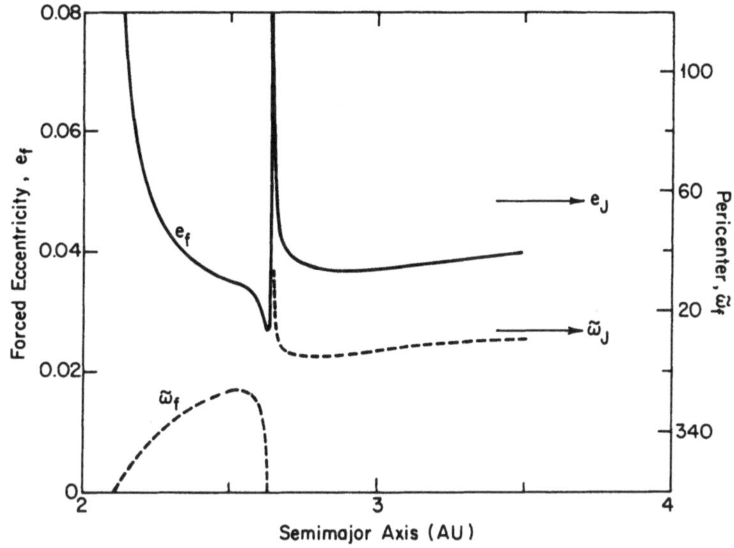

Fig. 2 Variation of the forced eccentricity e_f and the corresponding longitude of pericenter $\tilde{\omega}_f$ with semimajor axis in the main asteroid belt. The elements with subscript J are the present elements of Jupiter.

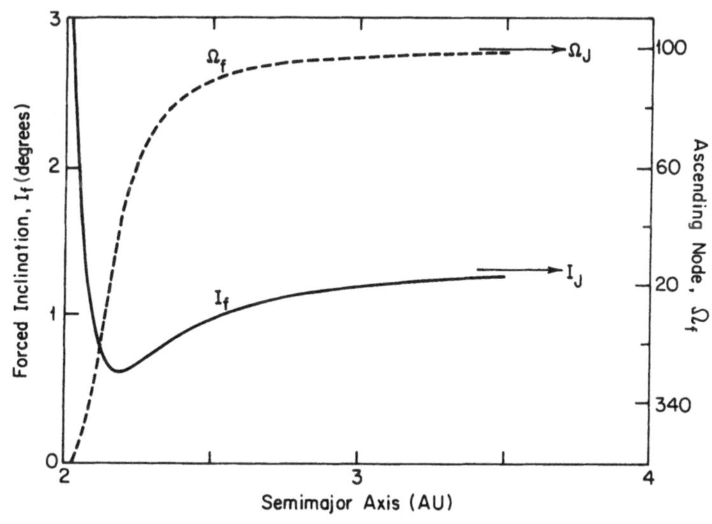

Fig. 3 Variation of the forced inclination I_f and the corresponding longitude of ascending node Ω_f with semimajor axis in the main asteroid belt. The elements with subscript J are the present elements of Jupiter.

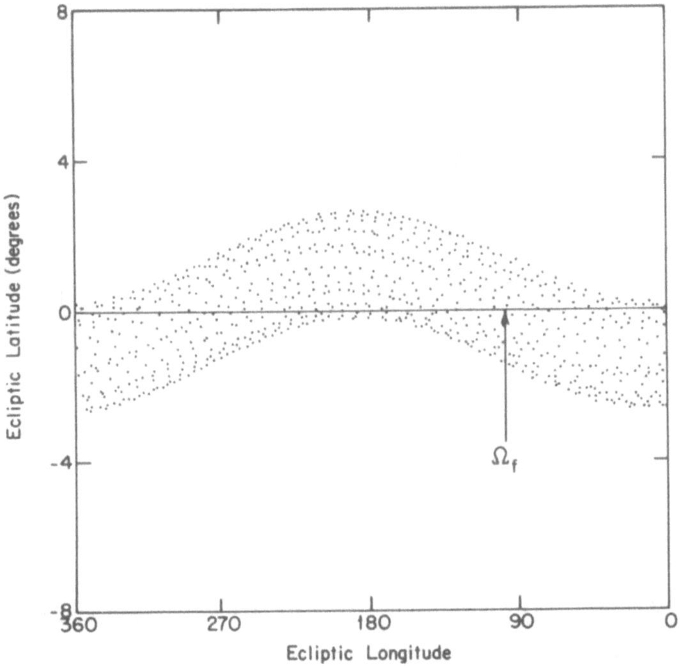

Fig. 4 Variation of ecliptic latitude with ecliptic longlitude, as measured in a sun-centered coordinate system, for the putative Themis dust band. The latitudinal width of the band is 2I' at all longitudes. The striations are an artifact of the small number of points used in the simulation and disappear as the number of points increases. (Figure copyright of Nature, Macmillan Journals Ltd.)

In closing this section, it should be mentioned that the above analysis is inadequate near resonances involving the mean motions of an asteroid and Jupiter, particularly the first-order and the lower-order resonances that dominate the orbital parameter space between the 2:1 resonance at 3.27 AU and the orbit of Jupiter (Dermott and Murray, 1983).

3. PARTICLE DISTRIBUTIONS

The existence of common forced inclinations and eccentricities gives rise to longitudinal variations in the latitude and the brightness of the bands that are diagnostic of their distances (Dermott et al., 1984). Geometrically, the effect of the forced inclination is to make all orbits with a particular value of semimajor axis precess about a common mean plane, whose inclination and ascending node are given by I_f and Ω_f, respectively. The peaks in the latitudinal distribution of an asteroidal family are symmetrically disposed with respect to this mean plane and, in a sun-centered coordinate system, form two parallel bands around the sky separated by 2I'. The appearance of the putative Themis dust band in a sun-centered ecliptic coordinate system is shown in Fig. 4. The sinusoidal variation of mean latitude with longitude has an amplitude I_f and the phase of the variation is determined by Ω_f. Variations in the apparent latitudinal separation of the bands as seen from the earth, corresponding to changes in telescope pointing, also give a method of determining the distances of the bands (Hauser et al., 1984). With this method, however, allowance must be made for the variation in apparent latitude corresponding to the changes in the distances of the bands associated with the forced eccentricity (q.v.).

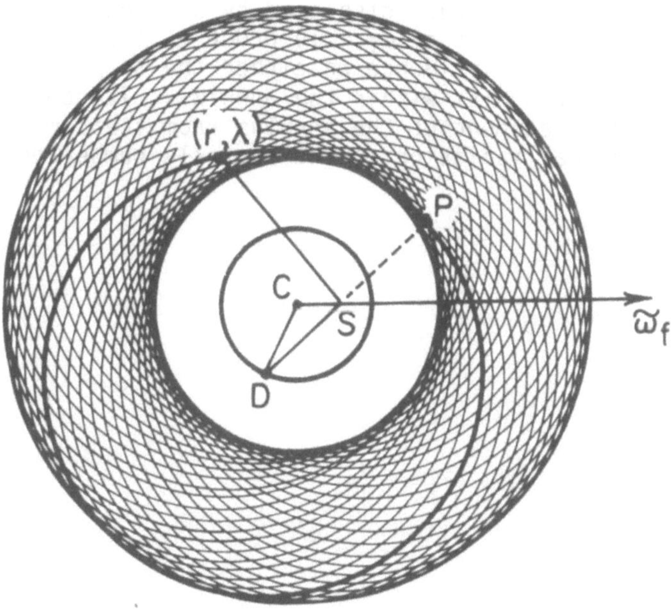

Fig. 5 Distribution of elliptical orbits. To order e, the figure is circularly symmetric about the point C which is displaced from the sun S by an amount ae_f in a direction opposite to that of $\tilde{\omega}_f$. The lines CS, DS and DC have lengths ae_f, ae and ae', respectively, and the triangle CDS is similar to the triangle CDS shown in Fig. 1. P is the pericenter of the highlighted orbit.

The effects of a forced eccentricity on the spatial distribution of an asteroidal family, or a dust band, are shown in Fig. 5. The distribution of the orbits depicted there can be understood from the properties of the vector diagram in Fig. 1. The distance of some point (r,λ) from the focus, S, in Fig. 5 varies, to lowest order in e, as

$$r = a[1 - e\cos(\lambda - \tilde{\omega}) + O(e^2)]. \tag{7}$$

However, the distance r' of that same point from the center of its elliptical orbit, D, varies as

$$r' = a[1 - \tfrac{1}{2}e^2\sin^2(\lambda - \tilde{\omega}) + O(e^3)] \; . \qquad (8)$$

Thus, to order e, an elliptical orbit can be represented by a circle of radius a centered on a point D which is displaced from its focus, S, by an amount ae in a direction opposite to that of its pericenter, P. Paradoxically, to order e, changing the eccentricity of an orbit does not result in a change in its shape: only its displacement from the focus changes. Since, in Fig. 1, the eccentricity vector e' rotates uniformly about C [cf. equations (1) and (2)], it follows that, in Fig. 5, the centers of the elliptical orbits are distributed uniformly on a circle of center C and radius ae' and that the orbits form an annulus of width 2ae' displaced by an amount ae_f in a direction opposite to that of $\tilde{\omega}_f$. To order e, the forced eccentricity e_f has no influence on either the shape, the size or the distribution of orbits within the annulus. It does, however, determine the displacement of the annulus from the common focus, S, and the longitudinal variation of the particle number density.

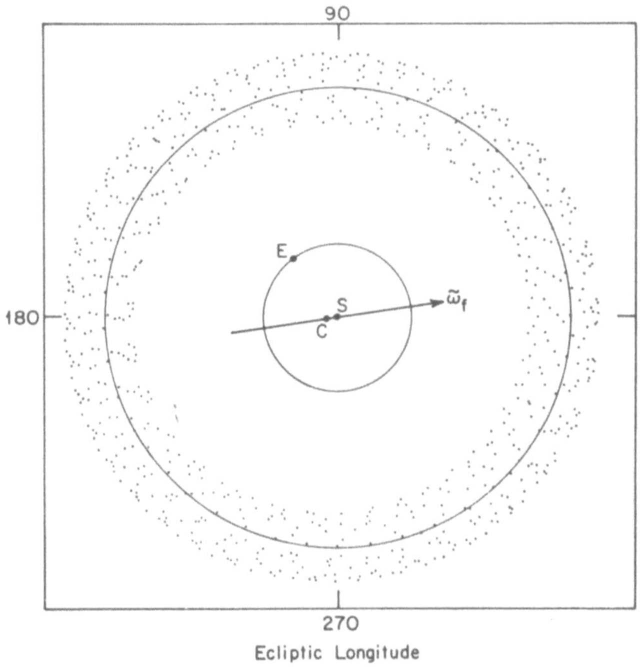

Fig. 6 A view of the putative Themis dust band in the ecliptic plane. The inner and the outer envelopes of the particle orbits are circularly symmetric about C (cf. Fig. 5). The radial width of the near-circular dust band is 2ae'. The solid circle within the dust band is centered on S and has a radius equal to the mean semimajor axis of the Themis family. The inner solid circle denotes the orbit of the earth E. (Figure copyright of Nature, MacMillan Journals Ltd.)

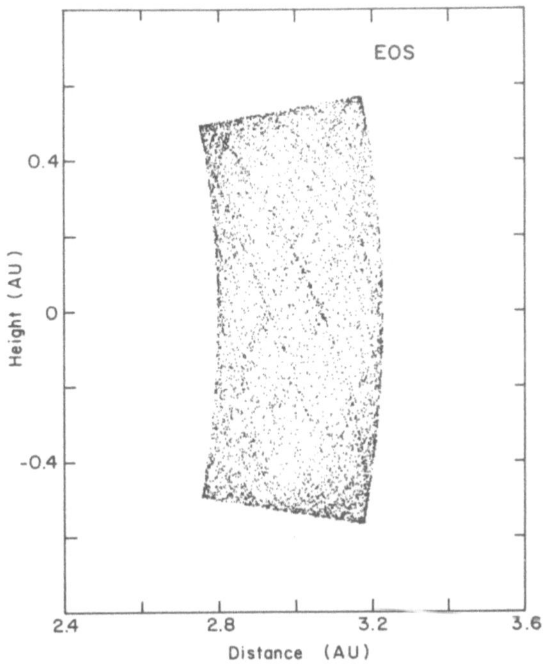

Fig. 7 A meridional cross-section of the putative Eos band. The particle number density is greatest at the corners of the distribution. Hence, from a point above the plane of the ecliptic an observer would see a pair of pericenter bands and a pair of apocenter bands.

The number of particles, $\delta N(\lambda,\tilde{\omega}')$, in a given section of the annulus on orbits with pericenters between $\tilde{\omega}'$ and $\tilde{\omega}' + \delta\tilde{\omega}'$, can be written

$$\delta N(\lambda,\tilde{\omega}') = \delta N_0 \delta N_p \ , \tag{9}$$

where δN_0 is the number of orbits in that section, as determined by $\delta\tilde{\omega}'$, and δN_p is the number of particles between λ and $\lambda + \delta\lambda$ on each orbit. Since $\tilde{\omega}'$ is distributed uniformly,

$$\delta N_0 = \frac{\delta\tilde{\omega}'}{2\pi} N_0 \ , \tag{10}$$

where N_0 is the total number of orbits in the annulus. From Kepler's law of areas,

$$\delta N_p \approx \frac{\delta\lambda}{2\pi}[1 - 2e \cos(\lambda - \tilde{\omega})]N_p \ , \tag{11}$$

where N_p is the total number of particles on each orbit. Hence, the variation of particle number density with longitude, λ, is given by

$$N(\lambda)\delta\lambda = \frac{N\delta\lambda}{4\pi^2}\int_0^{2\pi}[1 - 2e\cos(\lambda - \tilde{\omega})]d\tilde{\omega}' , \qquad (12)$$

where N (= $N_0 N_p$) is the total number of particles. By using

$$\lambda - \tilde{\omega} = (\lambda - \tilde{\omega}_f) - (\tilde{\omega} - \tilde{\omega}_f) , \qquad (13)$$

to expand $\lambda - \tilde{\omega}$ in equation (12) and by eliminating $\tilde{\omega}$ through the use of

$$e\sin(\tilde{\omega} - \tilde{\omega}_f) = e'\sin(\tilde{\omega}' - \tilde{\omega}_f) , \qquad (14)$$

and

$$e\cos(\tilde{\omega} - \tilde{\omega}_f) = e'\cos(\tilde{\omega}' - \tilde{\omega}_f) + e_f , \qquad (15)$$

obtained from triangle CDS in either Fig. 1 or Fig. 5, equation (12) reduces to

$$N(\lambda)\delta\lambda = \frac{N\delta\lambda}{2\pi}[1 - 2e_f\cos(\lambda - \tilde{\omega}_f)] \quad . \qquad (16)$$

Thus, the longitudinal variation of the particle number density is determined by e_f alone and is independent of e'. Two orthogonal views of typical particle distributions are shown in Figs. 6 and 7.

4. BRIGHTNESS VARIATIONS

Regardless of the source of the particles, longitudinal variations in brightness are predicted to exist in the IRAS data for two reasons. The particle number density is a minimum in the direction of $\tilde{\omega}_f$, but at $\lambda = \tilde{\omega}_f$ the particles are both closer to the earth's orbit and hotter. These latter effects more than compensate for the lower particle number density so that the dust bands are predicted to be brightest at pericenter (Dermott et al., 1984). The expected variations of brightness with ecliptic longitude, averaged over all latitudes, at the wavelengths that correspond to the four IRAS channels, are shown in Fig. 8. The phase of the longitudinal variation in brightness of each dust band is determined by $\tilde{\omega}_f$ and is therefore diagnostic of the dust band's semimajor axis, a, according to Fig. 2.

The relative amplitudes of the longitudinal variations in the four IRAS wavelength channels are determined by further properties of the dust grains, in particular, their albedos and radii. The particular calculation shown in Fig. 8 applies to the putative Themis band, but calculations for the Eos and the Koronis dust bands give almost identical results. For this heuristic model we have assumed (a) that

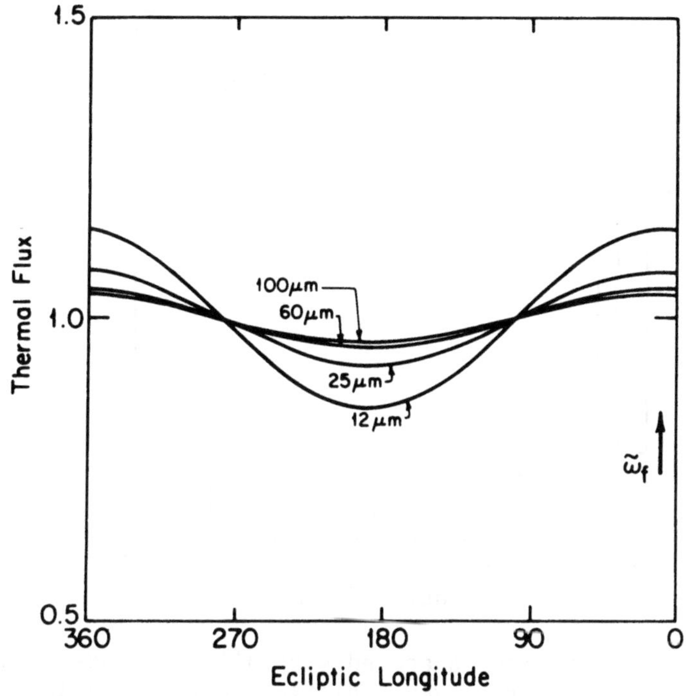

Fig. 8 Expected variation, with sun-centered ecliptic longitude, averaged over all latitudes, of the radiation received at the earth in the 12 μm, 25 μm, 60 μm, and 100 μm IRAS wavelength channels from dust particles associated with the Themis family. The arrow marks the longitude of perihelion of the forced eccentricity, $\tilde{\omega}_f$. (Figure copyright of Nature, MacMillan Journals Ltd.)

the radii of the dust particles are large (≫ the wavelengths of the IRAS observations), (b) that they have an emissivity of 0.9, a geometric albedo of 0.1 and a phase integral of 0.6, (c) that the earth's orbit is circular and (d) that the pointing direction of the IRAS telescope is tangential to the earth's orbit. The ecliptic longitude in Fig. 8 refers to the heliocentric longitude of the dust particles, rather than to their observed geocentric longitude, although the flux is that at the earth. The mean temperature of the dust particles associated with the Themis family is 160K. At this temperature, the Planck function peaks at a wavelength of 18 μm. The longitudinal variation in flux is strongest in the 12-μm channel because comparatively large changes in the thermal flux at wavelengths near 18 μm are produced by modest temperature changes.

Changes in the viewing aspect of the dust bands caused by longitudinal variations in their ecliptic latitude also result in longitudinal variations in brightness. A meridional cross section of a typical dust band is shown in Fig. 7. The harmonic oscillations of the

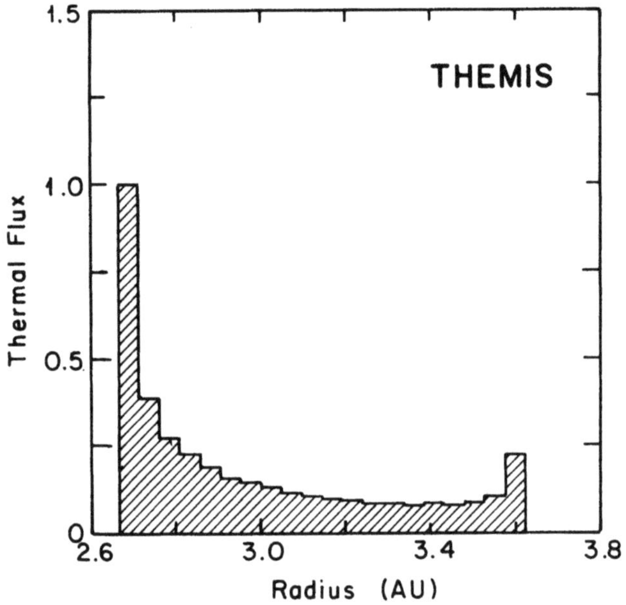

Fig. 9 The radial variation, averaged over all longitudes and latitudes, of the received radiation in the IRAS 12-μm channel, for the putative Themis dust band, as seen from its geometrical center, C. The inner edge of the band is four times as bright as the outer edge.

particles about their mean circular orbit result in a concentration of particles near the edges and, in particular, the corners of the cross section. From a viewpoint well above the plane of the ecliptic, an observer would see each of the two heliocentric bands split into a pair of bands, one band in each pair associated with the perihelia of the dust particle orbits and the other associated with their aphelia. The Voyager 1 spacecraft, which is at present leaving the solar system, may provide such a viewpoint (C. Porco, personal communication, 1984). These bands should also be sought in the data returned from the Pioneer integrated starlight experiment (Hong et al., 1985), which seems to find undulations in the zodiacal light brightness.

The perihelion and the aphelion bands, however, would not be equally bright. The expected radial variation, averaged over all longitudes and latitudes, of the thermal flux received in the IRAS 12-μm channel from the putative Themis dust band at its geometrical center, C, is shown in Fig. 9. The inner edge of the band is four times as bright as the outer edge and, thus, from a terrestrial viewpoint, we would expect the aphelia bands to be comparatively weak, particularly in the 12-μm channel. The difference between the perihelion and the aphelion bands is, however, less pronounced for the other families due to their smaller proper eccentricities.

The heliocentric distance of the prominent perhelia band varies as

$$r = a[1 - e' - e_f \cos(\lambda - \tilde{\omega}_f)] . \qquad (17)$$

The mean distance of the perihelia bands associated with the Koronis, Eos, and Themis families is ~ 2.73 AU (see Table 2). This distance compares favorably with the distance of 2.7 ± 0.5 AU estimated from the color temperatures of the bands (Low et al., 1984) and with the distance of 2.5 AU derived from parallax measurements (Hauser et al., 1985). However, as we have already remarked, the latter estimate needs to be corrected for the variation in the distance of the perihelia band given by equation (17).

For an annulus like that shown in Fig. 7, an observer at the sun would only see two bands separated by 2I' (Dermott et al., 1984), since the aphelion and the perihelion bands would be superposed. The ratio of the brightness extremes, at a given longitude, seen by such an observer, that is, the ratio, R, of the brightness at the edge of the band to that its center, is determined by the resolution of the telescope (or the angle over which the data are binned), δI, and by the dispersion, $\delta I'$, in the proper inclinations of the dust particle orbits. R cannot exceed $\sim(I'/\delta I')^{\frac{1}{2}}$. However, if $\delta I' \ll \delta I$, then

$$R = (2I'/\delta I)^{\frac{1}{2}} . \qquad (18)$$

Maximum values of R, from a solar viewpoint, for the Koronis, Eos, and Themis families are listed in Table 1.

Because of the non-heliocentric vantage point of the IRAS telescope, the observed values of R will be significantly less than the above maxima; moreover, due to the forced inclinations of the bands and the resulting changes in the viewing geometry, they will vary with ecliptic longitude. We deduce from Fig. 4 that, at a heliocentric ecliptic longitude, λ, of $\Omega_f + 90°$, for an earth-bound telescope pointing in a direction radially away from the sun, R would be a minimum in the Northern band and a maximum in the Southern band, whereas, at $\lambda = \Omega_f - 90°$ the reverse would apply. Since the pointing direction of the IRAS telescope was predominantly in, or close to, a plane perpendicular to the direction of the sun, the above deductions do not strictly pertain to the IRAS data. However, for a telescope pointing plane fixed with respect to the direction of the sun, and for a fixed pointing sense within that plane, we expect that the band brightness will vary sinusoidally with ecliptic longitude and that there will be a phase shift of 180° between the brightness variations in the Northern and in the Southern bands.

The two sources of longitudinal variation in brightness that we have identified are, in principle, separable. The sum of the brightnesses in the Northern and in the Southern bands is determined by e_f and $\tilde{\omega}_f$ and is independent of I_f and Ω_f, whereas, for the ratio of

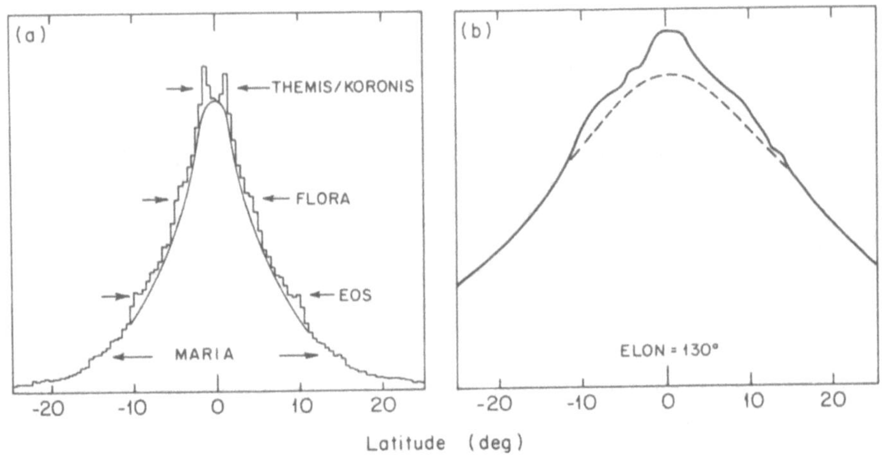

Fig. 10 A simulation of the the latitudinal distribution of dust particles associated with the ~ 2000 asteroids in the TRIAD file, (a), is compared with the IRAS flux data in the 60 μm wavelength channel (from Low et al., 1984), (b). In (a), latitude is measured with respect to the mean plane of the Eos family, whereas in (b) latitude is referred to the plane of the ecliptic. Deviations from the smooth curve in (a) are associated with the principal Hirayama families. In (b), ELON refers to the geocentric ecliptic longitude of the IRAS telescope pointing direction and the dashed curve represents the smooth zodiacal emission.

these brightnesses the opposite applies. All of these quantities are separately diagnostic of the bands' semimajor axes. Accurate predictions of R and its longitudinal variation for a given band, however, will require detailed modelling of the IRAS observations, beyond that attempted to date. The high-resolution IRAS data will be of particular interest, since R for this data will depend partly on the dispersion, $\delta I'$, in I', and partly on e'. Since $\delta I'$ for collision products is determined by the size of the parent body (collision products tend to disperse with relative velocities comparable to the escape velocity of the parent body), measurements of R should yield some information on the initial sizes of the dust particle sources.

5. PARTICLE SOURCES

Detailed analysis of the IRAS data should yield information on all the orbital elements of the dust particles and hence should disclose the source of the particles. Apart from the asteroid family model that we have discussed in detail above, and elsewhere (Dermott et al., 1984), the most plausible alternative model would involve the disintegration of one or more comets. In this case, however, the small-e and -I approximations would require some revision; nevertheless, our analytical approach should be equally suitable. In Fig. 10 we compare the IRAS flux data with the latitudinal distribution of dust particles associated with the known asteroidal orbits. It follows from this figure, and from a comparison of the relative volumes of the asteroids in the prominent families and those in the remainder of the belt, that, if the asteroid families provide the emitting area of the dust rings observed by IRAS, then the area of the entire asteroid belt must make a substantial contribution to the IRAS zodiacal signal.

This research was supported by grants from NASA and from NSF.

6. REFERENCES

Brouwer, D. and Clemence, G.M. Methods of Celestial Mechanics (Academic, New York, 1961).

Brouwer, D. and van Woerkom, A.J.J. (1950), Astr. Pap. U.S. Naval Obs. Naut. Almanac Off. Part II, 85-107.

Dermott, S. F., and Murray, C. D. (1983), "Nature of the Kirkwood gaps in the asteroid belt" Nature 301, 201-205.

Dermott, S. F., Nicholson, P. D., Burns, J. A., and Houck, J. R. (1984), "Origin of the solar system dust bands discovered by IRAS" Nature (in press).

Dohnanyi, J. S. (1976), "Sources of interplanetary dust: Asteroids" in Interplanetary Dust and Zodiacal Light (eds. Elsasser, H., and Fechtig, H.) 187-205 (Springer-Verlag, Berlin).

Hauser, M. G., Gautier, T. N., Good, J., and Low, F. J. (1985) "IRAS observations of interplanetary dust emission" (this conference).

Hong, F. F., Misconi, N. Y., Van Dijk, M.H.H., Weinberg, J. L., and Toller, G. N. (1985), "A search for small scale structure in the zodiacal light" (this conference).

Low, F. J., Beintema, D. A., Gautier, T. N., Gillet, F. C., Beichmann, C. A., Neugebauer, G., Young, E., Aumann, H. H., Boggess, N., Emerson, J. P., Habing, H. J., Hauser, M. G., Houck, J. R., Rowan-Robinson, M., Soifer, B. T., Walker, R. G., and Wesselius, P. R. (1984), "Infrared cirrus: New components of the extended infrared emission" Ap. J. 278, L19-L22.

Neugebauer, G., Beichmann, C. A., Soifer, B. T., Aumann, H. H., Chester, T. J., Gautier, T. N., Gillet, F. C., Hauser, M. G., Houck, J. R., Lonsdale, C. J., Low, F. J., and Young, E. (1984), "Early results from the Infrared Astronomical Satellite" Science 224, 14-21.

MASS INPUT INTO AND OUTPUT FROM THE METEORITIC COMPLEX

E. Grün[+], H.A. Zook*, H. Fechtig[+], and R.H. Giese[o]
+Max-Planck-Institut für Kernphysik, P.O. Box 10 39 80,
6900 Heidelberg, F.R.G.;
*SN3 NASA Johnson Space Center, Houston, Texas 77058, U.S.A.
oRuhr-Universität Bochum, Bereich Extraterrestrische Physik,
Universitätsstr. 150, 4630 Bochum-Querenburg, F.R.G.

ABSTRACT. Direct observations have established the size distribution of interplanetary meteoroids at 1 AU as well as the dependence of the spatial density with respect to the distance from the sun. After evaluating the consequences of mutual collisions and the effect of radiation pressure the following conclusions can be drawn: 1. Catastrophic collisions dominate the lifetimes of meteoroids with masses $m \gtrsim 10^{-5}$g. About 10 tons per second are lost within 1 AU (mostly in form of 10^{-4}g to 10^{-1}g particles). Under steady state conditions these meteor sized particles have to be replenished by other sources, e.g. comets. 2. After being crushed by collisions 70 to 85% of this mass will be in form of particles with masses $10^{-10}\text{g} \lesssim m \lesssim 10^{-5}$g. Part of these "zodiacal light" particles (about 0.3 tons per second) are transported by the Poynting Robertson effect towards the sun where they will evaporate. However, since the collisional production of these intermediate sized particles exceeds their losses this population is presently not in equilibrium. 3. 15 to 30% of the collisional fragments have masses $m \lesssim 10^{-10}$g. Most of these small particles will be injected into hyperbolic orbits by radiation pressure (β-meteoroids).

1. INTRODUCTION

Mutual collisions between interplanetary meteoroids, the Poynting Robertson effect, and radiation pressure ejection of small meteoroids give rise to some interesting consequences which are here examined. From lunar crater statistics and satellite data, the size distribution and flux of micrometeoroids at 1 AU distance are derived. We evaluate the relative rates of destruction and production of grains by mutual collisions and the operation of other processes that determine the stability of the interplanetary meteoritic population. From the comparison of collisional and radiation pressure effects conclusions are drawn on the present state of the meteoritic complex. Especially the mass input into and the mass output from three different mass intervals (meteor sized particles, zodiacal light particles and β-meteoroids) of the meteoritic cloud are discussed. A more detailed description of the methods employed

R. H. Giese and P. Lamy (eds.), Properties and Interactions of Interplanetary Dust, 411–415.

is given by Grün et al. (1984).

2. INTERPLANETARY METEOROID FLUX

Information on the interplanetary meteoroid flux is obtained from lunar crater statistics, in situ spacecraft measurements, meteor and zodiacal light observations. Grün et al. (1984) derive a flux model with the following characteristics:
Meteor observations (for a summary see Whipple, 1967) lead to a dependency of the cumulative flux on the meteoroid mass m according to $m^{-1.34}$ for masses $10^{-5}g \lesssim m \lesssim 10^{2}g$. The flux of smaller particles down to mass $m \sim 10^{-10}g$ is characterized by the size distribution of lunar microcraters (e.g. Morrison and Zinner, 1977). The absolute calibration of the fluxes at masses $10^{-7}g$ and $6x10^{-7}g$ is obtained from measurements of the Pegasus satellite (Naumann, 1966). At $m = 10^{-13}g$ and $10^{-12}g$ fluxes have been derived from the HEOS 2 experiment (Grün and Zook, 1980). These fluxes are below most lunar microcrater fluxes because the latter are dominated by secondary ejecta cratering (Zook et al., 1984) for particle masses $m \lesssim 10^{-10}g$. In the mass range $10^{-14}g \lesssim m \lesssim 10^{-9}g$ the slope of the cumulative meteoroid flux is ~ -0.36. The flux of smaller particles ($m \lesssim 10^{-14}g$) has been calculated from a collisional model assuming that all fragments of this size range which are produced inside 1 AU are pushed out of the solar system by radiation pressure and become β-meteoroids.
This interplanetary flux leads to a spatial mass density at 1 AU of $\sim 10^{-16}g/m^3$, where most mass per logarithmic mass interval is in meteoroids of masses $10^{-6}g$ to $10^{-4}g$. Measurements of the zodiacal light (Leinert et al., 1981) provide the radial dependence of the spatial number density n of interplanetary meteoroids: $n \sim r^{-1.3}$. The determination of the color of the zodiacal light (Pitz et al., 1979) shows some reddening compared to the solar spectral flux. This observation is compatible with the characteristics of the flux curve, i.e. most cross-sectional area per logarithmic mass interval originates from particles of masses $10^{-8}g$ to $10^{-5}g$. The total cross-sectional area of the interplanetary meteoroid cloud at 1 AU is $5x10^{-19}m^2/m^3$.

3. COLLISIONAL MODEL

The rate of catastrophic collisions is calculated using the size distribution described above. Further ingredients of the collisional model are:
- the effective mutual collision speed at 1 AU is 20 km/s and it varies with r as $r^{-0.5}$.
- the dependence of rupture energy upon particle size is given by Gault et al. (1972).
- the mass distribution of the fragments is taken from Fujiwara et al. (1977).

With these assumptions we have calculated the collisional lifetimes of interplanetary meteoroids (Fig. 1). At 1 AU the lifetimes are shortest

(10^4 years) for particles of mass 10^{-3}g to 1g. Both bigger and smaller particles have longer collisional lifetimes. For comparison we also show the Poynting Robertson lifetimes (Wyatt and Whipple, 1950). The efficiency factor used was that for olivine particles and the average initial eccentricity of the particle orbits at 1 AU is assumed to be 0.5. Collisions dominate the lifetimes of meteoroids with masses $m \gtrsim 10^{-5}$g. These large particles will not change their orbits significantly due to the Poynting Robertson effect before they are involved in a collision and fragmented into smaller particles. Only smaller particles will have their orbits circularized by the Poynting Robertson drag and will eventually spiral in towards the sun where they will evaporate. Using our collisional model the rate of catastrophic collisions is compared with the rate of production of fragments for logarithmic mass intervals in the range from 10^{-18}g to 10^{2}g. This comparison in terms of collisional loss and gain is shown in Fig. 2. We have also computed the radial loss due to the Poynting Robertson effect which is required in order to maintain a radial density distribution of $r^{-1.3}$. The result of this comparison is discussed in the next sections.

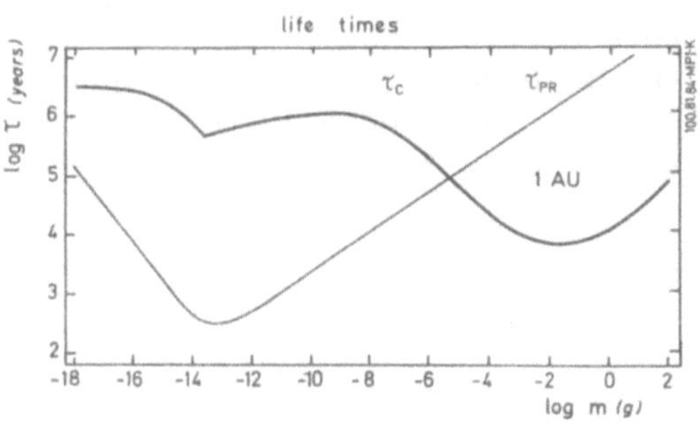

Fig. 1: Lifetimes of interplanetary meteoroids with respect to collisions τ_C and Poynting Robertson effect τ_{PR}. The sharp edge of the collisional lifetime at 10^{-14}g is caused by the artificial cut-off of the meteoroid distribution at $m = 10^{-18}$g.

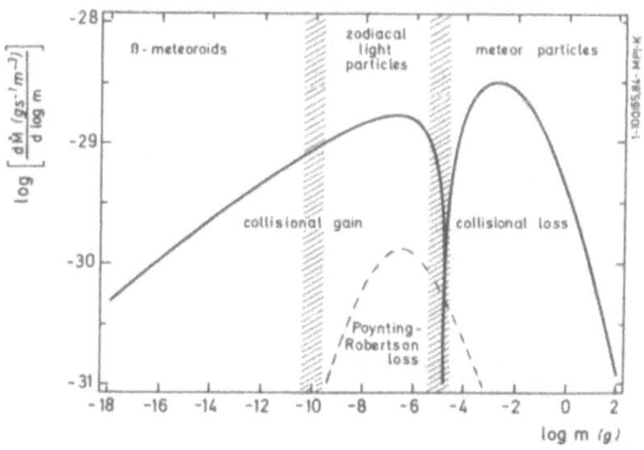

Fig. 2: Net mass loss and gain rates from collisions and transport losses due to Poynting Robertson effect at 1 AU. The total mass lost by collisions ($m \gtrsim 10^{-5}$g) and gained as fragments ($m \lesssim 10^{-5}$g) is 9×10^{-29} g/m³ s. The Poynting Roberston effect requires a loss of only 4×10^{-30} g/m³ s in order to maintain the spatial density $\sim r^{-1.3}$.

4. METEOR PARTICLES

Large meteor sized particles ($m \gtrsim 10^{-5}$g) are dominated by collisional fragmentation. Assuming a radial dependence according $r^{-1.3}$ and a filling factor $\varepsilon = 0.23$ (Leinert et al., 1983) then a total of 9 t/s is lost from this size range within 1 AU. This particle population would be depleted on a time scale of $\sim 10^4$ years without replenishment from cometary and asteroidal sources. Under steady state conditions most meteor particles are "young", i.e. they have not been fragmented by collisions and their initial orbits are not much altered by radiation pressure drag. Only planetary perturbations could distort the initial orbits significantly before the particles break up by catastrophic collisions. Observations of meteor streams support this finding.

5. ZODIACAL LIGHT PARTICLES

The optically active zodiacal light particles (10^{-10}g $\lesssim m \lesssim 10^{-5}$g) are dominated by radiation pressure drag and not by catastrophic disruption. Their lifetimes due to Poynting Robertson effect range from 10^5 years to 10^3 years for the smaller particles. However, many more particles are gained in this mass interval from collisional break-up of meteor-sized particles than are removed either by Poynting Robertson effect or by collisions. About 6 to 8 t/s of these particles are produced inside 1 AU. This compares to only $\sim$ 0.3 t/s which are lost by the Poynting Robertson effect. This situation is not stable but the zodiacal light particle population increases in time (on a time scale of about 10^4 years at 1 AU). Time stability of this particle population can only be maintained if we have overestimated the meteoroid flux by more than a factor 10 or if the break-up laws which we have applied are not at all representative for interplanetary meteoroids. Both alternatives are not supported by the data.

6. β-METEOROIDS

Small particles are affected by radiation pressure (see e.g. Burns et al., 1979) which reduces the solar gravitational attraction. A small fragment particle which is generated by a collision between a larger parent meteoroid and another meteoroid will move on an unbound trajectory if its reduced potential energy (gravitation minus radiation pressure) is exceeded by its kinetic energy which is derived from the parent particle. This is especially effective at the perihelion of an eccentric parent particle's orbit, where the kinetic energy and the collision rate are highest. Since the eccentricities of the parent particles are significant even fragment particles of masses as large as $m \sim 10^{-10}$g can get on hyperbolic trajectories and become β-meteoroids (Zook and Berg, 1975). This direct injection of fragment particles into hyperbolic orbits is a very efficient loss mechanism since the time these particles spend in the inner solar system is only order of 100 days. Therefore, most particles of masses $m \lesssim 10^{-10}$g which are produced from the disruption of larger meteoroids can efficiently be removed by this effect. Hence we conclude that particle population is in time stability. About 1 to 3 t/s of β-meteoroids pass the Earth's orbit.

REFERENCES

Burns, J.A., Lamy, P.L., and Soter, S. (1979). 'Radiation forces on small particles in the solar system', Icarus 40, 1-48.

Fujiwara, A., Kamimoto, G., and Tsukamoto, A. (1977). 'Destruction of basaltic bodies by high-velocity impact', Icarus 31, 277-288.

Gault, D.E., Hörz, F., and Hartung, J.B. (1972). 'Effects of microcratering on the lunar surface', Proc. Lunar Sci.Conf. 3rd, 2713-2734.

Grün, E. and Zook, H.A. (1980). 'Dynamics of micrometeoroids in the inner solar system', in: Solid Particles in the Solar System (Eds. I. Halliday and B.A. McIntosh) Reidel, Dordrecht, 293-298.

Grün, E., Zook, H.A., Fechtig, H., and Giese, R.H. (1984). 'Collisional balance of the meteoritic complex', submitted to Icarus.

Leinert, C., Richter, I., Pitz, E., and Planck, B. (1981). 'The zodiacal light from 1.0 to 0.3 AU', Astron.Astrophys. 103, 177-188.

Leinert, C., Röser, S., and Buitrago, J. (1983). 'How to maintain the spatial distribution of interplanetary dust', Astron.Astrophys. 118, 345-357.

Morrison, D.A., and Zinner, E. (1977). '12054 and 76215: New measurements of interplanetary dust and solar flare fluxes', Proc. Lunar Sci.Conf. 8th, 841-863.

Nauman, R.J. (1966). 'The near earth meteoroid environment', NASA TND 3717.

Pitz, E., Leinert, C., Schulz, A., and Link, H. (1979). 'Colour and polarization of the zodiacal light from the ultraviolet to the near infrared', Astron.Astrophys. 74, 15-20.

Whipple, F.L. (1967). 'On maintaining the meteoritic complex', in: Zodiacal Light and the Interplanetary Medium', NASA-SP 150, 409-426.

Wyatt, S.P., and Whipple, F.L. (1950). 'The Poynting-Robertson effect on meteor orbits', Astrophys.J. 111, 134-141.

Zook, H.A., and Berg, O.E. (1975). 'A source for hyperbolic cosmic dust particles', Planet. Space Sci. 23, 183-203.

Zook, H.A., Lange, G., Grün, E., and Fechtig, H. (1984). 'Lunar primary and secondary microcraters and the micrometeoroid flux', Lunar and Planet.Sci. XV, 965-966.

LORENTZ SCATTERING OF INTERPLANETARY DUST GRAINS

P. Barge*, R. Pellat**, J. Millet***

*Observatoire de Marseille, 13248 Marseille Cedex 4, France
**Laboratoire d'Astronomie Spatiale, F-13012 Marseille,France
***C.N.E.S. 129 rue de l'Universite, 75007 Paris, France

ABSTRACT. The scattering of dust grains orbits due to recurrent sectors of the interplanetary magnetic field is reinvestigated with a better formalism. Our method reveals the resonant character of the diffusion and is well suited for the problem. The spreads in the orbital parameters are found less important than believed untill now and to vary rapidly with eccentricity and semi-major axis. Only the small dielectric grains with size less than 0.5 µm may be scattered by the Lorentz force fluctuations; the main diffusion occurs in inclination and near the sun (20-60 $R_\odot$).

1. INTRODUCTION

The diffusion of charged dust grains by the fluctuations of the interplanetary magnetic field (IMF) is a mechanism often considered in the dynamics and the origin of the zodiacal cloud. The previous studies on this subject (Parker,1964; Consolmagno,1979; Morfill et al,1979) are either limited to simple cases or unsatisfactory. So, we have reexamined the problem with a formalism developped from the quasi-linear theory in Plasma Physics (Barge et al,1982a). The keplerian orbits of the dust grains are assumed weakly perturbed by the random Lorentz force and the diffusion time is taken much smaller than the Poynting-Robertson lifetime in such a way that a statistical approach with a Fokker-Planck approximation can be used. As explained in a previous paper (Barge et al, 1982b) the diffusion of keplerian orbits only occurs if the spectrum of the fluctuations encloses modes which are resonant with the orbital frequency. Therefore only the low frequency fluctuations associated with the large scale structure of the IMF have to be considered. Moreover it is clearly better to take three constants of the unperturbed motion (K_i) with their associated cyclic variables as new dynamical variables since , for example, the Fokker-Planck equation then becomes a simple diffusion equation. A convenient choice of the invariant is: $K_1 = E/2\omega_0$ where E is the energy and ω_0 the orbital frequency, $K_2 = h$ (the modulus of the angular momentum), $K_3 = h_z$ (the projection of $\vec{h}$ onto the spin axis).

R. H. Giese and P. Lamy (eds.), Properties and Interactions of Interplanetary Dust, 417–420.

2. SCATTERING BY MAGNETIC SECTORS

At heliocentric latitudes less than 15° the polarity of the interplanetary magnetic field and the solar wind speed are observed to be function of time; moreover the magnetic polarity appears divided in N sectors of opposite sign (N=2 or 4). Due to the random evolution of the large scale structure of the IMF the size of the sectors is stochastic and the periodicity of the field reversals have a gaussian like probability distribution P (Sawyer, 1976). The high-speed streams of the solar wind are reccurent with the same statistics because of their correlation with the IMF structure; in connection with the wind fluctuations, the charge of the grains may also vary stochastically. In fact, the various contributions to the Lorentz force give a diffusion which have grossly the same order of magnitude; so, we shall examine only the case of the diffusion by magnetic polarity fluctuations.

The following description of the IMF has been adopted:

$$B_r(\vec{r},t) = \langle B_r \rangle f_N(\vec{r},t)$$
$$B_\theta(\vec{r},t) = B_1 \; r_0/r \; f_N(\vec{r},t)$$
$$B_\phi(\vec{r},t) = \langle B_\phi \rangle f_N(\vec{r},t)$$

where $\langle B_r \rangle$ and $\langle B_\phi \rangle$ are the two components of the mean magnetic field of Parker, f_N is a function expressing the stochasticity of the magnetic polarity and B_1 is the magnetic perturbation associated with hydromagnetic fluctuations. The B_θ component has also been assumed to vary as 1/r (Consolmagno, 1979) and, from the power spectra of the IMF (Coleman, 1968), B_1 is deduced to be approximately 10^{-1} ($\sqrt{\varepsilon}$) smaller than the mean magnetic field B_0. We have also assumed f_N to be a function generated by a structure made up of N straight sectors rotating with the sun.

Then, after a Fourier expansion of the function f_N and of the orbital quantities (like the velocity), the diffusion coefficients $\langle \Delta K_i \Delta K_j \rangle / \Delta t$ can be derived from the motion equations (see Barge et al, 1982b, for detailled calculations). They are function of the power spectrum of f_N which is also related to the probability distribution P. As an example, for circular orbits, we have:

$$\frac{\langle \Delta K_1{}^2 \rangle}{\Delta t} = (qwr_0B_0)^2 \pi\varepsilon \int d\omega \sum_{m,l} C_l^2 \frac{2}{N} \langle f_N^2 (\omega,m) \rangle \; \delta\left[\omega+(m+l)\omega_0\right]$$

where w is the wind velocity, C_l is a Fourier coefficient, r_0 = 1 AU and δ is the Dirac distribution. See Barge et al (1982b) for the complete set of coefficients. Then, it is easy to derive the diffusion coefficients in the orbital parameters (semi-major axis a, eccentricity e, and inclination i) from the relations:

$$\dot{a} = -\frac{2}{\mu m} a^2 \omega_0 \dot{K}_1$$

$$\dot{e^2} = -\frac{2}{\mu m} a\omega_0 (1-e^2)^{1/2} \left[(1-e^2)^{1/2} \dot{K}_1 + \dot{K}_2\right]$$

$$\dot{i} = \frac{1}{\mu m} \frac{a\omega_0}{\sin i \, (1 - e^2)^{1/2}} \left[\cos i \, \dot{K}_2 - \dot{K}_3\right]$$

with $\mu = GM_\odot$.
The diffusion coefficients in a, e^2 and i have been numerically computed for a 2 sectors and a 4 sectors structure and for two values of the inclination. The square roots of the mean square changes of the orbital elements can be determined for a time interval of 10 years (approximately the duration of a solar cycle).
In the case of a 1 µm obsidian grain charged to a potential of 6 volts the spread in semi-major axis and in inclination are plotted as a function of the semi-major axis in figs 1a and 1b and as a function of the eccentricity in figs 2a and 2b, respectively.
(i) The most interesting feature on figs 1a and 1b is the wavy behaviour which is a direct consequence of the resonance between waves and orbital motions. The maxima of diffusion occur for orbits with semi-major axis satisfaying the relationship :

$$a = a_c \left(\frac{2n}{N}\right)^{2/3}$$

where $a_c = 0.165$ AU is the distance of corotation with the sun. On the other hand the general trend of the changes in the orbital elements shows a decrease of the diffusion with increasing semi-major axis.
(ii) The magnitude of the changes in the orbital elements begin to increase with increasing eccentricities (this behaviour is due to the increasing number of orbital harmonics); then, for high eccentricities, the general trend prevails with a behaviour in $(1 - e^2)^{1/2}$ for the dispersion in eccentricity and in $(1 - e^2)^{-1/2}$ for the dispersion in inclination. Initial inclination plays a minor role.

3. DISCUSSION AND CONCLUSION

The Poynting-Robertson drag force is known to take the particles inward to the sun with a decay time $(\Delta t)_{P-R}$ depending on the eccentricity of the orbit. From the work of Wyatt et al (1950) it is easy to show that $(\Delta t)_{P-R}$ is greater than the diffusion time (ie) that the diffusion can take place:
(i) if the eccentricities are less than 0.8
(ii) if the eccentricities are greater than 0.8 and if $a > 1$ AU.
On the other hand the dispersions in the orbital elements depend on the physical characteristics of the grains (material, size, charge).
For example if we consider two materials (iron and obsidian), only the small obsidian grains with size less than 0.5 µm and confined in the vicinity of the sun (20- 60 $R_\odot$) may be diffused. The main diffusion occurs in inclination: for 0.2 µm grains the dispersion is of the order of a few ten degrees near the first resonance of a 4 sectors structure (23 $R_\odot$) and falls down to only a few degrees at the next resonance.
Finally, we conclude that in the vicinity of the solar equatorial plane the stochastic Lorentz force have a negligible effect on the dynamics of most zodiacal dust grains.

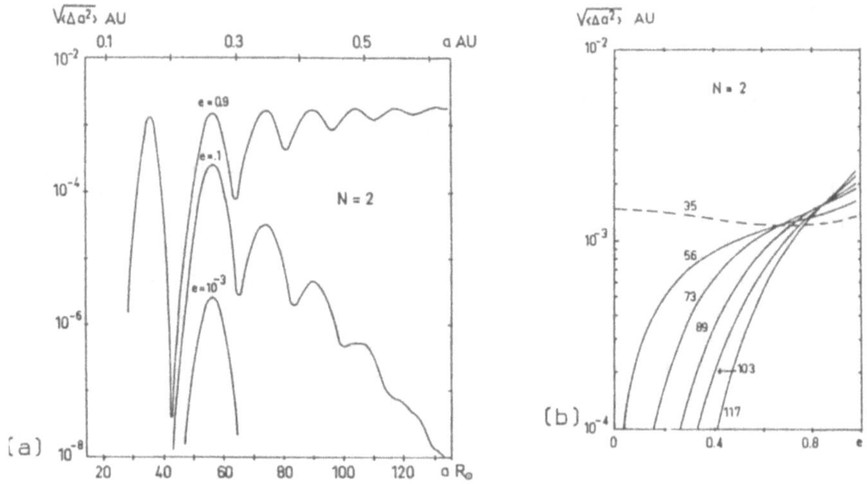

Figs. 1 Spread in semi-major axis (i = 0°): (a) versus semi-major axis, (b) versus eccentricity

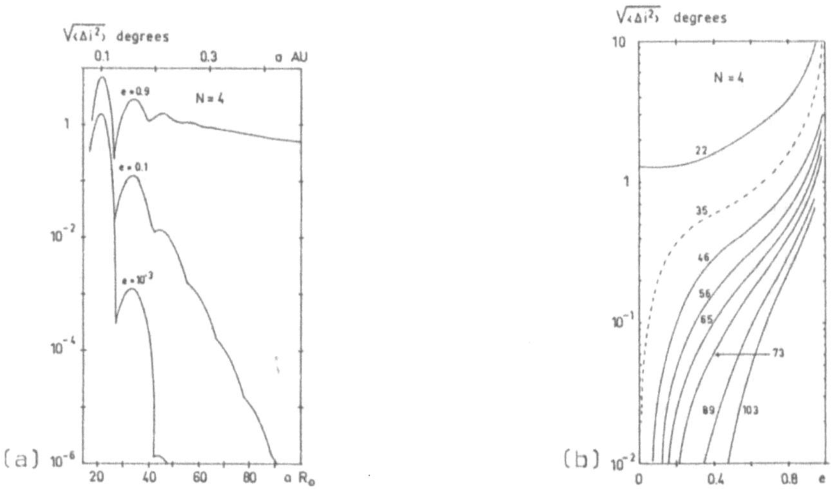

Figs. 2 Spread in eccentricity (i = 0°): (a) versus semi-major axis, (b) versus eccentricity

REFERENCES

Barge,P., Pellat,R., Millet,J.: 1982a, Astron. Astrophys. 109 , 228
Barge,P., Pellat,R., Millet,J.: 1982b, Astron. Astrophys. 115 , 8
Coleman,P.J.: 1968, Astrophys. J. 153 , 371
Consolmagno,G.: 1979, Icarus 38 , 398
Morfill,G.E., Grün,E.: 1979, Planetary Space Sci. 27 , 1269
Parker,E.N.: 1964, Astrophys. J. 139 , 951
Sawyer,C.: 1976, J. Geophys. Res. 81 , 2437
Wyatt,S.P., Whipple,F.L.: 1950, Astrophys. J. 111 , 134

POSSIBLE DISTRIBUTIONS OF SOME ORBITAL ELEMENTS OF INTERSTELLAR PARTICLES IN THE SOLAR SYSTEM

O.I. Belkovich*, I.N. Potapov**
* Engelhardt Astronomikal Observatory
Kazan, SU-422526
** Pedagogical Institute
Vitebsk, SU-210036
U.S.S.R.

ABSTRACT:

Distribution of eccentricities, perihelion distances of interstellar particles and their concentrations in the Solar system have been computed in the assumption that their velocities are similar to ones of the nearby stars of late spectral classes. More than 75% of orbits have eccentricities not exceeding the value 1.1 and concentration of particles at the Earth's orbit 3.5 times greater than in the interstellar space.

All meteor catalogues contain some quantity of hyperbolic orbits. The question is: Are some of them real interstellar particles or are they just due to inaccuracy of observations (Matsenko and Tkachuk 1982, Radzievsky 1967, Stohl 1970). That is why estimation of characteristics of orbital elements of particles possibly of interstellar origin would be useful.

We consider as the most probable hypothesis that particles with masses that can be detected by Earth-based optical or radar observations were born during formation of systems similar to our Solar system. Losses of these particles are possible from the peripheries of the systems due to perturbations by nearby stars or due to mass loss by the central star. In this case the most probable velocity is the minimal velocity (minimal energy) of the particles relative the central star. Then the velocity distribution of the lost (interstellar) particles must be similar to one of the nearby stars. This distribution found as from the catalogue (Wooley et al., 1970) for F and G classes of stars is

$$p^{(n)}(v_s) = 3.68\ v_s^2\ \exp(-1.38\ v_s^2), \qquad (1)$$

and the isotropic radiant distribution is

$$p^{(n)}(\varepsilon_s, \psi_s) = \frac{1}{4\pi}\ \sin\varepsilon_s, \qquad (2)$$

R. H. Giese and P. Lamy (eds.), Properties and Interactions of Interplanetary Dust, 421–424.

where v_s is the particle velocity, ε_s is the elongation of a particle radiant from the apex of the Sun and ψ_s is the second angle in the frame system of the Sun's centroid. Here and then all velocities are given in the units of the Earth's velocity and distances in astronomical units. The index (n) will point out velocity and radiant direction distributions of particles in some volume of space and the index (q) does the same for flux.

The three-dimensional distribution of radiants and velocities of the particles can be written as the product of the two previous equations because of the independence of the variables v_s, ε_s and ψ_s:

$$P^{(n)}(v_s,\varepsilon_s,\psi_s) = 0.29\ v_s^2\ \exp(-1.38\ v_s^2)\ \sin\varepsilon_s . \qquad (3)$$

The distribution $P^{(n)}(\psi_s) = P^{(n)}(\psi_h) = \frac{1}{2\pi}$ i.e. the same in the frame of reference of the centroid and in heliocentric one. Then in the future we can use the two-dimension distribution $P^{(n)}(v_s,\varepsilon_s)$ which is the same as the eq.(3) except the factor 1.89 instead of 0.29.

The distribution of the impact parameter a of the particles is

$$P^{(n)}(a) = C_1 a, \qquad (4)$$

where C_1 is the constant.

$$P^{(n)}(v_s,\varepsilon_s,a) = 1.84\ C_1\ a\ v_s^2\ \exp(-1.38\ v_s^2)\ \sin\varepsilon_s \qquad (5)$$

because of the independence of the variables v_s, ε_s and a.

Let ρ_s be the volume density of the particles outside the sphere of the Sun's gravitational influence and ρ_R the volume density at the distance R to the Sun. Than (Belkovich, 1983)

$$\rho_R P^{(n)}(v_R,\varepsilon_R) = \rho_s P^{(n)}(v_s,\varepsilon_s)\ \frac{v_R^2\ \sin\varepsilon_R}{v_s^2\ \sin\varepsilon_s}\ , \qquad (6)$$

Here v_s and ε_R are the variables in the heliocentric frame of reference at the distance R to the Sun. They are functions of v_s and ε_s.

From eqs. (3) and (6) we have:

$$\frac{\rho_R}{\rho_s} = 1.84 \int_0^\infty \int_0^\pi v_R^2\ \sin\varepsilon_R\ \exp(-1.38\ v_s^2)\ dv_R d\varepsilon_R . \qquad (7)$$

The following approximation can be used for eq. (7):

$$\frac{\rho_R}{\rho_s} = 0.38 + 0.62\ \exp\ (\frac{5}{R})^{0.3} . \qquad (8)$$

The flux density Q of the sporadic meteors according tò Belkowich (1983) is

$$Q_S P^{(q)}(v_S,\varepsilon_S) = \rho_S v_S P^{(n)}(v_S,\varepsilon_S)\left[\int_0^\infty \int_0^\pi v_S P^{(n)}(v_S,\varepsilon_S)dv_S d\varepsilon_S\right]^{-1} =$$

$$= 1.21\ \rho_S v_S^3\ \exp(-1.38\ v_S^2)\ \sin\varepsilon_S, \qquad (9)$$

and in the heliocentric system at the distance R to the Sun the flux is

$$Q_R P^{(q)}(v_R,\varepsilon_R) = Q_S P^{(q)}(v_S,\varepsilon_S)\ \frac{v_R^3\ \sin\varepsilon_R}{v_S^3\ \sin\varepsilon_S} =$$

$$= 1.21\ \rho_S v_R^3\ \exp(-1.38\ v_S^2)\ \sin\varepsilon_R. \qquad (10)$$

From (5) and (10) we have

$$P^{(q)}(v_R,\varepsilon_R,a) = C_2\ a\ v_R^3\ \exp(-1.38\ v_S^2)\ \sin\varepsilon_R\ , \qquad (11)$$

where C_2 is a constant.

The distribution of eccentricities of particles orbits P(e) is

$$P(e) = \int_{v_{Re}}^\infty \int_0^\pi P^{(q)}(v_R,\varepsilon_R,e)dv_R d\varepsilon_R = \int_{v_{Re}}^\infty \int_0^\pi P^{(q)}(v_R,\varepsilon_R,a)\ \frac{da}{de}\ dv_R d\varepsilon_R. (12)$$

$\frac{da}{de} = \frac{e}{v_h^4\ a}$ can be found from $a^2 = v_h^{-4}\ (e^2-1)$. Here v_h is the heliocentric velocity of a particle at the distance $R \to \infty$. v_{Re} is the minimal value of v_R.

$$v_{Re} = \sqrt{\frac{1}{D_o} + \frac{2}{R}}\ \text{for}\ e \le e_o\quad \text{and}\ v_{Re} = \sqrt{\frac{e+1}{R}}\ \text{for}\ e \ge e_o,$$

where D_o is the radius of the sphere of influence of the Sun taken to be equal to 10^5 a.u. and $e_o = \frac{R}{D_o} + 1$. The cumulative distribution of eccentricities was calculated from (12) for R = 1 and approximated as

$$P(e) = \begin{cases} \left(\frac{e-1}{e_o-1}\right)^{-0.145}, & (e_o \le e \le 1.14), \\ 0.1\ \ e^3(e-1)^{-0.6}, & (e \ge 1.14). \end{cases} \qquad (13)$$

This means that 75% of interstellar particle orbits have eccentricities within the limits 1.0 and 1.1 at the distance R=1 to the Sun and 8% only have eccentricities greater than 1.5.

The distribution of perihelions of the interstellar particles can be found in a similar way:

$$P(q) = \int_{\nu_{he}}^{\infty} \int_{0}^{\pi} p^{(q)}(v_h, \varepsilon_h, q)\, dv_h d\varepsilon_h =$$

$$= \int_{\nu_{he}}^{\infty} \int_{0}^{\pi} p^{(q)}(v_s, \varepsilon_s, a) \frac{v_h^3 \sin\varepsilon_h}{v_s^3 \sin\varepsilon_s} \frac{da}{dq}\, dv_h d\varepsilon_h = C_3 q, \qquad (14)$$

where C_3 is the constant, $\frac{da}{dq} = \frac{q}{a}(1 + \frac{2}{v_h^2})$, $v_{he} = D_o^{-0.5}$.

CONCLUSIONS

One must observe the strong concentration of the values of eccentricities of interstellar particle orbits to 1 that lead to the difficulties to select that particles from tose relating to the Solar system.

REFERENCES

Belkovich, O.I.: Astronomichesky vestnik, 1983, **17**, no.2, pp.108-115.
Matsenko, S.V., Tkachuk, A.A.: 1982, Meteornoe veschestvo v mejhplanetnom prostranstve, pp. 36-40.
Öpik, E.J.: 1951, Proc. Irish Acad., **A 54**, No. 12, pp. 165-199.
Radzievsky, V.V.: 1967, Astronomichesky J., **44**, No.1, pp. 166-177.
Stohl, J.: 1970, Bull. Astron. Inst. Cz., **21**, No.1, pp. 10-17.
Wooley, R. et al.: 1970, Roy. Observ. Ann., No.5.

F I N A L   D I S C U S S I O N

compiled by R.H. Giese

By the end of the meeting a final discussion took place to identify unsolved problems and to collect suggestions for future activities. To start this event a panel of few participants was asked to summarize personal views. Then the discussion was extended to involve the other outstanding experts fortunately being present in the auditory. Discussion sheets were kindly provided by Drs. Grün, Lamy, McDonnell, Parkin, Singer, Weinberg, and Zook. From these notes and records of many oral contributions hopefully the most important items were compiled in the following list. It is arranged in sections representing the different methods and within each section or subsection in an order ranging from near future to later and more demanding tasks. The list includes both, items, which have been already investigated but which still need more thorough treatment and projects which are far from implementation but strongly desired in the future. As a general trend it clearly turned out, that an interdisciplinary and synoptic view of the problems by combination of information from different fields is the most promising way to complete our knowledge of the interplanetary dust cloud and to relate it to the exciting problems of Astrophysics.

L I S T   O F   F U T U R E   T A S K S

I.   OBSERVATIONS

Structures in ZL (bands, rings, fine structure)
Time variation (short and long time) of Zodiacal Light (ZL)
More Infrared (IR) and Ultraviolet (UV) measurements (Balloons, Rockets, e.g. INTERZODIAK, Shuttle)
F-corona, Inner ZL
- 1988 solar eclipse (balloon, IR imaging)
Improvement of Doppler measurements

II.   SPACE INVESTIGATIONS

Large capture cells (e.g. Shuttle, Space Station)
Observations of comets, asteroids from orbiting telescope
Distribution of m, v and time variations of possible Beta meteoroids,

R. H. Giese and P. Lamy (eds.), Properties and Interactions of Interplanetary Dust, 425–427.

collisions in the inner solar system
Space coronograph (V,IR,Imaging, 20R , long time observations)
- brightness, color, polarization, thermal radiation
- time variations
- transient phenomena (transients, grazing comets)
Further flux- and/or optical measurements (e.g. cooperative NASA/ESA)
- out of ecliptic
- beyond Asteroid belt and Jupiters orbit
- in the inner solar system (0.3 AU)
- deep in planetary gravitational fields
Active space experiments for
- dust or plasma releases
- study of EM radiations effects (Poynting-Robertson, spin up)
Comet and/or asteroid rendezvous for
- investigation of production of submicron and meteoroid
  sized particles
- active experiments for in-situ analysis
- sample return
Use of spacecraft
- LDEF
- SHUTTLE (or free flying platforms)
- GIOTTO
- GALILEO
- ULYSSES
- SPACE TELESKOPE
Use of future space missions
- SPACE STATION
- GIOTTO 2
- CRAFT
- AGORA
- COBE
- CASSINI ORBITER
- SUN PROBE

III. LABORATORY

Analysis of collected grains (stratosphere, deep sea), and lunar samples
- analysis of more samples of extraterrestrial particles
- improvement of discrimination from space debris,
  how representative are collected particles (selection effects)?
- application of analytical methods used for atmospheric particles
  to deep sea spherules
- monitoring of variations of dust influx and big meteor events
  over millions of years by analysis of deep sea clay/antarctic
  ices
- same for past solar flare activity (flare tracks)
- search for products of astrophysical nuclear reactions
Scattering experiments (microwave, laser)
- further experiments to test theoretical approaches to agglomerates

- color effects (structure, microroughness)
- absorption and radiation pressure on irregular particles
- measurements on real interplanetary particles

Collisions of grains

Plasma-grain-interactions

IV. THEORY

Further analysis of existing data base (Pioneer 10/11, Helios 1/2, Hawaii, South Africa data, IRAS, Japanese baloon experiments)

Advanced analysis, including IR and UV data, to obtain
- variations with heliocentric distance
- size, material, number densitsy
- optical properties (albedo, polarization, colour)
- three dimensional structure of the zodiacal cloud (combination of optics and dynamics)

Analysis of inner zodiacal light

Explanation of optical properties (advanced theory of agglomerates)

Search for evidence of and interrelations to interstellar dust

Dust-plasma-gas interactions

Dynamics
- collisions (break up laws, production/destruction rates)
- destruction rates by Poynting-Robertson effect, erosion)
- production of meteoroid sized particles from comets
- evolution and dispersion of meteor streams to sporadic background
- planetary perturbations (including Jupiter as possible barrier)
- nongravitational forces changing orbital elements (e.g. inclination)
- explanation of belt structures (IRAS), swarms (Heos) and terrestrial dust belts (Electron 1,3)
- long time stability of zodiacal cloud

SUMMARY AND CONCLUSION OF THE COLLOQUIUM

P. L. LAMY
Laboratoire d'Astronomie Spatiale du C.N.R.S.
Traverse du Siphon - Les Trois Lucs
13012 MARSEILLE (FRANCE)

There is now a satisfactory convergence of the observational results for the absolute intensity of the zodiacal light for elongations larger than 20° to 30°. They are still discrepancies affecting the data for the F-corona but after carefully scrutinizing available data, Koutchmy and Lamy could obtain an axy-asymmetric non-spherical model for the equatorial and polar regions which satisfactorily bridges to the zodiacal light. However, a closer examination reveals several points of concern:

i) the gradient of the brightness which is of utmost importance for retrieving the phase function of interplanetary dust: the "Hawaï" data (Misconi and Weinberg), the PCN data (Nikolsky et al.) and the classical data of Dumont and Sanchez are in disagreement;

ii) a lack of data for the inner zodiacal light;

iii) insufficient out-of-ecliptic data; the on-going analysis of the "Hawaï" data (which benefits from an improved reduction method) and of the "South Africa" data will help partly remedy this situation.

A consensus appears to be reached on the near solar color of the zodiacal light or, to be more precise, a slight reddening down to 2000 Å. The inner zodiacal light and the F-corona are clearly redder than the Sun. Some structure near 2800 Å reported by Feldman et al. is certainly worth of further investigation. Ultraviolet observations of Comet Halley may well provide adequate opportunities. The situation of the polarization of the zodiacal light is less satisfactory as large discrepancies persist. Of major importance for the properties of the dust grains are the behaviours at small and large scattering angles - in particular, the shape of the negative branch - as well as the angle of maximum polarization and its magnitude. The polarization is certainly color dependent but the present data are insufficient for a clear-cut conclusion.

With the background zodiacal light on a firm basis, investigations have started, mainly by the Florida group, to search for fluctuations in

R. H. Giese and P. Lamy (eds.), Properties and Interactions of Interplanetary Dust, 429–434.

the brightness indicating small scale structures. Results from ground-based observations are inconclusive, their reliability is always a matter of concern as the fluctuating airglow poses a major problem. This is really a space program further motivated by the impressive results of the IRAS mission. Infrared observations were carried out in the past, but the tremondous achievement of IRAS represents a giant step in our probing the interplanetary dust cloud. Structures are definitively observed for the first time, the so-called interplanetary dust bands as well as irregularities of the profiles near the ecliptic plane. The fact that the data of the D2B ultraviolet satellite do not reveal these bands (Maucherat, private communication) confirmes that they are composed of large grains. If they result from a collision, small micronic and submicronic grains possibly created then have already been swept from the bands. Generally speaking, we must expect a quite different zodiacal cloud in the infrared than in the visible and ultraviolet (as examplified by Comet Iras-Araki-Alcock, 1983d) since they do not relate to the same grains. In particular, we may well expect a different plane of symmetry. When fully exploited, all these data will possibly help to differentiate the dynamical evolution of grains as a function of their size.

The balloon observation of the F-corona at the 1983 eclipse represents a major effort of the japanese astronomical community. Accurate photometry (and polarization) with a television camera is always a difficult task but the data are certainly worth further elaborated reduction. The excess emission observed in the infrared on the west side of the Sun does not really ressemble the "peaks" reported by Peterson and MacQueen. It is further puzzling that this excess tends to vanish beyond 2 microns. Although there probably exists a dust free zone (or zones), with edge enhancement(s) possibly connected to the reinforcements observed in the visible in the equatorial direction, it is not clear if they are permanent stationary features of a local component of the F-corona. Then the relative importance of this local component (scattering at large angles) and of the classical F-corona (diffraction by grains along the line-of-sight) is not clear either.

The investigation of the dynamics of the zodiacal cloud based on the measurement of the Doppler shift is actively pursued with a variety of technics. Fabry-Perot interferograms as obtained by Shcheglov et al. on the F-corona are a powerful technics which should probably be extended to the zodiacal light. It is of interest to note that these authors could well have observed the debris of a Sun-grazing comet. High-resolution spectrographic observations by Robley et al. reveal that the MgI lines may be contaminated by a weak OH system, a situation which could explain some puzzling aspects (e.g., the Gegenschein receding from the Earth at 4 km/sec) of the otherwise impressive results obtained by East and Reay who reject the hyperbolic trajectories found by Fried.

There is one field which does not benefit from the expanding activities of man in space, on the contrary! it is the in-situ sensing of meteroids in the near-Earth region. The reassessed importance of lunar secondary ejecta based on new laboratory simulations is a further drawback for the determination of the size distribution of interplanetary dust from lunar microcraters data. However, Zook concludes that in-situ experiments carried out in the early 1960's were not significantly contaminated by space debris and that the lunar microcraters distribution of impact pits larger than 7 microns satisfactorily represents the size distribution of interplanetary dust. If the "solar flare track" lunar clock is rejected as unreliable, then both sources of information are consistent as shown by Grün et al. But the size distribution for grains radii below ≃ 2 micron (or mass below ≃ 10^{-10} g) cannot be directly obtained. As pointed out by Walker during the discussion, one could compare the crater populations on different faces of a large glass deposit to determine the contribution of secondary craters. Alternatively, the deconvolution process worked out by the Canterbury group could be applied again using the new laboratory data on a sample (such as 12054) whose lunar orientation is well known. Because of so many years of effort and improved knowledge on processes taking place on the lunar surface, the true size distribution of impact pits smaller than 7 microns is certainly within reach.

The complexity of the near-Earth region is revealed by new analysis of in-situ observations realized in the sixties. While the 1962 Explorer 46 data lead Kessler to conclude that nearly all the penetrations were from exhaust product from solid rocket motors, Barsukov et al. consider that the 1964 Elektron 1 and 3 data were not severely contaminated and reveal structures in the distribution of meteoritic matter in orbit around the Earth, similar to the "swarms" registered by HEOS-2.

Deep-space, in-situ measurements are not (yet!) subject to contamination and the continuying analysis of the Helios data gives, for the first data, direct information on the radial variation of the micrometeoroid flux between 0.3 and 1 AU. Contrary to all expectations and to a straightforward conclusion from the results of the zodiacal light experiment, the "ecliptic" flux levels off at 0.5 to 0.6 AU or even decreases further in! Theoretical studies are now required to understand this result while we are expecting more surprises from the forthcoming Galileo/ISPM missions.

To the new and challenging problem of distinguishing interplanetary dust from man-made space debris (and eventually, from simulated interplanetary materials by Strong et al.), investigators are responding with a new generation of detectors of amazing ingeniosity which, of course, benefits from past experience and improved technology. Combining orbital, chemical and isotopic information should unambiguously allow to decipher the impacts on the LDEF detectors.

The field of laboratory measurements of interplanetary dust grains, which already produced significant astrophysical results, has now reached a phase of rapid development for, at least, three reasons:
i) the routine collection program managed by NASA; this will greatly increase the number of IDG available for analysis and improve the

statistical significance of the results;
ii) the parallel use of a large variety of complementary analysis techniques;
iii) the new methodology for the measurements which consists in studying each particle individually and establishing its extraterrestrial origin on the basis of several criteria, large D excesses, presence of solar flare tracks and solar-type gases.
As it is the case for a new, rapidly expanding field, more and more refined and detailed results have been presented which probably preclude a synthesis for the time-being. The limited subset of "chondritic" dust grains, which have been thoroughfully measured by the St Louis group, indicates from infrared transmission spectra, that these grains fall into three classes, "layer-latticed silicates", "pyroxenes" and "olivines". Worth mentioning is the presence, in spectra of the first class, of a band at 6.8 micron as seen in the spectra of some protostars and probably due to a carbonate mineral.

The impressive development of analysis techniques benefits equally to the renewal of deep-sea spherules studies, a field which have been absent from the interplanetary dust conferences for about a decade. Several criteria are proposed to test their origin such as the $^{26}Al/^{10}Be$ ratio, the chemical and isotopic compositions, the cosmic ray exposure age, the density of the internal metal globules. While Raisbeck et al. conclude for a cometary origin, Parkin favors their formation as splash ejecta in asteroidal collisions. Of importance is the ability of cosmic exposure ages to provide the first experimental checks of theoretical lifetimes of micrometeoroids in space.

The harmonious parallel development of both experimental and theoretical studies of the optical properties of irregular and rough dust grains is also one of the main achievements of these past years in our field. We have therefore available good laboratory measurements, especially from the Bochum group, and a variety of models (which are in fact complementary) cooperating in an improved understanding of the light scattering by non-spherical rough grains such as the negative branch of polarization resulting from preferential geometrical blocking of reflected rays. Likewise, efforts are being pursued to extract local optical properties of interplanetary dust grains from zodiacal light data especially by Dumont who has been fighting for years to retrieve non model dependent information which may be viewed as offering "absolute calibration" points for any model. Although, the out-of-ecliptic optical probing of the zodiacal cloud will not be possible, Giese et al. have shown that a clever analysis of available data can indeed help to confirm or reject three dimensional distribution models of interplanetary dust.

Combining all these results lead to an improved understanding of the main features of the zodiacal light. They (including the IRAS data) converge to a low albedo, of the order of a few percents. The increased flux model proposed by Grün et al. considerably eases the compatibility with zodiacal light data (especially, in view of the low albedo) but still, the contribution of submicronic grains is non-negligible as shown by Lamy and Perrin and probably required to explain the slight reddening of the zodiacal light down to 2000 Å. The question of the heliocentric variation of the spatial density of grains ($r^{-\nu}$) must remain opened as the result $\nu = 1.3$ derived from the Helios zodiacal light experiment is not compatible with the observed brightness of the inner zodiacal light and F-corona as noted by Leinert and Hong. It may be argued that part of the $r^{-1.3}$ variation can be attributed to the variation of the out-of-ecliptic "thickness" of the zodiacal cloud since the Helios photometers were pointing away from the ecliptic. Furthermore, the Helios impact experiment reveals that the "ecliptic" flux tends to level off near 0.5 AU. It is also likely that ν is not a constant, but a function of heliocentric distance. As an alternative view and if a bimodal population is considered, each population may have its own ν such as $\nu = 1$ for large grains and $\nu = 2$ for small grains. Taking into account the scattering efficiency of both populations, one could well produce an apparent, intermediate ν such as $\nu = 1.3$.

This, by no means, precludes investigating constraints on the dynamics of interplanetary grains to produce $\nu > 1$ (the Poynting-Robertson effect alone would result in $\nu = 1$). Three mechanism recently proposed were discussed and compared in a very enlighting way by Leinert. Catastrophic collisions of larger meteoritic particles indeed appear to be the most natural process and Grün et al. studied the collisional balance of the "meteoritic complex" and found that the population of "zodiacal light" particles is not presently in equilibrium.

The dynamics of grains larger than a few microns is essentially controlled by solar gravitational and radiation forces and gravitational disturbances by the major planets. When applicable, it can be studied by perturbation theory and Dermott et al. presented a thorough application to the case of the IRAS dust bands to predict longitudinal variations in their brightness and in their latitude so as to discriminate between cometary and asteroidal origins. Gustafson proposed a method to calculate and integrate the perturbations on ensembles of particles of low eccentricity and showed that dust close to the orbital plane of Venus is flattened while a dust band is created. The general case is investigated by Burkhardt who numerically integrated the equations of motion of test particles in three-dimensions, all planets from Mercury to Jupiter being included; although this kind of approach is always limited by the number of test particles preventing a global view, it does show some interesting behaviours not predicted by the planar treatment. Electromagnetic forces may become important for submicronic grains. Applying the quasi-linear theory, Barge et al. convincingly proved that only the low frequency fluctuations associated with the sector structure of the interplanetary magnetic field must be considered and that, introducing the cyclic variables associated with the three constants of the unperturbed motion, allows to obtain a simple diffusion equation. The diffusion, which has a resonant behaviour, mainly affects the inclination.

The general problem of connecting the interplanetary dust distribution with the dust dynamics and dust sources and sinks was addressed by Fahr and Ripken. The problem may be rigorously formulated in terms of the Boltzmann kinetic equation (or a Fokker-Planck equation when stochastic interactions are considered). In the case of loss-free motion and circular orbits, these authors could obtain an explicit solution for the intensity distribution function. They also explained how the dust dynamics can be traced by measuring the velocity distribution function of hydrogen and helium atoms which desorb from the grains surface and how it will be applied to a forthcoming rocket experiment.

Progress in the calculation of the charge of interplanetary dust grains were mainly achieved on both theoretical and experimental fronts by a french consortium. In particular, the effect of surface roughness on the charging process was investigated by Lafon and Millet. Such efforts, which are somehow parallel to those in the optical domain, must be further developped and backed by laboratory experiments which are not too difficult in principle. The case of energetic particles penetrating dust grains was examined by Rössler and Eich while Ibadov looked at the possibility of X-ray emission in dusty comets coming from a high-temperature plasma generated by collisions between cometary and interplanetary grains.

Looking in retrospect to past conferences in our field has led me to refrain from drawing definitive conclusions. The panel and I have pointed out directions where immediate or near-future progress can be achieved, domains of potential and promising results, studies and (space) experiments which are needed. In my opinion, the continuous vitality of our field and its perpetual renewal stems from the variety of techniques of probing the zodiacal dust cloud, each one involving in turn its part of observational, laboratory and theoretical studies. It is essential that this type of interdisciplinary approach be pursued to assure continuous progress in our understanding of the properties and interactions of interplanetary dust. The potential access to primitive interstellar matter offered by collected interplanetary grains and the recent observation of "zodiacal" dust cloud around stars will probably led us to expand our horizons and to cast a new light on our interplanetary medium.

Writing this "summary and conclusion" was the most demanding task of all of the organization of this colloquium. But it was also the most interesting and rewarding and I am most thankful to R. Giese for having asked me to do it.

INDEX OF AUTHORS

Ajmanov 77
Amari 185

Baba 49
Baggaley 299
Barge 417
Barsukov 99
Belkovich 421
Blank 157
Bliek 231
Bode 275
Borghesi 159
Brownlee 143, 169, 275
Bücher 85
Burkhardt 389
Burns 395
Bussoletti 159

Carey 131
Cebula 31
Colangeli 159
Courtes 231
Crifo 261

Dermott 395
Dixon 131
Dumont 207

East 81
Eich 351

Fahey 149
Fahr 305
Farnum 275
Fechtig 89, 105, 121, 411
Feldman 31
Fernandez 287

Gautier 43
Giese 219, 223, 255, 411
Giovane 39
Göller 113
Gonin 27
Good 43
Grün 89, 105, 113, 287, 411
Gustafson 227, 381, 385

Hasegawa 55
Hauser 43
Henry 31
Hirayama 49
Hiromoto 55
Hörz 121
Hong 33, 215
Honma 175
Houck 395
Huebner 275

Ibadov 365
Igenbergs 121, 129
Iglseder 129
Ip 325
Isobe 49

Jessberger 121, 157
Jockers 283

Kassel 117
Kessler 97
Kimura 279, 283, 325
Kinateder 255
Kissel 105
Klein 169
Kneissel 255
Koma 49
Koutchmy 7, 63, 85, 141
Kuczera 121, 129
Kunkel 275
Kviz 191

Lafon 335, 341, 347
Lamy 7, 63, 85, 137, 141, 231, 239, 245, 335
Lange 89, 121
Lefevre 329, 335
Leinert 369
Leuprecht 17
Levasseur-Regourd 207
Liu, C.P. 279, 283, 325
Liu, Z.K. 163
Llebaria 27
Low 43

Maihara 55
Mann 39
Maucherat 27
McDonnell 131
McKeegan 149
Middleton 169
Millet 335, 341, 347, 417
Misconi 11, 33, 377
Mizutani 55
Mukai, S. 235
Mukai, T. 59, 235

Nazarova 99
Nesmjanovich 7
Nicholson 395
Nikolsky 7

Oliver 39
Ozima 185

Pailer 121
Papoular 267
Parkin 183
Pegourie 267
Pellat 417
Peng 163
Pfleiderer 17
Perrin 239, 245
Potapov 421
Potter 293
Priol 329

Quemerias 329

Raisbeck 169
Reay 81
Ripken 305
Rittich 255
Robley 85
Rössler 351, 357
Rusk 377

Sandford 149, 273
Schiffer 249
Seignac 329
Shcheglov 77
Shestakova 75, 77
Singer 117
Soegijo 49
Stanley 117
Steel 299
Stephens 275
Strong 275
Sutton 121
Swan 121

Takami 55
Takayanagi 185
Tanabe 49
Tennyson 31
Thielheim 249
Toller 21, 33
Traxel 157

Van Dijk 33
Vernadsky 99

Walker 121, 127, 149, 273
Wallenwein 157
Weinberg 1, 11, 21, 33, 39
Weishaupt 129
Weiss-Wrana 219, 223
Wopenka 149

Yamakoshi 175, 179
Yiou 169
Yu 163

Zashu 185
Zerull 197, 219, 223
Zhuang 163
Zinner 121, 127, 149
Zook 89, 287, 293, 411

INDEX OF SUBJECTS

Ablation → meteors
Actinides 127
Airglow 12, 17, 34
Albedo 43, 235, 243
  infrared 45, 213, 405
  ultraviolet 31
Allende meteorite → meteorites
Alpha meteoroids → particles
Aluminium oxide 90, 97
Andesite 224
Asteroids 183, 312, 395
Atmosphere 35, 85, 306

Background (starlight, galactic) 2, 17, 21, 34, 45, 65, 212
Backscattering 210, 223, 235
Beta meteoroids → particles
Bumper → meteroid bumper
Boltzmann equation 311
Brightness integral (→ ZL) 210, 215, 257

Capture cells 121, 131 ff
Carbon, Carbonates 159, 175, 273, 307, 329
CCD image sensors 294
Centroid 381 ff
Charge of grains → particles
Chemical interactions 357
Chondritic material (→ particles) 139, 145, 149
Circular orbits 385 ff
Circumstellar dust 267
Collection of dust → particles
Collisions → particles
Color effects 197, 201, 219, 225, 227, 249
Comets
  coma 225, 288
  discovery 287
  dust from 80, 146, 169, 191 ff, 223, 279 ff, 280, 306, 325 ff
  dust striae 283 ff
  Encke 118
  Halley 118
  Ikeya-Seki 227
  IRAS-Araki-Alcock 279, 430
  Kohoutek 150, 273, 280
  Mrkos 283 ff
  nucleus 288
  optical detection 287 ff
  plasma 325 ff
  small comets 287 ff

tails 227
West 225, 283 ff
Complex angular momentum theory 261
Corona (solar) 55, 141
F-corona (→ also inner ZL) 56, 59, 63, 75, 77
K-corona 56, 80
Cosmic chemistry 351, 357
Cosmic rays 127, 145, 169, 179 ff
Cratering 91 ff, 131
Cross polarisation 198, 202, 220
Cylinders 227

Debris 90, 97, 431
Debye Sphere 316
Deep sea sediments (→ spherules) 163, 169, 179 ff, 185
DIDSY 114
Discharges 345
Doppler shift 83, 85, 212, 430
Double reflection 245
Dust → particles, micrometeoroids, ZL
Dust arcs → ZL
Dust bands 46, 387 ff, 395 ff, 433
Dust cloud → zodiacal dust cloud
Dust collections, detectors → particles
Dust envelope of the earth 99 ff
Dust free zone, ring → F-corona
D2B Satellite 27, 430

EDX 149
Electric field 325, 341
Electric suspension 200, 219
Electron 1,2 99 ff
Electron microscope 149, 164 ff
Electrostatic potential → particles
Enstatite 269
Excentricity (particle orbits) 379 ff, 391, 417 ff, 423 ff
Explorer
Explorer XVI 90
Explorer XXIII 90
Explorer 46 97, 117
Extinction 161
Extraterrestrial origin → particles

Fabry-Perot Interferometer 77, 81, 85, 430
Fan-Model → ZL
F-Corona (and inner ZL) 63 ff
brightness, polarization, color 7, 49, 55, 63 ff, 258
polarization of
dust free zone, dust ring 49, 55, 59, 75
infrared radiation 56, 61, 70
models of 63 ff, 258

Field emission 341
Fluff 341 ff
Fluffy → particles
Flux curves → particles
Flux models → particles
Focker Planck approximation 311 ff, 417
Foils, Films 131 ff, 139
Fosterite 269
Fresnel reflection 202, 236, 245
FSN grains 240

Galactic cosmic rays → cosmic rays
Galactic emission → background
Galileo 113 ff
Gegenschein 2, 30, 33, 39, 81, 83, 210, 249
Gemini 144
Geometrical optics 245, 249
Grains → particles
Gravitation 377 ff, 389

Helios 2, 21, 105 ff, 138, 207, 249, 311
Heliosphere 305 ff
Heliocentric dependence → ZL
Helium 308
Henyey-Greenstein function 215
High energy approximation 245
Hirayama families 47, 395
Hyperbolic orbits, trajectories 84, 105, 421 ff

Impacts (→ dust particles) 89, 100, 105 ff, 113, 117, 129, 365
IMP 7 311
Inclination (→ orbits) 105, 390, 418 ff
Interferometric observations 77, 81, 85
Interplanetary dust (→ ZL, micrometeoroids, particles) 89, 157, 197, 213, 235, 239, 255, 275, 412
Interplanetary dust emission (→ ZL, infrared) 43 ff
Interplanetary magnetic field 373, 417
Interstellar dust, particles 1, 121, 267, 351, 421 ff
Interstellar medium 267, 306
INTERZODIAK 319
Inversion (→ ZL) 207, 216
Ion implantation 357
IR, infrared (→ F-corona, ZL) 56, 61, 70, 149 ff, 212, 267, 273, 275
IRAS 1, 2, 43, 212, 279, 395 ff
Iridium 177
Iron 233, 271, 329, 419
Isotopes (ratio, composition) 121, 169 ff, 432
ISPM (International Solar Polar Mission) 113

K-Cornona (→ corona) 56, 80
Keplerian orbits, motion → orbits

Laboratory measurements of dust (→ scattering) 121, 197, 219, 223
Lagrangian points 5, 39
Laser experiments → scattering
LDEF 117 ff, 121 ff, 127
Lifetime of particles 118, 411 ff
Lorentz force 286, 305 ff, 373, 417, 418
Lunar
  crater statistics 91 ff, 412
  ejecta 89
  microcraters 89 ff, 137 ff, 239

Magellan dust collector 144
Magnetic perturbations → Lorentz force
Marlowe (computer code) 351
Mass input → particles
Mass → particles
Mass spectrocospy 113
Meteors 138, 299 ff, 421
  ablation products 169, 302
  fireballs 293
  showers 117, 299 ff
Meteorites
  Allende 223, 240
  carbonaceous chondrites 223, 243
  Murchison 157, 223
Meteoroids 293 ff, 411 ff
Meteoroid bumper 97
Mg I line 77, 81, 85
Micrometeorites, micrometeoroids → particles
Microwave measurements → particles
Mie-theory → scattering
Mueller Matrix 198
Murchison → meteorites

NI emission 86
Node (Martian, quasi radial) 209
Nodes (of symmetry plane, → ZL)
Nongravitational forces 389, → particles:electromagnetic effects

Obsidian 60, 417 ff
OH-spectrum 86
OI-emission 8
Olivine 55, 150, 273
Orbits, orbital elements of dust 105 ff, 386, 390, 398 ff, 421 ff
Organic molecules 351, 357
Origin of life 356
Osmium 177

Particles, dust
  absorbing 225
  aggregates, conglomerates, low density 99 ff, 105, 137, 240

albedo → albedo
alpha meteoroids 90
beta meteoroids 39, 89, 108, 411 ff
charge of 113, 308, 335, 418
chondritic 139, 145, 149
collection 121, 139, 143 ff
collisions 351 ff, 368, 371, 372, 411 ff
composition 121, 149 ff
density 105, 133, 137 ff, 239, 255
detectors 105, 113 ff, 117, 121
dielectric 225, 234, 417 ff
electromagnetic effects
Lorentz force 286, 305 ff, 373, 417, 418
→ Poynting-Robertson effect 39, 309 ff, 381, 389, 414, 417
→ radiation pressure 377, 386, 389
electrostatic charge, potential, fragmentation 325, 326, 335, 341 ff, 349, 418
elongated 227
extraterrestrial origin 143 ff, 186 ff
fluffy 203, 223
fluxes, flux models 89, 90, 105 ff, 255, 412
ice 261
interstellar (→ interstellar dust) 121, 267, 421
irregular → rough
laboratory measurements 149 ff, 432
lifetime of 411 ff
mass 90, 118, 137
mass loss, input 411 ff
optical properties 245, 268, 273, 329 ff
orbits, evolution, dynamics 100, 275, 386, 398 ff, 421 ff
physical properties 4, 59, 201, 219, 223
rough particles, roughness, asperities 203, 223, 235, 239, 245, 249, 341
scattering properties (→ scattering) 198, 201, 219, 223, 370
sizes, size distribution 45, 57, 137, 146, 227, 279, 396, 411 ff, 431
sources 371, 411 ff
spatial density, distribution 105, 225, 239 ff, 240, 369 ff, 371 ff, 400, 411 ff, 433
sublimation 60
submicron 89, 117, 159, 239, 327
temperature, thermal radiation 1, 45, 46, 55 ff, 58, 261
velocity of 75, 77, 81, 113, 129, 421 ff
Pegasus 90
Perturbations → also Lorentz, Radiation, Poynting Robertson forces
planetary 385 ff, 390, 396 ff
Jupiter 390, 396 ff
Mercury 386
Saturn 386
Venus 387, 390
Penetration by particles 90, 126, 131

Phase function → scattering function
Photoelectric emission, properties 329 ff, 335
Pioneer 2, 11, 21, 33, 39, 117
Plane of symmetry → ZL symmetry plane
Plasma 275, 305 ff, 307, 326
Polarization → scattering, ZL
Pole (ecliptic) → ZL
Porous materials 348
Potential → particles
Poynting-Robertson effect 59, 75, 179 ff, 183, 305 ff, 369 ff, 381, 389, 414, 417
Prograde motion 75, 77, 79, 85
Protostars 267
Pyroxene 150, 273
Pyrrhotite 240

Radiation forces, pressure (→ also particles) 39, 377 ff, 386, 389
Rainfall 191 ff
Raman spectroscopy 151
Rayleigh-Rice-approach 250
Rendezvous 426
Resonance 391, 398, 419
Resonance cell 320
Retrograde Motion 77
Rocket observations → ZL
Rough grains → particles

Saliout 7
Scattering 235, 245
- cometary dust 227
- function, cross section 215, 231 ff, 240, 249
  - empirical scattering function 215, 223, 370
  - laboratory measurements (laser, microwave) 200, 202, 219, 223, 227, 231, 432
  - phase function 29, 215
- Mie theory 198, 202, 241
- polarization 197 ff, 223 ff, 227, 235 ff, 245 ff, 247
- solar x-radiation by comets 365
- wavelength dependence (→ color, ZL) 197, 201, 219, 249

Schuerman dust arcs (→ ZL) 5, 39
Secondary emission 347 ff
Secondary ions 124
Semimajor axis 377 ff, 417 ff
SEM 144, 149
SESAME dust collector 144
Shuttle 275
Silicate 59, 137, 150, 224, 227, 233, 241, 267, 273, 275, 307, 351
Silicon carbide 159, 275
SIMS 121 ff, 149 ff
Size distribution → particles
Sky background → background

Skylab 4, 90, 140
Solar, wind 305 ff, 325 ff, 338, 418
Solar flares, flare tracks 89, 127, 145, 149 ff, 169
Solar system(s) 261, 421
Sombrero model → ZL
Sources → particles
Space debris 89, 97, 129
Spacelab, space station 127, 426
Spatial distribution → particles
Spectrum → ZL
Sphere of influence, action 377, 423
Spherules, spheres (ocean, ice, deep sea, metallic) 143, 169, 175, 179 ff, 181, 183
Sputtering 360
Starlight → background
Stokes parameters, Stokes vector 52, 198
Stratospheric dust collections 143
Sublimation 60
Submicron particles → particles
Swarms (of micrometeoroids) 100
Symmetrie plane → ZL

TEM 149 ff
Temperature, thermal radiation → particles, F-corona, ZL
Three body problem (→ dust arcs) 39
Titanium 90

Ulysses (→ = ISPM)
UV zodiacal light → ZL

Vela 311
Velocity of particles → particles
Venus 306
Venus flytrap 143
VSF (volume scattering function) (→ scattering) 215, 232, 240

X-rays 144, 365

Zodiacal dust cloud 207, 215, 305, 385
  models 7, 45, 256
Zodiacal light (ZL), Zodical Emission (ZE) 1, 89, 299
  bands 5, 46, 387 ff, 395 ff
  brightness (brightness distribution, isophotes) 8, 9, 11 ff, 17 ff, 22, 34, 45, 53, 210, 215, 240, 257, 429
  brightness integral 210, 215, 257
  color, spectrum, wavelength dependence, reddening 3, 7, 9, 31, 57, 61, 67, 85, 249, 412
    infrared 45, 56, 149, 212, 273
    inner (→ F-corona) 7, 49, 55, 61, 63, 258
    ultraviolet 31, 429, 430
  Doppler shift 83, 85, 212

dust arcs 5, 39
ellipsoid model 7, 256
fan model 7, 256
Gegenschein 2, 30, 33, 39, 81, 83, 210, 299
heliocentric, dependence 2, 4, 22
inner → F-corona
inversion 207, 216
measurements from balloons 49, 55
from ground 11, 19, 33, 40, 217, 255
from rockets 31, 255
from spacecraft 1 ff, 7, 23 ff
models 7, 45, 255, 256
out of ecliptic 9
photographs 7
photometry 8, 17, 255
photopolarimetry 49
polarization 3, 4, 56, 207 ff, 210, 218, 240
pole (ecliptic) 18
Schuerman dust arcs 5, 39
Sombrero model 255
structures 12, 22, 33, 36
symmetry plane, photometric axis 5, 7, 12, 29, 43, 45, 255, 385
variations 46
Zodiacal light particles (→ particles) 105, 157, 414
Zone of influence (→ sphere of influence, action) 377 ff, 423

MIX
Papier aus verantwortungsvollen Quellen
Paper from responsible sources
FSC® C105338

If you have any concerns about our products,
you can contact us on
ProductSafety@springernature.com

In case Publisher is established outside the EU,
the EU authorized representative is:
Springer Nature Customer Service Center GmbH
Europaplatz 3, 69115 Heidelberg, Germany

Printed by Libri Plureos GmbH
in Hamburg, Germany